Formulas

Rules of Exponents (p. 215)

1. $x^a x^b = x^{a+b}$
2. $(x^a)^b = x^{ab}$

2A. $(xy)^a = x^a y^a$

3. $\dfrac{x^a}{x^b} = x^{a-b} \qquad (x \neq 0)$

3A. $\left(\dfrac{x}{y}\right)^a = \dfrac{x^a}{y^a} \qquad (y \neq 0)$

4. $x^{-n} = \dfrac{1}{x^n} \qquad (x \neq 0)$
5. $x^0 = 1 \qquad (x \neq 0)$
6. $\left(\dfrac{x^a y^b}{z^c}\right)^n = \dfrac{x^{an} y^{bn}}{z^{cn}}$ **None of the letters can have a value that makes the denominator zero**

The Pythagorean Theorem (p. 204)

$$c^2 = a^2 + b^2$$

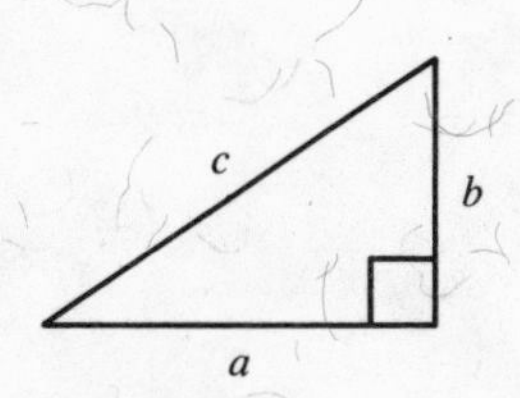

The Quadratic Formula (p. 269)

The solution for $ax^2 + bx + c = 0$ is: $x = \dfrac{-b \pm \sqrt{b^2 - 4ac}}{2a} \qquad (a \neq 0)$

The Logarithm of a number *N* is the exponent ℓ to which the base must be raised to give N *(p. 340)*

Logarithmic form		*Exponential form*
$\log_b N = \ell$	$\Longleftrightarrow$	$b^\ell = N$

The Rules of Logarithms (p. 347)

1. $\log_b MN = \log_b M + \log_b N$
2. $\log_b \left(\dfrac{M}{N}\right) = \log_b M - \log_b N$
3. $\log_b N^p = p \log_b N$
4. $\log_{10} 10^k = k$
5. $\log_b N = \dfrac{\log_a N}{\log_a b}$ (change of base)

Arithmetic Progressions (p. 372)

$$a_n = a_1 + (n-1)d$$

$$S_n = \frac{n(a_1 + a_n)}{2}$$

Geometric Progressions (p. 377, 382)

$$a_n = a_1 r^{n-1}$$

$$S_n = \frac{a_1(1 - r^n)}{1 - r} \qquad r \neq 1$$

$$S_\infty = \frac{a_1}{1 - r} \qquad |r| < 1$$

The Binomial Expansion (p. 386)

$$(a+b)^n = 1\,(a)^n (b)^0 + \frac{n}{1}(a)^{n-1}(b)^1 + \frac{n(n-1)}{1 \cdot 2}(a)^{n-2}(b)^2 + \frac{n(n-1)(n-2)}{1 \cdot 2 \cdot 3}(a)^{n-3}(b)^3 + \cdots$$

INTERMEDIATE ALGEBRA

INTERMEDIATE ALGEBRA

Alden T. Willis

Carol L. Johnston
East Los Angeles College

Wadsworth Publishing Company
Belmont, California

A Division of Wadsworth, Inc.

Mathematics Editor: Richard Jones
Editorial/Production services: Phoenix Publishing Services, San Francisco
Designer: Brenn Lea Pearson
Cover Photograph: © David Powers 1981

Library of Congress Cataloging in Publication Data

Willis, Alden T
Intermediate algebra.

Includes index.
1. Algebra. I. Johnston, Carol Lee, joint author.
II. Title.
QA154.2.J63 1981 512.9 80-24353
ISBN 0-534-00887-9

Printed in the United States of America 1 2 3 4 5 6 7 8 9 10—85 84 83 82 81

CONTENTS

PREFACE

In this book we use simple, direct language to present the essential content of intermediate algebra for students who have already had beginning algebra. It has been our experience that two kinds of students study intermediate algebra: students who have just completed a course in beginning algebra, and those who have not had algebra for several years. For the benefit of students who need a review, whenever possible we begin the discussion of a topic by using familiar material from beginning algebra. The discussion then quickly advances to the level of intermediate algebra.

Some of the major features of this book are:

1. The contents are arranged in **small sections,** each with its own examples and exercises. We use a one-step, one-concept-at-a-time approach.
2. After careful and detailed topic explanation, we use many concrete examples that lead up to general algebraic principles which can then be used to work the exercises.
3. Important concepts and algorithms are enclosed in **boxes** for easy identification and reference.
4. Common algebraic errors are clearly identified in special **"Words of Caution."** This helps to reduce the number of mistakes commonly made by the inexperienced student.
5. Liberal use is made of **visual aids** such as the number line, shading, and other graphics.
6. Special attention is given to the **operations with zero** in Chapter One, including a discussion of the common errors students make with zero.

7. Our approach to solving **word problems** is to provide a detailed method for changing a word statement into an algebraic equation. In addition, we emphasize word problems having wide application: Ratio, proportion, percent, and variation.

8. The importance of **checking solutions** is stressed throughout the book.

9. We have included **over 700 examples** which present orderly solutions clarifying points that often prove difficult for students. With few exceptions, there are examples illustrating solutions for all exercises other than the Brainteasers.

10. The even-numbered exercises parallel the odd exercises so that assignment of either set provides complete coverage of the material.

11. **Answers** to all **odd-numbered exercises,** as well as many selected solutions for these odd exercises, are provided in the back of this text. No answers for the **even-numbered exercises** are given in this textbook. A **Student's Solutions Manual** containing solutions for all even-numbered exercises is available for student use.

12. An **Instructor's Manual** is available that contains:
    i. Four different *tests for each chapter.*
    ii. Two *final examinations.*
    iii. A *pretest* for Intermediate Algebra.
    iv. *Keys* for all of the above tests.
    v. *Student answer forms* for all tests.

    All of the above items may be easily removed and duplicated for class use.

13. To prepare a student for taking a chapter test, a **diagnostic test** is included at the end of each chapter. Complete solutions to all problems in these diagnostic tests, together with section references, are given in the answer section.

14. A comprehensive **summary** is included at the end of each chapter.

We wish to thank our many friends for their valuable suggestions. In particular, we are deeply grateful to: William Cipriano, Roger Debelak, Tom Drouet, Robert Kester, Jeanne Lazaris, David Morin, Edward Rosenberg, Paddie Sampson, and Merwin Waite, of East Los Angeles College. ■

Alden Willis
Carol Johnston

INTERMEDIATE ALGEBRA

1 REVIEW OF ELEMENTARY TOPICS

Intermediate algebra is made up of two types of topics: (1) Those that were introduced in beginning algebra and are expanded in this course, and (2) new topics not covered in beginning algebra. With most topics we begin with the ideas learned in beginning algebra, and then develop these ideas further.

In Chapter One we review sets, the properties of real numbers, integral exponents, polynomials, and simplifying and evaluating algebraic expressions.

1–1 Sets

Ideas in all branches of mathematics—arithmetic, algebra, geometry, calculus, statistics, etc.—can be explained in terms of sets. For this reason you will find it helpful to have a basic understanding of sets.

Set

A **set** is a collection of objects or things.

Example 1 Sets

(a) The set of students attending college in the United States.

(b) The set *natural numbers:* 1, 2, 3, etc.

Element of a Set

The objects or things that make up a set are called its **elements** (or *members*). Sets are usually represented by listing their elements, separated by commas, within braces { }.

Example 2 Elements of sets

(a) Set {5, 7, 9} has elements 5, 7, and 9.

(b) Set $\{a, d, f, h, k\}$ has elements $a, d, f, h,$ and k.

A set may contain just a few elements, many elements, or no elements at all. A set is usually named by a capital letter such as A, N, W, etc. Letters used as elements of sets are usually written in lowercase letters. The expression "$A = \{m, t, c\}$" is read "A is the set whose elements are m, t, and c."

Roster Method

A class roster is a list of the members of the class. When we represent a set by $\{3, 8, 9, 11\}$, we are representing the set by a **roster** (or *list*) of its members.

Modified Roster Method. Sometimes the number of elements in a set is so large that it is not convenient or even possible to list all its members. In such cases we modify the roster notation. For example, the set of natural numbers can be represented as follows:

$$\{1, 2, 3, \ldots\}$$

This is read "The set whose elements are 1, 2, 3, and so on." The three dots to the right of the number 3 indicate that the remaining numbers are to be found by counting in the same way we have begun: Namely, by adding 1 to the preceding number to find the next number.

Equal Sets

Two sets are **equal** if they both have exactly the same members.

Example 3 Equal sets and unequal sets

(a) $\{1, 5, 7\} = \{5, 1, 7\}$. Notice that both sets have exactly the same elements, even though they are not listed in the same order.

(b) $\{1, 5, 5, 5\} = \{5, 1\}$. Notice that both sets have exactly the same elements. It is not necessary to write the same element more than once when writing the roster of a set.

(c) $\{7, 8, 11\} \neq \{7, 11\}$. These sets are not equal because they both do not have exactly the same elements.

Rule Method (Set-Builder)

A set can also be represented by giving a **rule** describing its members in such a way that we definitely know whether a particular element is in that set or is not in that set.

$$\{x \mid x \text{ is an even integer}\}$$

is read "The set of all x such that x is an even integer."

The rule

Example 4 Changing from set-builder notation to roster notation
Write $\{x \mid x + 2 = 5 \text{ and } x \text{ is a natural number}\}$ in roster notation.

$$\{x \mid x + 2 = 5 \text{ and } x \text{ is a natural number}\} = \{3\}$$

Example 5 Changing from roster notation to set-builder notation
Write $\{0, 3, 6, 9\}$ in set-builder notation.

$$\{0, 3, 6, 9\} = \{x \mid x \text{ is a digit exactly divisible by } 3\}$$

The Symbol $\in$

The expression $2 \in A$ is read "2 is an element of set A." If $A = \{2, 3, 4\}$, we can say: $2 \in A$, $3 \in A$, and $4 \in A$. If we wish to show that a number or object is *not* an element of a given set, we use the symbol $\notin$, which is read "is not an element of." If $A = \{2, 3, 4\}$, then $5 \notin A$, which is read "5 is not an element of set A." Notice that $\in$ looks like the first letter of the word *element*.

Example 6 The use of $\in$ and $\notin$
If $F = \{x \mid x \text{ is a negative integer}\}$, then $-7 \in F$, $-33 \in F$, and $6 \notin F$.

Empty Set (Null Set)

Set $B = \{1, 5\}$ has two elements. Set $C = \{5\}$ has only one element. Set $D = \{\ \}$ is empty, since it has no elements. A set having no elements is called the **empty set** (or *null set*). We use the symbol $\{\ \}$ or $\varnothing$ to represent the empty set.

Example 7 The empty set

(a) The set of all people in your math class who are 10 ft tall $= \varnothing$.

(b) The set of all digits that are greater than $10 = \{\ \}$.

Universal Set

A **universal set** is a set that consists of all the elements being considered in a particular problem. The universal set is represented by the letter U.

Example 8 Universal sets

(a) Suppose we are going to consider only digits. Then $U = \{0, 1, 2, 3, 4, 5, 6, 7, 8, 9\}$, since U contains *all* digits.

(b) Suppose we are going to consider the whole numbers. Then $U = \{0, 1, 2, 3, \ldots\}$, since U contains *all* whole numbers.

Notice that there can be different universal sets.

Finite Set

If in counting the elements of a set the counting comes to an end, the set is called a **finite set.**

Example 9 Finite sets

(a) $A = \{5, 9, 10, 13\}$

(b) $\varnothing = \{\ \}$

Infinite Set

If in counting the elements of a set the counting *never* comes to an end, the set is called an **infinite set**.

Example 10 Infinite sets

(a) $N = \{1, 2, 3, \ldots\}$ The natural numbers

(b) $J = \{x \mid x \text{ is an integer}\}$

Subsets

A set A is called a **subset** of set B if every member of A is also a member of B. "A is a subset of B" is written "$A \subseteq B$."

Note: The symbol $\subseteq$ is used to indicate that one *set* is a *subset* of another set. The symbol $\in$ is used to indicate that a particular *element* is a *member* of a particular set. ■

Example 11 Subsets

(a) $A = \{3, 5\}$ is a subset of $B = \{3, 5, 7\}$ because every member of A is also a member of B. Therefore $A \subseteq B$.

(b) $F = \{10, 7, 5\}$ is a subset of $G = \{5, 7, 10\}$ because every member of F is also a member of G. Every set is a subset of itself.

(c) $D = \{4, 7\}$ is *not* a subset of $E = \{7, 8, 5\}$ because $4 \in D$, but $4 \notin E$. Therefore $D \not\subseteq E$, which is read "D is not a subset of E."

(d) The empty set is a subset of every set. $\{\ \}$ is a subset of A because there is no member of $\{\ \}$ that is not also a member of A.

Example 12 All the subsets of the set $\{a, b, c\}$

$\{\ \}$	the subset having *no* elements	The empty set is a subset of every set
$\{a\}, \{b\}, \{c\}$	the subsets having *one* element	
$\{a, b\}, \{a, c\}, \{b, c\}$	the subsets having *two* elements	
$\{a, b, c\}$	the subset having *three* elements	Every set is a subset of itself

Venn Diagrams

A useful tool for helping you understand set concepts is the **Venn diagram.** A simple Venn diagram is shown in Figure 1-1A.

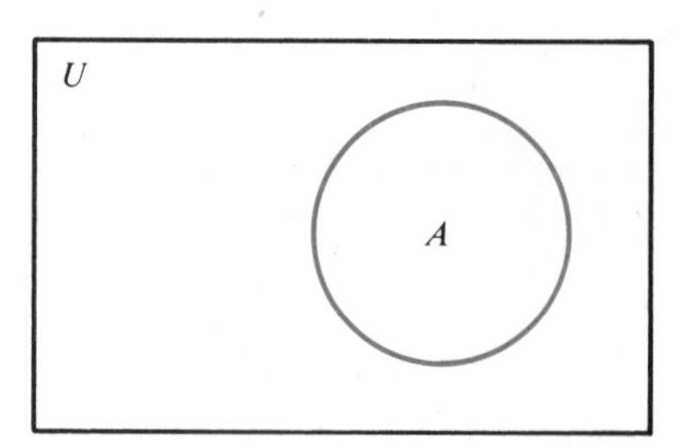

Figure 1–1A Venn Diagram

All elements of the universal set U are considered to lie in the rectangle labeled U. All elements of set A are considered to lie in the circle labeled A. Since all elements of A also lie in rectangle U, $A \subseteq U$.

Union of Sets

The **union of sets** A and B, written $A \cup B$, is the set that contains all the elements of A as well as all the elements of B. For an element to be in $A \cup B$, it must be in set A *or* set B.

In the Venn diagram (Figure 1-1B) $A \cup B$ is represented by the shaded area. In terms of set notation, suppose $A = \{b, c, g\}$ and $B = \{1, 2, 5, 7\}$. Then $A \cup B = \{b, c, g, 1, 2, 5, 7\}$.

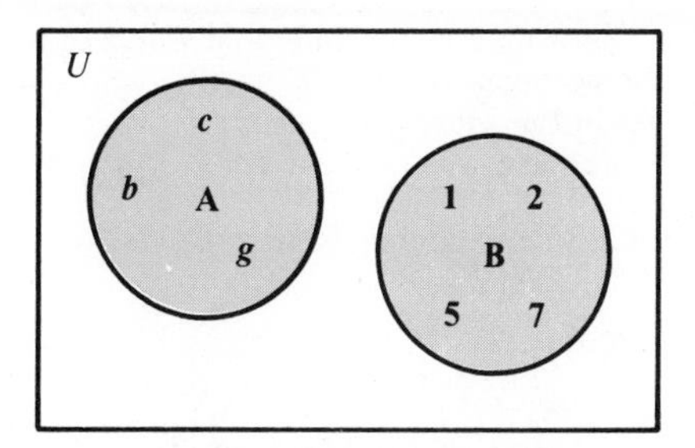

Figure 1–1B $A \cup B$

Intersection of Sets

The **intersection of sets** C and D, written $C \cap D$, is the set that contains only elements in *both* C and D. For an element to be in $C \cap D$, it must be in set C *and* set D.

Consider $C = \{g, f, m\}$ and $D = \{g, m, t, z\}$. Then $C \cap D = \{g, m\}$ because g and m are the only elements in both C and D. In the Venn diagram (Figure 1-1C), the shaded area represents $C \cap D$ because that area lies in both circles.

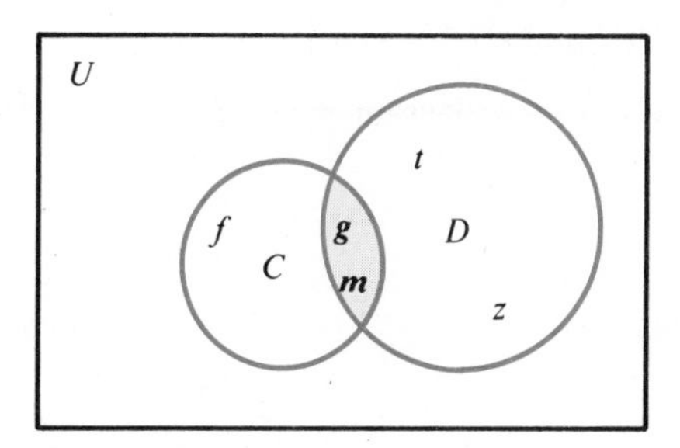

Figure 1–1C $C \cap D$

Disjoint Sets

Disjoint sets are sets whose intersection is the empty set. Figure 1-1B is an example of a Venn diagram showing disjoint sets A and B.

Example 13 Disjoint sets

(a) If $A = \{1, 2\}$ and $B = \{3, 4\}$, then $A \cap B = \varnothing$. Therefore A and B are disjoint sets.

(b) If $R = \{5, 7, 9\}$ and $T = \{9, 10, 12\}$, then $R \cap T = \{9\} \neq \varnothing$. Therefore R and T are *not* disjoint sets; they are intersecting sets.

EXERCISES 1-1

1. Are $\{2, 2, 7, 7\}$ and $\{7, 2\}$ equal sets?

2. State which of the following sets are finite and which are infinite:
(a) The set of whole numbers
(b) The set of books in the Library of Congress
(c) The set of fish in all the seas of the earth at this instant

3. State which of the following statements are true and which are false:
(a) If $B = \{x, y, z, w\}$, then $a \notin B$
(b) $0 \in \varnothing$
(c) $\varnothing \notin \{\ \}$

4. If $A = \{3, 5, 10, 11\}$, $B = \{3, 5, 12\}$, and $C = \{5, 3\}$, state which of the following statements are true and which are false:
(a) $B \subseteq A$ (b) $C \subseteq B$ (c) $C \nsubseteq A$

5. Write each of the given sets in roster notation.
(a) $\{x \mid x \text{ is an even digit}\}$
(b) $\{x \mid x - 1 = 3 \text{ and } x \text{ is a natural number}\}$

6. Write each of the given sets in set-builder notation.
(a) $\{0, 4, 8\}$ (b) $\{10, 20, 30, \ldots\}$

7. Given $X = \{2, 5, 6, 11\}$, $Y = \{5, 7, 11, 13\}$, and $Z = \{0, 3, 4, 6\}$:
(a) Write $X \cap Y$ in roster notation
(b) Write $X \cup Z$ in roster notation
(c) Find any two sets that are disjoint

8. Shade in $A \cup B$ in the Venn diagram given here.

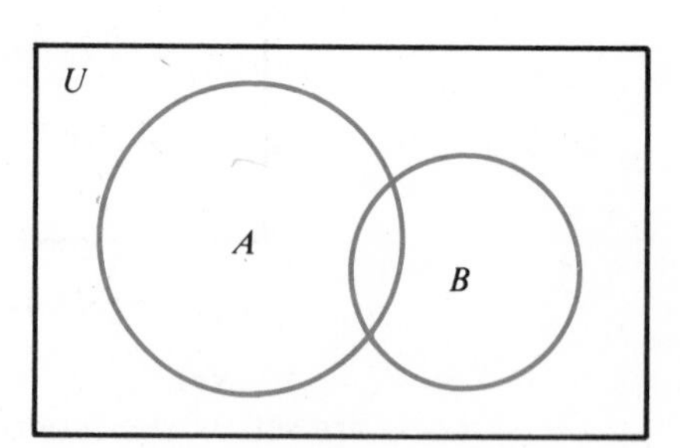

9. Write the name of the set represented by the shaded area.

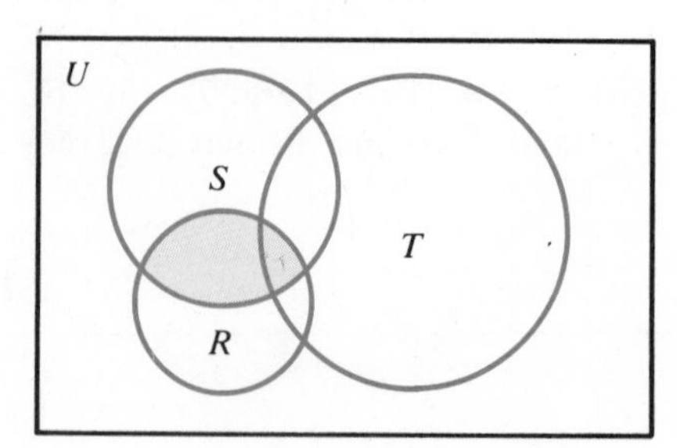

10. Write the name of the set represented by the shaded area.

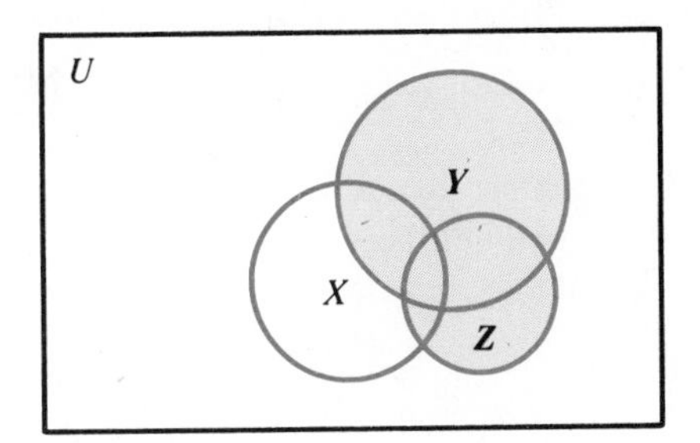

1–2 Real Numbers

Real Numbers

All the numbers that can be represented by points on the *number line* are called **real numbers** (Figure 1-2A). We represent the set of real numbers by R.

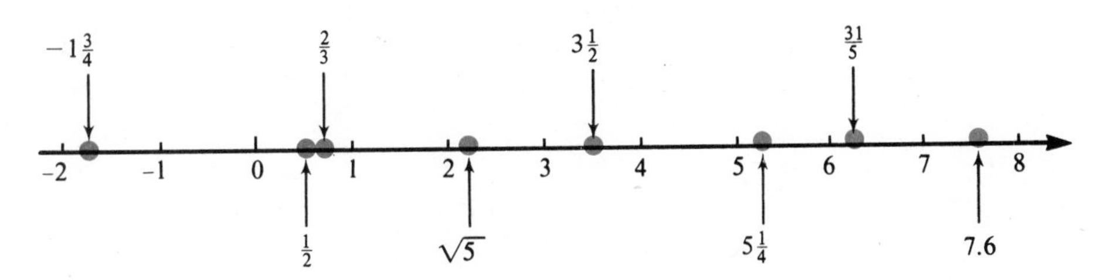

Figure 1–2A Number Line

With one exception, all the numbers used in this book are real numbers (see Complex Numbers, Section 6–8).

Some subsets of the real numbers are:

The set of digits, $D = \{0, 1, 2, 3, 4, 5, 6, 7, 8, 9\}$

Any number can be written by using some or all of these ten digits.

The set of natural numbers, $N = \{1, 2, 3, \ldots\}$

The set of whole numbers, $W = \{0, 1, 2, \ldots\}$

The set of integers, $J = \{\ldots, -3, -2, -1, 0, 1, 2, 3, \ldots\}$

The set of rational numbers, $Q = \left\{\frac{a}{b} \middle| a, b \in J, \quad b \neq 0\right\}$

The set of integers is a subset of the set of rational numbers ($J \subseteq Q$) because any integer can be written as a rational number. For example:

$$5 = \frac{5}{1}$$

The set of terminating decimals is a subset of the set of rational numbers Q because any terminating decimal can be written as a rational number. For example:

$$1.03 = \frac{103}{100}$$

The set of repeating, nonterminating decimals is a subset of the set of rational numbers Q because any nonterminating, repeating decimal can be written as a rational number. For example:

$$0.666\ldots = \frac{2}{3}$$

The set of mixed numbers is a subset of the set of rational numbers Q because any mixed number can be written as a rational number. For example:

$$2\frac{1}{3} = \frac{7}{3}$$

It is also true that $N \subseteq Q$ and $W \subseteq Q$.

The set of irrational numbers, $H = \{x \mid x \in R,\ x \notin Q\}$. This set of irrational numbers consists of those real numbers that are not rational numbers.

$$\sqrt{3}, \quad \sqrt[3]{5}, \quad \text{and} \quad \pi \quad \text{are examples of irrational numbers}$$

Any nonterminating, nonrepeating decimal is an irrational number.

Some of the relationships just discussed are shown in Figure 1-2B.

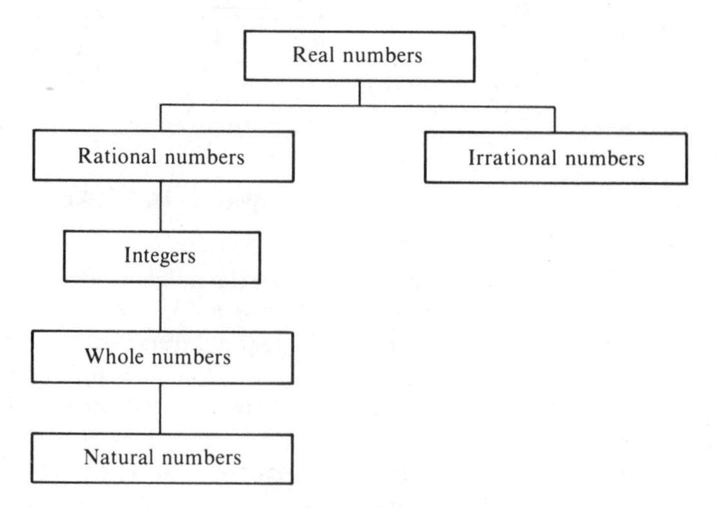

Figure 1–2B

The symbols $>$ and $<$ are called *inequality symbols*. Let x be any number on the number line. Then numbers to the right of x on the number line are said

Greater Than
Less Than

to be *greater than* x, written "$> x$." Numbers to the left of x on the number line are said to be *less than* x, written "$< x$." The arrowhead on the number line indicates the direction in which numbers get larger.

Example 1 $-1 > -4$, read "-1 is greater than -4"

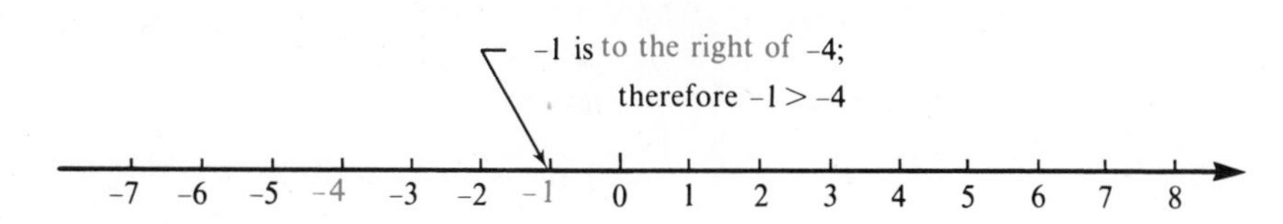

Example 2 $-2 < 1$, read "-2 is less than 1"

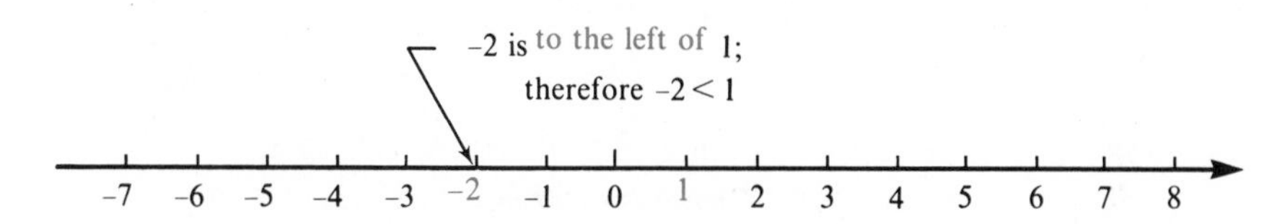

Example 3 "Greater than" and "less than" symbols

(a) $-\frac{1}{2} > -1$ because $-\frac{1}{2}$ is to the right of -1 on the number line.

(b) $-20 < -10$ because -20 is to the left of -10 on the number line.

Note: The statements $7 > 6$ and $6 < 7$ give the same information even though they are read differently. ■

Less Than
or Equal to

The inequality $a \leq b$ is read "a is less than *or* equal to b."

This means: if $\begin{Bmatrix} \text{either } a < b \\ \text{or } \quad a = b \end{Bmatrix}$ is true, then $a \leq b$ is true.

For example, $2 \leq 3$ is true because $2 < 3$ is true (even though $2 = 3$ is *not* true). Remember, only *one* of the two statements $\begin{Bmatrix} 2 < 3 \\ 2 = 3 \end{Bmatrix}$ need be true in order that $2 \leq 3$ be true.

Greater Than
or Equal To

The inequality $a \geq b$ is read "a is greater than *or* equal to b."

This means: if $\begin{Bmatrix} \text{either } a > b \\ \text{or } \quad a = b \end{Bmatrix}$ is true, then $a \geq b$ is true.

For example, $5 \geq 1$ is true because $5 > 1$ is true (even though $5 = 1$ is *not* true). Remember, only *one* of the two statements $\begin{Bmatrix} 5 > 1 \\ 5 = 1 \end{Bmatrix}$ need be true in order that $5 \geq 1$ be true.

Combined Inequalities

Example 4 Combined inequalities

(a) $4 < x < 9$ is read "4 is less than x is less than 9."

This means: $\left\{\begin{array}{r} 4 < x \\ \textit{and } x < 9 \end{array}\right\}$. Therefore x is between 4 and 9.

(b) $10 > x > -3$ is read "10 is greater than x is greater than -3."

This means: $\left\{\begin{array}{r} 10 > x \\ \textit{and } x > -3 \end{array}\right\}$. Therefore x is between 10 and -3.

Note: Even though the statement $10 > x > -3$ is correct as written, it is customary to write combined inequalities such as this one so that the numbers are in the same order that they appear on the number line: $-3 < x < 10$. ■

(c) $x \in J$ and $-3 < x \leq 2$

This means: $\left\{\begin{array}{l} \quad x \text{ is an integer} \\ \textit{and } x \text{ lies between } -3 \text{ and } 2 \\ \textit{or} \quad x \text{ can equal } 2 \end{array}\right\}$.

Therefore $x \in \{-2, -1, 0, 1, 2\}$.

Slash Line

A **slash line** drawn through a symbol puts a "not" in the meaning of the symbol.

Example 5 The use of the slash line

(a) $4 \neq 5$ is read "4 is *not* equal to 5."

(b) $3 \not< -2$ is read "3 is *not* less than -2."

(c) $-6 \not> -5$ is read "-6 is *not* greater than -5."

Absolute Value

The **absolute value** of a real number x is written $|x|$, where

$$|x| = \begin{cases} x & \text{if } x \geq 0 \\ -x & \text{if } x < 0 \end{cases}$$

x is greater than zero, or x equals 0

x is less than zero

$-x$ is *not* a negative number in this case because x is a negative number, and the negative of a negative number is positive; for example, if $x = -2$, then $-x = -(-2) = 2$

Example 6 The use of the absolute value sign

(a) $|15| = 15$

(b) $|0| = 0$

(c) $|-23| = \underbrace{-(-23)} = 23$

Because -23 is negative

The absolute value of a number is positive (or zero). The absolute value of a number is the number of units the point representing that number is from the point representing 0 on the number line. See Figure 1-2C.

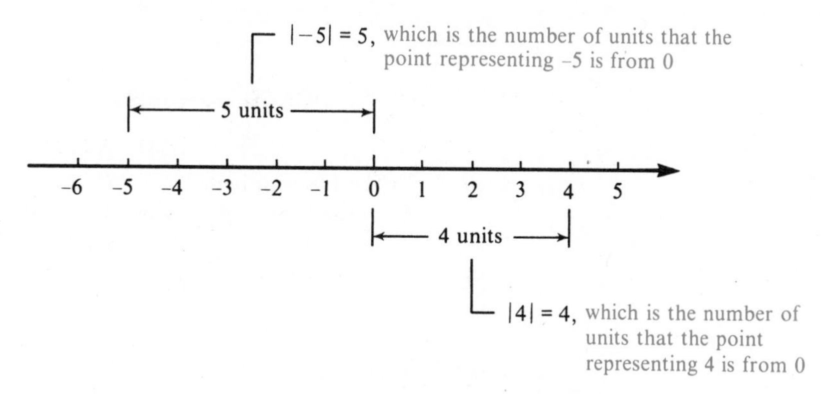

Figure 1-2C Absolute Value

EXERCISES 1-2

1. Which of the sets, R, D, N, W, J, Q, and H, have the given number as an element?
   (a) 7 (b) -5 (c) 30 (d) 2.5 (e) $\sqrt{6}$ (f) $-\frac{3}{4}$
2. Which of the two symbols, $<$ and $>$, should be used to make each statement true?
   (a) $9 \underline{\ ?\ } 4$ (b) $0 \underline{\ ?\ } -5$ (c) $-15 \underline{\ ?\ } -13$ (d) $-6 \underline{\ ?\ } -17$ (e) $-6 \underline{\ ?\ } |-17|$

In Exercises 3 and 4, what integers can be used in place of x to make the statement true? Use the roster method to express the answers.

3. $-8 < x < -5$
4. $-2 \leq x < 3$
5. Evaluate each expression.
   (a) $|5|$ (b) $|-3|$ (c) $|-14|$
6. Write each of the given sets in roster notation.
   (a) $\{x \mid 4 \leq x < 7 \text{ and } x \in W\}$ (b) $\{x \mid -4 < x < 0 \text{ and } x \in J$
7. Replace the question mark with either $\subseteq$ or $\in$ to form a true statement.
   (a) $\{1, 3\} \underline{\ ?\ } J$ (b) $\frac{3}{5} \underline{\ ?\ } R$ (c) $N \underline{\ ?\ } R$
8. Given $A = \{1, 3, 5\}$, $B = \{2, 4, 6\}$, and $C = \{5, 6, 7\}$. Write the set $(A \cup B) \cap C$ in roster notation.
9. Replace the question mark with either $\subseteq$ or $\in$ to form a true statement.
   (a) $\{0, 5\} \underline{\ ?\ } W$ (b) $J \underline{\ ?\ } R$ (c) $5 \underline{\ ?\ } J$
10. List all the subsets of $\{1, 2, 3\}$.

1-3 Operations with Signed Numbers

1-3A THE FOUR FUNDAMENTAL OPERATIONS

ADDITION

TO ADD TWO SIGNED NUMBERS

Case 1. When the numbers have the same sign	1st Add their absolute values. 2nd The sum has the same sign as both numbers.
Case 2. When the numbers have different signs	1st Subtract the smaller absolute value from the larger. 2nd The sum has the sign of the number that has the larger absolute value.

Example 1 Adding signed numbers

(a) $(3) + (5) = +8$ (Case 1)
- 8: Sum of absolute values
- +: Same sign as both numbers

(b) $(-7) + (-11) = -18$ (Case 1)
- 18: Sum of absolute values $(7 + 11 = 18)$
- −: Same sign as both numbers

(c) $(+18) + (-32) = -14$ (Case 2)
- 14: Difference of absolute values $(32 - 18 = 14)$
- −: Sign of number that has larger absolute value (-32)

(d) $(-9) + (25) = +16$ (Case 2)
- 16: Difference of absolute values $(25 - 9 = 16)$
- +: Sign of number that has larger absolute value (25)

Addition of Zero The sum of zero and any real number is the identical real number. For this reason zero is called the **additive identity.**

$$a + 0 = 0 + a = a$$

SUBTRACTION

To Find the Negative of a Number

1. Change the sign of the number.
   The negative of $b = -b$.
   The negative of $-b = -(-b) = b$.
2. The negative of $0 = 0$.

Note: Zero is the only real number whose negative is equal to the number itself. ■

The negative of any real number can also be found by multiplying it by (-1).

$$(-1) \cdot a = a \cdot (-1) = -a$$

Definition of Subtraction

To subtract b from a, add the negative of b to a.

$$a - b = a + (-b)$$

TO SUBTRACT ONE SIGNED NUMBER FROM ANOTHER

1. Change the sign of the number being subtracted.
2. Add the resulting signed numbers.

Example 2 Subtracting signed numbers

(a) Subtract 9 from 6. [This means $(6) - (9)$.]
$(6) - (9) = (6) + (-9) = -3$

(b) $(-7) - (-11) = (-7) + (11) = 4$

(c) $(+13) - (-14) = (13) + (14) = 27$

(d) $(-147) - (+59) = (-147) + (-59) = -206$

Subtractions Involving Zero

If a is any real number, then

1. $a - 0 = a$
2. $0 - a = 0 + (-a) = -a$

Subtraction is called the *inverse* of addition because it "undoes" addition.

Example 3

Note that 5 is where we begin and 5 is where we end

$5 + 4 = 9$ (Addition of 4) $\qquad$ $9 - 4 = 5$ (Subtraction of 4)

$$6 \times 2 = 12$$

Factors: 6, 2 — Product: 12

Factors

The numbers 6 and 2 are **factors** of 12; 12 is the **product** of 6 and 2. The numbers 3 and 4 are also factors of 12, because $3 \times 4 = 12$.

TO MULTIPLY TWO SIGNED NUMBERS

1. Multiply their absolute values.
2. The product is *positive* when the signed numbers have the same sign.

   The product is *negative* when the signed numbers have different signs.

Multiplicative Identity

Any number multiplied by 1 gives the identical number. For this reason 1 is called the **multiplicative identity.**

Example 4 Multiplying signed numbers

(a) $(-7)(4) = -28$

Product of their absolute values: $7 \times 4 = 28$
Product is negative because the numbers have different signs

(b) $(-14)(-10) = +140$

$14 \times 10 = 140$
Because the numbers have the same sign

Multiplication by Zero

The product of zero and any real number is zero.

$$a \cdot 0 = 0 \cdot a = 0$$

DIVISION

$$\begin{array}{r} \text{Quotient} \\ \text{Divisor}\overline{)\text{Dividend}} \end{array}$$

To check the answer in a division problem:

$$\text{Divisor} \times \text{Quotient} = \text{Dividend}$$

Division is called the *inverse* of multiplication because it "undoes" multiplication.

Example 5

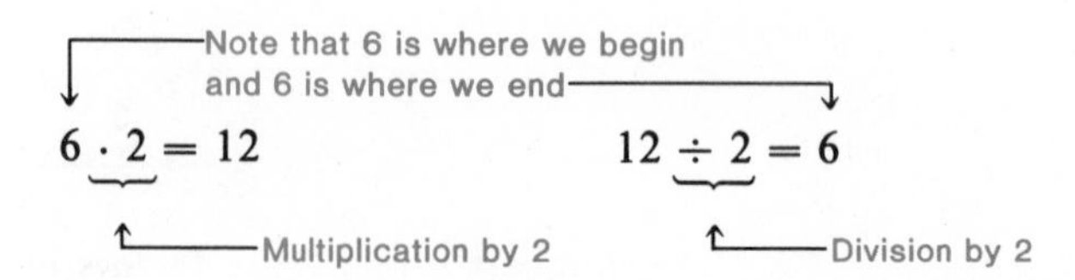

Because of this inverse relation between division and multiplication, the rules for finding the sign of a quotient are the same as the ones used for finding the sign of a product.

TO DIVIDE ONE SIGNED NUMBER BY ANOTHER

1. Divide their absolute values.
2. The quotient is *positive* when the signed numbers have the same sign.

   The quotient is *negative* when the signed numbers have different signs.

Example 6 Dividing signed numbers

(a) $(-30) \div (5) = -6$

- Quotient of absolute values: $30 \div 5 = 6$
- Negative because the numbers have different signs

(b) $\dfrac{-64}{-8} = +8$

- Quotient of absolute values: $\dfrac{64}{8} = 8$
- Positive because the numbers have the same sign

Divisions Involving Zero

DIVISIONS INVOLVING ZERO

1. Division of zero by any nonzero real number is zero.

   $\dfrac{0}{a} = 0 \quad (a \neq 0)$ See Example 7

2. Division of a nonzero real number by zero is not possible.

   $\dfrac{a}{0}$ is not possible $\quad (a \neq 0)$ See Example 8

3. *Division of zero by zero cannot be determined.*

   $\dfrac{0}{0}$ cannot be determined See Example 9

Example 7 Division of zero by any nonzero real number is zero

$$\frac{0}{a} = 0 \div a = a\overline{)0}^{\,0}$$

Because $a \cdot 0 = 0$
(Divisor × Quotient = Dividend)

Example 8 Division of a nonzero real number by zero is not possible

$$\frac{4}{0} = 4 \div 0 = 0\overline{)4}^{\,?}$$

Suppose the quotient is some unknown number we call x. Then

$$0\overline{)4}^{\,x} \text{ means } 0 \cdot x = 4, \text{ which is certainly false}$$

$$0 \cdot x \neq 4 \text{ because any number multiplied by zero} = 0$$

Therefore dividing any nonzero number by zero is impossible.

Example 9 Division of zero by zero cannot be determined

$$\frac{0}{0} = 0\overline{)0}^{\,1} \quad \text{means} \quad 0 \cdot 1 = 0, \text{ which is true}$$

$$\frac{0}{0} = 0\overline{)\;0}^{\,156} \quad \text{means} \quad 0 \cdot 156 = 0, \text{ which is true also!}$$

In other words, $0 \div 0 = 1$ and also 156. In fact, it can be any number. Therefore we do not know what answer to put for $0 \div 0$. For these reasons we say $0 \div 0$ cannot be determined.

EXERCISES 1-3A

Perform the indicated operations.

1. $(-5) + (-9)$
2. $(-10) + (-7)$
3. $(12) + (-7)$
4. $(17) + (-8)$
5. $(-74) + (35)$
6. $(-105) + (71)$
7. $(-17) + (0)$
8. $(0) + (-25)$
9. $(-13.5) + (-8.06)$
10. $(-24.08) + (-17.9)$
11. $(97.168) + (-39.779)$
12. $(-458.16) + (79.83)$
13. $(-15) - (11)$
14. $(-24) - (13)$
15. $(156) - (-97)$
16. $(284) - (-89)$
17. $(780) - (840)$
18. $(579) - (700)$
19. $(-18) - (0)$
20. $(0) - (-29)$
21. $(-8.96178) - (-3.75296)$
22. $(-6.48352) - (-2.79648)$
23. $(-26)(10)$
24. $(-20)(17)$
25. $(-11)(-7)$
26. $(-16)(-3)$
27. $(0)(-48)$
28. $(-31)(0)$
29. $(17.5)(-150)$
30. $(12.5)(-16)$

31. $(163.84)(-156.25)$

32. $(-655.36)(-78.125)$

33. $\frac{-150}{10}$

34. $\frac{-250}{100}$

35. $36 \div (-12)$

36. $56 \div (-8)$

37. $\frac{0}{-5}$

38. $\frac{0}{14}$

39. $\frac{-15}{-6}$

40. $\frac{-27}{-12}$

41. $\frac{24}{0}$

42. $\frac{-31}{0}$

43. $\frac{0}{0}$

44. $2 + \frac{0}{0}$

45. $\frac{7.5}{-0.5}$

46. $\frac{1.25}{-0.25}$

47. $(-10{,}240) \div (-655.36)$

48. $(-51{,}200) \div (-78.125)$

49. A team climbing Mt. Everest starts from a camp at 17,018 ft elevation. By evening they had climbed 2759 ft. What is their elevation at that time?

50. An airplane cruising at an altitude of 17,285 ft is ordered by a tower controller to climb 2500 ft. What will the plane's new altitude be?

51. At 6 AM the temperature in Hibbing, Minnesota, was $-35°$ F. If the temperature had risen 53° F by 2 PM, what was the temperature at that time?

52. At midnight in Billings, Montana, the temperature was $-50°$ F. By noon the temperature had risen 67° F. What was the temperature at noon?

53. A scuba diver descends to a depth of 141 ft below sea level. His buddy dives 68 ft deeper. What is his buddy's altitude at the deepest point of his dive?

54. When Fred checked his pocket altimeter at the seashore on Friday afternoon, it read -150 ft. Saturday morning it read 9650 ft when he checked it on the peak of a nearby mountain. Allowing for the obvious error in his altimeter reading, what is the correct height of that peak?

55. Mt. Everest (the highest known point on earth) has an altitude of 29,028 ft. The Mariana Trench in the Pacific Ocean (the lowest known point on earth) has an altitude of $-36{,}198$ ft. Find the difference in altitude of these two places.

56. An airplane is flying 75 ft above the level of the Dead Sea (elevation -1299 ft). How high must it climb to clear a 2573-ft peak by 200 ft?

57. A dune buggy starting from the floor of Death Valley (-282 ft) is driven to the top of a nearby mountain having an elevation of 5782 ft. What was the change in the dune buggy's altitude?

58. At 2 PM Cindy's temperature was 103.4° F. By 6 PM her temperature had dropped to 99.9° F. What was the drop in her temperature?

1-3B COMMUTATIVE AND ASSOCIATIVE PROPERTIES

Commutative Properties

Both addition and multiplication are commutative.

COMMUTATIVE PROPERTIES

If a and b are real numbers, then

1. Commutative Property of Addition

$$a + b = b + a$$

2. Commutative Property of Multiplication

$$a \cdot b = b \cdot a$$

Changing the order of the numbers in an addition or multiplication problem does not change the answer.

Example 1 Showing addition is commutative

$$(-6) + (2) \stackrel{?}{=} (2) + (-6) \quad \text{Order changed}$$
$$-4 = -4 \quad \text{Same answer}$$

Example 2 Showing multiplication is commutative

$$(-9)(3) \stackrel{?}{=} (3)(-9) \quad \text{Order changed}$$
$$-27 = -27 \quad \text{Same answer}$$

Neither subtraction nor division is commutative.

Example 3 Showing subtraction is *not* commutative

$$3 - 2 \stackrel{?}{=} 2 - 3 \quad \text{Order changed}$$
$$1 \neq -1 \quad \text{Different answers}$$

Example 4 Showing division is *not* commutative

$$10 \div 5 \stackrel{?}{=} 5 \div 10 \quad \text{Order changed}$$
$$2 \neq \frac{1}{2} \quad \text{Different answers}$$

Associative Properties

Both addition and multiplication are associative.

ASSOCIATIVE PROPERTIES

If a, b, and c are real numbers,

1. Associative Property of Addition

$$a + b + c = (a + b) + c = a + (b + c)$$

2. Associative Property of Multiplication

$$a \cdot b \cdot c = (a \cdot b) \cdot c = a \cdot (b \cdot c)$$

Changing the grouping of the numbers in an addition or multiplication problem does not change the answer.

Example 5 Showing addition is associative

$$(2 + 3) + 4 \stackrel{?}{=} 2 + (3 + 4) \quad \text{Grouping changed}$$
$$5 \quad + 4 \stackrel{?}{=} 2 + \quad 7$$
$$9 \quad = \quad 9 \quad \text{Same answer}$$

Example 6 Showing multiplication is associative

$$[(-6) \cdot (+2)] \cdot (-5) \stackrel{?}{=} (-6) \cdot [(+2) \cdot (-5)] \quad \text{Grouping changed}$$
$$[-12] \quad \cdot (-5) \stackrel{?}{=} (-6) \cdot \quad [-10]$$
$$60 \quad = \quad 60 \quad \text{Same answer}$$

Neither subtraction nor division is associative.

Example 7 Showing subtraction is *not* associative

$$(7 - 4) - 8 \stackrel{?}{=} 7 - (4 - 8) \quad \text{Grouping changed}$$
$$3 \quad - 8 \stackrel{?}{=} 7 - \quad (-4)$$
$$-5 \quad \neq \quad 11 \quad \text{Different answers}$$

Example 8 Showing division is *not* associative

$$[(-16) \div (4)] \div (-2) \stackrel{?}{=} (-16) \div [(4) \div (-2)] \quad \text{Grouping changed}$$
$$[-4] \quad \div (-2) \stackrel{?}{=} (-16) \div \quad [-2]$$
$$2 \quad \neq \quad 8 \quad \text{Different answers}$$

EXERCISES 1-3B

Tell whether each statement is true or false, and give the reason.

1. $(-5) + (10) = (10) + (-5)$

2. $x + y = y + x$

3. $(-3)[8+(-2)] = [(-2)+8](-3)$

4. $(-2)[(-5)+6] = [6+(-5)](-2)$

5. $8 \div (-2) = (-2) \div 8$

6. $(-5) \div 10 = 10 \div (-5)$

7. $(6)(-3)+(-4) = (-4)+(-3)(6)$

8. $(-8)(2) - 10 = 10 - (2)(-8)$

9. $(-7) - 4 = 4 - (-7)$

10. $x - y = y - x$

11. $8 + [(-2)+(-4)] = [(-4)+(-2)] + 8$

12. $[3+(-2)] + (-9) = (-9) + [3+(-2)]$

1-3C POWERS AND ROOTS OF SIGNED NUMBERS

Powers

In some products the same number is repeated as a *factor*.

$$\underbrace{3 \cdot 3 \cdot 3 \cdot 3} = 3^4 = 81$$

The 4 indicates that 3 is used as a factor 4 times

Exponential Notation

In the symbol 3^4, the 3 is called the **base.** The 4 is called the **exponent.** The entire symbol 3^4 is called the **fourth power of three** and is commonly read "three to the fourth power."

$$3^4 = 81$$

Exponent; Base; Fourth power of 3

A Word of Caution. Students often think that expressions such as $(-6)^2$ and -6^2 are the same. They are *not* the same. The exponent applies only to the symbol immediately preceding it.

In this statement $(-6)^2 = (-6)(-6) = 36$ the exponent applies to the ( )

In this statement $-6^2 = -(6 \cdot 6) = -36$ the exponent applies to the 6

Therefore, $-6^2 \neq (-6)^2$. ■

Even Power

If a base has an exponent that is exactly divisible by 2, we say that it is an **even power** of the base. For example, 3^2, 5^4, and $(-2)^6$ are even powers.

Odd Power

If a base has an exponent that is *not* exactly divisible by 2, we say that it is an **odd power** of the base. For example, 3^1, 10^3, and $(-4)^5$ are odd powers.

Example 1 Powers

(a) $1^4 = 1 \cdot 1 \cdot 1 \cdot 1 = 1$ — Any power of 1 is 1

(b) $(-2)^3 = (-2)(-2)(-2) = -8$ — An *odd* power of a negative number is *negative*

(c) $(-3)^2 = (-3)(-3) = 9$ — An *even* power of a negative number is *positive*

(d) $-3^2 = -(3 \cdot 3) = -9$ — Compare with (c)

(e) $0^2 = 0 \cdot 0 = 0$

POWERS OF ZERO

If a is any positive real number, then

$$0^a = 0$$

Cases where 0 is the *exponent* are discussed in Section 1–4B.

Example 2 Any positive power of zero is zero.

$$0^5 = 0 \cdot 0 \cdot 0 \cdot 0 \cdot 0 = 0$$

Square Roots

The **square root** of a number N is a number which, when squared, gives N. *Every positive number has both a positive and a negative square root.* The positive square root is called the **principal square root.** The principal square root of N is written $\sqrt{N}$. The negative square root of N is written $-\sqrt{N}$.

Example 3 The square roots of 9

3 is the principal square root of 9, written $3 = \sqrt{9}$

$$3^2 = 3 \cdot 3 = 9$$

It is also true that

-3 is a square root of 9, written $-3 = -\sqrt{9}$

$$(-3)^2 = (-3) \cdot (-3) = 9$$

$\sqrt{9}$ stands for 3, the principal root; therefore, $-\sqrt{9} = -3$

When the symbol $\sqrt{N}$ is used, it *always* represents the *principal square root of* N. Since the principal square root is positive, $\sqrt{N}$ is always positive (or zero).

Example 4 Finding principal square roots

(a) $\sqrt{36} = 6$ because $6^2 = 36$

(b) $\sqrt{0} = 0$ because $0^2 = 0$

(c) $\sqrt{1} = 1$ because $1^2 = 1$

In this section we discuss only the square roots of numbers greater than or equal to zero. Square roots of negative numbers will be discussed in Section 6–8.

Square Roots by Table (or Calculator)

Example 5 Find $\sqrt{94}$ using Table I (in the back of the book). Locate 94 in the column headed N. Then read the value of $\sqrt{94} = 9.695$ in the column headed $\sqrt{N}$. Table I gives square roots rounded off to three decimal places.

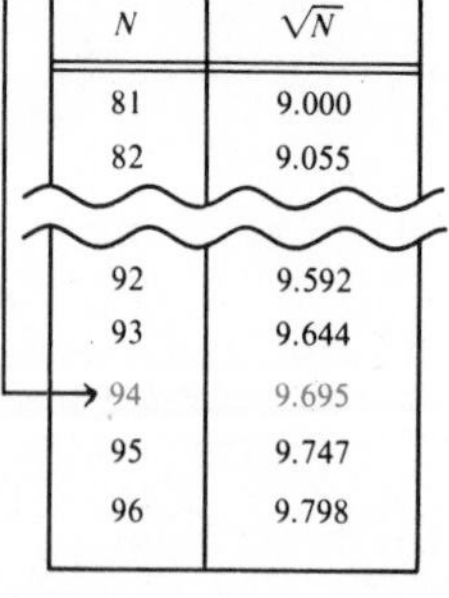

N	$\sqrt{N}$
81	9.000
82	9.055
92	9.592
93	9.644
94	9.695
95	9.747
96	9.798

$$\sqrt{94} \doteq 9.695 \text{ because } (9.695)^2 = 93.993025 \doteq 94$$

($\doteq$ Means "approximately equal to")

Further use of Table I for finding square roots is made in Section 8–4.

Square roots are often found by using calculators or computers. For example, a calculator with a square root key $\boxed{\sqrt{\ }}$ gives $\sqrt{94} = 9.695359714$. When 9.695359714 is rounded off to three decimal places, we get 9.695, which is the same number obtained from Table I.

Higher Roots

Roots other than square roots are called **higher roots.**

Example 6 Higher roots

(a) In the expression $(-2)^3$

Cube Roots

$$(-2)^3 = (-2)(-2)(-2) = -8$$

-2 is called the *principal** **cube root** of -8, written $\sqrt[3]{-8} = -2$

-8 is called the **cube** of -2, written $(-2)^3 = -8$ (*third power* of -2)

(b) In the expression 2^4

Fourth Root

$$2^4 = 2 \cdot 2 \cdot 2 \cdot 2 = 16$$

2 is called the *principal* **fourth root** of 16, written $\sqrt[4]{16} = 2$

16 is called the **fourth power** of 2, written $2^4 = 16$

ROOTS OF ZERO

If n is any positive integer, then

$$\sqrt[n]{0} = 0$$

Radicals

Indicated roots of numbers are called **radicals.** All the roots indicated in this section can be called radicals. Radicals are discussed in Chapter Six.

---

*More information about principal roots is given in Section 7–1.

Example 7 Showing the negative of a radical

(a) $-\sqrt{16} = -(4) = -4$ (b) $-\sqrt[3]{-8} = -(-2) = 2$

Inverse of a Root

Example 8 Finding the root of a number is the *inverse* of raising that number to a power because it "undoes" raising to that power.

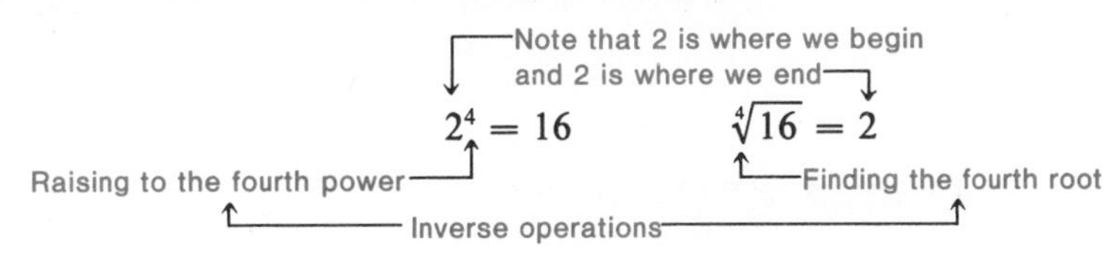

There is an inverse relation between any like power and root.

Real Numbers

All the numbers that can be represented by points on the number line are called **real numbers.** In Section 1–2 we showed that integers, fractions, mixed numbers, and decimals are real numbers. Principal roots are also real numbers and therefore can be represented by points on the number line (Figure 1–3A).

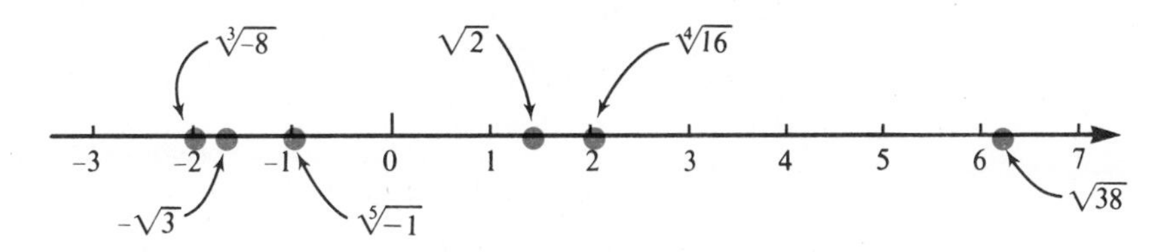

Figure 1–3A Representing Real Numbers on the Number Line

EXERCISES 1-3C

In Exercises 1–12, find the value of each expression.

1. $(-3)^4$	**2.** $(-2)^4$	**3.** -2^4	**4.** -3^4
5. 0^5	**6.** 0^6	**7.** $(-1)^{49}$	**8.** $(-1)^{50}$
9. 17.3^2	**10.** 9.2^3	**11.** $(-1.5)^5$	**12.** $(-2.5)^3$

In Exercises 13–16, find all square roots of each number.

13. 36	**14.** 49	**15.** 144	**16.** 256

In Exercises 17–26, evaluate each expression.

17. $\sqrt{25}$	**18.** $\sqrt{64}$	**19.** $-\sqrt{100}$	**20.** $-\sqrt{144}$
21. $\sqrt{81}$	**22.** $\sqrt{121}$	**23.** $-\sqrt{169}$	**24.** $-\sqrt{49}$
25. $\sqrt{225}$	**26.** $\sqrt{289}$		

In Exercises 27–38, find the square roots using Table I or a calculator.

27. $\sqrt{12}$	**28.** $\sqrt{17}$	**29.** $\sqrt{28}$	**30.** $\sqrt{58}$
31. $\sqrt{45}$	**32.** $\sqrt{73}$	**33.** $\sqrt{184}$	**34.** $\sqrt{191}$
35. $\sqrt{938}$	**36.** $\sqrt{85.2}$	**37.** $\sqrt{172.8}$	**38.** $\sqrt{517.6}$

In Exercises 39–56, evaluate each expression.

39. $\sqrt[3]{8}$ **40.** $\sqrt[5]{1}$ **41.** $-\sqrt[3]{27}$ **42.** $-\sqrt[4]{16}$
43. $\sqrt[3]{-64}$ **44.** $\sqrt[3]{-27}$ **45.** $-\sqrt[5]{32}$ **46.** $-\sqrt[3]{64}$
47. $\sqrt[3]{0}$ **48.** $-\sqrt[4]{0}$ **49.** $\sqrt[5]{-32}$ **50.** $\sqrt[4]{256}$
51. $-\sqrt[7]{-1}$ **52.** $\sqrt[3]{-125}$ **53.** $\sqrt[3]{-1000}$ **54.** $\sqrt[3]{-216}$
55. $-\sqrt[4]{81}$ **56.** $-\sqrt[6]{64}$

1–4 Integral Exponents

1–4A POSITIVE EXPONENTS

Multiplying Powers

RULE 1

> **MULTIPLYING POWERS**
>
> $$x^a \cdot x^b = x^{a+b}$$

Example 1 Using Rule 1

(a) $x^3 \cdot x^2 = (xxx)(xx) = xxxxx = x^5$

3 factors + 2 factors = 5 factors

Therefore, $x^3 \cdot x^2 = x^{3+2} = x^5$.

(b) $w \cdot w^7 = w^{1+7} = w^8$ When the base is written without an exponent, its exponent is understood to be 1

(c) $x^3y^2 = xxx \cdot yy = x^3y^2$ Rule 1 does not apply because the bases are different

(d) $10^9 \cdot 10^4 = 10^{9+4} = 10^{13}$

(e) $2^a \cdot 2^b = 2^{a+b}$

Power of a Power

RULE 2

> **POWER OF A POWER**
>
> $$(x^a)^b = x^{ab}$$

Example 2 Using Rule 2

(a) $(x^5)^3 = (x^5)(x^5)(x^5) = x^5 \cdot x^5 \cdot x^5 = x^{5+5+5} = x^{3 \cdot 5} = x^{5 \cdot 3} = x^{15}$

(b) $(x^5)^4 = x^{5 \cdot 4} = x^{20}$

(c) $(x)^4 = (x^1)^4 = x^{1 \cdot 4} = x^4$

(d) $(10^6)^2 = 10^{6 \cdot 2} = 10^{12}$

(e) $(2^a)^b = 2^{a \cdot b} = 2^{ab}$

An extension of Rule 2 is Rule 2A.

Power of a Product

RULE 2A

POWER OF A PRODUCT

$$(xy)^a = x^a y^a$$

Example 3 Using Rule 2A

(a) $(xy)^3 = (xy)(xy)(xy) = (xxx)(yyy) = x^3y^3$

(b) $(5a)^2 = 5^2a^2 = 25a^2$

Dividing Powers

RULE 3

DIVIDING POWERS

$$\frac{x^a}{x^b} = x^{a-b} \qquad (x \neq 0)^*$$

When using Rule 3 in Example 4 and Exercises 1–4A, we chose the exponent in the numerator larger than the exponent in the denominator in order to obtain a positive exponent in the quotient. Rule 3 is also true if the quotient has a negative exponent. We will discuss negative exponents in Section 1–4B.

Example 4 Using Rule 3

(a) $\dfrac{x^5}{x^3} = \dfrac{xxxxx}{xxx} = \dfrac{xxx \cdot xx}{xxx \cdot 1} = \dfrac{xxx}{xxx} \cdot \dfrac{xx}{1} = 1 \cdot \dfrac{xx}{1} = xx = x^2$

The value of this fraction is 1 (for $x \neq 0$)

Therefore, $\dfrac{x^5}{x^3} = x^{5-3} = x^2$.

(b) $\dfrac{r^{12}}{r^5} = r^{12-5} = r^7$

(c) $\dfrac{10^7}{10^3} = 10^{7-3} = 10^4$

(d) $\dfrac{y^3}{y} = \dfrac{y^3}{y^1} = y^{3-1} = y^2$

(e) $\dfrac{2^a}{2^b} = 2^{a-b}$

(f) $\dfrac{x^5}{y^2} = \dfrac{xxxxx}{yy} = \dfrac{x^5}{y^2}$ Rule 3 does not apply when the bases are different

---

*If $x = 0$ in Rule 3, it would mean dividing zero by zero, which cannot be determined. For example, $\dfrac{x^5}{x^2} = \dfrac{0^5}{0^2} = \dfrac{0 \cdot 0 \cdot 0 \cdot 0 \cdot 0}{0 \cdot 0} = \dfrac{0}{0}$, which cannot be determined (Section 1–3D). But by Rule 3, $\dfrac{0^5}{0^2} = 0^{5-2} = 0^3 = 0$, which is not correct.

A Word of Caution. An expression like $\frac{x^5 - y^3}{x^2}$ is usually left unchanged. It could be changed as follows:

$$\frac{x^5 - y^3}{x^2} = \frac{x^5}{x^2} - \frac{y^3}{x^2} = x^3 - \frac{y^3}{x^2}$$

A common mistake students make is to divide only x^5 by x^2, instead of dividing both x^5 and y^3 by x^2. Expressions of this form are discussed in Section 1–7C. ■

An extension of Rule 3 is Rule 3A.

Power of a Quotient

RULE 3A

POWER OF A QUOTIENT

$$\left(\frac{x}{y}\right)^a = \frac{x^a}{y^a} \qquad (y \neq 0)*$$

Example 5 Using Rule 3A

(a) $\left(\frac{x}{y}\right)^3 = \left(\frac{x}{y}\right)\left(\frac{x}{y}\right)\left(\frac{x}{y}\right) = \frac{xxx}{yyy} = \frac{x^3}{y^3}$

(b) $\left(\frac{a}{2}\right)^4 = \frac{a^4}{2^4} = \frac{a^4}{16}$

In this book, unless otherwise noted, none of the letters has a value that makes a denominator zero.

EXERCISES 1–4A

Use the rules of exponents to simplify each expression.

1. $10^2 \cdot 10^4$ **2.** $2^3 \cdot 2^2$ **3.** $x^2 \cdot x^5$ **4.** $y^6 \cdot y^3$

5. $\frac{a^8}{a^3}$ **6.** $\frac{x^5}{x^2}$ **7.** $2^x \cdot 2^y$ **8.** $3^m \cdot 3^n$

9. $5^z \cdot 5^{2z}$ **10.** $7^{3y} \cdot 7^y$ **11.** $(10^3)^2$ **12.** $(5^2)^4$

13. $(3^a)^b$ **14.** $(2^m)^n$ **15.** $(3^x)^x$ **16.** $(2^a)^a$

17. $(ab)^3$ **18.** $(ef)^5$ **19.** $(3x)^3$ **20.** $(5y)^2$

21. $(2y)^k$ **22.** $(10w)^m$ **23.** $\frac{x^5}{x^3}$ **24.** $\frac{y^7}{y^2}$

25. $\frac{10^b}{10^c}$ **26.** $\frac{2^f}{2^a}$ **27.** $\frac{a^{5x}}{a^{2x}}$ **28.** $\frac{x^{6a}}{x^{3a}}$

---

*If $y = 0$ and $x \neq 0$, it would mean division of a nonzero number by zero, which is not possible. If $y = 0$ and $x = 0$, it would mean dividing zero by zero, which cannot be determined.

29. $\dfrac{5^{3x}}{5^x}$ **30.** $\dfrac{6^{4y}}{6^y}$ **31.** $\dfrac{x^4}{y^2}$ **32.** $\dfrac{a^5}{b^3}$

33. $\left(\dfrac{w}{z}\right)^2$ **34.** $\left(\dfrac{h}{k}\right)^4$ **35.** $\left(\dfrac{x}{5}\right)^3$ **36.** $\left(\dfrac{z}{3}\right)^2$

37. $\left(\dfrac{2}{u}\right)^n$ **38.** $\left(\dfrac{4}{v}\right)^p$

1-4B NEGATIVE EXPONENTS

Rule 3 leads to the introduction of negative and zero exponents. Consider the expression $\dfrac{x^3}{x^5}$.

$$\frac{x^3}{x^5} = \frac{xxx}{xxxxx} = \frac{xxx \cdot 1}{xxx \cdot xx} = \frac{xxx}{xxx} \cdot \frac{1}{xx} = 1 \cdot \frac{1}{xx} = \frac{1}{x^2}$$

↑ The value of this fraction is 1

However, if we use Rule 3,

$$\frac{x^3}{x^5} = x^{3-5} = x^{-2}$$

For $x^{-2} = \dfrac{1}{x^2}$ to be true, we need the following definition:

Definition of a Negative Exponent

> **NEGATIVE EXPONENT**
>
> $$x^{-n} = \frac{1}{x^n} \qquad (x \neq 0)^*$$

Rule 4 can be used with negative or positive exponents.

Example 6 Using Rule 4 starting with a negative exponent

(a) $x^{-5} = \dfrac{1}{x^5}$ (b) $10^{-4} = \dfrac{1}{10^4}$

Example 7 Using Rule 4 starting with a positive exponent

(a) $x^4 = \dfrac{1}{x^{-4}}$ (b) $w^{11} = \dfrac{1}{w^{-11}}$

Example 8 Using Rule 4 reading from right to left

(a) $\dfrac{1}{x^{-4}} = x^4$ (b) $\dfrac{1}{y^3} = y^{-3}$

---

* If $x = 0$ in Rule 4, it would mean division of 1 by zero, which is not possible. For example, $0^{-3} = \dfrac{1}{0^3} = \dfrac{1}{0 \cdot 0 \cdot 0} = \dfrac{1}{0}$, which is not a number (Section 1-3D).

> A **factor** can be moved either from the numerator to the denominator or from the denominator to the numerator simply by changing the sign of its exponent.
>
> **Note:** This does not change the sign of the *expression.*

Example 9 Moving a factor from the numerator to the denominator of a fraction by changing the sign of its exponent

$$y^{-4}w^5z^{-2} = \frac{y^{-4}}{1} \cdot \frac{w^5}{1} \cdot \frac{z^{-2}}{1} = \frac{1}{y^4} \cdot \frac{w^5}{1} \cdot \frac{1}{z^2} = \frac{w^5}{y^4z^2}$$

Example 10 Moving a factor from the denominator to the numerator of a fraction by changing the sign of its exponent

(a) $\frac{h^5}{k^{-4}} = \frac{h^5}{1} \cdot \frac{1}{k^{-4}} = \frac{h^5}{1} \cdot \frac{k^4}{1} = h^5k^4$

(b) $\frac{a^3}{b^2} = \frac{a^3}{1} \cdot \frac{1}{b^2} = \frac{a^3}{1} \cdot \frac{b^{-2}}{1} = a^3b^{-2}$

Example 11 Writing expressions with only positive exponents

$$\frac{a^{-2}b^4}{c^5d^{-3}} = \frac{a^{-2}}{1} \cdot \frac{b^4}{1} \cdot \frac{1}{c^5} \cdot \frac{1}{d^{-3}} = \frac{1}{a^2} \cdot \frac{b^4}{1} \cdot \frac{1}{c^5} \cdot \frac{d^3}{1} = \frac{b^4d^3}{a^2c^5}$$

Example 12 Writing expressions without fractions, using negative exponents if necessary

(a) $\frac{y^3}{z^{-2}} = \frac{y^3}{1} \cdot \frac{1}{z^{-2}} = \frac{y^3}{1} \cdot \frac{z^2}{1} = y^3z^2$

(b) $\frac{a}{bc^2} = \frac{a^1}{1} \cdot \frac{1}{b^1} \cdot \frac{1}{c^2} = \frac{a^1}{1} \cdot \frac{b^{-1}}{1} \cdot \frac{c^{-2}}{1} = ab^{-1}c^{-2}$

A Word of Caution. An expression that is *not* a factor *cannot* be moved from the numerator to the denominator of a fraction simply by changing the sign of its exponent.

$$\frac{a^{-2} + b^5}{c^4} = \frac{\frac{1}{a^2} + b^5}{c^4}$$

a^{-2} *cannot* be moved to the denominator because it is not a *factor* of the numerator [the + sign indicates that a^{-2} is a *term* (see Section 1-6A) rather than a factor of the numerator].

Expressions of this kind will be simplified in Section 5-4. ■

Zero Exponents

In using Rule 3 when the exponent of the numerator is larger than that of the denominator, the answer has a *positive* exponent. When the exponent of the denominator is larger than that of the numerator, the answer has a *negative* exponent. When the exponents of the numerator and denominator are the same, we have:

$$\frac{x^4}{x^4} = \frac{xxxx}{xxxx} = 1$$ Because a number divided by itself is 1

and $$\frac{x^4}{x^4} = x^{4-4} = x^0$$ Using Rule 3

Therefore we define

RULE 5

ZERO EXPONENT

$$x^0 = 1 \qquad (x \neq 0)^*$$

Example 13 Showing zero as an exponent

(a) $a^0 = 1$ Provided $a \neq 0$

(b) $10^0 = 1$

(c) $6x^0 = 6 \cdot 1 = 6$ Provided $x \neq 0$ (the 0 exponent applies only to x)

EVALUATING EXPRESSIONS THAT HAVE NUMERICAL BASES

Example 14

(a) $10^3 \cdot 10^2 = 10^5 = 10 \cdot 10 \cdot 10 \cdot 10 \cdot 10 = \underbrace{100{,}000}_{\text{5 zeros}}$

(b) $10^{-2} = \dfrac{1}{10^2} = \dfrac{1}{10 \cdot 10} = \dfrac{1}{\underbrace{100}_{\text{2 zeros}}}$

(c) $(2^3)^{-1} = 2^{-3} = \dfrac{1}{2^3} = \dfrac{1}{2 \cdot 2 \cdot 2} = \dfrac{1}{8}$

(d) $\dfrac{5^0}{5^2} = \dfrac{1}{5 \cdot 5} = \dfrac{1}{25}$

A Word of Caution. A common mistake students make is shown by the following examples.

	Correct method	*Incorrect method*
(a)	$2^3 \cdot 2^2 = 2^{3+2}$	$2^3 \cdot 2^2 \neq (2 \cdot 2)^{3+2} = 4^5$
	$= 2^5 = 32$	$= 1024$

---

* If $x = 0$ in Rule 5, it would mean dividing zero by zero, which cannot be determined. For example, $x^0 = 0^0 = 0^{2-2} = \dfrac{0^2}{0^2} = \dfrac{0 \cdot 0}{0 \cdot 0} = \dfrac{0}{0}$, which cannot be determined.

(b) $10^2 \cdot 10 = 10^{2+1} = 10^3 = 1000$ $\quad$ $10^2 \cdot 10 \neq (10 \cdot 10)^{2+1} = 100^3 = 1{,}000{,}000$ ✗

In words: When multiplying powers of the same base, add the exponents; do *not* multiply the bases. ■

SIMPLIFIED FORM OF EXPRESSIONS THAT HAVE EXPONENTS

Simplifying Exponential Expressions

An expression with exponents is considered simplified when each different base appears only once, and its exponent is a single integer.

Example 15 Simplifying expressions that have exponents

(a) $x^2 \cdot x^7 = x^9$ $\quad$ (b) $\dfrac{x^5y^2}{x^3y} = x^2y$ $\quad$ (c) $(x^2)^3 = x^6$

Note: All the rules for positive exponents used in Section 1–4A can also be used with negative and zero exponents. ■

EXERCISES 1-4B

In Exercises 1–40 simplify each expression. Write answers using only positive exponents.

1. a^{-3} **2.** x^{-2} **3.** 10^{-3} **4.** 10^{-5}

5. $x^{-3}y^2$ **6.** r^3s^{-4} **7.** $xy^{-2}z^{-3}$ **8.** $a^{-4}bc^{-5}$

9. $\dfrac{a^3}{b^{-4}}$ **10.** $\dfrac{c^4}{d^{-5}}$ **11.** $\dfrac{x^{-3}}{y^{-2}}$ **12.** $\dfrac{P^{-2}}{Q^{-4}}$

13. $xy^{-2}z^0$ **14.** $x^{-3}y^0z^2$ **15.** $(a^3)^{-2}$ **16.** $(b^2)^{-4}$

17. x^6x^{-4} **18.** $y^{-5}y^3$ **19.** $\dfrac{e^4}{e^{-3}}$ **20.** $\dfrac{h^3}{h^{-5}}$

21. $(x^{2a})^{-3}$ **22.** $(y^{3c})^{-2}$ **23.** $\dfrac{x^{3a}}{x^{-a}}$ **24.** $\dfrac{a^{4x}}{a^{-2x}}$

25. $(3^x)^0$ **26.** $(2^a)^0$ **27.** $\dfrac{6x^2}{2x^{-3}}$ **28.** $\dfrac{10p^{-2}}{5p^{-4}}$

29. $(xy)^{-3}$ **30.** $(hk)^{-5}$ **31.** $(3x)^{-2}$ **32.** $(4y)^{-3}$

33. $\left(\dfrac{e}{f}\right)^{-1}$ **34.** $\left(\dfrac{c}{d}\right)^{-1}$ **35.** $\left(\dfrac{u}{5}\right)^{-2}$ **36.** $\left(\dfrac{v}{2}\right)^{-4}$

37. $\dfrac{15m^0n^{-3}}{5m^{-2}n^2}$ **38.** $\dfrac{16x^0y^{-2}}{12x^{-2}y^{-3}}$ **39.** $\dfrac{x^{-1}+y}{y}$ **40.** $\dfrac{a-b^{-1}}{a}$

In Exercises 41–48 evaluate each expression.

41. $10^5 \cdot 10^{-2}$ **42.** $2^4 \cdot 2^{-2}$ **43.** $(3^{-2})^{-2}$ **44.** $(10^{-1})^{-3}$

45. $\dfrac{10^2 \cdot 10^{-1}}{10^{-3}}$ **46.** $\dfrac{2^{-3} \cdot 2^2}{2^{-4}}$ **47.** $(10^0)^5$ **48.** $(3^0)^4$

In Exercises 49–54 write each expression without fractions, using negative exponents if necessary.

49. $\dfrac{y}{x^3}$ **50.** $\dfrac{x}{y^2}$ **51.** $\dfrac{x^3}{a^{-4}}$ **52.** $\dfrac{m^2}{n^{-3}}$

53. $\dfrac{x^4y^{-3}}{z^{-2}}$ **54.** $\dfrac{a^{-1}b^3}{c^{-4}}$

1-4C GENERAL RULE OF EXPONENTS

RULE 6

GENERAL RULE OF EXPONENTS

$$\left(\frac{x^a y^b}{z^c}\right)^n = \frac{x^{an} y^{bn}}{z^{cn}}$$

None of the letters can have a value that makes the denominator zero.

Exponent Rules Combined

When Rule 6 is used, we are using a combination of the preceding rules of exponents.

A Word of Caution. An exponent applies only to the symbol immediately preceding it. For example,

(a) Find the value of $3x^2$ if $x = 4$.

$$3x^2 = 3(4)^2 = 3(16) = 48$$

(b) $2x^3 \neq (2x)^3 = 2^3x^3 = 8x^3$

- The exponent 3 applies to *both* 2 and x (in $(2x)^3$)
- The exponent 3 applies *only* to x (in $2x^3$)

(c) Find the value of $-x^2$ when $x = 3$.

Correct method	*Incorrect method*
$-x^2 = -(3)^2 = -9$	$-x^2 \neq (-3)^2 = 9$

The exponent 2 applies *only* to x ■

When using Rule 6, notice the following:

1. x, y, and z are *factors* of the expression within the parentheses. They are *not* separated by + or − signs. See Example 16(i).
2. The exponent of each factor within the parentheses is multiplied by the exponent outside the parentheses.

Example 16 Using Rule 6, the general rule of exponents

(a) $(x^3y^{-1})^5 = x^{3\cdot 5}y^{(-1)5} = x^{15}y^{-5} = \dfrac{x^{15}}{y^5}$

The same rules of exponents apply to *numerical* bases as well as literal bases

(b) $\left(\dfrac{2a^{-3}b^2}{c^5}\right)^3 = \dfrac{2^{1\cdot 3}a^{(-3)3}b^{2\cdot 3}}{c^{5\cdot 3}} = \dfrac{2^3a^{-9}b^6}{c^{15}} = \dfrac{8b^6}{a^9c^{15}}$

(c) $\left(\frac{3^{-7}x^{10}}{y^{-4}}\right)^0 = 1$ The zero power of any nonzero expression is one.

(d) $\left(\frac{x^5y^4}{x^3y^7}\right)^2 = (x^2y^{-3})^2 = x^4y^{-6} = \frac{x^4}{y^6}$

Simplify the expression within the parentheses first whenever possible

(e) $(5^0h^{-2})^{-3} = (1h^{-2})^{-3} = (h^{-2})^{-3} = h^6$

(f) $\left(\frac{10^{-2} \cdot 10^5}{10^4}\right)^3 = \left(\frac{10^3}{10^4}\right)^3 = (10^{-1})^3 = 10^{-3} = \frac{1}{10^3}$

This exponent applies only to the symbol immediately preceding it, 5

(g) $\frac{-5^2}{(-5)^2} = \frac{-(5 \cdot 5)}{(-5)(-5)} = \frac{-25}{25} = -1$

This exponent applies only to the symbol immediately preceding it, (−5)

(h) $\frac{-2^3}{(-2)^3} = \frac{-(2 \cdot 2 \cdot 2)}{(-2)(-2)(-2)} = \frac{-8}{-8} = 1$

(i) $(x^2 + y^3)^4$ Rules 2 and 6 *cannot* be used here because the + sign means that x^2 and y^3 are *not* factors; they are *terms* of the expression being raised to the fourth power. See Section 11-5.

Even and Odd Powers of Negative Values

Compare Examples 16g and 16h and note the difference when negative values are raised to an even or an odd power.

EXERCISES 1-4C

Simplify each expression. Then write the answer using only positive signs in the exponents.

1. $(a^2b^3)^2$
2. $(x^4y^5)^3$
3. $(m^{-2}n)^4$
4. $(p^{-3}r)^5$
5. $(2x^3y^{-2})^4$
6. $(3h^{-2}k^{-1})^4$
7. $(x^{-2}y^3)^{-4}$
8. $(w^{-3}z^4)^{-2}$
9. $(5a^{-1}b^2)^{-3}$
10. $(4e^3f^{-4})^{-2}$
11. $(10^0k^{-4})^{-2}$
12. $(6^0z^{-5})^{-2}$
13. $\left(\frac{xy^4}{z^2}\right)^2$
14. $\left(\frac{a^3b}{c^2}\right)^3$
15. $\left(\frac{M^{-2}}{N^3}\right)^4$
16. $\left(\frac{R^5}{S^{-4}}\right)^3$
17. $\left(\frac{-x^{-5}}{y^4z^{-3}}\right)^{-2}$
18. $\left(\frac{-a^{-4}}{b^2c^{-5}}\right)^{-3}$
19. $\left(\frac{r^7s^8}{r^9s^6}\right)^0$
20. $\left(\frac{t^5u^6}{t^8u^7}\right)^0$
21. $\left(\frac{x^{-2}y^2}{2x^{-3}}\right)^3$
22. $\left(\frac{a^3b^{-1}}{5b^{-4}}\right)^2$
23. $\left(\frac{8ab^{-2}}{2a^2}\right)^{-3}$
24. $\left(\frac{4x^3}{12x^{-2}y^{-4}}\right)^{-1}$
25. $\left(\frac{u^{-4}}{uv^3}\right)^{-2}$
26. $\left(\frac{x^{-6}y}{x^{-4}}\right)^{-1}$
27. $\left(\frac{x^{3n}x^{2n}}{x^{4n}}\right)^2$
28. $\left(\frac{x^{2n-1}y^{3n}}{x^ny^{2n+2}}\right)^2$

29. $\dfrac{(x^{3n}y^{2n-1})^3}{(xy)^2}$ **30.** $\dfrac{-4(h^{2n}k^{3n-1})^2}{(-2hk)^3}$

1–5 Evaluating Algebraic Expressions

1-5A ORDER OF OPERATIONS

In Section 1–3 we discussed performing one operation with signed numbers at a time. When more than one operation occurs in the same expression, we need to know which operation must be performed first.

ORDER OF OPERATIONS

1. If there are operations within grouping symbols, that part of the expression within those grouping symbols is evaluated first. Then the entire expression is evaluated.
2. Any evaluation always proceeds in three steps:

   First: Powers and roots are done in any order.

   Second: Multiplication and division are done in order *from left to right.*

   Third: Addition and subtraction are done in order *from left to right.*

Example 1 Showing how grouping symbols change the usual order of operations

$5 + 4 \cdot 6 = 5 + 24 = 29$ Usual order—multiplication done before addition

$(5 + 4) \cdot 6 = 9 \cdot 6 = 54$ Because of parentheses—addition done before multiplication

Example 2 Showing the correct order of operations

(a) $(7 + 3) \cdot 5$ Do the part in parentheses first
$= 10 \cdot 5 = 50$

(b) $7 + 3 \cdot 5$ Do multiplication before addition
$7 + 15 = 22$

(c) $4^2 + \sqrt{25} - 6$ Do powers and roots first
$= 16 + 5 - 6$
$= 21 - 6 = 15$

(d) $16 \div 2 \cdot 4$
$= 8 \cdot 4 = 32$

Do division first because in reading from left to right the division comes before the multiplication

(e) $\sqrt{16} - 4(2 \cdot 3^2 - 12 \div 2)$

First evaluate the expression inside the parentheses

$= \sqrt{16} - 4(2 \cdot 9 - 12 \div 2)$ Do the power inside ( )

$= \sqrt{16} - 4(18 - 6)$ Do the × and ÷ inside ( )

$= \sqrt{16} - 4(12)$ Do the − inside ( )

$= 4 - 4(12)$ Do the root next

$= 4 - 48 = -44$

(f) $\sqrt[3]{-8}(-3)^2 - 2(-6)$ Do roots and powers first

$= -2(9) - 2(-6)$ Do multiplication before subtraction

$= -18 - (-12)$

$= -18 + 12 = -6$

(g) $\dfrac{(-4) + (-2)}{8 - 5}$

$= \dfrac{-6}{3} = -2$

This bar is a grouping symbol for both $(-4) + (-2)$ and $8 - 5$; notice that the bar can be used either above or below the numbers being grouped

---

EXERCISES 1-5A

Evaluate each expression.

1. $10 \div 2 \cdot 5$
2. $20 \cdot 15 \div 5$
3. $12 \div 6 \div 2$
4. $18 \div 3 \div 2$
5. $14 \cdot 2 \div 4$
6. $15 \cdot 3 \div 5$
7. $3(2^4)$
8. $5(3^2)$
9. $4 \cdot 3 + 15 \div 5$
10. $2 \cdot 5 + 18 \div 6$
11. $10 - 3 \cdot 2$
12. $12 - 2 \cdot 4$
13. $10 \cdot 15^2 - 4^3$
14. $3 \cdot 4^2 - 2^4$
15. $(785)^3(0) + 1^5$
16. $(456)^4(0) - 1^2$
17. $(10^2)\sqrt{16} \cdot 5$
18. $(5)\sqrt{25} + 4(6) - 6$
19. $28 \div 4 \cdot 2(6)$
20. $48 \div 16 \cdot 2(-8)$
21. $(8 - 2) \cdot 6$
22. $(10 - 6) \cdot 4$
23. $2\sqrt{9}(2^3 - 5)$
24. $5\sqrt{36}(4^2 - 8)$
25. $(3 \cdot 5^2 - 15 \div 3) \div (-7)$
26. $(3 \cdot 4^3 - 72 \div 6) \div (-9)$
27. $(-10) \div 2 - (-3)$
28. $(-18) \div (-3) \cdot (-6)$
29. $(-10)^3 - 5(10^2)\sqrt{100}$
30. $(-10)^2\,10 + 10^0(-20)$
31. $-18 \div (-\sqrt{81})$
32. $-30 \div (-\sqrt{100})$
33. $\dfrac{7 + (-12)}{8 - 3}$
34. $\dfrac{(-14) + (-2)}{9 - 5}$

35. $\dfrac{8-10}{12-6} - \dfrac{9-(-3)}{-8+4}$

36. $\dfrac{(-2)+(-7)}{18+(-15)} - \dfrac{7-(+5)}{14-11}$

37. $8 - [5(-2)^3 - \sqrt{16}]$

38. $10 - [3(-3)^2 - \sqrt{25}]$

39. $(667.5) \div (25.8) \cdot (2.86)$

40. $96.64 \div 12.08(11.94)$

1-5B EVALUATING ALGEBRAIC EXPRESSIONS

Algebraic Expressions

An **algebraic expression** consists of numbers, letters, signs of operation ($+$, $-$, $\times$, $\div$, powers, roots), and signs of grouping.

Constants

A **constant** is an object or symbol that does not change its value in a particular problem or discussion. It is usually represented by a number symbol. In the algebraic expression $4x^2 - 3y$, the constants are 4, 2, and -3.

Variables

A **variable** is an object or symbol that can change its value in a particular problem or discussion. It is represented by a letter. In the algebraic expression $2x - 3y$, the variables are x and y.

Domain of a Variable

The set of all numbers that can be used in place of a variable is called the **domain** (set) of that variable.

Evaluating Algebraic Expressions

> **TO EVALUATE AN ALGEBRAIC EXPRESSION**
>
> 1. Replace each letter by its number value.
> 2. Carry out all operations in the correct order (Section 1–5A).

Example 1 Evaluate $\dfrac{2a-b}{10c}$ when $a = -1$, $b = 3$, and $c = -2$

Remember that this bar is a grouping symbol

$$\frac{2a-b}{10c} = \frac{2(-1)-(3)}{10(-2)} = \frac{-2-3}{-20} = \frac{-5}{-20} = \frac{1}{4} = 0.25$$

Example 2 Evaluate $b - \sqrt{b^2 - 4ac}$ when $a = 3$, $b = -7$, and $c = 2$

This bar is a grouping symbol for $b^2 - 4ac$

$$\begin{aligned} b - \sqrt{b^2 - 4ac} & \\ &= (-7) - \sqrt{(-7)^2 - 4(3)(2)} \\ &= (-7) - \sqrt{49 - 24} \\ &= (-7) - \sqrt{25} \\ &= (-7) - \quad 5 = -12 \end{aligned}$$

Formulas

Students will encounter **formulas** in many courses, as well as in real-life situations. In the examples and exercises in this section we have listed the subject areas where the formulas are used.

Formulas are evaluated in the same way any other expression having numbers and letters is evaluated.

Example 3 Given the formula $A = P(1 + rt)$, find A when $P = 1000$, $r = 0.08$, and $t = 1.5$. (Business)

$$\begin{aligned} A &= P(1 + rt) \\ &= 1000\,[1 + (0.08)(1.5)] \\ &= 1000\,[1 + 0.12] \\ &= 1000\,[1.12] = 1120 \end{aligned}$$

Notice that [ ] were used in place of ( ) to clarify the grouping

Example 4 Given the formula $s = \frac{1}{2}gt^2$, find s when $g = 32$ and $t = 5\frac{1}{2}$.

$$\begin{aligned} s &= \frac{1}{2}gt^2 \\ &= \frac{1}{2}(32)\left(\frac{11}{2}\right)^2 \\ &= \frac{1}{2}\left(\frac{32}{1}\right)\left(\frac{121}{4}\right) = 484 \end{aligned}$$

Example 5 Given the formula $T = \pi\sqrt{\frac{L}{g}}$, find T when $\pi \doteq 3.14$, $L = 96$, and $g = 32$. (Physics)

This symbol means "approximately equal to"

$$\begin{aligned} T &= \pi\sqrt{\frac{L}{g}} \\ &\doteq (3.14)\sqrt{\frac{96}{32}} = (3.14)\sqrt{3} \doteq (3.14)(1.732) \\ &\doteq 5.44 \end{aligned}$$

Rounded off to two decimal places

EXERCISES 1-5B

In Exercises 1–6, evaluate the expression when $a = 3$, $b = -5$, $c = -1$, $x = 4$, and $y = -7$.

1. $3b - ab + xy$
2. $4 + a(x + y)$
3. $2(a - b) - 3c$
4. $ab^2 - xy$
5. $b^2 - 4ac$
6. $5c^2 + a(x - y)$

In Exercises 7–9, find the value of the expression when $E = -1$, $F = 3$, $G = -5$, and $H = -4$.

7. $\dfrac{3E}{G - 2H}$
8. $G - \sqrt{G^2 - 4EH}$
9. $\dfrac{\sqrt{5.6F - 1.7G}}{0.78H^2}$

10. Evaluate $\dfrac{-b + \sqrt{b^2 - 4ac}}{2a}$ when $a = 5$, $b = -6$, and $c = -1$.

In Exercises 11–22, evaluate each formula using the values of the letters given with the formula.

(Nursing)	**11.** $q = \dfrac{DQ}{H}$	$D = 5, H = 30, Q = 420$
(Business)	**12.** $I = prt$	$p = 600, r = 0.09, t = 4.5$
(Chemistry)	**13.** $K = PV$	$P = 3\frac{1}{2}, V = 13\frac{1}{3}$
(Business)	**14.** $A = P(1 + rt)$	$P = 500, r = 0.09, t = 2.5$
(Science)	**15.** $C = \dfrac{5}{9}(F - 32)$	$F = -10$
(Geometry)	**16.** $A = \dfrac{1}{2}bh$	$b = 15, h = 14$
(Electricity)	**17.** $I = \dfrac{E}{R}$	$E = 110, R = 22$
(Electricity)	**18.** $Z = \dfrac{Rr}{R + r}$	$R = 22, r = 8$
(Geometry)	**19.** $A = \pi R^2$	$\pi \doteq 3.14, R = 10$
(Geometry)	**20.** $V = \dfrac{4}{3}\pi R^3$	$\pi \doteq 3.14, R = 3$
(Mathematics)	**21.** $S = \dfrac{a(1 - r^n)}{1 - r}$	$a = 18, r = 0.06, n = 4$
(Business)	**22.** $R = \dfrac{Ai(1 + i)^n}{(1 + i)^n - 1}$	$A = 240, i = 0.0079, n = 4$

1-6 Simplifying Algebraic Expressions

1-6A DISTRIBUTIVE RULE

Terms

The + and − signs in an algebraic expression break it up into smaller pieces called **terms.**

Example 1 Identifying terms

(a) $3 + 2x - y$ — 3 terms

(b) $3 + (2x - y)$ — 2 terms

(c) $\dfrac{2 - x}{xy} + 5(2x^2 - y)$ — 2 terms

An expression with grouping symbols is considered as a single piece even though it may contain + and − signs.

Factors

The numbers that are multiplied together to give a product are called the **factors** of that product. One (1) is a factor of any number.

Coefficients

In a term that has *two* factors, the **coefficient** of one factor is the other factor. In a term that has *more than two* factors, the coefficient of each factor is the product of all the other factors in that term.

A **numerical coefficient** is a coefficient that is a number. If we say "*the* coefficient" of a term, it is understood to mean the *numerical* coefficient of that term.

A **literal coefficient** is a coefficient that is a letter or product of letters.

Example 2 Showing numerical and literal coefficients

(a) $12xy^2$

xy^2 is the literal coefficient of 12

12 is the numerical coefficient of xy^2

(b) $3x^2 - 9x(2y + 5z)$

-9 is the numerical coefficient of $x(2y + 5z)$

Distributive Rule

THE DISTRIBUTIVE RULE(S)

$$a(b + c) = ab + ac$$

$$(b + c)a = ba + ca$$

These rules may be extended to have any number of terms within the parentheses.

Meaning of the Distributive Rule. When a factor is multiplied by an expression enclosed within grouping symbols, the factor must be multiplied by *each term* within the grouping symbol; then the products are added.

Example 3 Using the distributive rule

(a) $5(2x - 3y) = 5(2x) + 5(-3y) = 10x - 15y$

(b) $4x(x^2 - 2xy + y^2) = (4x)(x^2) + (4x)(-2xy) + (4x)(y^2)$
$= 4x^3 - 8x^2y + 4xy^2$

(c) $(-2x^2 + xy - 5y^2)(-3xy)$
$= (-2x^2)(-3xy) + (xy)(-3xy) + (-5y^2)(-3xy)$
$= 6x^3y - 3x^2y^2 + 15xy^3$

A Word of Caution. A common mistake students make is to think that the distributive rule applies to expressions such as $2(3 \cdot 4)$.

The distributive rule only applies when this is an addition (or subtraction)

$$2(3 \cdot 4) \neq (2 \cdot 3)(2 \cdot 4)$$

$$2(12) \neq 6 \cdot 8$$

$$24 \neq 48$$

■

Example 4 **Using the distributive rule to remove grouping symbols**

Multiplying a number by 1 does not change its value

(a) $(3x - 5) = 1(3x - 5) = (1)(3x) + (1)(-5) = 3x - 5$

(b) $+(4y + 7) = +1(4y + 7) = (+1)(4y) + (+1)(+7) = +4y + 7$

(c) $-(8 - 6z) = -1(8 - 6z) = (-1)(8) + (-1)(-6z) = -8 + 6z$
(d) $-2x(4 - 5x) = (-2x)(4) + (-2x)(-5x) = -8x + 10x^2$

The applications of the distributive rule shown in Example 4 lead to the rules for removing grouping symbols given in the following box.

Removing Grouping Symbols

REMOVING GROUPING SYMBOLS

First, write a + sign for any enclosed term that has no written sign. Then:

1. **When removing a grouping symbol preceded by a + sign (or no sign):**
   Leave the enclosed terms unchanged. Drop the grouping symbol and the + sign preceding it (if there is one).
2. **When removing a grouping symbol preceded by a − sign:**
   Change the sign of each enclosed term. Drop the grouping symbol and the − sign preceding it.
3. **When removing a grouping symbol preceded by a factor:**
   Multiply each enclosed term by the factor and add these products. Drop the grouping symbol and the factor.
4. **When grouping symbols occur within other grouping symbols:**
   It is usually easier to remove the *innermost* grouping symbols first.

Nested Grouping Symbols

Example 5 Removing grouping symbols

(a) $5z + (4y + 7)$
$= 5z + (+4y + 7)$ Write + sign when enclosed term has no sign
$= 5z \quad +4y + 7$ Drop the ( ) and the + sign preceding the ( )

(b) $2x - (8 - 6z)$
$= 2x - (+8 - 6z)$ Write + sign when enclosed term has no sign
$= 2x \quad -8 + 6z$ Change the sign of each enclosed term, *and* drop the ( ) and the − sign preceding the ( )

(c) $-2x(4 - 5x)$
$= -2x(+4 - 5x)$ Write + sign when enclosed term has no sign
$= (-2x)(+4) + (-2x)(-5x)$ Multiply each enclosed term by the factor and add the products
$= \quad -8x \quad + \quad 10x^2$

(d) $(3a - b) - 2\{x - [(y - 2) - z]\}$
$= (3a - b) - 2\{x - [y - 2 - z]\}$ — Remove ( ) preceded by no sign
$= (3a - b) - 2\{x - y + 2 + z\}$ — Remove [ ] preceded by − sign
$= (3a - b) - 2x + 2y - 4 - 2z$ — Remove { } using the distributive rule
$= 3a - b - 2x + 2y - 4 - 2z$ — Remove ( ) preceded by no sign; these ( ) could have been removed as early as the first step

A Word of Caution. The following are some of the common errors made in removing parentheses:

	Correct form	Common error
(a)	$-(x - 2y) = -x + 2y$	$-(x - 2y) \neq -x - 2y$
(b)	$6(y - 3) = 6y - 18$	$6(y - 3) \neq 6y - 3$

■

EXERCISES 1-6A

In each of the Exercises 1–4: (a) determine the number of terms; (b) write the second term.

1. $3x^2y + \dfrac{2x + y}{3xy} + 4(3x^2 - y)$
2. $5xy^2 + \dfrac{5x - y}{7xy} + 3(x^2 - 4y)$
3. $(R + S) - 2(x + y)$
4. $(z + y) - 5(x + y)$

In Exercises 5 and 6: (a) write the numerical coefficient of the first term; (b) write the literal part of the second term.

5. $2R^2 - 5RT + 3T^2$
6. $4x^2 - 3xy + y^2$

In Exercises 7–16, find each of the following products.

7. $x(xy - 3)$
8. $a(ab - 4)$
9. $(x - 5)(-4)$
10. $(y - 2)(-5)$
11. $-3(x - 2y + 2)$
12. $-2(x - 3y + 4)$
13. $(3x^3 - 2x^2y + y^3)(-2xy)$
14. $(4z^3 - z^2y - y^3)(-2yz)$
15. $(-2ab)(3a^2b \cdot 6abc^3)$
16. $(5x^2y)(-2xy^3 \cdot 3xyz^2)$

In Exercises 17–30, remove the grouping symbols.

17. $(x - y) + 10$
18. $(3a - b) + 8$
19. $7 - (-4R + S)$
20. $9 - (-3m - n)$
21. $6 - 2(a - 3b)$
22. $12 - 3(2R - S)$
23. $(-2x)(x - 4y) + 3$
24. $(-5x)(2x - 3y) + 2$
25. $-(x - y) + (2 - a)$
26. $-(a - b) + (x - 3)$
27. $5 - 3[a - 4(2x - y)]$
28. $7 - 5[x - 3(2a - b)]$
29. $6 - 4[2x - 3(a - 2b)]$
30. $9 - 2[-3a - 4(2x - y)]$

1-6B SIMPLIFYING ALGEBRAIC EXPRESSIONS

Like Terms Terms that have *identical literal parts* are called **like terms.**

Unlike Terms Terms that have *different literal parts* are called **unlike terms.**

Example 6 Like terms

(a) $3x, 4x, x, \frac{1}{2}x, 0.7x$ are like terms. They are called "x-terms."

(b) $2x^2, 10x^2, \frac{3}{4}x^2, 2.3x^2$ are like terms. They are called "x^2-terms."

(c) $4x^2y, 8x^2y, x^2y, \frac{1}{5}x^2y, 2.8x^2y$ are like terms. They are called "x^2y-terms."

Example 7 Unlike terms

(a) $5x^2$ and $7x$ are unlike terms.

$$\left.\begin{aligned} 5x^2 &= 5(xx) \\ 7x &= 7(x) \end{aligned}\right\}$$ The literal parts are different

(b) $4x^2y$ and $10xy^2$ are unlike terms.

$$\left.\begin{aligned} 4x^2y &= 4(xxy) \\ 10xy^2 &= 10(xyy) \end{aligned}\right\}$$ The literal parts are different

Combining Like Terms

TO COMBINE LIKE TERMS

1. Identify the like terms by their identical literal parts.
2. Find the sum of each group of like terms by adding their numerical coefficients.

When we combine like terms, we are usually changing the grouping and the order in which the terms appear. The commutative and associative properties of addition guarantee that when we do this, the sum remains unchanged.

Example 8 Combining like terms

(a) $3x + 5x = (3 + 5)x = 8x$

This is an application of the distributive rule

$-ba = -1ab$

(b) $4ab - ba + 6ab = (4 - 1 + 6)ab = 9ab$

(c) $12a - 7b - 9a + 4b$

$= (12a - 9a) + (-7b + 4b)$ Only like terms may be combined

$= (12 - 9)a + (-7 + 4)b$

$= 3a + (-3)b = 3a - 3b$

TO SIMPLIFY AN ALGEBRAIC EXPRESSION

1. Remove grouping symbols.
2. Combine like terms.

It often helps to underline like terms before combining them.

Example 9 Simplifying algebraic expressions

(a) $3(5x - 2y) - 4(3x - 6y)$

$= \underline{15x} - 6y - \underline{12x} + 24y$ Remove grouping symbols

$= \underline{15x} - \underline{12x} - 6y + 24y$ Collect like terms

$= 3x + 18y$ Combine like terms

(b) $2x^2y(4xy - 3xy^2) - 5xy(3x^2y^2 - 2x^2y)$

$= 8x^3y^2 - \underline{6x^3y^3} - \underline{15x^3y^3} + 10x^3y^2$

$= 8x^3y^2 + 10x^3y^2 - \underline{6x^3y^3} - \underline{15x^3y^3}$

$= 18x^3y^2 - 21x^3y^3$

(c) $-8[-5(3x - 2) + 13] - 11x$

$= -8[-15x + \underline{10} + \underline{13}] - 11x$ Remove innermost grouping symbol first

$= -8[-15x + 23] - 11x$

$= +120x - 184 - 11x$

$= 109x - 184$

EXERCISES 1-6B

Simplify each algebraic expression.

1. $-3(a - 2b) + 2(a - 3b)$
2. $-2(m - 3n) + 4(m - 2n)$
3. $2h(3h^2 - k) - k(h - 3k^3)$
4. $4x(2y^2 - 3x) - x(2x - 3y^2)$
5. $x^2y(3xy^2 - y) - 2xy^2(4x - x^2y)$
6. $ab^2(2a - ab) - 3ab(2ab - ab^2)$
7. $3x(4y^2 - 2x) - x(5x - y^2)$
8. $x^2(x^2 + y^2) - y^2(x^2 + y^2)$
9. $2x(3x^2 - 5x + 1) - 4x(2x^2 - 3x - 5)$
10. $3x(4x^2 - 2x - 3) - 2x(3x^2 - x + 1)$
11. $u(u^2 + 2u + 4) - 2(u^2 + 2u + 4)$
12. $x(x^2 - 3x + 9) + 3(x^2 - 3x + 9)$
13. $2x - [3a + (4x - 5a)]$
14. $2y - [5c + (3y - 4c)]$
15. $5x + [-(2x - 10) + 7]$
16. $4x + [-(3x - 5) + 4]$
17. $-10[-2(3x - 5) + 17] - 4x$
18. $-20[-3(2x - 4) + 20] - 5x$
19. $25 - 2[3g - 5(2g - 7)]$
20. $40 - 3[2h - 8(3h - 10)]$

21. $(2x - y) - \{[3x - (7 - y)] - 10\}$

22. $(3a - b) - \{[2a - (10 - b)] - 20\}$

23. $3xy^2(2x - xy + 4) + x(3xy^3 - 12y^2)$

24. $5r^2s(3r - 5 - 2s) + 5r^2s(5 + 2s)$

25. $100 - \{2z - [3z - (4 - z)]\}$

26. $50 - \{-2m - [5m - (6 - 2m)]\}$

27. $60x - 2\{-3[-5(-5 - x) - 6x]\}$

28. $100n - 3\{-4[-2(-4 - n) - 5n]\}$

29. $-2\{-3[-5(-4 - 3z) - 2z] + 30z\}$

30. $-3\{-2[-3(-5 - 2z) - 3z] - 40z\}$

1-7 Polynomials

Polynomial in x

A **polynomial in x** is an algebraic expression that has only terms of the form ax^n, where a is any real number and n is a positive integer (or zero).

Example 1 Polynomials in one letter

(a) $3x$ — A polynomial of one term is called a **monomial**

(b) $4x^4 - 2x^2$ — A polynomial of two unlike terms is called a **binomial**

(c) $7x^2 - 5x + 2$ — A polynomial of three unlike terms is called a **trinomial**

(d) 5 — This is a polynomial of one term (monomial) because its only term has the form $5x^0 = 5 \cdot 1 = 5$

(e) $6z^3 - 3z + 1$ — Polynomials can be in any letter; this is a polynomial in z

Example 2 Algebraic expressions that are not polynomials

(a) $4x^{-2}$ — This expression is *not* a polynomial because the exponent -2 is not a positive integer (or zero)

(b) $\dfrac{2}{x - 5}$ — This is *not* a polynomial because the term does not have the form ax^n

Polynomials often have terms that contain more than one variable.

Polynomial in x and y

A **polynomial in x and y** is an algebraic expression that has only terms of the form ax^ny^m, where a is any real number and n and m are positive integers (or zero).

Example 3 Polynomials in more than one letter

(a) $x^3y^2 - 2x + 3y^2 - 1$

$-1 = -1x^0y^0$

$3y^2 = 3x^0y^2$

$-2x = -2x^1y^0$

(b) $7uv^4w - 5u^2v + 2uw^2$ — A polynomial in u, v, and w

Degree of Term

The **degree of a term** in a polynomial is the *sum of the exponents* of its variables.

Example 4 Degree of a term

(a) 5^2x^3 — 3rd degree—only exponents of *variables* determine the degree of the term

(b) $-2u^3vw^2$ 6th degree because $-2u^3vw^2 = -2u^3v^1w^2$
$3 + 1 + 2 = 6$

(c) 14 0 degree because $14 = 14x^0$

Degree of Polynomial

The **degree of a polynomial** is the same as that of its highest-degree term (provided like terms have been combined).

Example 5 Degree of a polynomial

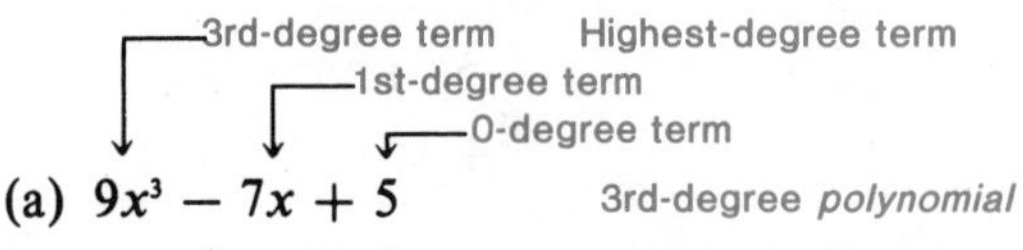

(a) $9x^3 - 7x + 5$ 3rd-degree *polynomial*

(b) $14xy^3 - 11x^5y + 8$ 6th-degree *polynomial*

Descending Powers

Polynomials are usually written in **descending powers** of one of the letters. For example,

Exponents get smaller from left to right

$$8x^3 - 3x^2 + 5x^1 + 7$$

When a polynomial has more than one letter, it can be arranged in descending powers of any one of its letters.

Example 6 Arrange $3x^3y - 5xy + 2x^2y^2 - 10$: (a) in descending powers of x, (b) in descending powers of y.

(a) $3x^3y + 2x^2y^2 - 5xy - 10$ Descending powers of x

(b) $2x^2y^2 + 3x^3y - 5xy - 10$ Descending powers of y

Since y is the same power in both terms, the higher-degree term is written first

Polynomial Equations

A **polynomial equation** is a polynomial set equal to zero. *The degree of the equation is the degree of the polynomial.*

Example 7 Polynomial equations

(a) $5x - 3 = 0$ 1st-degree polynomial equation in one variable

(b) $2x^2 - 4x + 7 = 0$ 2nd-degree polynomial equation in *one* variable (also called *quadratic* equation—see Chapter Eight)

(c) $2x^2y - 3xy + 5y^2 = 0$ 3rd-degree polynomial equation in *two* variables

1-7A ADDITION AND SUBTRACTION OF POLYNOMIALS

Polynomials are algebraic expressions. Therefore, the same rules used in Section 1–6 for algebraic expressions can be used for polynomials.

Adding Polynomials

TO ADD POLYNOMIALS

1. Remove grouping symbols.
2. Combine like terms.

Example 8 Adding polynomials

(a) $(3x^2 + 5x - 4) + (2x + 5) + (x^3 - 4x^2 + x)$
$= 3x^2 + 5x - 4 + 2x + 5 + x^3 - 4x^2 + x$
$= x^3 - x^2 + 8x + 1$

(b) $(5x^3y^2 - 3x^2y^2 + 4xy^3) + (4x^2y^2 - 2xy^2) + (-7x^3y^2 + 6xy^2 - 3xy^3)$
$= 5x^3y^2 - 3x^2y^2 + 4xy^3 + 4x^2y^2 - 2xy^2 - 7x^3y^2 + 6xy^2 - 3xy^3$
$= -2x^3y^2 + x^2y^2 + xy^3 + 4xy^2$

Subtracting Polynomials

TO SUBTRACT POLYNOMIALS

1. Change the sign of *all* terms in the polynomial being subtracted.
2. Combine like terms.

Example 9 Subtracting polynomials

(a) Subtract $(-4x^2y + 10xy^2 + 9xy - 7)$ from $(11x^2y - 8xy^2 + 7xy + 2)$.

$(11x^2y - 8xy^2 + 7xy + 2) - (-4x^2y + 10xy^2 + 9xy - 7)$
$= 11x^2y - 8xy^2 + 7xy + 2 + 4x^2y - 10xy^2 - 9xy + 7$
$= 15x^2y - 18xy^2 - 2xy + 9$

(b) Subtract $(2x^2 - 5x + 3)$ from the sum of $(8x^2 - 6x - 1)$ and $(4x^2 + 7x - 9)$.

$(8x^2 - 6x - 1) + (4x^2 + 7x - 9) - (2x^2 - 5x + 3)$
$= 8x^2 - 6x - 1 + 4x^2 + 7x - 9 - 2x^2 + 5x - 3$
$= 10x^2 + 6x - 13$

EXERCISES 1-7A

1. If the expression is a polynomial, find its degree.
   (a) $2x^2 + \frac{1}{3}x$ (b) $x^{-2} + 5x^{-1} + 4$

(c) 10

(d) $x^3y^3 - 3^2x^2y + 3^4xy^2 - y^3$

(e) $\dfrac{1}{2x^2 - 5x}$

2. Write each polynomial in descending powers of the indicated letter.
(a) $3x^2y + 8x^3 + y^3 - xy^5$ powers of y
(b) $6xy^3 + 7x^2y - 4y^2 + y$ powers of y

3. $(2m^2 - m + 4) + (3m^2 + m - 5)$

4. $(5n^2 + 8n - 7) + (6n^2 - 6n + 10)$

5. $(y^2 - 3y + 12) - (8y^3 - 3y)$

6. $(-3z^2 - z + 9) - (9 - z + z^2)$

7. $(6a - 5a^2 + 6) + (4a^2 + 6 - 3a)$

8. $(2b + 7b^2 - 5) + (4b^2 - 2b + 8)$

9. $(-3x^4 - 2x^3 + 5) - (4x^4 + 2x^3 - 4)$

10. $(5y^3 - 6y - 7) - (3y^3 - 4y + 7)$

11. $(7 - 8v^3 + 9v^2 + 4v) + (9v^3 - 8v^2 + 4v + 6)$

12. $(15 - 10w + w^2 - 3w^3) + (18 + 4w^3 + 7w^2 + 10w)$

13. Subtract $(6 + 3x^5 - 4x^2)$ from $(4x^3 + 6 + x)$.

14. Subtract $(7 - 4x^4 + 3x^3)$ from $(x^3 + 7 - 3x)$.

15. Add:
$$\begin{array}{r} 4x^3 + 7x^2 - 5x + 4 \\ 2x^3 - 5x^2 + 5x - 6 \\ \hline \end{array}$$

16. Add:
$$\begin{array}{r} 3y^4 - 2y^3 + 4y + 10 \\ -5y^4 + 2y^3 + 4y - 6 \\ \hline \end{array}$$

17. Subtract $(3x^4 + 2x^3 - 5)$ from $(2x^4 + 4x^3 + 8)$.

18. Subtract $(2y^3 - 3y^2 + 4)$ from $(7y^3 + 5y^2 - 5)$.

19. $(6r^3t + 14r^2t - 11) - (19 - 8r^2t + r^3t) + (8 - 6r^2t)$

20. $(13m^2n^2 + 4mn + 23) - (17 + 4mn - 9m^2n^2) + (-29 - 8mn)$

21. Subtract $(-3m^2n^2 + 2mn - 7)$ from the sum of $(6m^2n^2 - 8mn + 9)$ and $(-10m^2n^2 + 18mn - 11)$.

22. Subtract $(-9u^2v + 8uv^2 - 16)$ from the sum of $(7u^2v - 5uv^2 + 14)$ and $(11u^2v + 17uv^2 - 13)$.

23. Given the polynomials $(2x^2 - 5x - 7)$, $(-4x^2 + 8x - 3)$, and $(6x^2 - 2x + 1)$, subtract the sum of the first two from the sum of the last two.

24. Given the polynomials $(8y^2 + 10y - 9)$, $(13y^2 - 11y + 5)$, and $(-2y^2 - 16y + 4)$, subtract the sum of the first two from the sum of the last two.

25. Subtract the sum of $(x^3y + 3xy^2 - 4)$ and $(2x^3y - xy^2 + 5)$ from the sum of $(5 + xy^2 + x^3y)$ and $(-6 - 3xy^2 + 4x^3y)$.

26. Subtract the sum of $(2m^2n - 4mn^2 + 6)$ and $(-3m^2n + 5mn^2 - 4)$ from the sum of $(5 + m^2n - mn^2)$ and $(3 + 4m^2n + 2mn^2)$.

27. $(8.586x^2 - 9.030x + 6.976) - [1.946x^2 - 41.45x - (7.468 - 3.914x^2)]$

28. $(24.21 - 35.28x - 73.92x^2) - [82.04x - 53.29x^2 - (64.34 - 19.43x^2)]$

1-7B MULTIPLICATION OF POLYNOMIALS

MULTIPLYING A POLYNOMIAL BY A MONOMIAL

To multiply a polynomial by a monomial multiply *each* term in the polynomial by the monomial; then add the resulting products.

Example 10 Multiplying a polynomial by a monomial

(a) $5x(3x^2 - 2x + 6) = (5x)(3x^2) + (5x)(-2x) + (5x)(6)$
$= \underbrace{15x^3 - 10x^2 + 30x}$

These terms *cannot* be combined because they are not *like* terms

(b) $(2a^3bc^2 - 3ac + b^3)(-5ab^2)$
$= (2a^3bc^2)(-5ab^2) + (-3ac)(-5ab^2) + (b^3)(-5ab^2)$
$= -10a^4b^3c^2 + 15a^2b^2c - 5ab^5$

MULTIPLYING A POLYNOMIAL BY A POLYNOMIAL

Consider the product $(x^2 - 3x + 2)(x - 5)$. By the distributive rule, this is equal to $(x^2 - 3x + 2)(x) + (x^2 - 3x + 2)(-5)$. This means that the first polynomial $(x^2 - 3x + 2)$ must be multiplied by each term of the second polynomial and the results added. This is conveniently arranged as follows:

$$\begin{array}{rrrr} & x^2 & -\ 3x & +\ 2 \\ & & x & -\ 5 \\ \hline & -\ 5x^2 & +\ 15x & -\ 10 \\ x^3 & -\ 3x^2 & +\ 2x & \\ \hline x^3 & -\ 8x^2 & +\ 17x & -\ 10 \end{array}$$

Product $(x^2 - 3x + 2)(-5)$

Product $(x^2 - 3x + 2)(x)$

Notice that the second line is moved one place to the left so we have like terms in the same vertical line

Example 11 Multiply: $(2m + 2m^4 - 5 - 3m^2)(2 + m^2 - 3m)$

Note that $0m^3$ was written in to save a place for the m^3 terms that arise in the multiplication

$$\begin{array}{rrrrrrr} & & 2m^4 & +\ 0m^3 & -\ 3m^2 & +\ 2m & -\ 5 \\ & & & & m^2 & -\ 3m & +\ 2 \\ \hline & & 4m^4 & & -\ 6m^2 & +\ 4m & -\ 10 \\ & -\ 6m^5 & & +\ 9m^3 & -\ 6m^2 & +\ 15m & \\ 2m^6 & & -\ 3m^4 & +\ 2m^3 & -\ 5m^2 & & \\ \hline 2m^6 & -\ 6m^5 & +\ m^4 & +\ 11m^3 & -\ 17m^2 & +\ 19m & -\ 10 \end{array}$$

Multiplication is simplified by first arranging the polynomials in descending powers of m

Example 12 Multiply: $(3a^2b - 6ab^2)(2ab - 5)$

$$\begin{array}{rrrr} & & 3a^2b & -\ 6ab^2 \\ & & 2ab & -\ 5 \\ \hline & & -\ 15a^2b & +\ 30ab^2 \\ 6a^3b^2 & -\ 12a^2b^3 & & \\ \hline 6a^3b^2 & -\ 12a^2b^3 & -\ 15a^2b & +\ 30ab^2 \end{array}$$

Note that the second line is moved over far enough so only like terms are in the same vertical line

---

EXERCISES 1-7B

Find the products.

1. $6x(5x^2 - 3x - 7)$
2. $8y(3y^2 + 4y - 9)$
3. $(2m^3 + m^2 - 6m)(-4m^2)$
4. $(7h^3 - 5h - 11)(-3h^3)$
5. $-3z^2(5z^3 - 4z^2 + 2z - 8)$
6. $-4m^3(2m^3 - 3m^2 + m - 5)$
7. $(4y^4 - 7y^2 - y + 12)(-5y^3)$
8. $(2z^5 + z^3 - 8z^2 - 12)(-7z^2)$
9. $(2h - 3)(4h^2 - 5h + 7)$
10. $(5k - 6)(2k^2 + 7k - 3)$
11. $(4 + a^4 + 3a^2 - 2a)(a + 3)$
12. $(3b - 5 + b^4 - 2b^3)(b - 5)$
13. $(4 - 3z^3 + z^2 - 5z)(4 - z)$
14. $(3 + 2v^2 - v^3 + 4v)(2 - v)$
15. $(3u^2 - u + 5)(2u^2 + 4u - 1)$
16. $(2w^2 + w - 7)(5w^2 - 3w - 1)$
17. $-2xy(-3x^2y + xy^2 - 4y^3)$
18. $-3xy(-4xy^2 - x^2y + 3x^3)$
19. $(a^3 - 3a^2b + 3ab^2 - b^3)(5a^2b)$
20. $(m^3 - 3m^2p + 3mp^2 - p^3)(-4mp^2)$
21. $(x^2 + 2x + 3)^2$
22. $(z^2 - 3z - 4)^2$
23. $[(x + y)(x^2 - xy + y^2)][(x - y)(x^2 + xy + y^2)]$
24. $[(a - 1)(a^2 + a + 1)][(a + 1)(a^2 - a + 1)]$
25. $(2.56x^2 - 5.03x + 4.21)(3.05x - 9.28)$
26. $(8.04y^2 + 4.13y - 1.28)(5.03y - 7.22)$

1-7C DIVISION OF POLYNOMIALS

DIVIDING A POLYNOMIAL BY A MONOMIAL

To divide a polynomial by a monomial, divide *each* term in the polynomial by the monomial, then add the resulting quotients.

Example 13 Dividing a polynomial by a monomial

(a) $$\frac{9x^3 - 6x^2 + 12x}{3x} = \frac{9x^3}{3x} + \frac{-6x^2}{3x} + \frac{12x}{3x} = 3x^2 - 2x + 4$$

(b) $$\frac{4x^2y - 8xy^2 + 3y}{-4xy} = \frac{4x^2y}{-4xy} + \frac{-8xy^2}{-4xy} + \frac{3y}{-4xy}$$

$$= -x + 2y - \frac{3}{4}x^{-1} = -x + 2y - \frac{3}{4x}$$

DIVIDING A POLYNOMIAL BY A POLYNOMIAL

The method used to divide a polynomial by a polynomial is like long division of whole numbers in arithmetic. It is based on the fact that division is repeated subtraction.

Example 14 $(x^2 - 3x - 10) \div (x + 2)$

First term in quotient $= \dfrac{\text{First term of dividend}}{\text{First term of divisor}} = \dfrac{x^2}{x} = x$

Second term in quotient $= \dfrac{-5x}{x} = -5$

$$
\begin{array}{rl}
 & x - 5 \\
x + 2 \,\big) & x^2 - 3x - 10 \\
 & \underline{+x^2 + 2x} \qquad \text{Subtracting } (x + 2)x = x^2 + 2x \\
 & -5x - 10 \\
 & \underline{-5x - 10} \qquad \text{Subtracting } (x + 2)(-5) = -5x - 10 \\
 & 0
\end{array}
$$

Example 15 $(17ab^2 + 12a^3 - 8b^3 - 11a^2b) \div (3a - 2b)$

Arrange the terms of the dividend and divisor in descending powers of a before beginning the division.

$$
\begin{array}{rl}
 & 4a^2 - ab + 5b^2 \text{ R } 2b^3 \\
3a - 2b \,\big) & 12a^3 - 11a^2b + 17ab^2 - 8b^3 \\
 & \underline{12a^3 - 8a^2b} \\
 & -3a^2b + 17ab^2 \\
 & \underline{-3a^2b + 2ab^2} \\
 & 15ab^2 - 8b^3 \\
 & \underline{15ab^2 - 10b^3} \\
 & 2b^3 \qquad \text{Remainder (R)}
\end{array}
$$

This quotient could also be written with the remainder expressed as a fraction: $4a^2 - ab + 5b^2 + \dfrac{2b^3}{3a - 2b}$

Example 16 $(2x^4 + x^3 - 8x^2 + 2) \div (x^2 - x - 2)$

When the divisor is a polynomial of more than two terms, exactly the same procedure is used.

$$
\begin{array}{rl}
 & 2x^2 + 3x - 1 \text{ R } 5x \quad \text{or } 2x^2 + 3x - 1 + \dfrac{5x}{x^2 - x - 2} \\
x^2 - x - 2 \,\big) & 2x^4 + x^3 - 8x^2 - 0x + 2 \\
 & \underline{2x^4 - 2x^3 - 4x^2} \\
 & 3x^3 - 4x^2 \\
 & \underline{3x^3 - 3x^2 - 6x} \\
 & -x^2 + 6x + 2 \\
 & \underline{-x^2 + x + 2} \\
 & 5x + 0
\end{array}
$$

Note that $0x$ was written in to save a place for the x-terms that arise in the division

Checking Division

> **TO CHECK DIVISION**
>
> Dividend = Divisor × Quotient + Remainder

Checking Example 16 by this method, we have

$$
\begin{array}{rrrrrl}
 & & 2x^2 & +\ 3x & -\ 1 & \quad \text{Quotient} \\
 & & x^2 & -\ \ x & -\ 2 & \quad \text{Divisor} \\
\hline
 & & -4x^2 & -\ 6x & +\ 2 & \\
 & -2x^3 & -\ 3x^2 & +\ \ x & & \\
2x^4 & +\ 3x^3 & -\ \ x^2 & & & \\
\hline
2x^4 & +\ \ x^3 & -\ 8x^2 & -\ 5x & +\ 2 & \\
 & & & +\ 5x & & \quad \text{Remainder} \\
\hline
2x^4 & +\ \ x^3 & -\ 8x^2 & & +\ 2 & \quad \text{Dividend}
\end{array}
$$

Dividend = Divisor × Quotient + Remainder

$$2x^4 + x^3 - 8x^2 + 2 = (x^2 - x - 2)(2x^2 + 3x - 1) + 5x$$

EXERCISES 1-7C

Perform the indicated divisions.

1. $\dfrac{18x^5 - 24x^4 - 12x^3}{6x^2}$

2. $\dfrac{16y^4 - 36y^3 + 20y^2}{-4y^2}$

3. $\dfrac{55a^4b^3 - 33ab^2}{-11ab}$

4. $\dfrac{26m^2n^4 - 52m^3n}{-13mn}$

5. $\dfrac{-15x^2y^2z^2 - 30xyz}{-5xyz}$

6. $\dfrac{-24a^2b^2c^2 - 16abc}{-8abc}$

7. $\dfrac{5x^3 - 4x^2 + 10}{-5x^2}$

8. $\dfrac{7y^3 - 5y^2 + 14}{-7y^2}$

9. $\dfrac{13x^3y^2 - 26xy^3 + 39xy}{13x^2y^2}$

10. $\dfrac{21m^2n^3 - 35m^3n^2 - 14mn}{7m^2n^2}$

11. $\dfrac{6a^2bc^2 - 4ab^2c^2 + 12bc}{6abc}$

12. $\dfrac{8a^3b^2c - 4a^2bc - 10ac}{4abc}$

13. Divide and check: $(6x^2 + 5x - 6) \div (3x - 2)$
14. Divide and check: $(20x^2 + 13x - 15) \div (5x - 3)$
15. Divide and check: $(15v^2 + 19v + 10) \div (5v - 7)$
16. Divide and check: $(15v^2 + 19v - 4) \div (3v + 8)$
17. $(6x^3 + 7x^2 - 11x - 12) \div (2x + 3)$
18. $(6z^3 - 13z^2 - 4z + 15) \div (3z - 5)$
19. $(8x - 4x^3 + 10) \div (2 - x)$

20. $(12x - 15 - x^3) \div (3 - x)$
21. $(17ab^2 + 12a^3 - 10b^3 - 11a^2b) \div (3a - 2b)$
22. $(10h^3 + 13h^2k - 11hk^2 - 12k^3) \div (2h + 3k)$
23. $(x^4 + 2x^3 - x^2 - 2x + 1) \div (x^2 + x - 1)$
24. $(x^4 - 2x^3 + 3x^2 - 2x + 1) \div (x^2 - x + 1)$
25. $(2u^4 + u^3v - 8u^2v^2 - 5uv^3 - 2v^4) \div (u^2 - uv - 2v^2)$
26. $(3m^4 - 7m^3n + m^2n^2 - 7mn^3 - 4n^4) \div (m^2 - 2mn + n^2)$

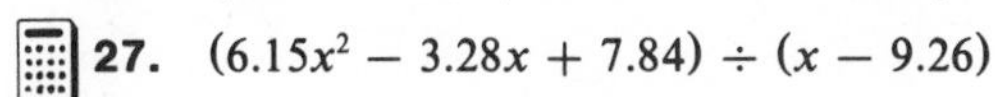

27. $(6.15x^2 - 3.28x + 7.84) \div (x - 9.26)$
28. $(81.3x^2 - 19.7x - 43.5) \div (x + 26.2)$

Chapter One Summary

Sets. (Section 1–1). A *set* is a collection of objects or things. The *elements of a set* are the objects that make up the set.
Set A is a subset of set B, written $A \subseteq B$, if every element of A is also an element of B.
The union of sets A and B, written $A \cup B$, is the set that contains all the elements of A, as well as all the elements of B.
The intersection of sets A and B, written $A \cap B$, is the set that contains all the elements that are in *both* A and B.

Kinds of Numbers. (Section 1–2). All the numbers that can be represented by points on the number line are *real numbers*. Some subsets of the real numbers are:
The set of digits, $D = \{0, 1, 2, 3, 4, 5, 6, 7, 8, 9\}$
The set of natural numbers, $N = \{1, 2, 3, \ldots\}$
The set of whole numbers, $W = \{0, 1, 2, \ldots\}$
The set of integers, $J = \{\ldots, -3, -2, -1, 0, 1, 2, 3, \ldots\}$
The set of rational numbers, $Q = \left\{ \frac{a}{b} \,\middle|\, a, b \in J; b \neq 0 \right\}$
The set of irrational numbers, $H = \{x \mid x \in R, x \notin Q\}$

Absolute Value. (Section 1–2). The absolute value of a number is always positive (or zero).

$$|x| = \begin{cases} x & \text{if } x \geq 0 \quad x \text{ greater than zero, or } x = 0 \\ -x & \text{if } x < 0 \quad x \text{ less than zero} \end{cases}$$

Inverse Operations. (Section 1–3).
Subtraction is the inverse of addition (Section 1–3A).
Division is the inverse of multiplication (Section 1–3A).
Finding the root of a number is the inverse of raising that number to a power (Section 1–3C).

Commutative Properties. (*Order Changed*) (Section 1–3B)

$$\left.\begin{array}{ll} \textit{Addition:} & a + b = b + a \\ \textit{Multiplication:} & a \cdot b = b \cdot a \end{array}\right\} \quad a, b \in R$$

Subtraction and *division* are *not* commutative.

Associative Properties. (*Grouping Changed*) (Section 1–3B)

$$\left.\begin{array}{ll} \textit{Addition:} & a + b + c = (a + b) + c = a + (b + c) \\ \textit{Multiplication:} & a \cdot b \cdot c = (a \cdot b) \cdot c = a \cdot (b \cdot c) \end{array}\right\} \quad a, b, c \in R$$

Subtraction and *division* are *not* associative.

Operations with Zero. (Sections 1–3A, 1–3B)
If a is any real number,

1. $a + 0 = 0 + a = a$
2. $a - 0 = a$
3. $0 - a = 0 + (-a) = -a$
4. $a \cdot 0 = 0 \cdot a = 0$

If a is any real number *except* 0,

5. $\frac{0}{a} = 0$
6. $\frac{a}{0}$ is not possible $\left(\frac{a}{0} \text{ is not a real number}\right)$
7. $\frac{0}{0}$ cannot be determined
8. $0^a = 0$
9. $a^0 = 1$

The Rules of Exponents. (Section 1–4)
None of the letters can have a value that makes any denominator zero.

1. $x^a x^b = x^{a+b}$
2. $(x^a)^b = x^{ab}$

2A. $(xy)^a = x^a y^a$

3. $\frac{x^a}{x^b} = x^{a-b} \quad x \neq 0$
4. $x^{-n} = \frac{1}{x^n} \quad x \neq 0$

3A. $\left(\frac{x}{y}\right)^a = \frac{x^a}{y^a} \quad y \neq 0$

5. $x^0 = 1 \quad x \neq 0$
6. $\left(\frac{x^a y^b}{z^c}\right)^n = \frac{x^{an} y^{bn}}{z^{cn}}$

Order of Operations. (Section 1–5A)

1. If there are operations within grouping symbols, that part of the expression within those grouping symbols is evaluated first; then the entire expression is evaluated.
2. Any evaluation always proceeds in three steps:
   First: Powers and roots are done in any order.
   Second: Multiplication and division are done in order from left to right.
   Third: Addition and subtraction are done in order from left to right.

Evaluating Algebraic Expressions. (Section 1–5)

1. Replace each letter by its number value.
2. Then carry out all operations in the correct order.

Distributive Rule(s). (Section 1–6A)

$a(b + c) = ab + ac$
$(b + c)a = ba + ca$

These rules can be extended to have any number of terms within the parentheses

Removing Grouping Symbols. (Section 1–6A)
In any of the following four cases, first write a + sign for any enclosed term that has no written sign:

1. *Grouping symbol preceded by + sign (or no sign):* Leave the enclosed terms unchanged. Drop the grouping symbol and the + sign (if there is one) preceding the grouping symbol.

2. *Grouping symbol preceded by − sign:* Change the sign of each enclosed term. Drop the grouping symbol and the − sign preceding the grouping symbol.

3. *Grouping symbol preceded by factor:* Multiply each enclosed term by the factor and add these products. Drop the grouping symbol and the factor.

4. *Grouping symbols within grouping symbols:* It is usually easier to remove the innermost grouping symbols first.

Simplifying Algebraic Expressions. (Section 1–6B)

1. Remove the grouping symbols.

2. Then combine like terms.

Polynomials. (Section 1–7)
A polynomial in x is an algebraic expression that has only terms of the form ax^n, where a is any real number and n is a positive integer (or zero).

A polynomial in x and y is an algebraic expression that has only terms of the form ax^ny^m, where a is any real number and n and m are positive integers (or zero).

The degree of a term in a polynomial is the sum of the exponents of its variables.

The degree of a polynomial is the same as that of its highest-degree term (provided like terms have been combined).

Chapter One Diagnostic Test or Review

Allow yourself about 50 minutes to do these problems. Complete solutions for every problem, together with section references, are given in the answer section at the end of the book.

---

1. Are $\{5, 3, 3, 5\}$ and $\{3, 5\}$ equal sets?

2. Write $\{x \mid 6 < x < 10, x \in N\}$ in roster notation.

3. State which of the following sets are finite and which are infinite.
(a) The set of integers (b) $\{x \mid 1 < x < 5, x \in J\}$
(c) $\{x \mid 1 < x < 5, x \in R\}$

4. Given: $A = \{x, z, w\}$, $B = \{x, y, w\}$, $C = \{y, r, s\}$. Find:
(a) $A \cup C$ (b) $B \cap C$ (c) $A \cap B$ (d) $B \cup C$

5. Given the numbers -3, 2.4, 0, $\sqrt{3}$, 5, and $\frac{1}{2}$:

(a) Which are real numbers?
(b) Which are integers?
(c) Which are natural numbers?
(d) Which are irrational numbers?
(e) Which are rational numbers?

In Problems 8–34 find the value of each expression (if it has one).

6. $|-17|$

7. $(-11) + (15)$

8. $(14)(-2)$

9. $(-5)^2$

10. $(30) \div (-5)$

11. $|0|$

12. $\dfrac{9}{0}$

13. $(-35) - (2)$

14. $(-27) - (-17)$

15. $(-9)(-8)$

16. $|-3|$

17. $(-19)(0)$

18. $|5|$

19. $\dfrac{-40}{-8}$

20. $(-9) + (-13)$

21. 0^4

22. -6^2

23. $(-2)^0$

24. $\dfrac{0}{-5}$

25. $-|-3|$

26. $\sqrt[3]{-27}$

27. $\sqrt[7]{-1}$

28. $\sqrt[4]{16}$

29. $\sqrt{81}$

30. $(3^{-2})^{-1}$

31. $10^{-3} \cdot 10^5$

32. $\dfrac{2^{-4}}{2^{-7}}$

33. $\left(\dfrac{1}{10^{-3}}\right)^2$

34. (a) $16 \div 4 \cdot 2$ (b) $3 + 2 \cdot 5$
(c) $2\sqrt{9} - 5$ (d) $16 \div (-2)^2 - \dfrac{7-1}{2}$

In Problems 35–38 simplify each expression. Write the answer using only positive exponents.

35. $x^2 \cdot x^{-5}$

36. $(N^2)^4$

37. $\left(\dfrac{2x^3}{y}\right)^2$

38. $\left(\dfrac{xy^{-2}}{y^{-3}}\right)^{-1}$

In Problems 39 and 40 simplify each expression. Write the answer without fractions, using negative exponents if necessary.

39. $\dfrac{1}{a^{-3}}$

40. $\dfrac{x^2y^{-1}}{y^3z^4}$

In Problems 41 and 42 simplify each expression. Write the answer without fractions.

41. $\dfrac{x^{4a}}{x^{7a}}$

42. $\left(\dfrac{x^{-a}}{2}\right)^{-3}$

43. Given the formula $C = \dfrac{5}{9}(F - 32)$, find C when $F = -4$.

In Problems 44–50 perform the indicated operations.

44. $7x - 2(5 - x)$

45. $y - [2(x - y) - 3(1 - y)]$

46. $6x(2xy^2 - 3x^3) - 3x^2(2y^2 - 6x^2)$

47. $7x - 2\{6 - 3[8 - 2(x - 3) - 2(6 - x)]\}$

48. $\dfrac{42u^4 - 7u^2 + 28}{-14u^2}$

49. $(8 + 2z^4 - z^2 - 9z)(z - 4)$

50. $(20a^3 - 23a^2 - 29a + 14) \div (5a - 2)$

2 FIRST-DEGREE EQUATIONS AND INEQUALITIES IN ONE UNKNOWN

Most problems in algebra are solved by the use of equations or inequalities. In this chapter we show how to solve equations and inequalities that have only one unknown. We discuss other types of equations and inequalities in later chapters.

2–1 Solving First-Degree Equations that Have Only One Unknown

2-1A SOLVING AND CHECKING EQUATIONS

The Parts of an Equation

An equation has three parts:

$$7x - 11 \quad = \quad 2x + 5$$

Left side: $7x - 11$ Equal sign: $=$ Right side: $2x + 5$

A **first-degree equation that has only one unknown** is an equation with only one letter, in which the highest power of that letter is the first power.

Solution of an Equation

A **solution of an equation** that has only one unknown is a number which, when put in place of the letter, makes the two sides of the equation equal. A solution of an equation is also called a **root** of the equation.

Solution Set

The **solution set** of an equation that has only one unknown is the set of *all* numbers that are solutions of that equation. For example,

3 is the *solution* of $x + 2 = 5$ because $(3) + 2 = 5$;
$\{3\}$ is the *solution set* of $x + 2 = 5$.

Principles of Equality

The following **principles of equality** are used in solving equations.

Addition Rule: The same number may be added to both sides.
Subtraction Rule: The same number may be subtracted from both sides.
Multiplication Rule: Both sides may be multiplied by the same number.
Division Rule: Both sides may be divided by the same nonzero number.

When the above rules are used, the two sides of the equation remain equal. The key is to do the same thing to each side of the equation.

TO SOLVE A FIRST-DEGREE EQUATION THAT HAS ONLY ONE UNKNOWN

1. *Remove fractions* by multiplying both sides by the lowest common denominator (LCD).
2. *Remove grouping symbols.*
3. *Combine like terms* on each side of the equation.
4. If the unknown appears on both sides, use addition or subtraction to remove it from one side.
5. Remove all numbers from the side of the equation that has the unknown.

   First: Remove the numbers being added or subtracted.
   Second: Divide both sides by the coefficient of the unknown.
6. Check the solution.

To Check the Solution of an Equation

1. Replace the unknown letter in the given equation by the number found in the solution.
2. Perform the indicated operations on both sides of the = sign.
3. If the resulting number on each side of the = sign is the same, the solution is correct.

Example 1 Solve the equation $7y - 3(2y - 5) = 6(2 + 3y) - 31$

$$
\begin{aligned}
7y - 3(2y - 5) &= 6(2 + 3y) - 31 \\
7y - 6y + 15 &= 12 + 18y - 31 && \text{Removed grouping symbols} \\
y + 15 &= 18y - 19 && \text{Combined like terms} \\
\underline{-y \quad} &\quad \underline{-y \quad} && \text{To remove } y \text{ from left side} \\
15 &= 17y - 19 \\
\underline{+19} &\quad \underline{+19} && \text{To remove } -19 \text{ from right side} \\
34 &= 17y \\
2 &= y && \text{Divided both sides by 17}
\end{aligned}
$$

The *solution* of the equation is 2.
The *solution set* of the equation is {2}.

Check for $y = 2$.

$$
\begin{aligned}
7y - 3(2y - 5) &= 6(2 + 3y) - 31 \\
7(2) - 3[2(2) - 5] &\stackrel{?}{=} 6[2 + 3(2)] - 31 && \text{Replaced } y \text{ by 2} \\
14 - 3[4 - 5] &\stackrel{?}{=} 6[2 + 6] - 31 \\
14 - 3[-1] &\stackrel{?}{=} 6[8] - 31 \\
14 + 3 &\stackrel{?}{=} 48 - 31 \\
17 &= 17
\end{aligned}
$$

Example 2 Solve: $\dfrac{x - 4}{2} - \dfrac{x}{5} = \dfrac{1}{10}$

LCD = 10

$$\frac{\overset{5}{\cancel{10}}}{1} \cdot \left(\frac{x - 4}{\cancel{2}}\right) - \frac{\overset{2}{\cancel{10}}}{1} \cdot \left(\frac{x}{\cancel{5}}\right) = \frac{\cancel{10}}{1} \cdot \left(\frac{1}{\cancel{10}}\right) \qquad \text{Multiplied by LCD 10}$$

$$
\begin{aligned}
5(x - 4) - 2x &= 1 \\
5x - 20 - 2x &= 1 && \text{Removed grouping symbols} \\
3x - 20 &= 1 && \text{Combined like terms} \\
3x &= 21 && \text{Added 20 to both sides} \\
x &= 7 && \text{Divided both sides by 3}
\end{aligned}
$$

The *solution* of the equation is 7.
The *solution set* of the equation is {7}.

Check for $x = 7$.

$$\frac{x-4}{2} - \frac{x}{5} = \frac{1}{10}$$

$$\frac{(7)-4}{2} - \frac{(7)}{5} \stackrel{?}{=} \frac{1}{10} \qquad \text{Replaced } x \text{ by 7}$$

$$\frac{3}{2} - \frac{7}{5} \stackrel{?}{=} \frac{1}{10}$$

$$\frac{\overset{5}{\cancel{10}}}{1} \cdot \frac{3}{\cancel{2}} - \frac{\overset{2}{\cancel{10}}}{1} \cdot \frac{7}{\cancel{5}} \stackrel{?}{=} \frac{\cancel{10}}{1} \cdot \frac{1}{\cancel{10}} \qquad \text{Multiplied by the LCD 10}$$

$$15 - 14 \stackrel{?}{=} 1$$

$$1 = 1$$

GRAPHING SOLUTIONS OF EQUATIONS ON THE NUMBER LINE

Domain of Variable The solution set can contain numbers only from the *domain of the variable. If the domain of the variable is not mentioned, we assume it to be the set of real numbers, R.*

Example 3 Solve $2x - 5 = 7$ and plot its solution set on the number line.

$$\begin{aligned} 2x - 5 &= 7 \\ +5 &\quad +5 \\ \hline 2x &= 12 \\ x &= 6 \qquad \text{Solution} \end{aligned}$$

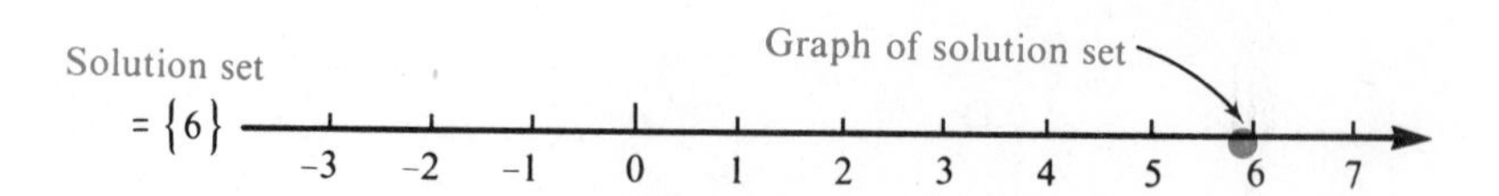

Example 4 Plot $\{x \mid 2(3x + 5) = 14, x \in J\}$ on the number line

$$\begin{aligned} 2(3x + 5) &= 14 \\ 6x + 10 &= 14 \\ -10 &\quad -10 \\ \hline 6x &= 4 \end{aligned}$$

$$x = \frac{4}{6} = \frac{2}{3}$$

Since $\frac{2}{3}$ is not an integer, then $\{x \mid 2(3x + 5) = 14, x \in J\} = \{\ \}$.

This means that the equation $2(3x + 5) = 14$ has *no solution that is an integer.* Since there is no solution, there are no points to plot on the number line.

EXERCISES 2-1A

In Exercises 1–10 solve and check each equation; then graph the solution set on the number line.

1. $5x - 3(2 + 3x) = 6$

2. $7x - 2(5 + 4x) = 8$

3. $7x + 5 = 3(3x + 5)$

4. $8x + 6 = 2(7x + 9)$

5. $3y - 2(2y - 7) = 2(3 + y) - 4$

6. $4a - 3(5a - 14) = 5(7 + a) - 9$

7. $2(3x - 6) - 3(5x + 4) = 5(7x - 8)$

8. $4(7z - 9) - 7(4z + 3) = 6(9z - 10)$

9. $7(2 - 5x) + 27 = 18x - 3(8 - 4x)$

10. $5(3 - 2k) = 8(3k - 4) - 4(1 + 7k)$

In Exercises 11–33 solve each equation.

11. $\frac{a}{2} - \frac{a}{5} = 6$

12. $\frac{b}{3} - \frac{b}{7} = 12$

13. $7 = \frac{x}{3} + \frac{x}{4}$

14. $8 = \frac{x}{5} + \frac{x}{3}$

15. $\frac{M - 2}{5} + \frac{M}{3} = \frac{1}{5}$

16. $\frac{y + 2}{4} + \frac{y}{5} = \frac{1}{4}$

17. $\frac{2x - 1}{3} + \frac{3x}{4} = \frac{5}{6}$

18. $\frac{3}{4} = \frac{3z - 2}{4} + \frac{3z}{8}$

19. $6(3 - 4x) + 12 = 10x - 2(5 - 3x)$

20. $6(5 - 4h) = 3(4h - 2) - 7(6 + 8h)$

21. $2[3 - 5(x - 4)] = 10 - 5x$

22. $3[2 - 4(x - 7)] = 26 - 8x$

23. $6(3h - 5) = 3[4(1 - h) - 7]$

24. $4(2 - 6x) - 6 = 8x + [-(3x - 11) + 20]$

25. $\frac{2(m - 3)}{5} - \frac{3(m + 2)}{2} = \frac{7}{10}$

26. $\frac{5(x - 4)}{6} - \frac{2(x + 4)}{9} = \frac{5}{18}$

27. $5(3 - 2x) - 10 = 4x + [-(2x - 5) + 15]$

28. $-2\{5 - [6 - 3(4 - x)] - 2x\} = 13 - [-(2x - 1)]$

29. $12 = -\{-3[4z - 2(z - 2)]\}$

30. $-3\{10 - [7 - 5(4 - x) - 8]\} = 11 - [-(5x - 4)]$

31. $6.23x + 2.5(3.08 - 8.2x) = -14.7$

32. $9.84 - 4.6x = 5.17(9.01 - 8.23x)$

33. $7.02(5.3x - 4.28) = 11.6 - 2.94x$

34. *Brainteaser.* If Sam adds \$5 to what he had before he spent \$5, he finds that he would have had five times as much as he has now. How much did he have originally?

2-1B CONDITIONAL EQUATIONS, IDENTITIES, AND EQUATIONS THAT HAVE NO SOLUTION

In general, when first-degree equations in one unknown are solved, there are three possible results.

1. A single solution is obtained.
2. Both sides of the equation reduce to the same expression.
3. The two sides of the equation reduce to unequal expressions.

A method of determining which of these three cases you have is given in the following box.

Conditional, Identical, and Unsolvable Equations

WHEN SOLVING A FIRST-DEGREE EQUATION THAT HAS ONLY ONE UNKNOWN

1. Attempt to solve the equation by the method given on page 58.

2. There are three possible results:

(a) **Conditional Equation:** If a *single solution* is obtained, the equation is a conditional equation.

(b) **Identity:** If the two sides of the equation *reduce to the same expression,* the equation is an identity. In this case every permissible* real number is a solution.

(c) **No Solution:** If the two sides of the equation reduce to unequal expressions, the equation has no solution. In this case the solution set is the empty set.

Example 5 Solve:

$$\begin{aligned} 4x - 2(3 - x) &= 12 \\ 4x - 6 + 2x &= 12 \\ 6x &= 18 \\ x &= 3 \end{aligned}$$

Conditional equation (Single solution)

*A permissible number is any number in the domain of the unknown letter (Section 5–1).

Example 6

$$\begin{aligned} 4x - 2(3 + 2x) &= -6 \\ 4x - 6 - 4x &= -6 \\ -6 &= -6 \end{aligned}$$

Identity
(Both sides the same)

Example 7

$$\begin{aligned} 4x - 2(3 + 2x) &= 8 \\ 4x - 6 - 4x &= 8 \\ -6 &\neq 8 \end{aligned}$$

No solution
(Sides unequal)

EXERCISES 2-1B

Identify each equation as a conditional equation, an identity, or an equation that has no solution. Find the solution of each conditional equation.

1. $5x - 2(4 - x) = 6$

2. $4x - 2(6 + 2x) = -12$

3. $6x - 3(5 + 2x) = -15$

4. $6x - 3(5 + 2x) = -12$

5. $4x - 2(6 + 2x) = -15$

6. $8x - 3(5 - x) = 7$

7. $7(2 - 5x) - 32 = 10x - 3(6 + 15x)$

8. $2(2x - 5) - 3(4 - x) = 7x - 20$

9. $6(3 - 4x) + 10 = 8x - 3(2 - 3x)$

10. $3(x - 4) - 5(6 - x) = 2(4x - 21)$

11. $2[3 - 4(5 - x)] = 2(3x - 11)$

12. $3[5 - 2(7 - x)] = 6(x - 7)$

2-2 Solving First-Degree Inequalities that Have Only One Unknown

The Parts of an Inequality

An inequality has three parts:

$$5x + 3 \quad < \quad 2x - 1$$

Left side (5x + 3); Inequality symbol (<); Right side (2x − 1)

Inequality Symbols

The **inequality symbol** may be any one of the following:

$$<, >, \leq, \geq, \neq, \not<, \not>, \not\leq, \not\geq$$

In this text we will discuss only inequalities that have the symbols $<$, $>$, $\leq$, or $\geq$.

Solution of an Inequality

A **solution of an inequality** is a number that, when put in place of a letter, makes the inequality a true statement.

The **solution set of an inequality** is the set of *all* numbers that are solutions of the inequality.

Sense of Inequality

The **sense** of an inequality symbol refers to the direction the symbol points.

Same sense:
$$a > b$$
$$c > d$$

Opposite sense:
$$a < b$$
$$c > d$$

The method of solving inequalities is very much like the method used for solving equations. We show how the methods are alike or different in the following box.

Comparing the Solution of Equations and Inequalities

IN SOLVING EQUATIONS	IN SOLVING INEQUALITIES
Addition Rule: The same number may be added to both sides.	**Addition Rule:** The same number may be added to both sides.
Subtraction Rule: The same number may be subtracted from both sides.	**Subtraction Rule:** The same number may be subtracted from both sides.
Multiplication Rule: Both sides may be multiplied by the same number.	**Multiplication Rule:** **1.** Both sides may be multiplied by the same *positive* number. **2.** When both sides are multiplied by the same *negative* number, *the sense must be changed.*
Division Rule: Both sides may be divided by the same nonzero number.	**Division Rule:** **1.** Both sides may be divided by the same *positive* number. **2.** When both sides are divided by the same *negative* number, *the sense must be changed.*

Example 1 Illustrating the use of the inequality rules

(a) Using the addition rule

$$\begin{array}{rcr} 10 & > & 5 \\ +6 & & +6 \\ \hline 16 & > & 11 \end{array}$$

Added 6 to both sides

Sense is not changed

(b) Using the subtraction rule

$$\begin{array}{rcl} 7 & < & 12 \\ \underline{-2} & & \underline{-2} \\ 5 & < & 10 \end{array}$$

Subtracted 2 from both sides

Sense is not changed

(c) Using the multiplication rule (positive multiplier)

$$3 < 4$$
$$2\,(3) < 2\,(4) \quad \text{Multiplied both sides by 2}$$
$$6 < 8 \quad \textit{Sense is not changed}$$

(d) Using the multiplication rule (negative multiplier)

$$3 < 4$$
$$(-2)(3) > (-2)(4)$$

Multiplied both sides by the same *negative* number, -2

$$-6 > -8$$

Sense is changed

(e) Using the division rule (positive divisor)

$$9 > 6$$
$$\frac{9}{3} > \frac{6}{3} \quad \text{Divided both sides by 3}$$
$$3 > 2$$

Sense is not changed.

(f) Using the division rule (negative divisor)

$$9 > 6$$
$$\frac{9}{-3} < \frac{6}{-3}$$

Divided both sides by the same *negative* number, -3

$$-3 < -2$$

Sense is changed

Example 2 Solve: $3x - 2(2x - 7) \le 2(3 + x) - 4$

$$3x - 2(2x - 7) \le 2(3 + x) - 4$$
$$3x - 4x + 14 \le 6 + 2x - 4 \quad \text{Removed grouping symbols}$$
$$-x + 14 \le 2 + 2x$$
$$-3x \le -12 \quad \text{Subtracted } 2x \text{ from both sides}$$
$$\frac{-3x}{-3} \ge \frac{-12}{-3} \quad \text{Divided both sides by } -3$$
$$x \ge 4 \quad \textit{Sense is changed}$$

The *solution* of the inequality is $x \ge 4$.
The *solution set* of the inequality is $\{x \mid x \ge 4\}$.

Example 3 Solve: $\dfrac{y+3}{4} \le \dfrac{y-2}{3} + \dfrac{1}{4}$

LCD = 12

$$\frac{\overset{3}{\cancel{12}}}{1}\left(\frac{y+3}{\cancel{4}}\right) \le \frac{\overset{4}{\cancel{12}}}{1}\left(\frac{y-2}{\cancel{3}}\right) + \frac{\overset{3}{\cancel{12}}}{1}\left(\frac{1}{\cancel{4}}\right) \qquad \text{Multiplied by LCD 12}$$

$$3(y+3) \le 4(y-2) + 3(1)$$
$$3y + 9 \le 4y - 8 + 3$$
$$3y + 9 \le 4y - 5$$
$$14 \le y$$

Which is the same as $y \ge 14$.

The *solution* of the inequality is $y \ge 14$.
The *solution set* of the inequality is $\{y \mid y \ge 14\}$.

> **TO SOLVE AN INEQUALITY**
>
> Proceed in the same way used to solve equations, with the *exception* that *the sense must be changed when multiplying or dividing both sides by a negative number.*

A Word of Caution. We solve an inequality by a method very much like the method used for solving an equation. For this reason some students confuse the *solution* of an inequality with that of an equation. For example:

$$\begin{array}{rl} 3x < & 2x + 6 \\ \underline{-2x} & \underline{-2x \quad\;\;} \\ x = & \quad\;\; 6 \end{array} \qquad \text{Incorrect}$$

↑ This = sign is the error

$$x < \quad 6 \qquad \text{Correct}$$

↑ Instead of an = sign this must be a < symbol

Infinitely many numbers satisfy this inequality, since any real number less than 6 is the solution. For example, 3, −6, and −201 are all solutions. ■

GRAPHING SOLUTIONS OF INEQUALITIES ON THE NUMBER LINE

Example 4 Solve $5x + 3 < 13$ and plot its solution set on the number line

$$\begin{array}{rl} 5x + 3 < & 13 \\ \underline{-3} & \underline{-3} \\ 5x \quad < & 10 \\ x < & 2 \end{array} \qquad \text{Solution}$$

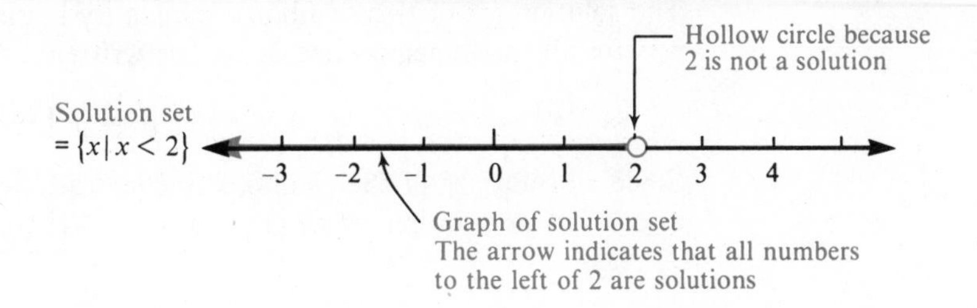

Example 5 **Solve $3x - 2 \geq -14$ and plot its solution set on the number line**

$$
\begin{aligned}
3x - 2 &\geq -14 \\
+2 &\quad +2 \\
\hline
3x &\geq -12 \\
x &\geq -4 \quad \text{Solution}
\end{aligned}
$$

The arrow on the number line together with the solid circle indicates that -4 and all numbers to the right of -4 are solutions

Solution set $= \{x \mid x \geq -4\}$

-5 -4 -3 -2 -1 0 1 2 3 4 5

Solid circle because -4 is a solution

Combined Inequalities

We sometimes have two inequality symbols in the same statement. For example, $2 < x + 5 < 9$. In such **combined inequalities** (page 10) we actually have three inequalities:

1. $2 < x + 5$

$$
\begin{aligned}
2 &< x + 5 \\
-5 &\quad -5 \\
\hline
-3 &< x \quad \textit{Solution}
\end{aligned}
$$

$\{x \mid -3 < x\}$ *Solution set*

2. $x + 5 < 9$

$$
\begin{aligned}
x + 5 &< 9 \\
-5 &\quad -5 \\
\hline
x &< 4 \quad \textit{Solution}
\end{aligned}
$$

$\{x \mid x < 4\}$ *Solution set*

3. $2 < 9$ This inequality is always true, therefore *Solution set* $= R$

The solution set of this combined inequality is the *intersection* of the solution sets for all three inequalities. It can be written

$$\{x \mid -3 < x\} \cap \{x \mid x < 4\} \cap R = \{x \mid -3 < x\} \cap \{x \mid x < 4\}$$

Since all solutions of the combined inequalities lie between -3 and 4, we can also write the solution set as $\{x \mid -3 < x < 4\}$ (see Figure 2–2).

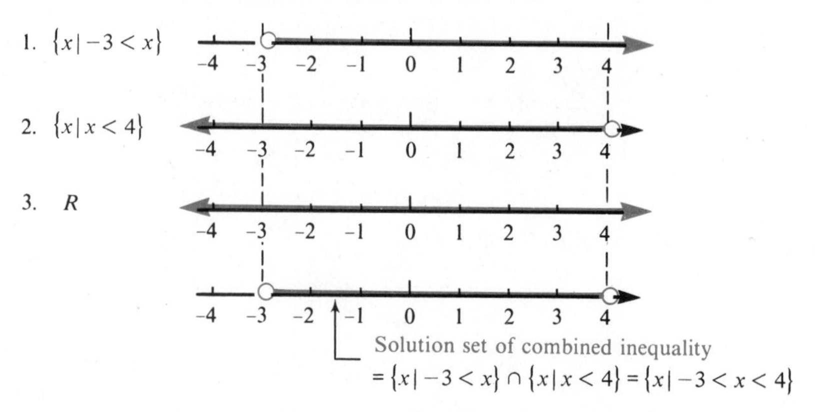

Figure 2–2 Inequalities Combined on Number Line

The solutions of combined inequalities may be conveniently arranged as shown in Examples 6 and 7.

Example 6 Solve $2 < x + 5 < 9$ and plot its solution set

$$\begin{array}{rcccl} 2 < & x + 5 & < & 9 & \\ \underline{-5} & \underline{-5} & & \underline{-5} & \\ -3 < & x & < & 4 & \text{Solution} \end{array}$$

$\{x \mid -3 < x < 4\}$

-4 -3 -2 -1 0 1 2 3 4 5 6

Example 7 Plot $\{x \mid 1 > x - 2 \geq -3, x \in J\}$

J is the domain of variable x

x is an integer

$$\begin{array}{rcccl} 1 > & x - 2 & \geq & -3 & \\ \underline{+2} & \underline{+2} & & \underline{+2} & \\ 3 > & x & \geq & -1 & \end{array}$$

and x is an integer

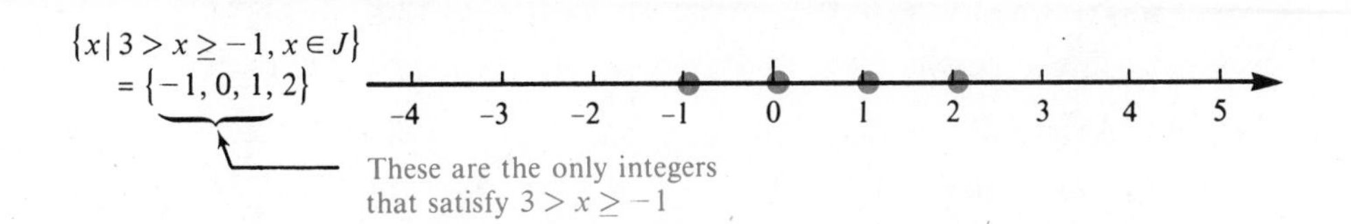

A Word of Caution. Special care must be taken when writing combined inequalities. For example,

(a) $-8 > x > 2$

Incorrect because $-8 \not> 2$

This incorrect combined inequality *cannot* be written correctly by reversing the sense of the two inequalities.

(b) $2 < x < -8$

Incorrect because $2 \not< -8$

(c) Combined inequalities should not be written with different senses for the two inequality symbols.

$0 < x > 1$

Incorrect because $0 <> 1$ is a contradiction ■

EXERCISES 2-2

Solve the inequalities.

1. $3x - 1 < 11$

2. $7x - 12 < 30$

3. $17 \geq 2x - 9$

4. $33 \geq 5 - 4x$

5. $-3 \leq x + 4$

6. $18 - 7y > -3$

7. $2y - 16 > 17 + 5y$

8. $6y + 7 > 4y - 3$

9. $4z - 22 < 6(z - 7)$

10. $8(a - 3) > 15a - 10$

11. $11a - 7 < 5a - 13$

12. $3(2 + 3m) \geq 5m - 6$

13. $9(2 - 5m) - 4 \geq 13m + 8(3 - 7m)$

14. $18k - 3(8 - 4k) \leq 7(2 - 5k) + 27$

15. $10 - 5x > 2[3 - 5(x - 4)]$

16. $3[2 + 4(y + 5)] < 30 + 6y$

17. $6(10 - 3k) + 25 \geq 4k - 5(3 - 2k)$

18. $6z < 2 - 4[2 - 3(z - 5)]$

In Exercises 19–32, solve each inequality and graph its solution on the number line.

19. $\frac{z}{3} > 7 - \frac{z}{4}$

20. $\frac{t}{5} - 8 > -\frac{t}{3}$

21. $\frac{1}{3} + \frac{w + 2}{5} \geq \frac{w - 5}{3}$

22. $\frac{u - 2}{3} - \frac{u + 2}{4} \geq -\frac{2}{3}$

23. $\frac{w}{3} > 12 - \frac{w}{6}$

24. $\frac{1}{2} + \frac{u + 9}{5} \geq \frac{u + 1}{2}$

25. $7(4x + 10) > 14$

26. $9 \geq 2x - 1 \geq -9$

27. $-3 \leq x + 1 < 2$

28. $5 > x - 2 \geq 3$

29. $\{x \mid 4 \geq x - 3 \geq -5, x \in N\}$

30. $\{x \mid -3 \leq 2x + 1 \leq 7, x \in N\}$

31. $\{x \mid -1 < x + 2 < 4, x \in J\}$

32. $\{x \mid -13 < 2x - 3 < 7, x \in N\}$

33. $14.73(2.65x - 11.08) - 22.51x \geq 13.94x(40.27)$

34. $1.065 - 9.801x \leq 5.216x - 2.740(9.102 - 7.641x)$

35. $54.7x - 48.2(20.5 - 37.6x) \leq 81.9(60.3x - 19.1) + 97.4$

36. $1.203(4.071x) \leq 9.214(52.18 - 6.022x) - 2.947$

2–3 Equations and Inequalities that Have Absolute Value Signs

EQUATIONS

An equation that has an absolute value sign is equivalent to *two* equations without absolute value signs.

Definition of Absolute Value

For any algebraic expression *N*, (see Section 1–2)

$$|N| = N \quad \text{if } N \geq 0$$
$$|N| = -N \quad \text{if } N < 0$$

Example 1 Applying the definition of absolute value

$$\left.\begin{aligned} |2y + 5| &= 2y + 5 && \text{if } 2y + 5 \geq 0 \\ |2y + 5| &= -(2y + 5) && \text{if } 2y + 5 < 0 \end{aligned}\right\} \quad \text{Definition of absolute value}$$

Note that $|2y + 5|$ can be replaced by either $(2y + 5)$ or $-(2y + 5)$, depending upon whether $(2y + 5) \geq 0$ or $(2y + 5) < 0$. When we do not know the value of y, we must allow for both possibilities.

Example 2 Solve $|2y + 5| = 3$

$|2y + 5|$ can be replaced by either

$$\begin{aligned} |2y + 5| &= 3 \\ 2y + 5 &= 3 \\ 2y &= -2 \\ y &= -1 \end{aligned} \quad \text{or} \quad \begin{aligned} |2y + 5| &= 3 \\ -(2y + 5) &= 3 \\ 2y + 5 &= -3 \\ 2y &= -8 \\ y &= -4 \end{aligned}$$

Therefore the *solution* is $y = -1$ *or* $y = -4$.
The *solution set* is $\{-1, -4\}$.

Check for $y = -1$:

$$|2y + 5| = 3$$
$$|2(-1) + 5| \stackrel{?}{=} 3$$
$$|-2 + 5| \stackrel{?}{=} 3$$
$$|3| \stackrel{?}{=} 3$$
$$3 = 3$$

Check for $y = -4$:

$$|2y + 5| = 3$$
$$|2(-4) + 5| \stackrel{?}{=} 3$$
$$|-8 + 5| \stackrel{?}{=} 3$$
$$|-3| \stackrel{?}{=} 3$$
$$3 = 3$$

The method used in Example 2 is summarized in the following box.

> **TO SOLVE AN EQUATION THAT HAS AN ABSOLUTE VALUE SIGN**
>
> *If* $|N| = a$ where N is any algebraic expression and $a \geq 0$
>
> *then* solve *two* equations:
>
> (1) $N = a$ and (2) $-N = a$
>
> The solution set of $|N| = a$ is the union of the solution sets of equations (1) and (2).

If $a < 0$, $|N| = a$ has no solution since the absolute value of an expression must be non-negative.

Example 3 Solve $\left|\dfrac{3 - 2x}{5}\right| = 2$ and plot its solution on the number line

In this example the N referred to in the preceding box is replaced by $\dfrac{3 - 2x}{5}$.

Solve: (1) $N = 2$ and (2) $-N = 2$, $N = -2$

$$\frac{3 - 2x}{5} = 2 \qquad\qquad \frac{3 - 2x}{5} = -2$$

$$3 - 2x = 10 \qquad\qquad 3 - 2x = -10$$

$$-2x = 7 \qquad\qquad -2x = -13$$

$$x = -\frac{7}{2} = -3\tfrac{1}{2} \qquad\qquad x = \frac{13}{2} = 6\tfrac{1}{2}$$

Therefore the *solution set* is $\{-3\frac{1}{2}, 6\frac{1}{2}\}$.

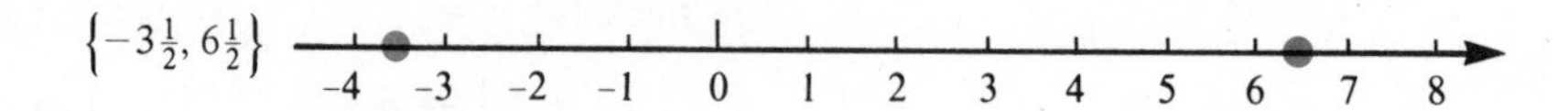

INEQUALITIES

An inequality that has an absolute value sign is equivalent to *two* inequalities without absolute value signs.

Consider $|x| \leq 5$. If $x < 0$, then $|x|$ is replaced by $-x$. If $x \geq 0$, then $|x|$ is replaced by x. Thus we have two cases to consider.

	$|x| \leq 5$		$|x| \leq 5$
<u>If $x < 0$,</u> then	$-x \leq 5$	<u>If $x \geq 0$,</u> then	$x \leq 5$
	$x \geq -5$	Since $x \geq 0$, and	$x \leq 5$
or	$-5 \leq x$		
Since $-5 \leq x$ and	$x < 0$,	(2)	$0 \leq x \leq 5$
(1)	$-5 \leq x < 0$		

Combining (1) and (2), we have

$$-5 \leq x \leq 5$$

The graph of $-5 \leq x \leq 5$ is shown in Figure 2–3A.

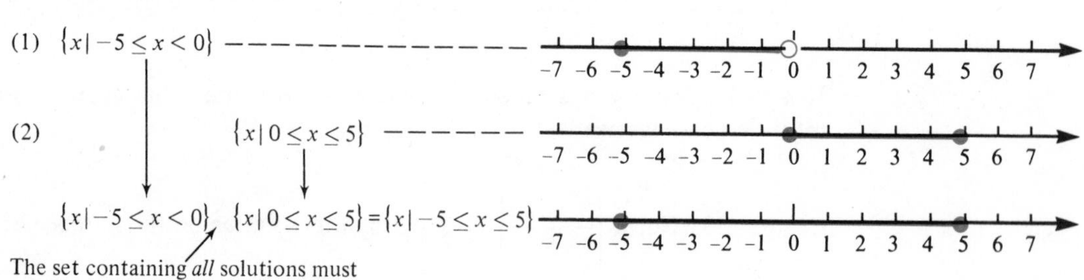

Figure 2–3A Graph of $-5 \leq x \leq 5$

These results may be generalized as follows:

> **THEOREM 1**
>
> If $|N| \leq a$, then $-a \leq N \leq a$ $(a \geq 0)$
>
> If $|N| < a$, then $-a < N < a$
>
> where N is any algebraic expression

Note: The value of a cannot be negative in Theorem 1 because if $a < 0$, then $|N| \leq a$ becomes $|N| < 0$, which is impossible. (The definition of $|N|$ requires that $|N|$ never be negative.) In this case the solution set is the empty set. ■

Consider $|x| \geq 3$. If $x < 0$, then $|x|$ is replaced by $-x$. If $x \geq 0$, then $|x|$ is replaced by x. Therefore we have two cases to consider.

$\|x\| \geq 3$	$\|x\| \geq 3$
If $x < 0$, then $-x \geq 3$	If $x \geq 0$, then $x \geq 3$ (2)
(1) $x \leq -3$	

Combining (1) and (2), we have

$$x \leq -3 \quad \text{or} \quad x \geq 3$$

The graph of $x \leq -3$ or $x \geq 3$ is shown in Figure 2–3B.

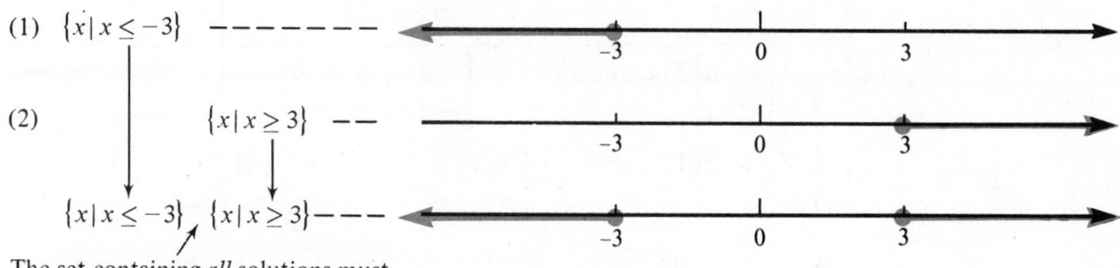

Figure 2–3B Graph of $x \leq -3$ or $x \geq 3$

These results may be generalized as follows:

> **THEOREM 2**
>
> If $|N| \geq a$, then $N \geq a$ or $N \leq -a$ $(a \geq 0)$
>
> If $|N| \geq a$, then $N > a$ or $N < -a$
>
> where N is any algebraic expression

Note: If $a < 0$ in Theorem 2, then the solution set is R. For example, if $N = x$ and $|x| \geq -2$, then (1) $x \geq -2$ *or* (2) $x \leq -(-2)$ by Theorem 2.

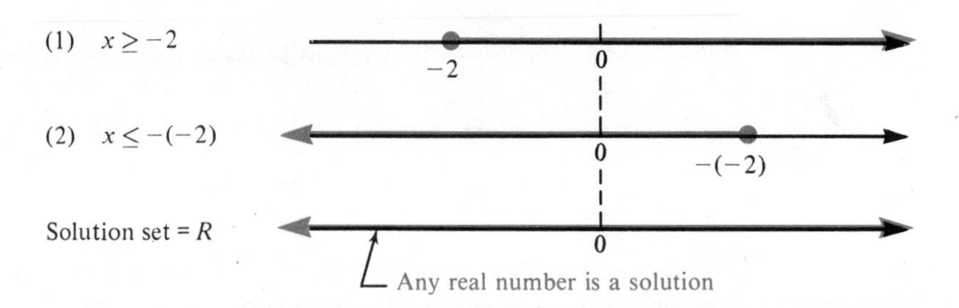

We solve inequalities that have absolute value signs by using either Theorem 1 or Theorem 2, as explained in the following box:

TO SOLVE AN INEQUALITY THAT HAS AN ABSOLUTE VALUE SIGN

Type of Inequality	Equivalent Statement to Solve	General Appearance of Graph
$\left.\begin{matrix}\lvert N\rvert < a \\ \lvert N\rvert \leq a\end{matrix}\right\}$ Theorem 1	$\left\{\begin{matrix}-a < N < a \\ -a \leq N \leq a\end{matrix}\right.$	
$\left.\begin{matrix}\lvert N\rvert > a \\ \lvert N\rvert \geq a\end{matrix}\right\}$ Theorem 2	$\left\{\begin{matrix}N > a \quad \text{or} \quad N < -a \\ N \geq a \quad \text{or} \quad N \leq -a\end{matrix}\right.$	

Example 4 Solve $\left|\dfrac{4-3x}{2}\right| \leq 6$ and plot its solution set

Use Theorem 1 with $N = \dfrac{4-3x}{2}$

Replace $\left|\dfrac{4-3x}{2}\right| \leq 6$ by $-6 \leq \dfrac{4-3x}{2} \leq 6$.

$$-6 \leq \frac{4-3x}{2} \leq 6$$

$$-12 \leq 4 - 3x \leq 12 \qquad \text{Multiplied by 2}$$

$$-16 \leq -3x \leq 8 \qquad \text{Subtracted 4}$$

$$\frac{-16}{-3} \geq \frac{-3x}{-3} \geq \frac{8}{-3} \qquad \text{Divided both sides by } -3$$

Sense is changed

$$\frac{16}{3} \geq x \geq -\frac{8}{3} \qquad \text{Solution}$$

$$\left\{x \,\middle|\, -\frac{8}{3} \leq x \leq \frac{16}{3}\right\} \qquad \text{Solution set}$$

$\left\{x \mid -2\frac{2}{3} \le x \le 5\frac{1}{3}\right\}$ −4 −3 −2 −1 0 1 2 3 4 5 6

Example 5 Solve $|5 - 2x| > 3$ and plot its solution set

Use Theorem 2 with $N = 5 - 2x$

Replace $|5 - 2x| > 3$ by $5 - 2x > 3$ *or* $5 - 2x < -3$.

$$\begin{array}{rcl} 5 - 2x > 3 & \textit{or} & 5 - 2x < -3 \\ -5 \quad\;\; -5 & & -5 \quad\;\; -5 \\ \hline -2x > -2 & & -2x < -8 \\ \dfrac{-2x}{-2} < \dfrac{-2}{-2} & & \dfrac{-2x}{-2} > \dfrac{-8}{-2} \end{array}$$

Divided both sides by −2
Sense is changed

$x < 1$ *or* $x > 4$ — Solution

$\{x \mid x < 1$ *or* $x > 4\}$ — Solution set

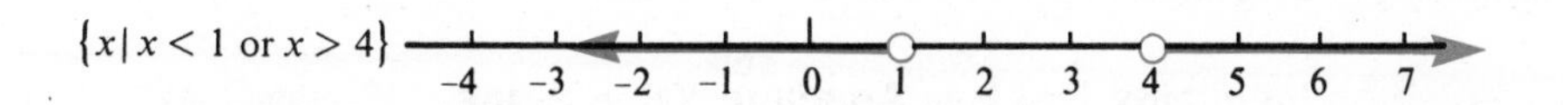

EXERCISES 2-3

Solve the equations and inequalities and graph their solution sets.

1. $|x| = 3$
2. $|3x| = 12$
3. $4|x| = 8$
4. $5|x| = 15$
5. $|x| < 2$
6. $|2x| < 8$
7. $3|x| \le 9$
8. $4|x| \ge 12$
9. $|x - 3| = 4$
10. $3|x - 3| = 9$
11. $|x - 5| \ge 4$
12. $2|x + 3| \le 4$
13. $|2x + 1| = 5$
14. $|3x - 2| = 7$
15. $|2x - 3| = -5$
16. $|7 - 5x| = -2$
17. $2|x - 4| = 6$
18. $3|x - 4| \ge 6$
19. $|1 - 3x| < 10$
20. $|5 - 2x| > 3$
21. $\left|\dfrac{1 - x}{2}\right| = 6$
22. $\left|\dfrac{5x - 6}{3}\right| = 8$
23. $\left|\dfrac{3x - 5}{2}\right| \le 10$
24. $\left|\dfrac{5 - x}{2}\right| < 7$
25. $\left|\dfrac{4 - x}{3}\right| > 2$
26. $\left|\dfrac{2x + 4}{3}\right| \ge 2$
27. $\left|3 - \dfrac{x}{2}\right| > 4$
28. $\left|5 - \dfrac{x}{3}\right| < 2$
29. $\left|4 - \dfrac{x}{3}\right| < 1$
30. (calculator) $|74.7 - 35.2x| = 19.1$
31. (calculator) $\left|\dfrac{2.94x - 8.08}{5.15}\right| > 4.42$
32. (calculator) $\left|\dfrac{89.1 - 27.4x}{35.2}\right| = 62.4$

Chapter Two Summary

When Solving a First-Degree Equation that Has Only One Unknown. (Section 2–1)
There are three possible results:

1. A single solution is obtained (*conditional equation*).
2. The two sides are identical (*identity*). Every permissible real number is a solution.
3. The two sides are unequal (*no solution*).

To Solve an Equation that Has Absolute Value Signs. (Section 2–3)

If $|N| = a$ where N is any algebraic expression and $a \geq 0$

then solve *two* equations:

$$(1)\ N = a \quad \text{and} \quad (2)\ -N = a$$

The solution set of $|N| = a$ is the union of the solution sets of equations (1) and (2).

When Solving a First-Degree Inequality that Has Only One Unknown. (Section 2–2)
Proceed in the same way used to solve equations, with the *exception* that the *sense* must be changed when *multiplying or dividing* both sides by a *negative* number.

To Solve an Inequality that Has Absolute Value Signs. (Section 2–4)
For $a > 0$:

If $|N| < a$, then $-a < N < a$

If $|N| > a$, then $N > a$ or $N < -a$

For $a < 0$: There is no solution.

Chapter Two Diagnostic Test or Review

Allow yourself about 50 minutes to do these problems. Complete solutions for every problem, together with section references, are given in the answer section at the end of the book.

In Problems 1–4 identify each equation as a conditional equation, an identity, or an equation that has no solution. Find and check the solution of each conditional equation.

1. $8x - 4(2 + 3x) = 12$

2. $3(x - 6) = 5(1 + 2x) - 7(x - 4)$

3. $4(3x - 8) - 2(-13 - 7x) = 5(4x + 6)$

4. $2[7x - 4(1 + 3x)] = 5(3 - 2x) - 23$

In Problems 5–7 solve each inequality.

5. $5w + 2 \leq 10 - w$

6. $13h - 4(2 + 3h) \geq 0$

7. $2[-5y - 6(y - 7)] < 6 + 4y$

In Problems 8–12 solve each inequality or equation and graph the solution set.

8. $\{x \mid 4 \geq 3x + 7 > -2,\ x \in R\}$

9. $\left\{x \,\middle|\, \frac{5(x - 2)}{3} + \frac{x}{4} \leq 12,\ x \in R\right\}$

10. $|7 - 3x| \geq 6$

11. $\{x \mid |2x - 5| < 11,\ x \in J\}$

12. $\left\{x \,\middle|\, \left|\frac{2x + 3}{5}\right| = 1,\ x \in J\right\}$

3 WORD PROBLEMS

The main reason for studying algebra is to equip oneself with the tools necessary to solve problems. Most problems are expressed in words. In this chapter we show methods for solving some traditional word problems. The skills learned in this chapter can be applied to solving mathematical problems encountered in many fields of learning as well as in real-life situations.

3–1 Method of Solving Word Problems

In this section we show how to change the words of a written problem into an equation. The equation can then be solved by the methods learned in Chapter Two.

Example 1 Seven increased by three times an unknown number is thirteen. What is the unknown number?

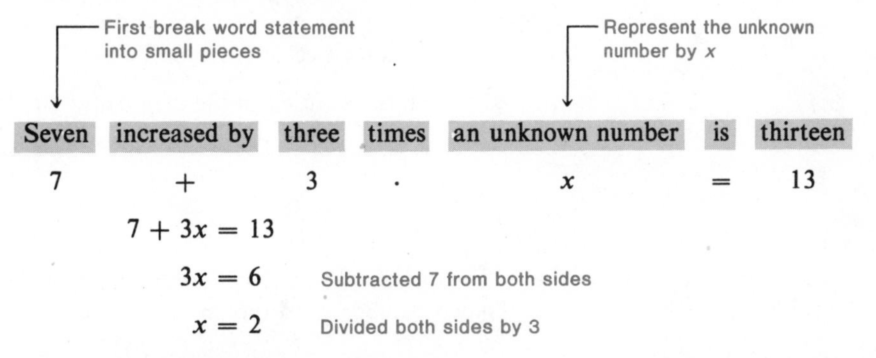

Checking Word Problems

To check a word problem, *the solution must be checked in the word statement.* An error may have been made in writing the equation which would not be discovered if you substitute the solution into the equation.

Check for Example 1.

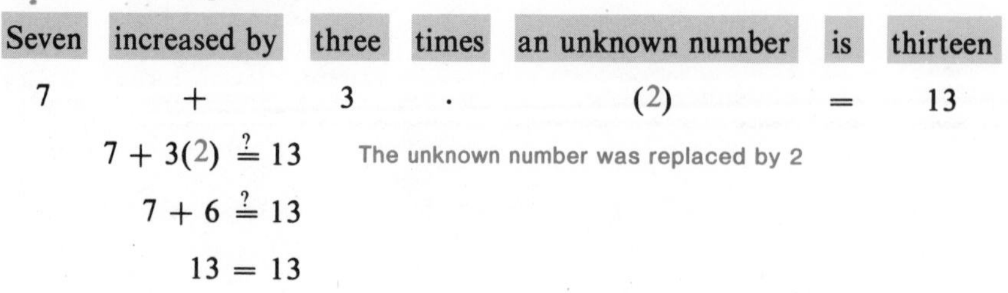

$$7 + 3(2) \stackrel{?}{=} 13$$ The unknown number was replaced by 2

$$7 + 6 \stackrel{?}{=} 13$$

$$13 = 13$$

We summarize the method for solving word problems in the following box.

METHOD FOR SOLVING WORD PROBLEMS

1. To solve a word problem, first read the problem very carefully *to determine what is unknown.* What is being asked for? *Don't* try to solve the problem at this time.
2. Represent one unknown number by a letter. Then reread the problem to see how you can represent any other unknown numbers in terms of that same letter.
3. Reread the entire word problem, breaking it up into small pieces that can be represented by algebraic expressions. Since every unknown number has been represented in terms of the same letter, there is nothing left in the word problem that cannot be represented by algebraic symbols.
4. After each of the pieces has been written as an algebraic expression, fit them together into an equation.
5. Solve the equation for the unknown letter by the methods learned in Chapter Two.
6. Check the solution in the *original word* statement.

Example 2 Four times an unknown number is equal to twice the sum of five and that unknown number. Find the unknown number.

Four times	an unknown number	is equal to	twice	the sum of five and that unknown number
$4\ \cdot$	x	$=$	$2\ \cdot$	$(5 + x)$

$$4x = 2(5 + x)$$

$$4x = 10 + 2x \quad \text{Distributive rule}$$

$$2x = 10 \quad \text{Subtracted } 2x \text{ from both sides}$$

$$x = 5 \quad \text{Divided both sides by 2}$$

Checking the solution in the word statement is left to the student.

Example 3 When seven is subtracted from one-half of an unknown number, the result is eleven. What is the unknown number?

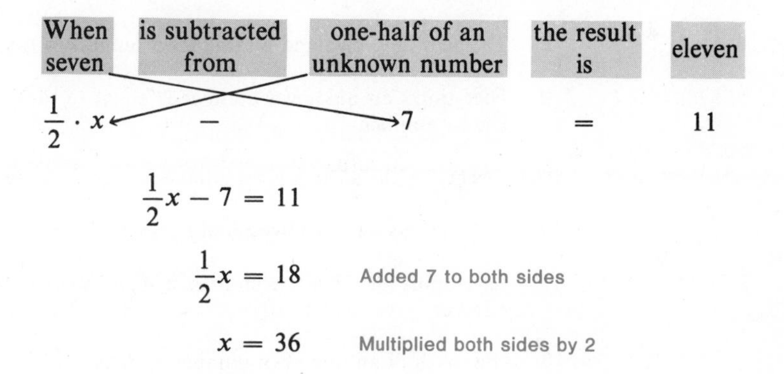

$$\frac{1}{2}x - 7 = 11$$

$$\frac{1}{2}x = 18 \quad \text{Added 7 to both sides}$$

$$x = 36 \quad \text{Multiplied both sides by 2}$$

Example 4 When three times an unknown number is subtracted from thirteen and the result is divided by nine, the quotient is four times the unknown number.

When three times an unknown number is subtracted from thirteen and the result is divided by nine,	the quotient is	four times the unknown number
$\dfrac{13 - 3x}{9}$	$=$	$4x$

$$\frac{13 - 3x}{9} = 4x$$

$$13 - 3x = 36x \quad \text{Multiplied both sides by 9}$$

$$13 = 39x \quad \text{Added } 3x \text{ to both sides}$$

$$\frac{1}{3} = x \quad \text{Divided both sides by 39}$$

EXERCISES 3-1

Solve each word problem for the unknown number. Check each solution in the word statement.

1. Seven more than twice an unknown number is twenty-three.
2. Eleven more than three times an unknown number is thirty-eight.
3. Nine more than four times an unknown number is thirty-three.
4. Five times an unknown number, decreased by eight, is twelve.
5. Four times an unknown number, decreased by seven, is twenty-five.
6. Five times an unknown number, decreased by six, is forty-nine.
7. Three times an unknown number is equal to the sum of twelve and the unknown number.
8. The sum of twelve and an unknown number is four times the unknown number.
9. Four times an unknown number is equal to the sum of thirty-three and the unknown number.
10. Twenty minus an unknown number is equal to the unknown number plus four.
11. Eight minus an unknown number is equal to the unknown number plus two.
12. Seven plus an unknown number is equal to seventeen decreased by the unknown number.
13. One-sixth of an unknown number is three.
14. One-seventh of an unknown number is one hundred five.
15. An unknown number divided by seven is eight.
16. When an unknown number is decreased by twelve, the difference is one-third the unknown number.
17. When an unknown number is decreased by seven, the difference is half the unknown number.
18. When an unknown number is subtracted from sixteen, the difference is one-third the unknown number.
19. When the sum of an unknown number and itself is multiplied by three, the result is forty-two.
20. Four times the sum of an unknown number and itself is seventy-two.
21. When the sum of an unknown number and itself is multiplied by five, the result is two hundred.
22. Four times the sum of six and an unknown number is equal to three times the sum of the unknown number and ten.

23. Three times the sum of eight and an unknown number is equal to twice the sum of the unknown number and seven.

24. Four times the sum of five and an unknown number is equal to three times the sum of the unknown number and nine.

25. When twice the sum of five and an unknown number is subtracted from eight times the unknown number, the result is equal to four times the sum of eight and twice the unknown number.

26. When seven times the sum of two and an unknown number is subtracted from four times the sum of three and twice the unknown number, the result is equal to zero.

27. When three times the sum of four and an unknown number is subtracted from eight times the unknown number, the result is equal to four times the sum of three and twice the unknown number.

3–2 Ratio and Proportion Problems

3-2A RATIO PROBLEMS

Ratio is another word for *fraction,* so in algebra fractions are often called *ratio*nal numbers.

The ratio of a to b is written $\frac{a}{b}$ or $a : b$

The ratio of b to a is written $\frac{b}{a}$ or $b : a$

Terms of a Ratio

We call a and b the **terms of the ratio.**

The terms of a ratio may be any kind of number; the only restriction is that the denominator cannot be zero.

The key to solving ratio problems is to use the given ratio to help represent the unknown numbers.

TO REPRESENT THE UNKNOWNS IN A RATIO PROBLEM

1. Multiply each term of the ratio by x.
2. The resulting products are used to represent the unknowns.

Example 1 Two numbers are in the ratio 3 to 5. Their sum is 32. Find the numbers.

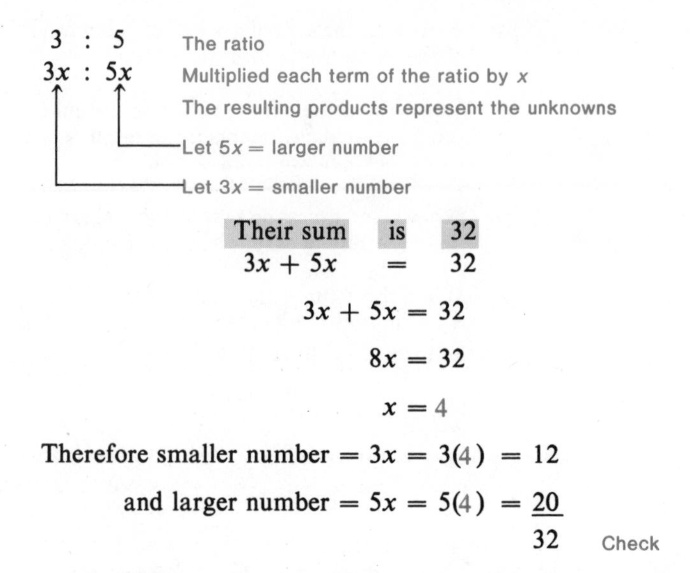

Example 2 The three sides of a triangle are in the ratio 2 : 3 : 4. The perimeter* is 63. Find the three sides.

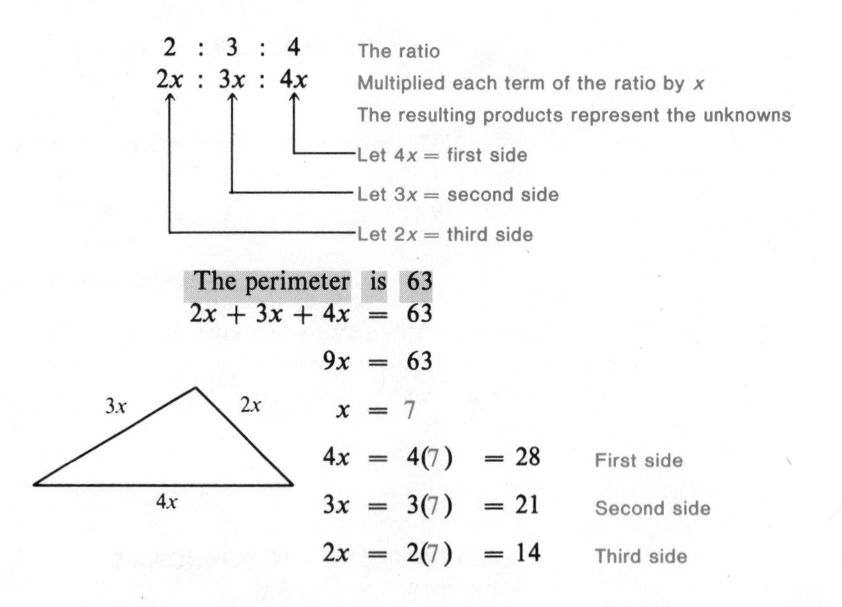

* The *perimeter* of a geometric figure is the sum of the lengths of all its sides. The word perimeter means "the measure around a figure."

EXERCISES 3-2A

1. Two numbers are in the ratio of 4 to 5. Their sum is 81. Find the numbers.
2. Two numbers are in the ratio of 8 to 3. Their sum is 77. Find the numbers.
3. Two numbers are in the ratio of 2 to 7. Their sum is 99. Find the numbers.
4. The three angles of a triangle are in the ratio 5 : 6 : 7. Their sum is 180 degrees. Find the angles.
5. The three sides of a triangle are in the ratio 3 : 4 : 5. The perimeter is 108. Find the three sides.
6. The three sides of a triangle are in the ratio 4 : 5 : 6. The perimeter is 120. Find the three sides.
7. Fifty-four hours of a student's week are spent in study, class, and work. The times spent in these activities are in the ratio 4 : 2 : 3. How many hours are spent in each activity?
8. The amounts Staci spent for food, rent, and clothing are in the ratio 3 : 5 : 2. She spends an average of $400 a month for these items. What can she expect to spend for each of these items in a year?
9. The formula for a particular shade of green paint calls for mixing 3 parts of blue paint with 1 part of yellow paint. Find the number of liters of blue and the number of liters of yellow needed to make 14 liters of the desired shade of green paint.
10. In Mrs. Sampson's bread recipe, flour and water are to be mixed in the ratio of 3 parts water to 5 parts flour. How many parts of flour are needed to make a mixture of 64 parts of water and flour?
11. The length and width of a rectangle are in the ratio of 7 to 6. The perimeter is 78. Find the length and width.
12. The length and width of a rectangle are in the ratio of 4 to 3. The area is 192. Find the length and width.
13. An aunt divided $28,000 among her three nieces in the ratio 6 : 5 : 3. How much did each receive?
14. A farmer plants 2200 acres of corn and soybeans in the ratio of 5 to 3. How many acres of each does he plant?
15. A man cuts a 78-inch board into two parts whose ratio is 9 : 4. Find the lengths of the pieces.
16. The gold solder used in a crafts class has gold, silver, and copper in the ratio 5 : 3 : 2. How much gold is there in 25 grams of this solder?
17. To maintain the ratio 7 : 9 of employed people over 40 to employed people under 40, how many of 224 people hired would have to be over 40?
18. The numerical ratio of the perimeter of a rectangle to its area is 5 : 11. Find the length of the rectangle if its width is 6 cm.

IN CLASS

3-2B PROPORTION PROBLEMS

A **proportion** is a statement that two ratios are equal.

Common notation for a proportion:

$$\frac{a}{b} = \frac{c}{d}$$ Read: "a is to b as c is to d"
or: "a over b equals c over d"

Another notation for a proportion:

$a : b :: c : d$ Read: "a is to b as c is to d"

Terms of a Proportion

The **terms of a proportion:**

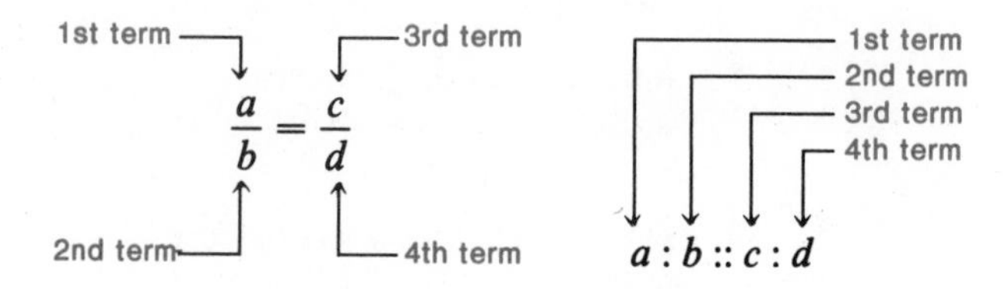

Means and Extremes

The **means and extremes** of a proportion:

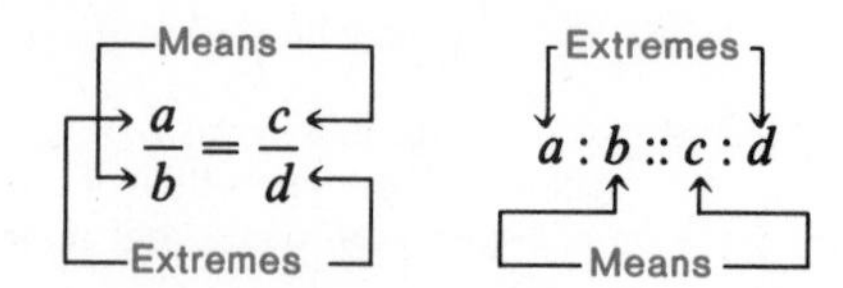

The *means* of this proportion are b and c.
The *extremes* of this proportion are a and d.

Cross-Multiplication

The product of the means equals the product of the extremes. This is sometimes called the *cross-multiplication rule* and these products called *cross-products*. This rule can be justified by multiplying both sides of the proportion by bd.

$$\frac{a}{b} = \frac{c}{d}$$

$$\not{b}d\left(\frac{a}{\not{b}}\right) = b\not{d}\left(\frac{c}{\not{d}}\right)$$

$$da = bc$$

$$\underbrace{b \cdot c}_{\text{Product of means}} = \underbrace{a \cdot d}_{\text{Product of extremes}}$$

Solving a Proportion

Solving a proportion means finding the value of the unknown that makes the two ratios of the proportion equal.

When three of the four terms of a proportion are known, it is always possible to find the value of the unknown term.

> **TO SOLVE A PROPORTION**
>
> **1.** Set the product of the means equal to the product of the extremes.
>
> **2.** Solve the resulting equation for the unknown.

Example 3 Solve $\frac{x}{25} = \frac{18}{15}$ for x

$$15 \cdot x = 18 \cdot 25$$ Product of extremes = Product of means

$$15x = 450$$

$$\frac{\overset{1}{\cancel{15}}x}{\underset{1}{\cancel{15}}} = \frac{\overset{30}{\cancel{450}}}{\underset{1}{\cancel{15}}}$$ Divided both products by the coefficient of the unknown

$$x = 30$$

$$\frac{(30)}{25} \stackrel{?}{=} \frac{18}{15}$$ x is replaced by 30

$$30 \cdot 15 \stackrel{?}{=} 25 \cdot 18$$

$$450 = 450$$

Often the work in solving a proportion can be simplified by reducing a ratio to its lowest terms. We use this procedure to simplify the solution of Example 3.

$$\frac{x}{25} = \frac{18}{15}$$

$$\frac{x}{25} = \frac{6}{5}$$ Because $\frac{18}{15} = \frac{6}{5}$ when reduced to lowest terms

$$5 \cdot x = 25 \cdot 6$$

$$\frac{\cancel{5} \cdot x}{\cancel{5}} = \frac{\overset{5}{\cancel{25}} \cdot 6}{\cancel{5}}$$

$$x = 30$$

Example 4 Solve: $\dfrac{x+1}{x-2} = \dfrac{7}{4}$

$$4(x+1) = 7(x-2)$$
$$4x + 4 = 7x - 14$$
$$18 = 3x$$
$$6 = x$$

Check: $\dfrac{x+1}{x-2} = \dfrac{7}{4}$

$$\frac{(6)+1}{(6)-2} \stackrel{?}{=} \frac{7}{4}$$

$$\frac{7}{4} = \frac{7}{4}$$

EXERCISES 3-2B

In Exercises 1–10 solve and check each proportion.

1. $\dfrac{x}{12} = \dfrac{3}{8}$
2. $\dfrac{4}{7} = \dfrac{x}{21}$
3. $\dfrac{x}{18} = \dfrac{12}{15}$
4. $\dfrac{22}{15} = \dfrac{33}{x}$
5. $\dfrac{x}{14} = \dfrac{1.5}{6}$
6. $\dfrac{32}{4.8} = \dfrac{12}{x}$
7. $\dfrac{P}{3} = \dfrac{\frac{5}{6}}{5}$
8. $\dfrac{P}{16} = \dfrac{\frac{3}{4}}{6}$
9. $\dfrac{2x+7}{3x+10} = \dfrac{3}{4}$
10. $\dfrac{5}{3x+4} = \dfrac{26}{8x-2}$

In Exercises 11–18 solve each proportion.

11. $\dfrac{3x-7}{2x+2} = \dfrac{2}{3}$
12. $\dfrac{4x-6}{3} = \dfrac{3x-12}{2}$
13. $\dfrac{12}{6-11x} = \dfrac{4}{5x}$
14. $\dfrac{3}{-7} = \dfrac{9x}{12-7x}$
15. $\dfrac{4(2x+7)}{3(3x+10)} = 1$
16. $\dfrac{2(2-8x)}{3(3x+4)} = -1$
17. $\dfrac{2x}{35.4} = \dfrac{17.5}{24.8}$
18. $\dfrac{A}{35.4} = \dfrac{17.6}{54.8}$

3-2C SOLVING WORD PROBLEMS USING PROPORTIONS

METHOD FOR SOLVING WORD PROBLEMS USING PROPORTIONS

1. Represent the unknown quantity by a letter.
2. Write the appropriate unit of measure next to each number in your proportion.
3. The same units must occupy corresponding positions in the two ratios of your proportion.

Correct Arrangements	*Incorrect Arrangements*
$\dfrac{\text{miles}}{\text{hours}} = \dfrac{\text{miles}}{\text{hours}}$	$\dfrac{\text{dollars}}{\text{weeks}} = \dfrac{\text{weeks}}{\text{dollars}}$
$\dfrac{\text{hours}}{\text{miles}} = \dfrac{\text{hours}}{\text{miles}}$	$\dfrac{\text{dollars}}{\text{week}} = \dfrac{\text{dollars}}{\text{days}}$
$\dfrac{\text{miles}}{\text{miles}} = \dfrac{\text{hours}}{\text{hours}}$	

4. Once the numbers have been correctly entered in the proportion by using the units as a guide, drop the units when cross-multiplying to solve for the unknown.

Example 5 There are 25 men in a college class of 38 students. Assuming this is a typical class, how many of the college's 7500 students are men?

There are 25 men in a college class of 38 students.	Assuming this is a typical class, how many of the college's 7500 students are men?

$$\frac{25 \text{ men}}{38 \text{ students}} = \frac{x \text{ men}}{7500 \text{ students}}$$

$$\frac{25}{38} = \frac{x}{7500}$$

$$38 \cdot x = 25(7500)$$

$$\frac{\cancel{38} \cdot x}{\cancel{38}} = \frac{25(\overset{3750}{\cancel{7500}})}{\underset{19}{\cancel{38}}}$$

$$x = \frac{25(3750)}{19} = \frac{93{,}750}{19} \doteq 4934 \text{ men}$$

Example 6 If a 6-ft man casts a $4\frac{1}{2}$-ft shadow, how tall is a tree that casts a 30-ft shadow?

A 6-ft man casts a $4\frac{1}{2}$-ft shadow.	How tall is a tree that casts a 30-ft shadow?

$$\begin{array}{ccc} \textit{Man} & & \textit{Tree} \\ \dfrac{6\text{-ft height}}{4\frac{1}{2}\text{-ft shadow}} & = & \dfrac{x\text{-ft height}}{30\text{-ft shadow}} \end{array}$$

$$\frac{6}{4\frac{1}{2}} = \frac{x}{30}$$

$$4\tfrac{1}{2} \cdot x = 6 \cdot 30$$

$$\frac{9}{2} \cdot x = 180$$

$$\frac{\frac{9}{2} \cdot x}{\frac{9}{2}} = \frac{180}{\frac{9}{2}}$$

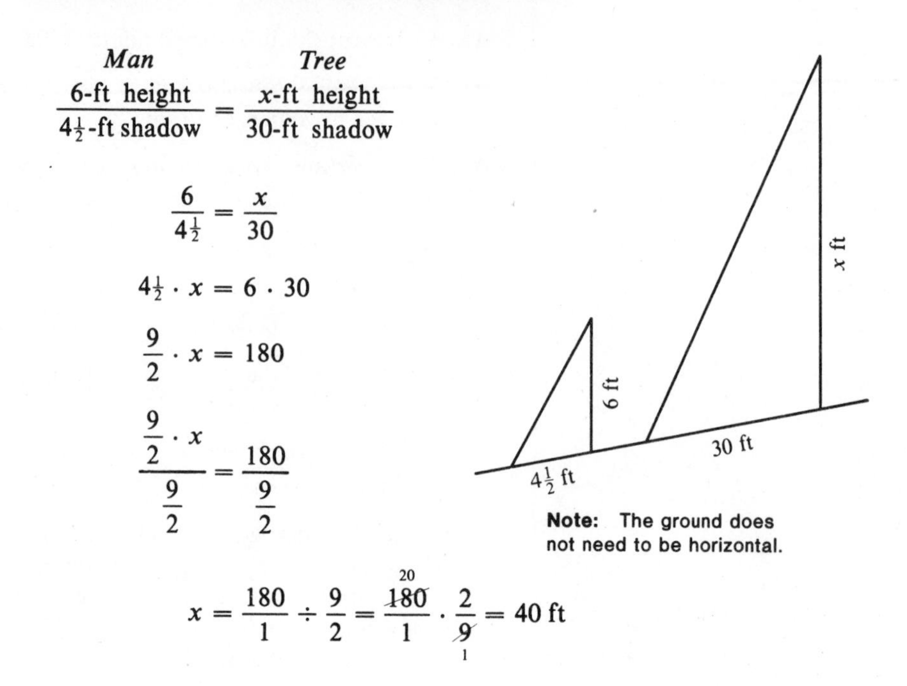

Note: The ground does not need to be horizontal.

$$x = \frac{180}{1} \div \frac{9}{2} = \frac{\overset{20}{\cancel{180}}}{1} \cdot \frac{2}{\underset{1}{\cancel{9}}} = 40 \text{ ft}$$

EXERCISES 3-2C

1. A baseball team wins 7 of its first 12 games. How many would you expect it to win out of 36 games if the team continues to play with the same degree of success?
2. There are 18 women in a college class containing 34 students. Assuming this is typical of all classes, how many of the college's 5100 students are women?
3. A 5-ft woman has a 3-ft shadow when a tree casts a 27-ft shadow. How tall is the tree?
4. A 6-ft man has a 4-ft shadow when a tree casts a 22-ft shadow. How tall is the tree?
5. If an investment of $7000 earns $1200 in 18 months, how much must be invested at the same rate to earn $1500 in the same length of time?
6. The property tax on a $155,000 home is $547.68. At this rate what would be the tax on a $135,000 home?
7. The scale in an architectural drawing is 1 in = 8 ft. What are the dimensions of a room that measures $1\frac{1}{2}$ by $2\frac{1}{4}$ in on the drawing?
8. Brian drives 720 miles in $2\frac{1}{2}$ days. At this rate, how far should he be able to drive in 4 days?

9. The ratio of a person's weight on earth compared to that same person's weight on the moon is 6 : 1. How much would a 126-lb woman weigh on the moon?

10. The ratio of a person's weight on Mars compared to that same person's weight on earth is 2 : 5. How much would a 180-lb man weigh on Mars?

11. One particular car model experienced 28 steering column failures in the first 40,000 cars sold. How many replacements could we expect to make if 1,500,000 cars of that model are recalled and inspected?

12. Ed's car used 12 quarts of oil on a 6,400-mile trip. What should he expect to spend for oil on a 9,600-mile trip if the oil costs $1.10 per quart?

13. A 340-gram box of corn flakes sells for 82 cents. A 510-gram box of corn flakes costs $1.15. For 340 grams, how much money is saved by buying the larger box?

14. An 18-oz jar of orange drink mix is marked $1.31. A 27-oz jar of the same mix is marked $1.82. For 18 oz, how much money is saved by buying the larger jar?

15. The ratio of the weight of aluminum and an equal volume of iron is 9 : 25. If a casting made of aluminum weighs 45 lb, what would the same casting weigh if it were made of iron?

16. The ratio of the weight of gold to an equal volume of copper is 19.3 : 8.9. If a bracelet of copper weighs 64.6 g, what would an identical bracelet made of gold weigh?

17. A man 6 feet tall is 20 feet from the point directly beneath a street light. What is the length of his shadow if the light is 22 feet above the street?

3–3 Percent Problems

Meaning of Percent

Percent means *hundredths*. Consider a 5-gal can containing 2 gal of paint. We can say "the can is $\frac{2}{5}$ full."

$$\text{Since } \frac{2}{5} = \frac{2 \cdot 20}{5 \cdot 20} = \frac{40}{100} = 0.40 = 40 \text{ hundredths} = 40 \text{ percent}$$

we can describe the amount of paint in the can in three different ways:

1. The can is $\frac{2}{5}$ full. — **Fraction** representation
2. The can is 0.40 (40 hundredths) full. — **Decimal** representation
3. The can is 40 percent full. — **Percent** representation

In the example of the paint can we have talked about three quantities:

1. 40 **percent** (***P***).
2. 5, which represents the **whole thing,** and is called the **base** (***B***).
3. 2, which is called the **amount** (***A***).

Percent as a Proportion

These three numbers are related by the proportion

$$\frac{2}{5} = \frac{40}{100}$$

When the numbers are replaced by the corresponding letters A, B, and P, we have:

THE PERCENT PROPORTION

$$\frac{A}{B} = \frac{P}{100}$$

where A = Amount, B = Base, P = Percent

There are three letters in the percent proportion. If any two are known, the proportion can be solved for the remaining letter by the method given in Section 3–2.

TO IDENTIFY THE NUMBERS IN A PERCENT PROBLEM

1. P is the number followed by the word "percent" (or %).
2. B is the number that follows the words "percent of."
3. A is the number left after P and B have been identified.

Example 1 What is 12% of 85?
First, identify the numbers.

What	is	12%	of	85	?
A		P		B	

$$\frac{A}{B} = \frac{P}{100}$$

$$\frac{A}{85} = \frac{12}{100}$$

$$\frac{A}{85} = \frac{\overset{3}{\cancel{12}}}{\underset{25}{\cancel{100}}}$$

$$25 \cdot A = 85 \cdot 3$$

$$\frac{\cancel{25} \cdot A}{\cancel{25}} = \frac{255}{25}$$

$$A = 10.2$$

Therefore 10.2 is 12% of 85.

Alternate Solution. $A = 12\%$ of 85
$= 0.12 \times 85 = 10.2$

$12\% = \frac{12}{100} = 0.12$ — "of" means "to multiply" in problems of this type

5 is what percent of 20?
First, identify the numbers.

5	is	what percent	of	20	?
A		P		B	

$$\frac{A}{B} = \frac{P}{100}$$

$$\frac{5}{20} = \frac{P}{100} \qquad \text{Reduce } \frac{5}{20} = \frac{1}{4}$$

$$\frac{1}{4} = \frac{P}{100}$$

$$4 \cdot P = 100$$

$$P = 25$$

Therefore 5 is 25% of 20.

Example 3 Ms. Delgado, a salesperson, makes a 6% commission on all items she sells. One week she made \$390. What were her gross sales for the week?

In this problem we are saying:

6%	of	some number	is	\$390
P		B		A

$$\frac{A}{B} = \frac{P}{100}$$

$$\frac{390}{B} = \frac{6}{100} \qquad \text{Reduce } \frac{6}{100} = \frac{3}{50}$$

$$\frac{390}{B} = \frac{3}{50}$$

$$3 \cdot B = 50 \cdot 390$$

$$\frac{3 \cdot B}{3} = \frac{19{,}500}{3} = 6500$$

$$B = \$6500$$

Therefore her gross sales for the week were \$6500.

EXERCISES 3-3

Solve the following percent problems.

1. What is 15% of \$62.40?
2. What is 13% of \$165?
3. 12 is what percent of 60?
4. 150 is what percent of 120?
5. What is 130% of \$480?
6. 15 is what percent of 120?
7. Ms. Salazar makes an 8% commission on all items she sells. One week she made \$328.56. What were her gross sales for the week?
8. Mrs. Clark makes a 9% commission on all items she sells. One week she made \$369.81. What were her gross sales for the week?
9. Mr. Lee makes a 6% commission on all items he sells. One week he made \$369.72. What were his gross sales for the week?
10. A bank pays 5% on regular passbook savings accounts and loans money at 12%. What is the bank's yearly profit made by lending \$150,000,000 from savings accounts?
11. If a business pays \$80 for an item, what is the selling price* of the item if it is marked up 60%?
12. If a car costs a dealer \$3450 and the markup is 22%, what is the selling price of the car?
13. If a business pays \$75 for an item, what is the selling price of the item if it is marked up 40%?
14. If a business pays \$125 for an item, what is the selling price of the item if it is marked up $37\frac{1}{2}$%? (Express the answer to the nearest cent.)
15. A company withholds 6.05% of an employee's salary for Social Security tax. How much is deducted from a salary of \$340 for Social Security?
16. Mr. Walker purchased 300 shares of stock at \$8 a share and then sold the stock for \$9.20 a share one year later. What percent gain did he make on this investment?
17. A shipping clerk receives an 8% reduction in his annual salary. After the reduction, his new salary is \$8832. What was his previous salary?
18. Nickel coinage used to be composed of copper and nickel in the ratio of 3 to 1. How much copper had to be melted with 20 lb of a copper-nickel alloy containing 60% copper to obtain the alloy for nickel coinage?
19. Tammy has an average grade of 83% on all work up to the final examination. What grade must she make on the final to bring her term average up to 85% if the final examination counts 20% of the course grade?

3–4 Variation Problems

Variation

A **variation** is an equation that relates one variable to one or more other variables by means of multiplication, division, or both.

---

*Merchandise must be sold at a price high enough to cover what the merchant pays for it, the expenses, and the profit. To accomplish this, the cost of each item must be *marked up* before it is sold. We use *markup* based on cost. Selling price, cost, and markup are related by the following formula:

$$\text{Selling price} = \text{Cost} + \text{Markup}$$
$$S = C + M$$

DIRECT VARIATION

Direct variation is a type of variation that relates one variable to another by the formula

Constant of proportionality

$$y = kx$$

Dependent variable — Independent variable

Dependent Variable
Independent Variable
Subscripts

In the formula $y = kx$, the value of y depends upon the value of x. For this reason y is called the **dependent variable,** and x is called the **independent variable.**

Subscripts make it possible to indicate *corresponding values* of variables. For example, x_1 and y_1 are a pair of corresponding variables. *Condition 1* lists all corresponding values of the variables with the subscript 1. *Condition 2* lists all corresponding values of the variables with the subscript 2.

$$\text{Condition 1} \quad \left\{\begin{matrix} \text{If } x = x_1 \\ \text{and } y = y_1 \end{matrix}\right\} \text{ then } y_1 = kx_1 \qquad (1)$$

Equation (1) results from substituting x_1 and y_1 into the formula $y = kx$.

$$\text{Condition 2} \quad \left\{\begin{matrix} \text{If } x = x_2 \\ \text{and } y = y_2 \end{matrix}\right\} \text{ then } y_2 = kx_2 \qquad (2)$$

Equation (2) results from substituting x_2 and y_2 into the formula $y = kx$.

Dividing Equation (1) by Equation (2), we have $\dfrac{y_1}{y_2} = \dfrac{\not{k}x_1}{\not{k}x_2} = \dfrac{x_1}{x_2}$.

Related Proportion

Therefore $\dfrac{y_1}{y_2} = \dfrac{x_1}{x_2}$ This proportion is called the **related proportion for direct variation**

Note that in *direct* variation, the subscripts of the letters in *both* ratios are in the *same* order

Example 1 The circumference C of a circle varies directly with the diameter D according to the formula $C = \pi D$. If $C = 12.56$ when $D = 4$, find C when $D = 7$.

$$\text{Condition 1} \quad \begin{cases} D_1 = 4 \\ C_1 = 12.56 \end{cases} \qquad \text{Condition 2} \quad \begin{cases} D_2 = 7 \\ C_2 = ? \end{cases}$$

$$\frac{C_1}{C_2} = \frac{D_1}{D_2}$$ The related proportion for direct variation

$$\frac{12.56}{C_2} = \frac{4}{7}$$ Substituted the given values in the related proportion

$$4C_2 = 12.56(7)$$ Product of means = Product of extremes

$$C_2 = \frac{12.56(7)}{4} = 21.98$$ Divided each side by 4

INVERSE VARIATION

Inverse variation is a type of variation that relates one variable to another by the formula

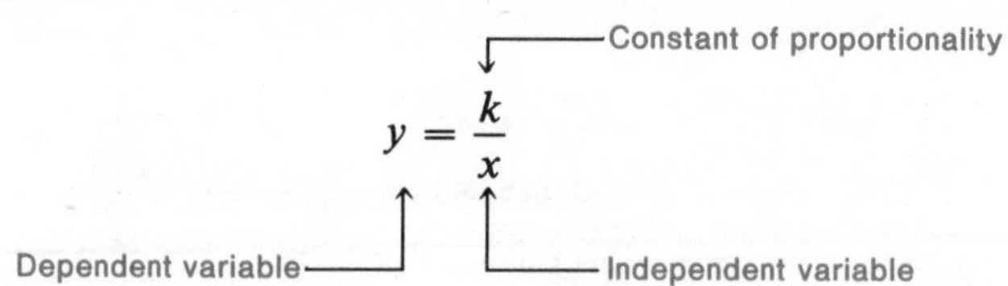

$$\left.\begin{array}{l}\text{If } x = x_1\\ \text{and } y = y_1\end{array}\right\} \text{ then } y_1 = \frac{k}{x_1} \quad (1)$$

By substituting x_1 and y_1 in the formula $y = \frac{k}{x}$

$$\left.\begin{array}{l}\text{If } x = x_2\\ \text{and } y = y_2\end{array}\right\} \text{ then } y_2 = \frac{k}{x_2} \quad (2)$$

By substituting x_2 and y_2 in the formula $y = \frac{k}{x}$

Dividing Equation (1) by Equation (2), we have

$$\frac{y_1}{y_2} = \frac{\frac{k}{x_1}}{\frac{k}{x_2}} = \frac{k}{x_1} \cdot \frac{x_2}{k} = \frac{x_2}{x_1}$$

Therefore $\frac{y_1}{y_2} = \frac{x_2}{x_1}$ This proportion is called the **related proportion for inverse variation.**

Note that in inverse variation, the subscripts of the letters in the two ratios are in *opposite* order.

Example 2 The pressure P varies inversely with the volume V according to the formula $PV = k$. If $P = 30$ when $V = 500$, find P when $V = 200$.

$$PV = k$$

$$P = \frac{k}{V}$$ Divided both sides by V

Condition 1 $\begin{cases} P_1 = 30 \\ V_1 = 500 \end{cases}$ Condition 2 $\begin{cases} P_2 = ? \\ V_2 = 200 \end{cases}$

$$\frac{P_1}{P_2} = \frac{V_2}{V_1}$$ The related proportion for inverse variation

$$\frac{30}{P_2} = \frac{200}{500} = \frac{2}{5}$$ Substituted the given values and reduced the fraction

$$2P_2 = 30(5)$$ Product of means = Product of extremes

$$P_2 = \frac{30(5)}{2} = 75$$

JOINT VARIATION

Joint variation is a type of variation that relates one variable to the product of two other variables by an equation of the following form:

$$z = kxy$$

Constant of proportionality (k); Dependent variable (z); Independent variables (x, y)

To derive the related proportion, we set up

$$z_1 = kx_1y_1 \qquad (1)$$

$$z_2 = kx_2y_2 \qquad (2)$$

Dividing Equation (1) by Equation (2), we have

$$\frac{z_1}{z_2} = \frac{\not{k}x_1y_1}{\not{k}x_2y_2} = \frac{x_1}{x_2} \cdot \frac{y_1}{y_2}$$

Therefore $\frac{z_1}{z_2} = \frac{x_1}{x_2} \cdot \frac{y_1}{y_2}$ This proportion is called the **related proportion for joint variation.**

Subscripts in *same* order

Example 3 The heat H generated by an electric heater involves a joint variation according to the formula $H = kI^2R$. If $H = 1200$ when $I = 8$ and $R = 15$, find H when $I = 5.5$ and $R = 20$.

Condition 1 $\begin{cases} I_1 = 8 \\ R_1 = 15 \\ H_1 = 1200 \end{cases}$ Condition 2 $\begin{cases} I_2 = 5.5 \\ R_2 = 20 \\ H_2 = ? \end{cases}$

$$\frac{H_1}{H_2} = \frac{I_1^2}{I_2^2} \cdot \frac{R_1}{R_2}$$

$$\frac{1200}{H_2} = \frac{8^2}{(5.5)^2} \cdot \frac{\overset{3}{\not{15}}}{\underset{4}{\not{20}}} = \frac{\overset{16}{\not{64}}(\not{3})}{30.25(\not{4})} = \frac{48}{30.25}$$

$$48H_2 = 1200(30.25)$$

$$H_2 = \frac{1200(30.25)}{48} = 756.25$$

COMBINED VARIATION

A **combined variation** is a variation that relates one variable to two or more other variables.

Example 4 Examples of combined variation, with k the constant of proportionality in each equation

(a) $w = \frac{kx}{y}$ *w* is directly proportional to *x*; *w* is inversely proportional to *y*

(b) $R = \dfrac{kL}{d^2}$ *R* is directly proportional to *L*; *R* is inversely proportional to d^2

(c) $F = \dfrac{kMm}{d^2}$ *F* is directly proportional to *M* and *m*; *F* is inversely proportional to d^2

The method for writing the related proportion for *any* variation with a known formula is given in the following box.

Related Proportions

WRITING THE RELATED PROPORTION

1. The related proportion has a ratio for each different variable that appears in the formula.
2. The subscripts in each ratio are in the *same* order for *direct* variation, and in *opposite* order for *inverse* variation.
3. The ratio for a particular letter is raised to the same power that letter has in the formula.

Example 5 The strength of a rectangular beam varies according to the formula $S = \dfrac{kbd^2}{L}$, which is a combined variation.

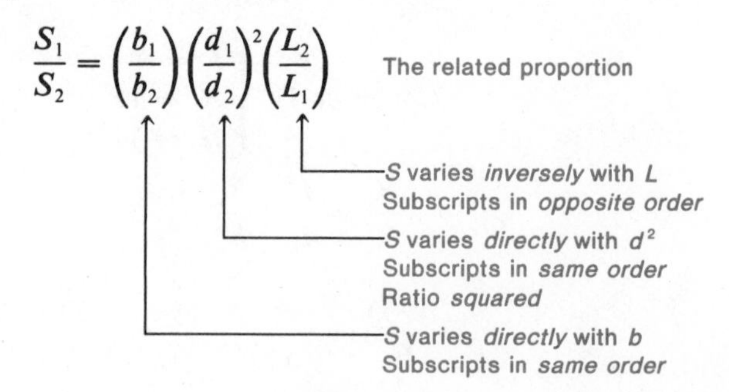

If $S = 2000$ when $b = 2$, $d = 10$, and $L = 15$, find S when $b = 4$, $d = 8$, and $L = 12$.

Condition 1 $\begin{cases} b_1 = 2 \\ d_1 = 10 \\ L_1 = 15 \\ S_1 = 2000 \end{cases}$ Condition 2 $\begin{cases} b_2 = 4 \\ d_2 = 8 \\ L_2 = 12 \\ S_2 = ? \end{cases}$

$$\frac{2000}{S_2} = \left(\frac{2}{4}\right)\left(\frac{10}{8}\right)^2\left(\frac{12}{15}\right) = \frac{5}{8}$$

$$5S_2 = 2000(8)$$

$$S_2 = \frac{2000(8)}{5} = 3200$$

EXERCISES 3-4

1. The pressure P per square inch in water varies directly with the depth d according to the formula $P = kd$. If $P = 8.66$ when $d = 20$, find P when $d = 50$.

2. The surface area S of a sphere varies directly with the square of the radius r according to the formula $S = 4\pi r^2$. If $S = 12.56$ when $r = 2$, find S when $r = 4$.

3. The distance d in miles that a person can see to the horizon from a point h ft above the surface of the earth varies approximately as the square root of the height h. If, for a height of 600 ft, the horizon is 30 miles distant, how far is the horizon from a point that is 1174 ft high?

4. The amount of sediment a stream will carry is directly proportional to the sixth power of its speed. If a stream carries 1 unit of sediment when the speed of the current is 2 mph, how many units of sediment will it carry when the current is 4 mph?

5. Given that F varies inversely with d^2 according to the formula $F = \frac{k}{d^2}$. If $F = 3$ when $d = 4$, find the constant of proportionality k.

6. The intensity I of light received from a light source varies inversely with the square of the distance d from the source according to the formula $I = \frac{k}{d^2}$. If the light intensity is 15 candelas at a distance of 10 ft from the light source, what is the light intensity at a distance of 15 ft?

7. The gravitational attraction F between two bodies varies inversely with the square of the distance d separating them. If the attraction measures 36 when the distance is 4 cm, find the attraction when the distance is 80 cm.

8. The simple interest I earned in a given time t varies jointly as the principal P and the interest rate r. If $I = 255$ when $P = 1250$ and $r = 0.0975$, find I when $P = 1500$ and $r = 0.0825$.

9. The pressure P in a liquid varies jointly with the depth h and density D of the liquid. If $P = 204$ when $h = 163.2$ and $D = 1.25$, find P when $h = 182.5$ and $D = 13.56$.

10. The wind force F on a vertical surface varies jointly as the area A of the surface and as the square of the wind velocity V. When the wind is blowing 20 mph, the force on 1 sq ft of surface is 1.8 lb. Find the force exerted on a surface 2 sq ft when the wind velocity is 60 mph.

11. Given that H varies jointly with R and the square of I according to the formula $H = kI^2R$. If $H = 1458$ when $I = 4.5$ and $R = 24$, find H when $I = 5.5$ and $R = 22$.

12. On a certain truck line, it costs \$56.80 to send 5 tons of goods 8 miles. How much will it cost to send 14 tons a distance of 15 miles?

13. Given that z varies according to the formula $z = \frac{kx}{y}$. If $z = 12$ when $x = 6$ and $y = 2$, find z when $x = -8$ and $y = -4$.

14. The electrical resistance R of a wire varies according to the formula $R = \frac{kL}{d^2}$. If $R = 2$ when $L = 8$ and $d = 4$, find R when $L = 10$ and $d = 5$.

15. The elongation e of a wire varies according to the formula $e = \frac{kPL}{A}$. If $e = 3$ when $L = 45$, $P = 2.4$, and $A = 0.9$, find e when $L = 40$, $P = 1.5$, and $A = 0.75$.

16. Given that W varies according to the formula $W = \frac{kxy}{z^2}$. If $W = 1200$ when $x = 8$, $y = 6$, and $z = 2$, find W when $x = 5.6$, $y = 3.8$, and $z = 1.5$.

17. The gravitational attraction F between two masses varies according to the formula $F = \frac{kMm}{d^2}$. If $F = 1000$, $d = 100$, $m = 50$, and $M = 2000$, find F when $d = 66$, $m = 125$, and $M = 1450$.

18. When a horizontal beam with rectangular cross section is supported at both ends, its strength S varies jointly as the breadth b and the square of the depth d and inversely as the length L. A 2- by 4-in beam 8 ft long resting on the 2-in side will safely support 600 lb. What is the safe load when the beam is resting on the 4-in side?

19. The centrifugal force F of an object following a curved course is given by the formula $F = \frac{kwv^2}{r}$, where w is weight in pounds, v is velocity in feet per second, and r is radius in feet. If $F = 200$ lb when $w = 3200$ lb, $v = 22$ ft per sec, and $r = 242$ ft, find the centrifugal force when a 3850 lb car traveling 30 mph (44 ft per sec) rounds a curve of 275 ft radius.

20. Power is the rate of doing work. One formula for horsepower is $P = \frac{kFs}{t}$, where P is horsepower, F is force in pounds, s is distance in feet, and t is time in minutes. When you climb a hill, you are exerting a vertical force equal to your weight. You exert this force through the vertical distance climbed. If $P = 0.2$ when $F = 171$ lb, $s = 2200$ ft, and $t = 51$ min, find the horsepower developed by a 180 lb man in climbing a 3400 ft mountain in 1 hr and 36 min.

21. *Brainteaser.* Two coins have a value of 60¢. One is not a dime. What are the coins?

3–5 Other Kinds of Word Problems

In this section we discuss other types of word problems that you may encounter in other courses.

3-5A DISTANCE-RATE-TIME PROBLEMS (Used in any field involving motion)

A physical law relating *distance* traveled d, *rate* of travel r, and *time* of travel t is

Distance Formula

$$d = r \cdot t$$

For example, you know that if you are driving your car at an average speed of 50 mph, then

	$d = r \cdot t$
you travel a distance of 100 miles in 2 hr:	$100 = 50(2)$
you travel a distance of 150 miles in 3 hr:	$150 = 50(3)$

and so on.

Example 1 A carload of campers leaves Los Angeles for Lake Havasu at 8:00 AM. A second carload of campers leaves Los Angeles at 8:30 AM and drives 10 mph faster over the same road. If the second car overtakes the first at 10:30 AM, what is the average speed of each car?

Let $x =$ Rate of first car
then $x + 10 =$ Rate of second car

	d =	r •	t	
First Car	d_1	x	$2\frac{1}{2}$	This car leaves at 8:00 AM and is overtaken at 10:30 AM
Second Car	d_2	$x + 10$	2	This car leaves at 8:30 AM and overtakes the first car at 10:30 AM

Use the formula $d = r \bullet t$ to fill in the two empty boxes.

	d =	r •	t	
First Car	$x\left(\frac{5}{2}\right)$	x	$\frac{5}{2}$	$d = r \bullet t$ $d_1 = x\left(\frac{5}{2}\right)$
Second Car	$(x + 10)\,2$	$x + 10$	2	$d = r \bullet t$ $d_2 = (x + 10)\,2$

Since both cars start at the same place and end at the same place, they have both traveled the same distance. Therefore

$$d_1 = d_2$$
$$\frac{5}{2}x = 2(x + 10)$$
$$5x = 4(x + 10) \quad \text{Multiplied both sides by 2}$$
$$5x = 4x + 40$$
$$x = 40 \text{ mph} \quad \text{First car}$$
$$x + 10 = 40 + 10 = 50 \text{ mph} \quad \text{Second car}$$

Example 2 A boat cruises downstream for 4 hr before heading back. After traveling upstream for 5 hr, it is still 16 miles short of the starting point. If the speed of the stream is 4 mph, find the speed of the boat in still water.

Let $x =$ Speed of boat in still water
then $x + 4 =$ Speed of boat downstream
and $x - 4 =$ Speed of boat upstream

$$\boxed{\text{Distance traveled downstream}} = 16 + \boxed{\text{Distance traveled upstream}}$$

$$(x + 4)4 = 16 + (x - 4)5$$
$$4x + 16 = 16 + 5x - 20$$
$$20 = x$$
$$x = 20 \text{ mph} \quad \text{Speed of boat in still water}$$

	d	= r	• t
Downstream	$(x + 4)\,4$	$x + 4$	4
Upstream	$(x - 4)\,5$	$x - 4$	5

$d = r \cdot t$

$d_1 = (x + 4)\,4$

$d = r \cdot t$

$d_2 = (x - 4)\,5$

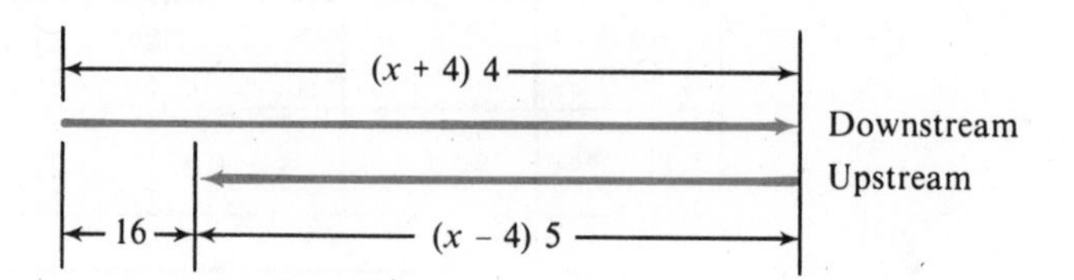

METHOD FOR SOLVING DISTANCE-RATE-TIME PROBLEMS

1. Draw the blank chart:

	d	= r	• t

2. Fill in four of the boxes using a single letter and the given information.
3. Use the formula $d = r \cdot t$ to fill in the remaining two boxes.
4. Write the equation by using information in the chart along with an unused fact given in the problem.
5. Solve the resulting equation.

EXERCISES 3-5A

1. The Malone family left San Diego by car at 7 AM, bound for San Francisco. Their neighbors the King family left in their car at 8 AM, also bound for San Francisco. By traveling 9 mph faster, the Kings overtook the Malones at 1 PM.
   (a) Find the average speed of each car.
   (b) Find the total distance traveled by each car before they met.
2. The Duran family left Ames, Iowa, by car at 6 AM, bound for Yellowstone National Park. Their neighbors the Silva family left in their car at 8 AM, also bound for Yellowstone. By traveling 10 mph faster, the Silvas overtook the Durans at 4 PM.
   (a) Find the average speed of each car.
   (b) Find the total distance traveled before they met.
3. Eric hiked from his camp to a lake in the mountains and returned to camp later in the day. He walked at a rate of 2 mph going to the lake and 5 mph coming back. If the trip to the lake took 3 hr longer than the trip back:
   (a) How long did it take him to hike to the lake?
   (b) How far is it from his camp to the lake?
4. Lee hiked from her camp up to an observation tower in the mountains and returned to camp later in the day. She walked up at the rate of 2 mph and jogged back at the rate of 6 mph. The trip to the tower took 2 hr longer than the return trip.
   (a) How long did it take her to hike to the tower?
   (b) How far is it from her camp to the tower?
5. Fran and Ron live 54 miles apart. Both leave their homes at 7 AM by bicycle, riding toward one another. They meet at 10 AM. If Ron's average speed is four-fifths of Fran's, how fast does each cycle?
6. Danny and Cathy live 60 miles apart. Both leave their homes at 10 AM by bicycle, riding toward one another. They meet at 2 PM. If Cathy's average speed is two-thirds of Danny's, how fast does each cycle?
7. Colin paddles a kayak downstream for 3 hr. After having lunch, he paddles upstream for 5 hr. At that time he is still 6 miles short of getting back to his starting point. If the speed of the stream is 2 mph, how fast does Colin row in still water? How far downstream did he travel?
8. The Wright family sails their houseboat upstream for 4 hr. After lunch they motor downstream for 2 hr. At that time they are still 12 miles away from the marina where they began. If the speed of the houseboat in still water is 15 mph, what is the speed of the stream? How far upstream did the Wrights travel?
9. Mr. Zaleva flew his private plane from his office to his company's storage facility bucking a 20-mph head wind all the way. He flew home the same day with the same wind at his back. The round trip took 10 hr of flying time. If the plane makes 100 mph in still air, how far is the storage facility from his office?
10. Mr. Summers drove his motor boat upstream a certain distance while pulling his son Brian on a water ski. He returned to the starting point pulling his other son Derek. The round trip took 25 min of skiing time. On both legs of the trip, the speedometer read 30 mph. If the speed of the current is 6 mph, how far upstream did he travel?
11. *Brainteaser.* A train 1 km long travels at 1 km per min through a tunnel 1 km long. How long will it take the train to pass completely through the tunnel?

3-5B MIXTURE PROBLEMS (Business)

Mixture problems involve mixing two or more ingredients.

TWO IMPORTANT FACTS NECESSARY TO SOLVE MIXTURE PROBLEMS

1. $\begin{pmatrix}\textit{Amount}\text{ of}\\ \text{ingredient A}\end{pmatrix} + \begin{pmatrix}\textit{Amount}\text{ of}\\ \text{ingredient B}\end{pmatrix} = \begin{pmatrix}\textit{Amount}\text{ of}\\ \text{mixture}\end{pmatrix}$
2. $\begin{pmatrix}\textit{Cost}\text{ of}\\ \text{ingredient A}\end{pmatrix} + \begin{pmatrix}\textit{Cost}\text{ of}\\ \text{ingredient B}\end{pmatrix} = \begin{pmatrix}\textit{Cost}\text{ of}\\ \text{mixture}\end{pmatrix}$

Example 3 A wholesaler makes up a 50-lb mixture of two kinds of coffee, one costing \$2.50 per pound, and the other costing \$2.60 per pound. How many pounds of each kind must be used if the mixture is to cost \$2.57 per pound?

Let $x =$ Amount of \$2.50 coffee
then $50 - x =$ Amount of \$2.60 coffee

		Ingredient A		Ingredient B		Mixture
	Unit Cost	2.50		2.60		2.57
(1)	Amount	x	+	$50 - x$	=	50
(2)	Total Cost	$2.50x$		$2.60\,(50 - x)$	=	$2.57\,(50)$

In each vertical line, the total cost is found by multiplying the unit cost by the amount.

$$\boxed{\begin{matrix}\text{Cost of}\\ \text{ingredient A}\end{matrix}} + \boxed{\begin{matrix}\text{Cost of}\\ \text{ingredient B}\end{matrix}} = \boxed{\begin{matrix}\text{Cost of}\\ \text{mixture}\end{matrix}}$$

$$(2) \qquad 2.50x + 2.60(50 - x) = 2.57(50)$$

$$250x + 260(50 - x) = 257(50) \qquad \text{Multiplied both sides by 100}$$

$$250x + 13000 - 260x = 12850$$

$$-10x = -150$$

$$x = 15 \text{ lb} \qquad (\$2.50 \text{ coffee})$$

$$50 - x = 50 - 15 = 35 \text{ lb} \qquad (\$2.60 \text{ coffee})$$

Check:

$$\begin{aligned} \text{Cost of \$2.50 coffee} &= 15(2.50) = \$\ 37.50 \\ \text{Cost of \$2.60 coffee} &= 35(2.60) = \underline{+91.00} \\ & \qquad\qquad\quad \$128.50 \\ \text{Cost of mixture} &= 50(2.57) = \$128.50 \end{aligned} \Bigg\} \text{ Check}$$

EXERCISES 3-5B

1. A dealer makes up a 100-lb mixture of Colombian coffee costing \$1.85 per pound and Brazilian coffee costing \$1.75 per pound. How many pounds of each kind must be used in order for the mixture to cost \$1.78 per pound?

2. A dealer makes up a 50-lb mixture of cashews and peanuts. If the cashews cost 80¢ a pound and the peanuts cost 65¢ a pound, how many pounds of each kind of nut must be used in order for the mixture to cost 74¢ a pound?

3. Mrs. Martinez mixes 15 lb of English toffee candy costing \$1.25 a pound with caramels costing \$1.50 a pound. How many pounds of caramels must she use to make a mixture costing \$1.35 a pound?

4. A 5-lb mixture of caramels and nougats costs \$7.35. If nougats cost \$2.10 a pound and caramels \$1.20 a pound, how many pounds of each kind are there?

5. A 50-lb mixture of Delicious and Jonathan apples costs \$14.50. If the Delicious apples cost 30¢ a pound and the Jonathan apples cost 20¢ a pound, how many pounds of each kind are there?

6. A 100-lb mixture of Valencia and Navel oranges costs \$22. If the Valencia oranges cost 20¢ a pound and the Navel oranges cost 25¢ a pound, how many pounds of each kind are there?

7. Mr. Wong wants to mix 30 bushels of soybeans with corn to make a 100-bushel mixture costing \$4.85 a bushel. How much can he afford to pay for each bushel of corn if soybeans cost \$8.00 a bushel?

8. Mrs. Lavalle wants to mix 6 pounds of Brand A with Brand B to make a 10-pound mixture costing \$11.50. How much can she afford to pay per pound for Brand B if Brand A costs \$1.23 per pound?

3-5C SOLUTION PROBLEMS (Chemistry and Nursing)

Solution problems involve the mixing of liquids. They can be solved by the same method we used for mixture problems. However, we think the following method makes it easier to visualize the mixing of liquids.

Example 4 How many liters of a 20% alcohol solution must be added to 3 liters of a 90% alcohol solution to make an 80% solution?

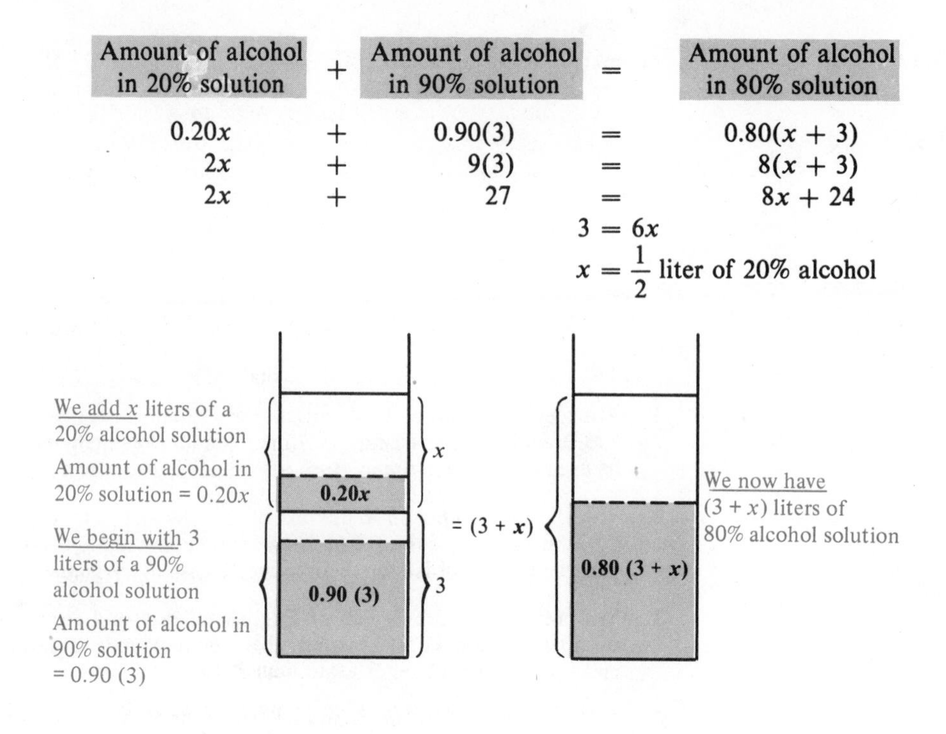

EXERCISES 3-5C

1. How many milliliters of water must be added to 500 ml of a 40% solution of sodium bromide to reduce it to a 25% solution?
2. How many liters of water must be added to 10 liters of a 30% solution of formaldehyde to reduce it to a 20% solution?
3. How many liters of pure alcohol must be added to 10 liters of a 20% solution of alcohol to make a 50% solution?
4. How many milliliters of pure alcohol must be added to 1000 ml of a 10% solution of alcohol to make a 40% solution?
5. How many cubic centimeters (cc) of a 20% solution of sulfuric acid must be mixed with 100 cc of a 50% solution to make a 25% solution of sulfuric acid?
6. How many pints of a 2% solution of disinfectant must be mixed with 5 pints of a 12% solution to make a 4% solution of disinfectant?
7. If 100 gal of 75% glycerin solution is made up by combining a 30% glycerin solution with a 90% glycerin solution, how much of each solution must be used?
8. If 1600 cc of 10% dextrose solution is made up by combining a 20% dextrose solution with a 4% dextrose solution, how much of each solution must be used?
9. A chemist has two solutions of hydrochloric acid. One is a 40% solution, and the other is a 90% solution. How many liters of each should be mixed to get 10 liters of a 50% solution?
10. Jan needs a 40% mixture of antifreeze and water in her 20-quart car radiator system before going on a weekend ski trip. How much of the present 25% mixture should be drained from her full radiator and replaced by pure antifreeze?

3-5D MISCELLANEOUS WORD PROBLEMS

The following exercise set includes word problems that have not been discussed. We leave it to students to devise their own solutions, guided by the general method given in Section 3–1.

EXERCISES 3-5D

1. Doris has 17 coins that have a total value of $1.15. If all the coins are nickels or dimes, how many of each does she have?
2. Dianne has $3.20 in nickels, dimes, and quarters. If there are 7 more dimes than quarters and 3 times as many nickels as quarters, how many of each kind of coin does she have?
3. The sum of two consecutive integers is thirty-three. What are the integers?
4. The sum of the first two of three consecutive odd integers added to the sum of the last two is 140. Find the integers.
5. The sum of the digits of a two-digit number is thirteen. The number formed by reversing the digits is 27 more than the original number. Find the original number.
6. The sum of the digits of a two-digit number is seven. The number formed by reversing the digits is 27 less than the original number. Find the original number.
7. How many minutes after 8 o'clock will the hands of a clock first be together?
8. At what time between 8 and 9 o'clock will the hands of a clock be opposite one another?
9. A drugstore buys eighteen cameras. If it had bought ten cameras of a higher quality, it would have paid $48 more per camera for the same total expenditure. Find the price of each type of camera.
10. The denominator of a fraction exceeds the numerator by twenty. If seven is added to the numerator and subtracted from the denominator, the resulting fraction equals $\frac{6}{7}$. Find the fraction.
11. Venus and the earth revolve around the sun in approximately 224.7 and 365.3 days, respectively. Find, to the nearest day, the time it takes Venus to gain one revolution on the earth.

Homework

Chapter Three Summary

Method for Solving Word Problems. (Section 3–1)

1. Read the problem and determine what is unknown.
2. Represent each unknown in terms of the same letter.
3. Break up the word statement into small pieces and represent each piece by an algebraic expression.
4. After each of the pieces has been written as an algebraic expression, fit them together into an equation.
5. Solve the equation for the unknown letter.
6. Check the solution in the *original word* statement.

Chapter Three Diagnostic Test or Review

Allow yourself about one hour to do these problems. Complete solutions for every problem, together with section references, are given in the answer section at the end of the book.

Solve each of the following word problems.

1. When twenty-three is added to four times an unknown number, the sum is thirty-one. Find the unknown number.
2. A 6-ft man has a 4-ft shadow when a building casts a 24-ft shadow. How tall is the building?
3. The length and width of a rectangle are in the ratio of 5 to 4. The area is 180. Find the length and width.
4. Linda has 14 coins that have a total value of \$2.15. If all the coins are dimes or quarters, how many of each kind does she have?
5. A grocer makes up a 60-lb mixture of cashews and peanuts. If the cashews cost \$1.00 a pound and the peanuts cost 80¢ a pound, how many pounds of each kind of nut must be used in order for the mixture to cost 85¢ a pound?
6. How many cubic centimeters of water must be added to 600 cc of a 20% solution of potassium chloride to reduce it to a 15% solution?
7. Mary's weekly salary is \$240. If her deductions amount to \$57.60, what percent of her salary is take-home pay?
8. The units digit of a two-digit number is twice the tens digit. If the digits were reversed, its value would be increased by twenty-seven. Find the original number.
9. The value of w varies with x, y, and z according to the formula $w = \dfrac{kxy}{z^2}$.

   If $w = 20$ when $x = 8$, $y = 6$, and $z = 12$, find w when $x = 6$, $y = 10$, and $z = 5$.
10. Roy hikes from his camp to a mountain lake one day and returns the next day. He walks up to the lake at the rate of 3 mph and returns to his camp at a rate of 5 mph. If the hike to the lake takes Roy 2 hr longer than the return trip:
    (a) How long does it take him to hike to the lake?
    (b) How far is the lake from his camp?

4 FACTORING AND SPECIAL PRODUCTS

In this chapter we review the methods of factoring and the special products covered in beginning algebra, and then discuss additional types of factoring and special products.

4–1 Prime Factorization and Greatest Common Factor (GCF)

PRIME FACTORIZATION OF POSITIVE INTEGERS

In this book we consider only factors (or divisors) that are integers.

Prime Numbers

A **prime number** is a positive integer greater than 1 that can be exactly divided only by itself and 1. A prime number has no factors other than itself and 1.

Composite Numbers

A **composite number** is a positive integer that can be exactly divided by some integer other than itself and 1. A composite number has factors other than itself and 1.

Note that 1 is neither prime nor composite.

Example 1 Prime and composite numbers

(a) 31 is a prime number because 1 and 31 are the only integral factors of 31.

(b) 45 is a composite number because it has factors 3, 5, 9, and 15 other than 1 and 45.

Prime Factorization

The **prime factorization** of a positive integer is the indicated product of all its factors that are themselves prime numbers. For example,

$$\left.\begin{array}{l} 18 = 2 \cdot 9 \\ 18 = 3 \cdot 6 \end{array}\right\}$$ These are *not prime* factorizations because 9 and 6 are not prime numbers

$$\left.\begin{array}{l} 18 = 2 \cdot 9 = 2 \cdot 3 \cdot 3 = 2 \cdot 3^2 \\ 18 = 3 \cdot 6 = 3 \cdot 2 \cdot 3 = 2 \cdot 3^2 \end{array}\right\}$$ These *are prime* factorizations because all the factors are prime numbers

Note that the two ways we factored 18 led to the *same* prime factorization ($2 \cdot 3^2$). The prime factorization of *any* positive integer (greater than 1) is unique.

Method for Finding the Prime Factorization. A partial list of prime numbers is: 2, 3, 5, 7, 11, 13, 17, 19, 23, 29, ... The smallest prime is 2; the next smallest is 3; the next smallest is 5; etc.

Note: The only even prime number is 2. All other even numbers are composite numbers with 2 among their factors. ■

The work of finding the prime factorization of a number can be conveniently arranged as shown in Examples 2 and 3.

Example 2 Find the prime factorization of 24

The smallest prime that divides 24 (↓)

$$\begin{array}{r|l} 2 & 24 \\ \hline 2 & 12 \\ \hline 2 & 6 \\ \hline & 3 \end{array}$$

Therefore $24 = 2 \cdot 2 \cdot 2 \cdot 3 = 2^3 \cdot 3$ where $2^3 \cdot 3$ is the prime factorization of 24

Example 3 Find the prime factorization of 315

The smallest prime that divides 315 (↓)

$$\begin{array}{r|l} 3 & 315 \\ \hline 3 & 105 \\ \hline 5 & 35 \\ \hline & 7 \end{array}$$

Prime factorization of $315 = 3 \cdot 3 \cdot 5 \cdot 7 = 3^2 \cdot 5 \cdot 7$

When trying to find the prime factors of a number, no prime that has a square greater than that number need be tried (Example 4).

Example 4 Find the prime factorization of 97

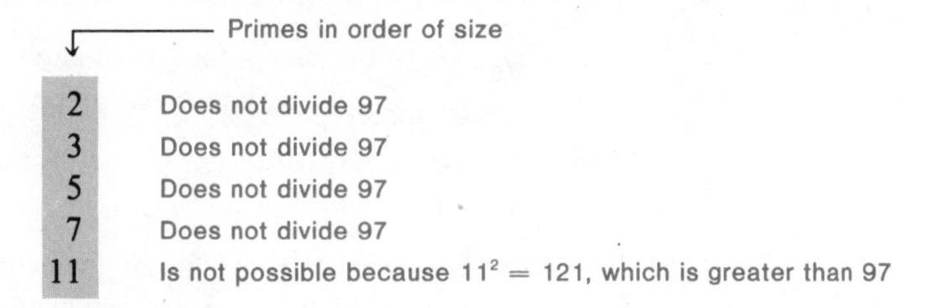

Therefore 97 is prime.

GREATEST COMMON FACTOR (GCF)

GCF

The **greatest common factor** (GCF) of two integers is the greatest integer that is a factor of both integers.

Example 5 Find the GCF of 12 and 16

$$12 \qquad\qquad 16$$

$$\underbrace{2 \cdot 2} \cdot 3 \qquad \underbrace{2 \cdot 2} \cdot 2 \cdot 2$$

Therefore, the greatest factor common to both 12 and 16 is $2 \cdot 2 = 4$

We can also find the GCF of terms in an algebraic expression.

Example 6 Find the GCF for the terms of $6y^3 - 21y$

$$6y^3 - 21y = 2 \cdot \underbrace{3 \cdot y} \cdot y \cdot y - 7 \cdot \underbrace{3 \cdot y}$$ Prime factorization of terms

GCF = $3y$

Polynomial Factor

$$6y^3 - 21y = 3y \ (2y^2 - 7)$$

GCF — $3y$; Polynomial factor — $(2y^2 - 7)$

Factoring

In Example 6, we found the factors from the terms of the product. The process of finding the factors when the product is already known is called **factoring.**

Finding the Product (Distributive rule) →

$$3y(2y^2 - 7) = 6y^3 - 21y$$

← Finding the Factors (Factoring)

TO FIND THE GREATEST COMMON FACTOR (GCF)

1. Write the *prime* factors of each term. Repeated factors should be expressed as powers.
2. Write each different prime factor that is common to all terms.
3. Raise each prime factor (selected in Step 2) to the *lowest* power it occurs anywhere in the expression.
4. The greatest common factor is the product of all the powers found in Step 3.

TO FIND THE POLYNOMIAL FACTOR

5. Divide the expression being factored by the greatest common factor found in Step 4.

Check. Find the product of the greatest common factor and the polynomial factor by using the distributive rule.

Example 7 Factor $15x^3 + 9x$

Finding the GCF.

Step 1: $15x^3 + 9x = 3 \cdot 5x^3 + 3^2x$ Each term in prime factored form

Steps 2 & 3:
- 3 is common to both terms (3^1 is the *lowest* power of 3 that occurs in any term)
- x is common to both terms (x^1 is the *lowest* power of x that occurs in any term)

Step 4: $\text{GCF} = 3^1x^1 = 3x.$

Finding the Polynomial Factor

Step 5:

$$15x^3 + 9x$$

$$3x(\quad + \quad)$$ The polynomial factor has as many terms as the original expression

$$3x(5x^2 + 3)$$

This term (3) is $\frac{9x}{3x} = 3$ ($3x$ must be multiplied by 3 to give $9x$)

This term ($5x^2$) is $\frac{15x^3}{3x} = 5x^2$ ($3x$ must be multiplied by $5x^2$ to give $15x^3$)

Therefore the factors of $15x^3 + 9x$ are $3x$ and $(5x^2 + 3)$. So that

$$15x^3 + 9x = \underbrace{3x}_{\text{GCF}}\ (\underbrace{5x^2 + 3}_{\text{Polynomial factor}})$$

Check. $3x(5x^2 + 3) = (3x)(5x^2) + (3x)(3) = 15x^3 + 9x$

Example 8 Factor $6a^3b^3 - 8a^2b^2 + 10a^3b$

$$6a^3b^3 - 8a^2b^2 + 10a^3b$$
$$= 2 \cdot 3a^3b^3 - 2^3a^2b^2 + 2 \cdot 5a^3b^1$$

The GCF $= 2^1 \cdot a^2 \cdot b^1 = 2a^2b$.

The polynomial factor $= \dfrac{6a^3b^3}{2a^2b} - \dfrac{8a^2b^2}{2a^2b} + \dfrac{10a^3b}{2a^2b}$

$$= 3ab^2 - 4b + 5a$$

Therefore, $6a^3b^3 - 8a^2b^2 + 10a^3b = 2a^2b(3ab^2 - 4b + 5a)$.

Example 9 Factor $2(a + b) - 8(a + b)^2$

$(a + b)$ is a common factor that is not a monomial.

When part of an algebraic expression is repeated, it is often helpful to replace the repeated part by a single letter.

$$2(a + b) - 8(a + b)^2$$

Let $(a + b) = x$

Then $2(a + b) - 8(a + b)^2 = 2x - 8x^2$

$= 2x(1 - 4x)$

$= 2(a + b)[1 - 4(a + b)]$ Since $x = (a + b)$

$= 2(a + b)(1 - 4a - 4b)$

EXERCISES 4-1

In Exercises 1–10 state whether the number is prime or composite.

1. 13 **2.** 15 **3.** 12 **4.** 11

5. 18 **6.** 19 **7.** 51 **8.** 42

9. 111 **10.** 101

In Exercises 11–22 find the prime factorization of each number.

11. 28 **12.** 30 **13.** 32 **14.** 33

15. 34 **16.** 35 **17.** 84 **18.** 75

19. 144 **20.** 180 **21.** 156 **22.** 221

Factor each expression.

23. $9x^2 + 3x$

24. $8y^2 - 4y$

25. $10a^3 - 25a^2$

26. $27b^2 - 18b^4$

27. $2a^2b + 4ab^2$

28. $3mn^2 + 6m^2n^2$

29. $12c^3d^2 - 18c^2d^3$

30. $15ab^3 - 45a^2b^4$

31. $4x^3 - 12x - 24x^2$

32. $18y - 6y^2 - 30y^3$

33. $24a^4 + 5a^2 - 40$

34. $45b^3 - 15b^4 - 22$

35. $6my + 15mz - 5n - 4n$

36. $4nx + 8ny + 16z - 4z$

37. $-14x^8y^9 + 42x^5y^4 - 28xy^3$

38. $-21u^7v^8 - 63uv^5 + 35u^2v^5$

39. $-44a^{14}b^7 - 33a^{10}b^5 + 22a^{11}b^4$

40. $-26e^8f^6 + 13e^{10}f^8 - 39e^{12}f^5$

41. $18u^{10}v^5 + 24 - 14u^{10}v^6$

42. $30a^3b^4 - 15 + 45a^8b^7$

43. $18x^3y^4 - 12x^2z^3 - 48x^4y^3$

44. $32m^5n^7 - 24m^8p^9 - 40m^3n^6$

45. $-35r^7s^5t^4 - 55r^8s^9u^4 + 40p^8r^9s^8$

46. $-120a^8 - 120a^8b^7c^5 + 40a^4c^3d^9 - 80a^5c^5$

47. $-24x^8y^3 - 12x^7y^4 + 48x^5y^5 + 60x^4y^6$

48. $64y^9z^5 + 48y^8z^6 - 16y^7z^7 - 80y^4z^8$

49. $5(x - y) - 15(x - y)^2$

50. $4(a - b) - 20(a - b)^2$

51. $3(x - y)^2 + 9(x - y)^3$

52. $4(a + b)^2 - 12(a + b)^3$

53. $3(a + b)(x - y) + 6(a + b)(x - y)^2$

54. $2(m - 4)^2(m + 2) + 8(m - 4)(m + 2)$

4–2 Factoring the Difference of Two Squares

Each type of factoring depends upon a particular special product. Consider the product $(a + b)(a - b)$.

$$\begin{aligned}(a + b)(a - b) &= (a + b)a + (a + b)(-b) && \text{Distributive Rule}\\ &= a^2 + ba - ab - b^2\\ &= a^2 - b^2\end{aligned}$$

Product of Sum and Difference

THE PRODUCT OF THE SUM AND DIFFERENCE OF TWO TERMS

Is equal to the square of the first term minus the square of the second term:

$$(a + b)(a - b) = a^2 - b^2$$

Example 1 The product of the sum and difference of two terms

(a) $(2x + 3y)(2x - 3y) = (2x)^2 - (3y)^2 = 4x^2 - 9y^2$

(b) $(5a^3b^2 + 6cd^4)(5a^3b^2 - 6cd^4) = (5a^3b^2)^2 - (6cd^4)^2$

$= 25a^6b^4 - 36c^2d^8$

Factoring $a^2 - b^2$ depends on the product $(a + b)(a - b) = a^2 - b^2$.

Finding the Product

$$(a + b)(a - b) = a^2 - b^2$$

Finding the Factors

Therefore $a^2 - b^2$ *factors into* $(a + b)(a - b)$.

When factoring the difference of two squares, we must find the principal square root of each term.

Example 2 Find the principal square root of each term.

(a) $\sqrt{25x^2} = 5x$ Because $(5x)^2 = 25x^2$

(b) $\sqrt{100a^4b^6} = 10a^2b^3$ Because $(10a^2b^3)^2 = 100a^4b^6$

TO FACTOR THE DIFFERENCE OF TWO SQUARES

1. Find a, the principal square root of the first term and b, the principal square root of the second term.
2. One factor is the sum of a and b, and the other factor is the difference of a and b.

$$a^2 - b^2 = (a + b)(a - b)$$

Example 3 Factor $25y^4 - 9z^2$

$$25y^4 - 9z^2 = (5y^2)^2 - (3z)^2 = (5y^2 + 3z)(5y^2 - 3z)$$

$\sqrt{9z^2}$ (points to $3z$); $\sqrt{25y^4}$ (points to $5y^2$)

Complete Factorization

An expression is **completely factored** when all its factors are prime. We will consider a factor prime if it cannot be factored by any method we have discussed.

Example 4 Factor $a^4 - b^4$

This is not a prime factor (points to $a^2 - b^2$)

$$a^4 - b^4 = (a^2 + b^2)(a^2 - b^2) = (a^2 + b^2)(a + b)(a - b)$$

Always remove a common factor first when possible (see Example 5).

Example 5 Factor $27x^4 - 12y^2$

Notice that neither $27x^4$ nor $12y^2$ is a perfect square. However, they have a common factor of 3 which can be removed.

$$27x^4 - 12y^2 = 3(9x^4 - 4y^2)$$

The factor $9x^4 - 4y^2$ is now the difference of two squares, and can therefore be factored.

$$27x^4 - 12y^2 = 3\underbrace{(9x^4 - 4y^2)} = 3\underbrace{(3x^2 + 2y)(3x^2 - 2y)}$$

Sometimes the quantities that are squared are not monomials (see Example 6).

Example 6 Factor $(x - y)^2 - (a + b)^2$

$(x - y)^2 - (a + b)^2$ Difference of two squares

$= [\quad + \quad][\quad - \quad]$

$= [(x - y) + (a + b)]\,[(x - y) - (a + b)]$

$= (x - y + a + b)(x - y - a - b)$

EXERCISES 4-2

In Exercises 1–14 find each product by inspection.

1. $(x + 3)(x - 3)$

2. $(z + 4)(z - 4)$

3. $(2u + 5v)(2u - 5v)$

4. $(3m - 7n)(3m + 7n)$

5. $(2x^2 - 9)(2x^2 + 9)$

6. $(10y^2 - 3)(10y^2 + 3)$

7. $(x^5 - y^6)(x^5 + y^6)$

8. $(a^7 + b^4)(a^7 - b^4)$

9. $(7mn + 2rs)(7mn - 2rs)$

10. $(8hk + 5ef)(8hk - 5ef)$

11. $(12x^4y^3 + u^7v)(12x^4y^3 - u^7v)$

12. $(11a^5b^2 + 9c^3d^6)(11a^5b^2 - 9c^3d^6)$

13. $(a + b + 2)(a + b - 2)$

14. $(x^2 + y + 5)(x^2 + y - 5)$

In Exercises 15–34 factor each expression completely.

15. $4c^2 - 1$

16. $16d^2 - 1$

17. $2x^2 - 8y^2$

18. $3x^2 - 27y^2$

19. $49u^4 - 36v^4$

20. $81m^6 - 100n^4$

21. $a^2b^2 - c^2d^2$

22. $m^2n^2 - r^2s^2$

23. $x^4 - y^4$

24. $a^4 - 16$

25. $4h^4k^4 - 1$

26. $9x^4y^4 + 1$

27. $25a^4b^2 - c^2d^4$

28. $x^2y^4 - 100x^4y^2$

29. $(x + y)^2 - 4$

30. $(a + b)^2 - 9$

31. $(a + b)x^2 - (a + b)y^2$

32. $x(a + b)^2 - x(a - b)^2$

33. $(3{,}146{,}721)^2 - (3{,}146{,}720)^2$

34. $9x^2 + y^2$

4–3 Factoring Trinomials

4-3A THE PRODUCT OF TWO BINOMIALS

In Section 4–2 when we multiplied two binomials, we obtained a binomial.

$$(a + b)(a - b) = a^2 - b^2$$

In general, however, when two binomials are multiplied, we obtain a trinomial.

Example 1 $\underbrace{(x + 2)(x + 3)}_{\text{product of two binomials}} = x(x + 3) + 2(x + 3)$

$$= x^2 + 3x + 2x + 6$$

$$= \underbrace{x^2 + 5x + 6}_{\text{trinomial}}$$

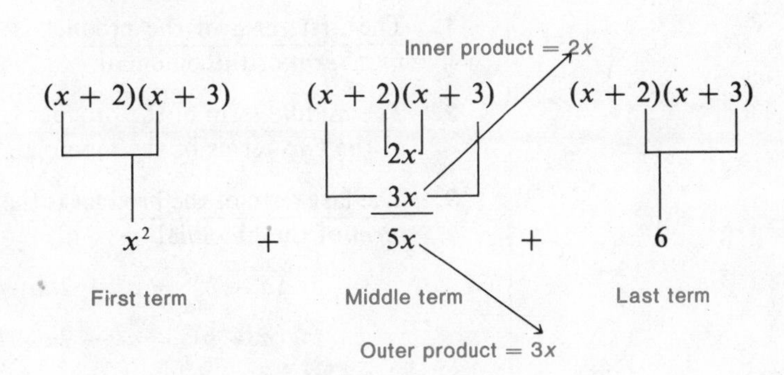

Multiplying Two Binomials

> **TO MULTIPLY TWO BINOMIALS**
>
> 1. The first term of the product is the *product* of the first terms of the binomials.
> 2. The middle term of the product is the *sum* of the inner and outer products.
> 3. The last term of the product is the *product* of the last terms of the binomials.

Example 2 $(3a - 8b) \cdot (4a - 5b)$

$$-32ab$$

$$-15ab$$

$$12a^2 - 47ab + 40b^2$$

Example 3 (a) $(a+b)^2 = (a+b) \cdot (a+b) = a^2 + 2ab + b^2$

$$\begin{array}{c} ab \\ \underline{ab} \\ a^2 + 2ab + b^2 \end{array}$$

(b) $(a-b)^2 = (a-b) \cdot (a-b) = a^2 - 2ab + b^2$

$$\begin{array}{c} -ab \\ \underline{-ab} \\ a^2 - 2ab + b^2 \end{array}$$

The two special products shown in Example 3 occur so often that they are worth remembering.

Squaring a Binomial

TO SQUARE A BINOMIAL

1. The first term of the product is *the square of the first term* of the binomial.
2. The middle term of the product is *twice the product* of the two terms of the binomial.
3. The last term of the product is the *square of the last term* of the binomial.

$$(a+b)^2 = a^2 + 2ab + b^2$$

$$(a-b)^2 = a^2 - 2ab + b^2$$

Example 4 The square of a binomial

(a) $(2x+3)^2 = (2x)^2 + 2(2x)(3) + (3)^2$

$= 4x^2 + 12x + 9$

(b) $(5m - 2n^2)^2 = (5m)^2 + 2(5m)(-2n^2) + (-2n^2)^2$

$= 25m^2 - 20mn^2 + 4n^4$

A Word of Caution. Students remember that

$(ab)^2 = a^2b^2$ From using the rules for exponents

↑ Here, a and b are *factors*

They try to apply this rule of exponents to the expression $(a+b)^2$. But $(a+b)^2$ cannot be found simply by squaring a and b.

↑ Here, a and b are *terms*

Correct Method

$$(a+b)^2 = (a+b)(a+b)$$
$$= a^2 + 2ab + b^2$$

A common error is to omit this term

Incorrect Method

$$(a+b)^2 \neq a^2 + b^2$$

EXERCISES 4-3A

Find each product by inspection.

1. $(a+5)(a+2)$
2. $(a+7)(a+1)$
3. $(y+8)(y-9)$
4. $(z-3)(z+10)$
5. $(x+3)^2$
6. $(x+5)^2$
7. $(b-4)^2$
8. $(b-6)^2$
9. $(3x+4)(2x-5)$
10. $(2y+5)(4y-3)$
11. $(7z-2)(8z+3)$
12. $(9z+1)(4z-5)$
13. $(2a+5b)(a+b)$
14. $(3c+2d)(c+d)$
15. $(4x-y)(2x+7y)$
16. $(3x-2y)(4x+5y)$
17. $(3x+4)^2$
18. $(2x+5)^2$
19. $(7x-10y)^2$
20. $(4u-9v)^2$
21. $(2a-3b)^2$
22. $(5m-2n)^2$
23. $(3x^2-2y)(4x^2+3y)$
24. $(4x-3y^2)(2x+5y^2)$

4-3B FACTORING A TRINOMIAL THAT HAS A LEADING COEFFICIENT OF 1

Leading Coefficient The **leading coefficient** of a polynomial is the numerical coefficient of its highest-degree term.

Example 5 (a) The leading coefficient of $4y^2 - 6y + 7$ is 4.

(b) The leading coefficient of $x^2 - 2x + 8$ is 1.

TO FACTOR A TRINOMIAL THAT HAS A LEADING COEFFICIENT OF 1

1. First, arrange the trinomial in descending powers.
2. Select as the first term of each binomial factor, expressions whose product is the first term of the trinomial.
3. List all pairs of factors of the coefficient of the last term of the trinomial.
4. Select the particular pair of factors whose sum equals the coefficient of the middle term of the trinomial.

Example 6 Factor $z^2 + 8z + 12$

The first term of each binomial factor is z.

List all pairs of factors of the last term of the trinomial.

$$\begin{aligned} 12 &= 1 \cdot 12 \\ &= 2 \cdot 6 \\ &= 3 \cdot 4 \end{aligned}$$

$2 \cdot 6$: Product = Last term $(+2)(+6) = +12$; Sum = Middle coefficient $(+2) + (+6) = +8$

$$z^2 + 8z + 12 = (z + 2)(z + 6)$$

Example 7 Factor $x^2 - 7xy - 30y^2$

$$\begin{aligned} 30 &= 1 \cdot 30 \\ &= 2 \cdot 15 \\ &= 3 \cdot 10 \\ &= 5 \cdot 6 \end{aligned}$$

$3 \cdot 10$: Product = Last coefficient $(+3)(-10) = -30$; Sum = Middle coefficient $(+3) + (-10) = -7$

$$x^2 - 7xy - 30y^2 = (x + 3y)(x - 10y)$$

Each last term must contain y

Sometimes the terms of the trinomial are not monomials (see Example 8).

Example 8 Factor $(x - y)^2 - 2(x - y) - 8$

Let $(x - y) = a$

Then $(x - y)^2 - 2(x - y) - 8$

$$= a^2 - 2a - 8 = (a + 2) \cdot (a - 4)$$

$$\begin{aligned} 8 &= 1 \cdot 8 \\ &= 2 \cdot 4 \end{aligned}$$

Therefore

$$\begin{aligned} (x - y)^2 - 2(x - y) - 8 &= [(x - y) + 2][(x - y) - 4] \\ &= (x - y + 2)(x - y - 4) \end{aligned}$$

EXERCISES 4-3B

Factor each expression completely.

1. $k^2 + 7k + 6$
2. $k^2 + 5k + 6$
3. $b^2 - 9b + 14$
4. $b^2 - 15b + 14$
5. $x^2 - 7x + 18$
6. $x^2 - 3x + 18$
7. $z^2 - z - 6$
8. $m^2 + 5m - 6$
9. $m^2 + 13m + 12$
10. $m^2 + 7m + 12$
11. $h^2 + 2h - 8$
12. $k^2 - 7k - 8$
13. $t^2 + 7t - 30$
14. $m^2 - 13m - 30$
15. $u^2 + 12u - 64$
16. $v^2 - 30v - 64$
17. $3x + x^2 + 2$
18. $7a + a^2 + 10$
19. $x^2 + xy - 2y^2$
20. $x^2 + 6xy + 9y^2$
21. $z^2 - 12za + 20a^2$
22. $x^2 + 9xy - 10y^2$
23. $f^2 - 7fg - 18g^2$
24. $v^2 - 10vw + 16w^2$
25. $(a + b)^2 + 6(a + b) + 8$
26. $(m + n)^2 + 9(m + n) + 8$
27. $(x + y)^2 - 13(x + y) - 30$
28. $(x + y)^2 - 10(x + y) - 24$
29. $(x + y)^2 - 2(x + y) - 15$
30. $(a + b)^2 + 6(a + b) + 8$

4-3C FACTORING A TRINOMIAL THAT HAS A LEADING COEFFICIENT UNEQUAL TO 1

When the leading coefficient equals 1, only the factors of the last term need to be considered. When the leading coefficient is not equal to 1, the factors of the first term as well as the last term must be considered.

TO FACTOR A TRINOMIAL THAT HAS A LEADING COEFFICIENT NOT EQUAL TO 1

1. Before factoring, arrange the trinomial in descending powers if the highest-degree term is positive. Arrange it in ascending powers if the highest-degree term is negative.
2. Make a blank outline and fill in all obvious information (Example 9).
3. List all pairs of factors of the coefficient of the first term and the last term of the trinomial.
4. Select the correct pairs of factors so that the sum of the inner and outer products equals the middle term of the trinomial.

Check your factoring by multiplying the binomial factors to see if their product is the correct trinomial.

Example 9 Factor $5x^2 + 13x + 6$

Step 1. The terms are already arranged in descending powers of x.

Step 2. Make a blank outline and fill in all the obvious information.

$$5x^2 + 13x + 6$$

$$(\ x + \quad)(\ x + \quad)$$

The letter in each binomial must be x so that the first term of the trinomial contains x^2

The sum of the inner and outer products must equal the middle term of the trinomial; find the correct pair by trial

$+13x$

Middle term

Step 3. Next, list the factors of the first and last coefficients of the trinomial.

Factors of first coefficient	*Factors of last coefficient*
$5 = 1 \cdot 5$	$6 = 1 \cdot 6$
	$= 2 \cdot 3$

Step 4. Select the pairs of factors so that:

1. The sum of the inner and outer products equals the middle coefficient of the trinomial.
2. The product of the last terms of the binomials equals the last term of the trinomial.
3. The product of the first terms of the binomials equals the first term of the trinomial.

$$(1x + 2) \cdot (5x + 3)$$

$$+10x$$

$$+\ 3x$$

$$5x^2 + 13x + 6$$

Therefore $5x^2 + 13x + 6$ factors into $(x + 2)(5x + 3)$.

Example 10 Factor $3m^2 - 2mn - 8n^2$

$3 = 1 \cdot 3$ | $8 = 1 \cdot 8$, $= 2 \cdot 4$

$$(m - 2n) \cdot (3m + 4n)$$

$$-6mn$$

$$+4mn$$

$$3m^2 - 2mn - 8n^2$$

Therefore $3m^2 - 2mn - 8n^2 = (m - 2n)(3m + 4n)$.

Sometimes the first term is a constant and the last term is the one with the letter. When this is the case, we proceed in almost the same way (see Example 11).

Example 11 Factor $3 - 2x - 5x^2$

$5 = 1 \cdot 5$ | $3 = 1 \cdot 3$

$$(\quad + x) \cdot (\quad - x)$$

$$-2x$$

$$(1 + 1x) \cdot (3 - 5x)$$

$$+3x$$

$$-5x$$

$$-2x$$

Therefore $3 - 2x - 5x^2 = (1 + x)(3 - 5x)$.

Sometimes the terms of the trinomial are not monomials (see Example 12).

Example 12 Factor $5(2y - z)^2 + 12(2y - z) + 4$

Let $(2y - z) = a$

Then

$$5(2y - z)^2 + 12(2y - z) + 4 = 5a^2 + 12a + 4 = (\ a + \quad)\cdot(\ a + \quad)$$

$5 = 1 \cdot 5$ | $4 = 1 \cdot 4$
$= 2 \cdot 2$

$+12a$

$(1a + 2) \cdot (5a + 2)$

$+10a$
$+\ 2a$
$+12a$

Therefore $5a^2 + 12a + 4 = (a + 2)(5a + 2)$.

But since $a = (2y - z)$,

$$5(2y - z)^2 + 12(2y - z) + 4 = [(2y - z) + 2][5(2y - z) + 2]$$
$$= (2y - z + 2)(10y - 5z + 2)$$

Sometimes every term of the trinomial to be factored has a common factor. In this case first factor out the common factor before factoring the remaining trinomial (see Example 13).

Example 13 Factor $15x^2 - 25x - 10$

$5(3x^2 - 5x - 2)$ First, remove the common factor 5

$5(x - 2) \cdot (3x + 1)$ Then factor the remaining trinomial

$-6x$
$+1x$
$-5x$

Therefore $15x^2 - 25x - 10 = 5(x - 2)(3x + 1)$.

EXERCISES 4-3C

Factor each expression completely.

1. $3x^2 + 7x + 2$
2. $3x^2 + 5x + 2$
3. $4x^2 + 7x + 3$
4. $4x^2 + 13x + 3$
5. $4 + 21x + 5x^2$
6. $4 + 12x + 5x^2$
7. $5a^2 - 16a + 3$
8. $5m^2 - 8m + 3$
9. $7 - 22b + 3b^2$

10. $7 - 10u + 3u^2$
11. $5v^2 - 17v + 6$
12. $5v^2 - 11v + 6$
13. $4 + 7h^2 - 11h$
14. $4 + 7h^2 - 16h$
15. $3t^2 + 17t - 6$
16. $3t^2 - 7t + 6$
17. $8 + 12z - 8z^2$
18. $9 + 21z - 18z^2$
19. $12x^2 + 10xy - 8y^2$
20. $45x^2 - 6xy - 24y^2$
21. $3n^2 - 8nt - 5t^2$
22. $5k^2 + 2kh - 7h^2$
23. $4v^2 + 14vw - 8w^2$
24. $6v^2 - 27vz - 15z^2$
25. $8x^2 + 2xy - 15y^2$
26. $7(x + y)^2 - 11(x + y) + 4$
27. $3(a - b)^2 + 7(a - b) + 2$
28. $4(a + b)^2 + 7(a + b) + 3$
29. $5(x + y)^2 + 21(x + y) + 4$
30. $5(x - y)^2 + 12(x - y) + 4$

4-3D FACTORING TRINOMIALS THAT ARE PERFECT SQUARES

Trinomial Squares

A Trinomial is a Perfect Square if

1. **Its first and last terms are perfect squares, and**
2. **Its middle term is twice the product of the square roots of its first and last terms.**

Factoring trinomials that are perfect squares depends on the following two special products.

$$(a + b)^2 = a^2 + 2ab + b^2$$
$$(a - b)^2 = a^2 - 2ab + b^2$$

TO FACTOR A PERFECT SQUARE TRINOMIAL
($a^2 \pm 2ab + b^2$)

1. Find a, the square root of the first term, and b, the square root of the last term.
2. The factors of the trinomial are:
   $(a + b)^2$ if the middle term is $+2ab$
   $(a - b)^2$ if the middle term is $-2ab$

Example 14 Factoring trinomials that are perfect squares

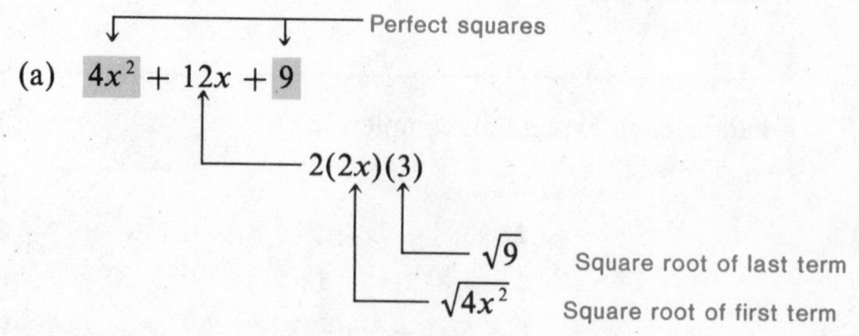

Therefore $4x^2 + 12x + 9 = (\sqrt{4x^2} + \sqrt{9})^2 = (2x + 3)^2$.

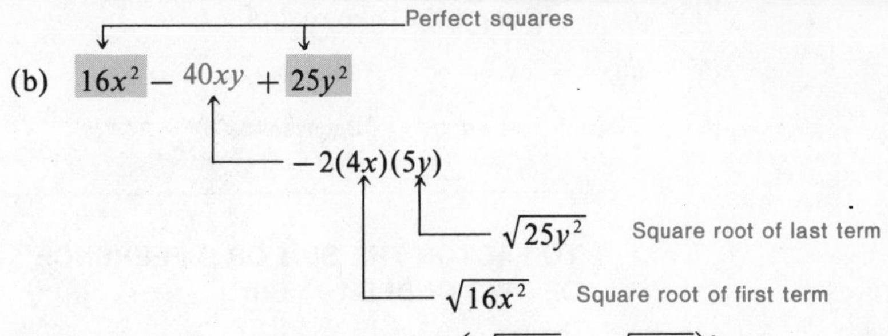

Therefore $16x^2 - 40xy + 25y^2 = (\sqrt{16x^2} - \sqrt{25y^2})^2 = (4x - 5y)^2$.

(c) Factor: $12x - 1 - 36x^2$

$= -(-12x + 1 + 36x^2)$ — Removed the common factor -1

$= -(1 - 12x + 36x^2)$ — Wrote the remaining trinomial in ascending powers of x

$= -[(1)^2 - 2(1)(6x) + (6x)^2]$

$= -(1 - 6x)^2$

EXERCISES 4-3D

Factor each expression completely.

1. $4a^2 + 4a + 1$
2. $9b^2 + 6b + 1$
3. $36x^2 + 12xy + y^2$
4. $16x^2 + 8xy + y^2$
5. $x^2 - 7xy + 49y^2$
6. $x^2 - 5xy + 25y^2$
7. $9 - 12m + 4m^2$
8. $25 - 30n + 9n^2$
9. $a^4 + 2a^2b^2 + b^4$
10. $x^4 - 2x^2y^2 + y^4$
11. $100m^2 + 180mn^2 + 81n^4$
12. $49R^4 + 28R^2T + 4T^2$
13. $2x^2y + 8xy^2 + 8y^3$
14. $3x^3 + 6x^2y + 3xy^2$
15. $3m^2n^2 - 6mn + 3$
16. $ab^2 - 2ab + a$
17. $x^4 - 6x^3y + 9x^2y^2$
18. $x^2y^2 - 8xy^3 + 16y^4$

4-4 Factoring the Sum or Difference of Two Cubes

Associated with factoring the sum or difference of two cubes are the following special products:

Finding the product →

$$(a + b)(a^2 - ab + b^2) = a^3 + b^3$$

← Finding the factors

Therefore $a^3 + b^3$ factors into $(a + b)(a^2 - ab + b^2)$

Finding the product →

$$(a - b)(a^2 + ab + b^2) = a^3 - b^3$$

← Finding the factors

$a^3 - b^3$ factors into $(a - b)(a^2 + ab + b^2)$.

Cube Root of a Term

When factoring the sum or difference of two cubes, we must find the cube root of each term.

Example 1 Finding the cube root of a term

(a) $\sqrt[3]{8x^3} = 2x$ Because $(2x)^3 = 8x^3$

(b) $\sqrt[3]{27a^6b^3} = 3a^2b$ Because $(3a^2b)^3 = 27a^6b^3$

Factoring Sum or Difference of Two Cubes

TO FACTOR THE SUM OR DIFFERENCE OF TWO CUBES ($a^3 \pm b^3$)

1. Find a, the cube root of the first term, and b, the cube root of the second term.
2. Substitute a and b in the factors given in the following formulas:

$$a^3 + b^3 = (a + b)(a^2 - ab + b^2)$$

$$a^3 - b^3 = (a - b)(a^2 + ab + b^2)$$

A Word of Caution. A common mistake students make is to think that the middle term of the trinomial factor is $2ab$ instead of ab.

$$a^3 + b^3 = (a + b)(a^2 - ab + b^2)$$

↑ *Not 2ab* ■

Example 2 Factor $x^3 + 8$

$\sqrt[3]{x^3} = x$ Because $(x)^3 = x^3$

$\sqrt[3]{8} = 2$ Because $2^3 = 8$

$$x^3 + 8 = (x)^3 + (2)^3$$

Same signs — Always +

$$= (\ \blacksquare + \blacksquare\ )(\ \blacksquare - \blacksquare + \blacksquare\ )$$

Always opposite signs

$$= (x + 2)[(x)^2 - (x)(2) + (2)^2]$$

$$= (x + 2)(x^2 - 2x + 4)$$

Example 3 Factor $27x^3 - 64y^3$

$$27x^3 - 64y^3 = (3x)^3 - (4y)^3$$

Same signs — Always +

$$= (\ \blacksquare - \blacksquare\ )(\ \blacksquare + \blacksquare + \blacksquare\ )$$

Always opposite signs

$$= (3x - 4y)[(3x)^2 + (3x)(4y) + (4y)^2]$$

$$= (3x - 4y)(9x^2 + 12xy + 16y^2)$$

Example 4 Factor $(x + 2)^3 - (x - 2)^3$

Let $(x + 2) = a$ and $(x - 2) = b$. Then

$$(x + 2)^3 - (x - 2)^3 = a^3 - b^3 = (a - b)(a^2 + ab + b^2)$$

But since $a = x + 2$ and $b = x - 2$,

$$\begin{aligned}
&(x + 2)^3 - (x - 2)^3 \\
&= [(x + 2) - (x - 2)][(x + 2)^2 + (x + 2)(x - 2) + (x - 2)^2] \\
&= [x + 2 - x + 2][x^2 + 4x + 4 + x^2 - 4 + x^2 - 4x + 4] \\
&= 4[3x^2 + 4]
\end{aligned}$$

EXERCISES 4-4

Factor each expression completely.

1. $x^3 - 8$
2. $x^3 - 27$
3. $64 + a^3$
4. $8 + b^3$
5. $a^3 - 1$
6. $1 - a^3$
7. $1 - 27a^3b^3$
8. $1 - 64a^3b^3$
9. $8x^3y^6 + 27$
10. $64a^6b^3 + 8$
11. $125x^6y^4 - 1$
12. $64a^6b^3 - 9$
13. $a^4 + ab^3$
14. $x^3y + y^4$
15. $16 - 2m^3$
16. $2x^3y^3 + 16z^6$
17. $(x + 1)^3 + 1$
18. $1 + (x - 1)^3$
19. $(a - 2)^3 + 8$
20. $8 - (x^2 + 2)^3$
21. $(x + 1)^3 - (x - 1)^3$
22. $(x - y)^3 - (x + y)^3$
23. $(a + b)^3 - (a - b)^3$

4–5 Factoring by Grouping

Sometimes it is possible to factor an expression by rearranging its terms into smaller groups.

Example 1 Factor $ax + ay + bx + by$

a is a common factor; *b* is a common factor

$$\begin{aligned}
&ax + ay + bx + by \\
&= a(x + y) + b(x + y) \quad (x + y) \text{ is a common factor} \\
&= (x + y)(a + b)
\end{aligned}$$

Therefore $ax + ay + bx + by = (x + y)(a + b)$.

Example 2 Factor $ab - b + a - 1$

Different Ways of Grouping

One Grouping

$$ab - b + a - 1$$
$$= b(a - 1) + 1(a - 1)$$
$$= (a - 1)(b + 1)$$

A Different Grouping

$$ab + a - b - 1$$
$$= a(b + 1) - 1(b + 1)$$
$$= (b + 1)(a - 1)$$

Same factors

Therefore $ab - b + a - 1 = (a - 1)(b + 1) = (b + 1)(a - 1)$.

Example 2 shows that it is sometimes possible to group terms differently and still be able to factor the expression. The same factors are obtained no matter what grouping is used.

Example 3 Factor $u^2 - v^2 + 3u + 3v$

$$u^2 - v^2 + 3u + 3v$$
$$= (u + v)(u - v) + 3(u + v)$$
$$= (u + v)(u - v + 3)$$

Therefore $u^2 - v^2 + 3u + 3v = (u + v)(u - v + 3)$.

Example 4 Factor $a^2 - a - 6 + ab - 3b$

$$a^2 - a - 6 + ab - 3b$$
$$= (a + 2)(a - 3) + b(a - 3)$$
$$= (a - 3)(a + 2 + b)$$

Therefore $a^2 - a - 6 + ab - 3b = (a - 3)(a + b + 2)$.

Example 5 Factor $a^3 + a^2 - b^3 - b^2$

$$a^3 - b^3 + a^2 - b^2$$ Terms have been rearranged

$$= (a - b)(a^2 + ab + b^2) + (a + b)(a - b)$$
$$= (a - b)(a^2 + ab + b^2 + a + b)$$

Therefore $a^3 + a^2 - b^3 - b^2 = (a - b)(a^2 + ab + b^2 + a + b)$.

A Word of Caution. An expression is *not* factored until it has been written as a single term that is a product of factors. To illustrate this, consider Example 1 again:

$$ax + ay + bx + by$$

$$= \boxed{a(x + y)} + \boxed{b(x + y)}$$

First term — Second term

This expression is *not* in factored form because it has *two* terms

$$= \boxed{(x + y)(a + b)}$$

Single term

Factored form of $ax + ay + bx + by$

EXERCISES 4-5

Factor each expression completely.

1. $am + bm + an + bn$
2. $cu + cv + du + dv$
3. $xy + x - y - 1$
4. $ad - d + a - 1$
5. $h^2 - k^2 + 2h + 2k$
6. $x^2 - y^2 + 4x + 4y$
7. $x^3 + 3x^2 - 4x + 12$
8. $a^3 + 5a^2 - 2a + 10$
9. $a^4 + 6a^3 + 6a^2 + 36a$
10. $x^3 - 3x^2 - 9x + 27$
11. $a^2 - b^2 + a^3 + b^3$
12. $a^3 - b^3 - a^2 + b^2$
13. $x^6 + x^3 - y^6 + y^3$
14. $a^8 + a^4 - b^8 + b^4$
15. $a^2 + b^2 + 2ab + 2b + 2a$
16. $m^2 + 4n^2 + 4mn + am + 2an$
17. $3x^2 - xy - 2y^2 + 3x + 2y$
18. $5h^2 - 7hk + 2k^2 + 5h - 2k$
19. $x^3 + x^2 - 2xy + y^2 - y^3$
20. $2x^4 + 2x^3 - 8x^2 - 8x$

4-6 Factoring by Completing the Square

Sometimes it is possible to add a term to a given binomial or trinomial to make it a trinomial that is a perfect square. Of course, *when a term is added, the same term must also be subtracted* so that the value of the original expression is unchanged.

Example 1 Factor $x^4 + x^2 + 1$

$(x^2 + 1)$ Form a binomial by taking the square root of the first and last terms of $x^4 + x^2 + 1$

$(x^2 + 1)^2$ Square the binomial just found

$= x^4 + 2x^2 + 1$ Notice that x^2 must be added to $x^4 + x^2 + 1$ to obtain $x^4 + 2x^2 + 1$

$$\begin{array}{r} x^4 + x^2 + 1 \qquad \\ + x^2 \qquad - x^2 \end{array}$$

Since x^2 is added *and* also subtracted, the value of $x^4 + x^2 + 1$ is unchanged

$$x^4 + x^2 + 1 = \underbrace{x^4 + 2x^2 + 1}_{\text{Trinomial square}} - x^2$$

$(x^2 + 1)^2 - (x)^2$ We now have the difference of two squares

$$[(x^2 + 1) + x][(x^2 + 1) - x]$$

$$(x^2 + 1 + x)(x^2 + 1 - x)$$

Therefore $x^4 + x^2 + 1 = (x^2 + x + 1)(x^2 - x + 1)$.

Example 2 Factor $a^4 + 4$

Found by the method shown in Example 1

$$\begin{array}{r} a^4 \qquad + 4 \qquad \\ + 4a^2 \qquad - 4a^2 \end{array}$$

Adding and subtracting $4a^2$ does not change the value of $a^4 + 4$

$$a^4 + 4 = \underbrace{a^4 + 4a^2 + 4}_{\text{Trinomial square}} - 4a^2$$

$(a^2 + 2)^2 - (2a)^2$ Difference of two squares

$$[(a^2 + 2) + 2a][(a^2 + 2) - 2a]$$

Therefore $a^4 + 4 = (a^2 + 2a + 2)(a^2 - 2a + 2)$.

Example 3 Factor $h^4 - 6h^2k^2 + 25k^4$

In order that the trinomial be a perfect square, the middle term must be $2\sqrt{h^4}\sqrt{25k^4} = 10h^2k^2$. Therefore we must add $16h^2k^2$ to the middle term $-6h^2k^2$ to get $10h^2k^2$. Since $16h^2k^2$ is a perfect square, the method of completing the square works.

Found by the method shown in Example 1

$$\begin{array}{rl} & h^4 - 6h^2k^2 + 25k^4 \\ & \quad + 16h^2k^2 \qquad\qquad - 16h^2k^2 \\ \hline h^4 - 6h^2k^2 + 25k^4 = & h^4 + 10h^2k^2 + 25k^4 - 16h^2k^2 \\ & (h^2 + 5k^2)^2 - (4hk)^2 \\ & [(h^2 + 5k^2) + 4hk][(h^2 + 5k^2) - 4hk] \end{array}$$

Therefore $h^4 - 6h^2k^2 + 25k^4 = (h^2 + 4hk + 5k^2)(h^2 - 4hk + 5k^2)$.

CONDITIONS NECESSARY FOR FACTORING BY COMPLETING THE SQUARE

1. The first and last terms must be perfect squares.
2. Any *literal* factor in the first and last terms must have an exponent divisible by four.
3. The trinomial square must be formed by adding a term.
4. The term added must itself be a perfect square.

EXERCISES 4-6

Factor each expression completely.

1. $x^4 + 3x^2 + 4$
2. $x^4 + 5x^2 + 9$
3. $u^4 + 4u^2 + 16$
4. $z^4 - 19z^2 + 25$
5. $4m^4 + 3m^2 + 1$
6. $9u^4 - 7u^2 + 1$
7. $64a^4 + b^4$
8. $a^4 + 4b^4$
9. $4a^4 + 1$
10. $a^4 + 64$
11. $x^4 - 3x^2 + 9$
12. $x^4 - x^2 + 16$
13. $a^4 - 17a^2b^2 + 16b^4$
14. $a^4 - 3a^2b^2 + 9b^4$
15. $16a^4 - 36a^2b^2 + 81b^4$
16. $9u^4 - 8u^2v^2 + 4v^4$
17. $9x^4 - 28x^2y^2 + 4y^4$
18. $4x^4 + 16x^2y^2 + 25y^4$
19. $9x^4 - 10x^2y^2 + y^4$
20. $4x^4 - 5x^2y^4 + y^8$

21. $50x^4 - 12x^2y^2 + 2y^4$

22. $32x^4 - 2x^2y^2 + 2y^4$

23. $8m^4n + 2n^5$

24. $50x^4y + 32x^2y^3 + 8y^5$

4–7 Factoring Using Synthetic Division

4–7A SYNTHETIC DIVISION

When a polynomial is divided by a binomial of the form $x - a$, the division process can be considerably shortened by a procedure called **synthetic division.**

Consider the following example:

$$
\begin{array}{r@{}l}
 & 2x^2 - 5x + 4 \quad \text{R } 7 \\
x - 3 & \overline{)2x^3 - 11x^2 + 19x - 5} \\
 & \underline{2x^3 - 6x^2} \\
 & -5x^2 + 19x \\
 & \underline{-5x^2 + 15x} \\
 & 4x - 5 \\
 & \underline{4x - 12} \\
 & 7
\end{array}
$$

Since we arrange the polynomials in descending powers of x, the position of a number tells us what power of x is associated with it. For this reason we can actually omit the letters when doing the division problem.

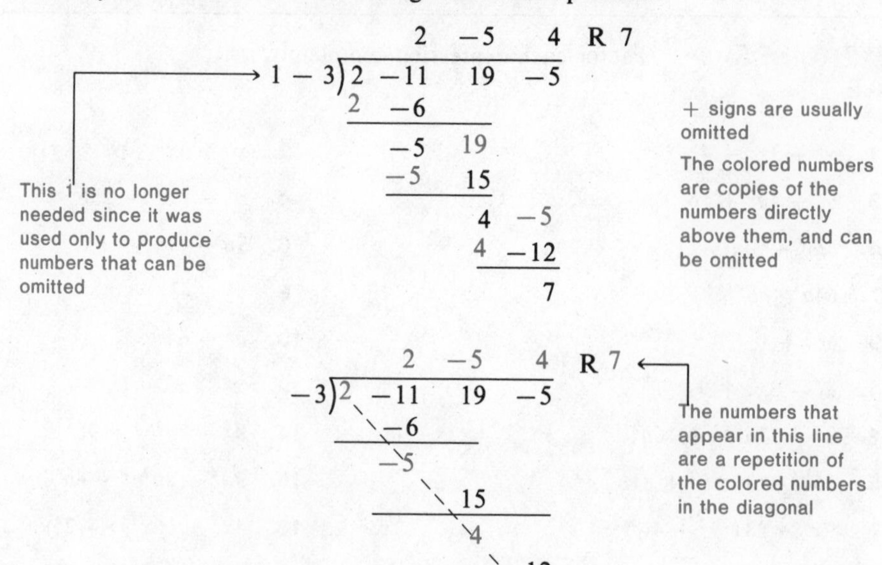

Spaces can be removed, and the work brought closer together as follows:

Changing the sign of the 3 in the divisor changes the signs of the colored numbers, making it possible to add instead of subtract

$$\begin{array}{r|rrrr} -3 & 2 & -11 & 19 & -5 \\ & & -6 & 15 & -12 \\ \hline & 2 & -5 & 4 & 7 \end{array}$$

$$\begin{array}{r|rrrr} & 2 & -11 & 19 & -5 \\ & & 6 & -15 & 12 \\ \hline 3 & 2 & -5 & 4 & 7 \end{array}$$

Replacing the variable, we have: $2x^2 - 5x + 4 \quad \text{R } 7$

Therefore $\dfrac{2x^3 - 11x^2 + 19x - 5}{x - 3} = 2x^2 - 5x + 4 + \dfrac{7}{x - 3}$.

Example 1 Divide $4x^3 + 7x^2 - 5x - 11$ by $x + 2$

Step 1 $x + 2\overline{)4x^3 + 7x^2 - 5x - 11}$

$$\begin{array}{r|rrrr} & 4 & 7 & -5 & -11 \\ \hline -2 & 4 & & & \end{array}$$

←Coefficients of dividend

Note sign change

Bring down leading coefficient of dividend

Step 2

$$\begin{array}{r|rrrr} & 4 & 7 & -5 & -11 \\ & & -8 & & \\ \hline -2 & 4 & & & \end{array}$$

Multiply

Step 3

Add

$$\begin{array}{r|rrrr} & 4 & 7 & -5 & -11 \\ & & -8 & & \\ \hline -2 & 4 & -1 & & \end{array}$$

Step 4

Add

$$\begin{array}{r|rrrr} & 4 & 7 & -5 & -11 \\ & & -8 & 2 & \\ \hline -2 & 4 & -1 & -3 & \end{array}$$

Multiply

Step 5

Add

$$\begin{array}{r|rrrr} & 4 & 7 & -5 & -11 \\ & & -8 & 2 & 6 \\ \hline -2 & 4 & -1 & -3 & -5 \end{array}$$

Multiply

$4x^2 \quad -1x \quad -3 \quad \text{R} \; -5$

Therefore $\dfrac{4x^3 + 7x^2 - 5x - 11}{x + 2} = 4x^2 - x - 3 - \dfrac{5}{x + 2}$.

Example 2 Divide $2x^4 - 3x^2 + 5$ by $x - 1$

$$
\begin{array}{r|rrrrr}
x - 1) & 2x^4 + & 0x^3 - & 3x^2 + & 0x + & 5 \\
 & 2 & 0 & -3 & 0 & 5 \\
 & & 2 & 2 & -1 & -1 \\
\hline
1 & 2 & 2 & -1 & -1 & 4 \\
 & 2x^3 + & 2x^2 & -1x & -1 & \text{R } 4
\end{array}
$$

Zeros are used for all missing powers

Sign changed

Therefore $\dfrac{2x^4 - 3x^2 + 5}{x - 1} = 2x^3 + 2x^2 - x - 1 + \dfrac{4}{x - 1}$.

Synthetic division can also be used with divisors of the type $ax + b$. In this book *we use synthetic division only when the divisor is of the form* $x - a$. The use of synthetic division in graphing is discussed in Section 7–7.

EXERCISES 4–7A

Use synthetic division to find the quotient and the remainder.

1. $(x^2 + 2x - 18) \div (x - 3)$
2. $(x^2 + 4x - 10) \div (x - 2)$
3. $(x^3 + 3x^2 - 5x + 6) \div (x + 4)$
4. $(x^3 + 6x^2 + 4x - 7) \div (x + 5)$
5. $(x^4 + 6x^3 - x - 4) \div (x + 6)$
6. $(2x^4 + 5x^3 + 10x - 2) \div (x + 3)$
7. $\dfrac{x^4 - 16}{x - 2}$
8. $\dfrac{x^7 - 1}{x - 1}$
9. $\dfrac{x^6 - 3x^5 - 2x^2 + 3x + 5}{x - 3}$
10. $\dfrac{x^6 - 3x^4 - 7x - 2}{x - 2}$
11. $(3x^4 - x^3 + 9x^2 - 1) \div \left(x - \dfrac{1}{3}\right)$
12. $(3x^4 - 4x^3 - x^2 - x - 2) \div \left(x + \dfrac{2}{3}\right)$
13. $(2.6x^3 + 1.8x - 6.4) \div (x - 1.5)$
14. $(3.8x^3 - 1.4x^2 - 23.9) \div (x - 2.5)$

4–7B FACTORING USING SYNTHETIC DIVISION

We divide synthetically only by binomials of the form $x - a$ where a is an *integer.*

A typical synthetic division problem is of the form

Quotient → $Q(x) + \text{R}$ ← Remainder

$$x - a \overline{)\, P(x)}$$

Divisor ↑ ↑ Dividend

This division is equivalent to

$$P(x) = (x - a) \cdot Q(x) + \text{R}$$

$$\text{Dividend} = (\text{Divisor})(\text{Quotient}) + \text{Remainder}$$

If the remainder R = 0, then

$$P(x) = (x - a) \cdot Q(x)$$

$$\text{Dividend} = (\text{Divisor})(\text{Quotient})$$

This means $(x - a)$ and $Q(x)$ are *factors* of $P(x)$. *If the remainder in a division problem is zero, the divisor and quotient are factors of the dividend.*

Consider the factored polynomial $(x - r_1)(x - r_2)(x - r_3)$.

If these factors are multiplied, the constant term in their product must be $r_1r_2r_3$, the product of the constant terms in each factor. For example:

$$(x - 5)(x + 2)(x - 3) = x^3 - 6x^2 - x + 30$$

$$(-5)(+2)(-3) = 30$$

Therefore *if a polynomial has a factor of the form $x - a$, then a must be factor of the constant term of the polynomial.* There can be no more *linear* factors than the degree of the polynomial being factored.

Factoring by Synthetic Division

TO FACTOR A POLYNOMIAL BY SYNTHETIC DIVISION

1. Find the factors of the constant term of the polynomial.
2. Use synthetic division to divide the polynomial by each of the factors found in Step 1. Stop when a remainder of zero is obtained.
3. If a remainder of zero is obtained for a, then $(x - a)$ and the quotient are factors of the polynomial.
4. Apply *any* method of factoring to see if the quotient can be factored further.

Example 3 Use synthetic division to factor $2x^3 - 3x^2 - 8x - 3$

Step 1. Factors of the constant term 3 are ± 1, ± 3.

Step 2. Divide: $2x^3 - 3x^2 - 8x - 3$

		2	−3	−8	−3	
			2	−1	−9	
Divide by +1	1	2	−1	−9	−12	Remainder *not* zero; therefore $(x - 1)$ is not a factor

	2	−3	−8	−3	
		−2	5	3	
Divide by −1 −1	2	−5	−3	0	Remainder *is* zero; therefore $(x + 1)$ is a factor

$2x^2 - 5x - 3$ ← Quotient is another factor

Step 3. $2x^3 - 3x^2 - 8x - 3 = (x + 1)(2x^2 - 5x - 3)$.

Step 4. See if the quotient $2x^2 - 5x - 3$ can be factored further.

$$2x^2 - 5x - 3$$
$$= (1x - 3)(2x + 1)$$

Factoring a trinomial that has a leading coefficient greater than 1

Therefore $2x^3 - 3x^2 - 8x - 3 = (x + 1)(x - 3)(2x + 1)$.

Example 4 Use synthetic division to factor $x^5 + 1$

Factors of the constant term $+1$ are ± 1.

$$1x^5 + 0x^4 + 0x^3 + 0x^2 + 0x + 1$$

	1	0	0	0	0	1	
		1	1	1	1	1	
1	1	1	1	1	1	2	Remainder is *not* zero; therefore $(x - 1)$ is not a factor

	1	0	0	0	0	1	
		−1	1	−1	1	−1	
−1	1	−1	1	−1	1	0	Remainder *is* zero; therefore $(x + 1)$ is a factor

$1x^4 - 1x^3 + 1x^2 - 1x + 1$ ← Quotient is another factor

See if the quotient can be factored further.

$$1x^4 - 1x^3 + 1x^2 - 1x + 1$$

	1	−1	1	−1	+1
		−1	2	−3	4
−1	1	−2	3	−4	5

± 1 are the only factors of the constant term 1; since we already know that $(x - 1)$ is not a factor, the only factor we need to try is $(x + 1)$

Therefore $x^5 + 1 = (x + 1)(x^4 - x^3 + x^2 - x + 1)$.

EXERCISES 4-7B

Factor each expression completely.

1. $x^3 + x^2 + x - 3$
2. $x^3 + x^2 - 5x - 2$
3. $x^3 - 3x^2 - 4x + 12$
4. $x^3 - 2x^2 - 5x + 6$

5. $6x^3 - 13x^2 + 4$

6. $6x^3 - 19x^2 + x + 6$

7. $x^3 - 4x^2 + x + 6$

8. $x^3 - 7x - 6$

9. $x^4 - 3x^2 + 4x + 4$

10. $x^4 + 4x^3 + 3x^2 - 4x - 4$

11. $x^3 - 2x^2 - 5x + 6$

12. $x^3 + 2x^2 - 2x + 3$

13. $2x^3 - 5x^2 - 2x + 8$

14. $2x^3 + 5x^2 - x + 6$

15. $x^4 - 13x^2 + 36$

16. $3x^3 + 3x^2 - 9x - 6$

17. $x^3 + x - y^3 - y$

18. $x^3 + x + y^3 - y$

19. $2x^4 + 3x^3 + 2x^2 + x + 1$

20. $3x^4 - 4x^3 - 1$

21. $x^4 - 4x^3 + 34x - 7x^2 - 24$

4–8 How to Select the Method of Factoring

With so many different kinds of factoring, you may be confused about which method of factoring to try first. The following is a procedure that will help you select the correct method for factoring a particular algebraic expression.

First, check for a common factor, no matter how many terms the expression has. If there is a common factor, factor it out. This makes the remaining polynomial easier to factor.

If the Expression to Be Factored Has 2 Terms:
1. Is it a *difference* of two *squares?* (Section 4–2)
2. Is it a *sum* or *difference* of two *cubes?* (Section 4–4)
3. Can it be factored by completing the square? (Section 4–6)
4. Can it be factored by synthetic division? (Section 4–7B)

If the Expression to be Factored Has 3 Terms:
1. Is the leading coefficient 1? (Section 4–3B)
2. Are the first and last terms perfect squares? If so, is the trinomial a perfect square? (Section 4–3D)
3. Is the leading coefficient unequal to 1? (Section 4–3C)
4. Can it be factored by completing the square? (Section 4–6)
5. Can it be factored by synthetic division? (Section 4–7B)

If the Expression to Be Factored Has 4 or More Terms:
1. Can it be factored by grouping? (Section 4–5)
2. Can it be factored by synthetic division? (Section 4–7B)

Check to see if any factor can be factored again. When the expression is *completely factored,* the same factors are obtained no matter what method is used.

Example 1 Selecting a method of factoring

	Method
(a) $6x^2y - 12xy + 4y$	Common factor
$2y(3x^2 - 6x + 2)$	$3x^2 - 6x + 2$ is not factorable

(b) $3x^3 - 27xy^2$ — Common factor
$3x(x^2 - 9y^2)$ — Difference of two squares
$3x(x + 3y)(x - 3y)$

(c) $2ac - 3ad + 10bc - 15bd$ — Grouping
$a(2c - 3d) + 5b(2c - 3d)$
$(2c - 3d)(a + 5b)$

(d) $3x^3 - 17x + 10$ — Synthetic division

$$\begin{array}{r|rrrr} & 3 & 0 & -17 & 10 \\ & & 6 & 12 & -10 \\ \hline 2 & 3 & 6 & -5 & 0 \end{array}$$

$(x - 2)(3x^2 + 6x - 5)$

(e) $2a^3b + 16b$ — Common factor
$2b(a^3 + 8)$ — Sum of two cubes
$2b(a + 2)(a^2 - 2a + 4)$

(f) $6x^2 + 9x - 10$ — Not factorable

(g) $1 \quad + 4x^4$ — Complete the square
$\quad + 4x^2 \quad - 4x^2$

$\underbrace{1 + 4x^2 + 4x^4}\; - 4x^2$

$(1 + 2x^2)^2 - (2x)^2$ — Difference of two squares

$[(1 + 2x^2) + 2x][(1 + 2x^2) - 2x]$
$(1 + 2x + 2x^2)(1 - 2x + 2x^2)$

(h) $12x^2 - 13x - 4$ — Trinomial that has a leading coefficient $\neq 1$
$(3x - 4)(4x + 1)$

(i) $16x^2 - 24xy + 9y^2$ — Trinomial is a perfect square
$(4x - 3y)^2$

(j) $2xy^3 - 14xy^2 + 24xy$ — Common factor
$2xy(y^2 - 7y + 12)$ — Trinomial that has a leading coefficient of 1
$2xy(y - 3)(y - 4)$

EXERCISES 4-8

Factor each expression completely.

1. $12e^2 + 13e - 35$
2. $30f^2 + 17f - 21$
3. $16y^3z - 4yz^3$
4. $5r^3t - 45rt^3$
5. $10x^2 + 28x - 21$
6. $w^2 + 2wk - 63k^2$
7. $6my - 4nz + 15mz - 5zn$
8. $10xy + 5mn - 6xy - nm$
9. $6ac - 6bd + 6bc - 6ad$
10. $10cy - 6cz + 5dy - 3dz$
11. $2xy^3 - 4xy^2 - 30xy$
12. $3yz^3 - 6yz^2 - 24yz$

13. $3x^3 + 24h^3$

14. $54f^3 - 2g^3$

15. $6x^2 + 4x - 10$

16. $2a^2mn - 18b^2mn$

17. $8ab + 9ac - 2ab - 3ac$

18. $4ac + 4bc - 8ad - 8bd$

19. $3x^3 + x^2 + 3x + 5$

20. $5x^3 - 7x^2 - 7x + 2$

21. $9e^2 - 30ef + 25f^2$

22. $16m^2 + 56mp + 49p^2$

23. $4 + k^4$

24. $64 + a^4$

25. $4x^3 + 32y^3$

26. $16 + 4a^4$

27. $x^3 - 2x^2 - 5x + 6$

4–9 Solving Equations by Factoring

Polynomial Equations

First-Degree

Second-Degree

A **polynomial equation** is a polynomial set equal to zero. The degree of the equation is the degree of the polynomial.

Polynomial equations with a first-degree term as the highest-degree term are called **first-degree equations.** Polynomial equations with a second-degree term as the highest-degree term are called **second-degree** or **quadratic equations.**

Example 1 Polynomial equations

(a) $5x - 3 = 0$ — First-degree equation in one variable

(b) $2x^2 - 4x + 7 = 0$ — Second-degree equation in one variable (also called *quadratic* equation)

The following fact is useful in solving some polynomial equations.

> *If* the product of two factors is zero, *then* one or both of the factors must be zero.
>
> If $a \cdot b = 0$, then $\begin{cases} a = 0 \\ \text{or } b = 0 \\ \text{or both } a \text{ and } b = 0 \end{cases}$

We illustrate the use of the above fact in the following example.

Example 2 Solve $(x - 1)(x - 2) = 0$

Since $(x - 1)(x - 2) = 0$

then $(x - 1) = 0$ or $(x - 2) = 0$ — Either factor may equal zero

If $x - 1 = 0$	If $x - 2 = 0$
then $x = 1$	then $x = 2$

Solution set = $\{1,2\}$

Check for $x = 1$:

$$(x-1)(x-2) = 0$$
$$(1-1)(1-2) \stackrel{?}{=} 0$$
$$(0)(-1) \stackrel{?}{=} 0$$
$$0 = 0$$

Check for $x = 2$:

$$(x-1)(x-2) = 0$$
$$(2-1)(2-2) \stackrel{?}{=} 0$$
$$(1)(0) \stackrel{?}{=} 0$$
$$0 = 0$$

TO SOLVE AN EQUATION BY FACTORING

1. Write all nonzero terms on one side of the equation by adding the same expression to both sides. Only zero must remain on the other side. Then arrange the polynomial in descending powers.
2. Factor the polynomial.
3. Set each factor equal to zero, and solve for the unknown letter.
4. Check apparent solutions in the *original* equation.

Example 3 Solve $4 - x = 3x^2$

$$0 = 3x^2 + x - 4$$
$$0 = (x-1)(3x+4)$$

In this case it is convenient to arrange all nonzero terms on the *right* side in descending order

$x - 1 = 0$	$3x + 4 = 0$
$x = 1$	$3x = -4$
	$x = -\dfrac{4}{3}$

Solution set $= \left\{1, -\dfrac{4}{3}\right\}$.

A Word of Caution. The product must equal *zero*, or no conclusions can be drawn about the factors.

Suppose $(x-1)(x-3) = 8$

No conclusion can be drawn because the product $\neq 0$

<u>*Some common mistakes.*</u> Students sometimes think that:

If $(x-1)(x-3) = 8$, then $x - 1 = 8$ *or* $x - 3 = 8$

Both these assumptions are incorrect, because:

If	$x - 1 = 8$	
then	$x = 9$	
Therefore	$x - 3 = 9 - 3$	$= 6$
Then	$(x - 1)(x - 3)$	
$=$	$8 \cdot 6$	$= 48$
		$\neq 8$

If	$x - 3 = 8$	
then	$x = 11$	
Therefore	$x - 1 = 11 - 1$	$= 10$
Then	$(x - 1)(x - 3)$	
$=$	$10 \cdot 8$	$= 80$
		$\neq 8$

The correct solution is:

$$(x - 1)(x - 3) = 8$$
$$x^2 - 4x + 3 = 8$$
$$x^2 - 4x - 5 = 0$$
$$(x - 5)(x + 1) = 0$$
$$x - 5 = 0 \quad | \quad x + 1 = 0$$
$$x = 5 \quad | \quad x = -1$$

■

EXERCISES 4-9 Solve each equation.

1. $3x(x - 4) = 0$
2. $5x(x + 6) = 0$
3. $x(x - 4) = 12$
4. $x(x - 2) = 15$
5. $4x(x - 2) = 0$
6. $x(3x - 1) = 2$
7. $x^2 - x - 12 = 0$
8. $x^2 + x - 12 = 0$
9. $x^2 - 18 = 9x$
10. $x^2 - 20 = 12x$
11. $x^2 + x = 12$
12. $6x^2 + 12 = 17x$
13. $6x^2 = 11x + 10$
14. $13x + 3 = -4x^2$
15. $5 - 16n = -3n^2$
16. $5k^2 = 34k + 7$
17. $5a^2 = 16a - 3$
18. $3z^2 = 22z - 7$
19. $4x^2 + 9 = 12x$
20. $25x^2 + 4 = 20x$
21. $4x(2x - 1)(3x + 7) = 0$
22. $5x(4x - 3)(7x - 6) = 0$
23. $2x^3 + x^2 = 3x$
24. $4x^3 = 10x - 18x^2$
25. $21x^2 + 60x = 18x^3$
26. $68x^2 = 30x^3 + 30x$
27. $z^3 - 2z^2 - z + 2 = 0$
28. $v^3 - 4v + v^2 - 4 = 0$
29. $2x^3 + 7x^2 + 2x - 3 = 0$
30. $3x^3 - 2x^2 - 19x - 6 = 0$

4–10 Solving Word Problems by Factoring

Some word problems lead to equations that can be solved by factoring.

Example 1 The difference of two numbers is 3. Their product is 10. What are the two numbers?

Let x = Smaller number
then $x + 3$ = Larger number
} Since their difference is 3

$$\begin{aligned} \text{Their product} &\quad \text{is} \quad 10 \\ x(x+3) &= 10 \end{aligned}$$

Remember, it is incorrect to say that $x = 10$ or $x + 3 = 10$

$$\begin{aligned} x^2 + 3x &= 10 \\ x^2 + 3x - 10 &= 0 \\ (x-2)(x+5) &= 0 \end{aligned}$$

$$\begin{array}{r|rl} x - 2 = 0 & x + 5 = & 0 \\ x = 2 & x = & -5 \\ x + 3 = 5 & x + 3 = & -2 \end{array}$$

Smaller number ($x = -5$)
Larger number ($x + 3 = -2$)

Therefore the numbers 2 and 5 are a solution and the numbers -5 and -2 are another solution.

Example 2 The first term of a proportion is 2 more than its fourth term. Write the proportion if its second term is 3 and its third term is 8.

Let $\quad x =$ Fourth term
then $x + 2 =$ First term

Then the proportion is

$$\frac{x+2}{3} = \frac{8}{x}$$

$$(x+2)x = 3(8) \qquad \text{Product of means = Product of extremes}$$

$$\begin{aligned} x^2 + 2x &= 24 \\ x^2 + 2x - 24 &= 0 \\ (x-4)(x+6) &= 0 \end{aligned}$$

$$\begin{array}{r|rl} x - 4 = 0 & x + 6 = & 0 \\ x = 4 & x = & -6 \end{array}$$

Therefore the proportions are:

For $x = 4$:

$$\frac{x+2}{3} = \frac{8}{x}$$

$$\frac{4+2}{3} = \frac{8}{4}$$

$$\frac{6}{3} = \frac{8}{4} \quad \text{True}$$

For $x = -6$:

$$\frac{x+2}{3} = \frac{8}{x}$$

$$\frac{-6+2}{3} = \frac{8}{-6}$$

$$\frac{-4}{3} = \frac{8}{-6} \quad \text{True}$$

In solving word problems about geometric figures, make a drawing of the figure and write the given information on it.

Example 3 One square has a side 3 ft longer than the side of a second square. If the area of the larger square is 4 times as great as the area of the smaller square, find the length of a side of each square.

Let $\quad x =$ side of smaller square
then $x + 3 =$ side of larger

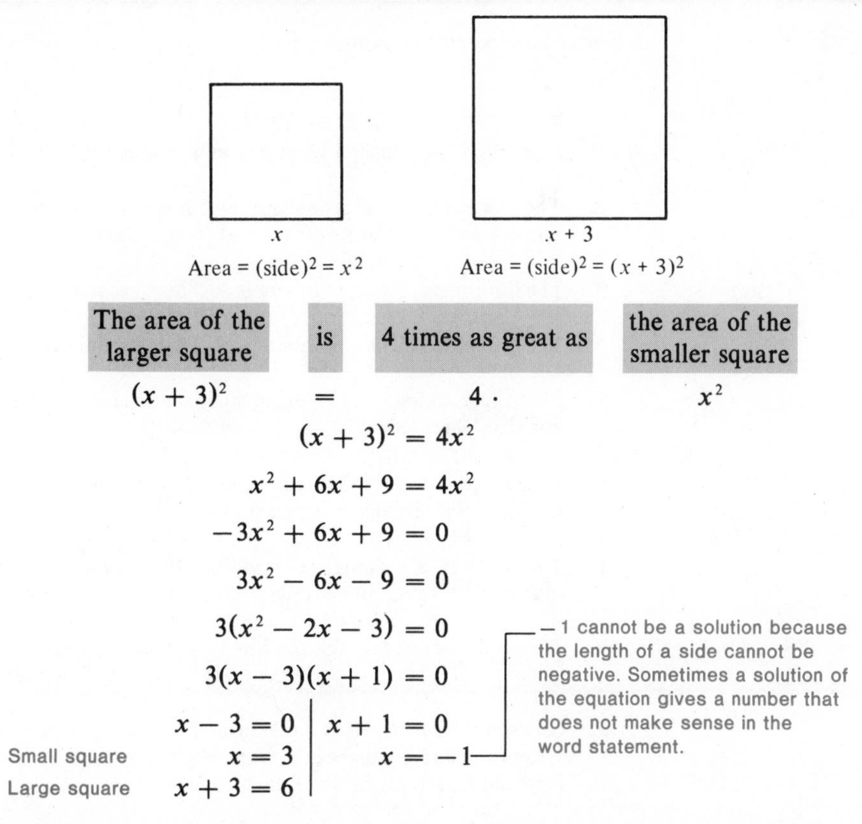

Example 4 The width of a rectangle is 5 cm less than its length. Its area is 10 more (numerically) than its perimeter. What are the dimensions of the rectangle?

Let L = Length
then $L - 5$ = Width

$L - 5$

L

$$\text{Area} = LW = L(L-5)$$
$$\text{Perimeter} = 2L + 2W = 2L + 2(L-5)$$

Its area	is	10 more than	its perimeter
$L(L-5)$	$=$	$10 +$	$2L + 2(L-5)$

$$L(L-5) = 10 + 2L + 2(L-5)$$
$$L^2 - 5L = 10 + 2L + 2L - 10$$
$$L^2 - 9L = 0$$
$$L(L-9) = 0$$

$L = 0$ (Not a solution) | $L - 9 = 0$
$L = 9$ Length
$L - 5 = 4$ Width

Therefore the rectangle has a length of 9 cm and a width of 4 cm.

EXERCISES 4-10

Solve each word problem.

1. The fourth term of a proportion is 7 more than its first term. Write the proportion if its second term is 3 and its third term is 6.
2. The third term of a proportion is 2 more than its second term. Write the proportion if its first term is 3 and its fourth term is 8.
3. The difference of two numbers is 5. Their product is 14. Find the numbers.
4. The difference of two numbers is 6. Their product is 27. Find the numbers.
5. Find three consecutive integers such that the product of the first two is 10 less than the product of the last two.
6. Find three consecutive integers such that the product of the first two is 20 less than the product of the last two.
7. The width of a rectangle is 4 yd less than its length. The area is 17 more (numerically) than its perimeter. What are the dimensions of the rectangle?
8. The width of a rectangle is 7 m less than its length. The area is 4 more (numerically) than its perimeter. What are the dimensions of the rectangle?
9. One square has a side 3 cm shorter than the side of a second square. The area of the larger square is 4 times as great as the area of the smaller square. Find the length of a side of each square.
10. One square has a side 2 m longer than the side of a second square. The area of the larger square is 16 times as great as the area of the smaller square. Find the length of a side of each square.
11. Forty square yards of carpet are laid in a rectangular room. The length is 2 yd less than three times the width. Find the dimensions of the room.
12. The area of a rectangle is 54 sq cm. Find its dimensions if its sides are in the ratio of 2 to 3.
13. A box with no top is to be formed from a rectangular sheet of metal by cutting 2 in. squares from the corners and folding up the sides. The length of the box is to be 3 in. more than its width, and its volume is to be 80 cu in.
    (a) Find the dimensions of the sheet of metal.
    (b) Find the dimensions of the box.
14. A box with no top is formed from a square sheet of metal by cutting 3 in. squares from the corners and folding up the sides. The box is to have a volume of 12 cu in.
    (a) Find the dimensions of the sheet of metal.
    (b) Find the dimensions of the box.
15. *Brainteaser.* Mr. Wiley divided $15 equally among his children. The product of all the amounts the children received from the $15 is the same as the number of square feet in a particular number of square yards. How much did each child receive, and how many children does Mr. Wiley have?

Chapter Four Summary

A Prime Number is a positive integer greater than 1 that can be exactly divided only by itself and 1. A prime number has no factors other than itself and 1. (Section 4–1)

A Composite Number is a positive integer that can be exactly divided by some integer other than itself and 1. A composite number has factors other than itself and 1. (Section 4–1)

The Prime Factorization of a Positive Integer is the indicated product of all its factors that are themselves prime numbers. (Section 4–1)

Special Products.

$$a(b + c) = ab + ac \qquad \text{Distributive rule}$$

$$(a + b)(a - b) = a^2 - b^2 \qquad \text{Sum and difference of two terms}$$

$$\left.\begin{aligned}(a + b)^2 &= a^2 + 2ab + b^2\\ (a - b)^2 &= a^2 - 2ab + b^2\end{aligned}\right\} \qquad \text{Square of a binomial}$$

Methods of Factoring.

1. Greatest common factor (GCF) (Section 4–1)
2. Difference of two squares (Section 4–2): $a^2 - b^2 = (a + b)(a - b)$
3. Sum of two cubes (Section 4–4): $a^3 + b^3 = (a + b)(a^2 - ab + b^2)$
   Difference of two cubes: $a^3 - b^3 = (a - b)(a^2 + ab + b^2)$
4. Trinomial square (Section 4–3D): $\begin{cases} a^2 + 2ab + b^2 = (a + b)^2 \\ a^2 - 2ab + b^2 = (a - b)^2 \end{cases}$
5. Trinomial $\begin{cases} \text{leading coefficient of 1 (Section 4–3B)} \\ \text{leading coefficient unequal to 1 (Section 4–3C)} \end{cases}$
6. Completing the square (Section 4–6)
7. Grouping (Section 4–5)
8. Synthetic division (Section 4–7B)

To Solve an Equation by Factoring. (Section 4–9)

1. Write *all* nonzero terms on one side of the equation by adding the same expression to both sides. *Only zero must remain on the other side.* Then arrange the polynomial in descending powers.
2. Factor the polynomial.
3. Set each factor equal to zero, and solve for the unknown letter.

Synthetic Division. (Section 4–7A). Synthetic division can be used when a polynomial is divided by a binomial of the form $x - a$.

Chapter Four Diagnostic Test or Review

Allow yourself about 55 minutes to do these problems. Complete solutions for every problem, together with section references, are given in the answer section at the end of the book.

---

1. Find the prime factorization of 126.

In Problems 2–5 find the products.

2. $2xy^3(4x^2y - 3xy - 5)$

3. $(2x^4 + 3)(2x^4 - 3)$

4. $(5m - 2)(3m + 4)$

5. $(3R^2 - 5)^2$

In Problems 6–12 factor the expressions completely.

6. $4x - 16x^3$

7. $y^3 - 1$

8. $x^2 - 2x - 15$

9. $6x^2 - 5x - 6$

10. $3ac + 6bc - 5ad - 10bd$

11. $x^4 + 5x^2 + 9$

12. $x^3 - 8x + 3$

13. Solve: $2x^2 + x = 15$

14. Solve: $8y^2 = 4y$

15. The length of a rectangle is 11 more than its width. Its area is 60. Find the length and width of the rectangle.

16. The third term of a proportion is 7 more than its second term. Write the proportion if its first term is 5 and its fourth term is 6.

5 FRACTIONS

In this chapter we define algebraic fractions, how to perform necessary operations with them, and how to solve equations and word problems involving them. A knowledge of the different methods of factoring is essential in your work with *algebraic* fractions.

5–1 Simplifying Algebraic Fractions

Rational Expression

A **simple algebraic fraction** (also called a **rational expression**) is an algebraic expression of the form

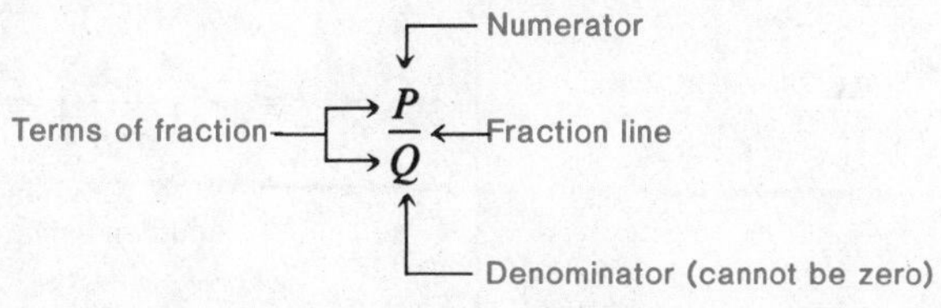

where P and Q are polynomials.

Excluded Values

Any value of the letter (or letters) which makes the denominator Q equal to zero must be excluded from the domain of the variable (Section 1–3D).

Example 1 Examples of algebraic fractions [some show excluded value(s) of the letter]

(a) $\dfrac{x}{3}$

No value of x is excluded because no value of x makes the denominator zero

(b) $\dfrac{5}{x}$ — x cannot be 0

(c) $\dfrac{2x-5}{x-1}$ — x cannot be 1, because that would make the denominator 0

(d) $\dfrac{x^2+2}{x^2-3x-4} = \dfrac{x^2+2}{(x-4)(x+1)}$ — x cannot be 4 or -1, because either value makes the denominator 0

(e) $\dfrac{2}{3}$ — Arithmetic fractions are also algebraic fractions; here, 2 and 3 are polynomials of degree 0

Note: After this section, whenever a fraction is written, it will be understood that the value(s) of the variable(s) that make the denominator zero are excluded. ■

The Three Signs of a Fraction

Every fraction has three signs associated with it.

$$+\frac{-2}{+3}$$

Sign of fraction → $+$; Sign of numerator → $-$ (of 2); Sign of denominator → $+$ (of 3)

> **If any two of the three signs of a fraction are changed, the value of the fraction is unchanged.**

Example 2 **Changing two signs of a fraction to obtain an equivalent fraction**

(a) $+\dfrac{+8}{+4} =$

$+\dfrac{-8}{-4} = +\left(\dfrac{-8}{-4}\right) = +(+2) = 2$

Here, we changed the signs of numerator and denominator

$-\dfrac{-8}{+4} = -\left(\dfrac{-8}{+4}\right) = -(-2) = 2$

Here, we changed the signs of fraction and numerator

$-\dfrac{+8}{-4} = -\left(\dfrac{+8}{-4}\right) = -(-2) = 2$

Here, we changed the signs of fraction and denominator

(b) $-\dfrac{1}{2-x} = +\dfrac{1}{-(2-x)} = \dfrac{1}{x-2}$

Here, we changed sign of fraction and denominator

The simplification in Example 2(b) can be explained as follows:

$$-(2-x) = (-1)(2-x) = -2 + x = x - 2 \quad \text{Distributive rule}$$

Reducing Fractions to Lowest Terms

We reduce fractions to lowest terms in algebra for the same reason we do in arithmetic: It makes them simpler and easier to work with. After this section, it is understood that *all fractions are to be reduced to lowest terms* (unless otherwise indicated).

The procedure for reducing fractions in algebra is the same as the method used in arithmetic.

TO REDUCE A FRACTION TO LOWEST TERMS

1. Factor the numerator and denominator completely.
2. Divide the numerator and denominator by all factors common to both.

Example 3 Reduce to lowest terms

(a) $\dfrac{4x^2y}{2xy} = \dfrac{\overset{2}{\cancel{4}}x^{\cancel{2}}\cancel{y}}{\underset{1}{\cancel{2}}\cancel{xy}} = 2x$ Here, the literal parts of the fraction are already factored

(b) $\dfrac{3x^2 - 5xy - 2y^2}{6x^3y + 2x^2y^2} = \dfrac{(x-2y)\cancel{(3x+y)}}{2x^2y\cancel{(3x+y)}} = \dfrac{x-2y}{2x^2y}$

(c) $\dfrac{x-y}{y-x} = -\dfrac{x-y}{-(y-x)} = -\dfrac{\overset{1}{\cancel{(x-y)}}}{\underset{1}{\cancel{(x-y)}}} = -1$

Changing sign of fraction and denominator

(d) $\dfrac{(b-a)(2b+3a)}{(a-b)(4a-5b)} = \dfrac{(-1)(b-a)\ (2b+3a)}{(-1)(a-b)(4a-5b)} = \dfrac{(a-b)\ (2b+3a)}{(-1)\ (a-b)\ (4a-5b)}$

$(-1)(b-a) = (a-b)$

Changing the sign of both numerator and denominator is equivalent to multiplying each by -1

$$= \frac{(2b+3a)}{(-1)(4a-5b)} = \frac{2b+3a}{5b-4a}$$

(e) $\frac{\overset{1}{\cancel{z}}}{2\underset{1}{\cancel{z}}} = \frac{1}{2}$ **Note:** A factor of 1 will always remain in the numerator and denominator after they have been divided by factors common to both. ■

(f) $\frac{x+3}{x+6}$ Cannot be reduced {Neither x nor 3 is a *factor* of numerator or denominator

(g) $\frac{x+y}{x}$ Cannot be reduced (See the following word of caution)

← x is *not* a *factor* of the numerator

A Word of Caution. A common error made in reducing fractions is to forget that the number the numerator and denominator are divided by *must* be a *factor* of *both* [see Examples 3(f) and 3(g)].

Error:

← 3 is *not* a factor of the numerator

$$\frac{3+2}{3} \neq 2 \quad \textit{Incorrect reduction}$$

The above reduction is incorrect because

$$\frac{3+2}{3} = \frac{5}{3} \neq 2$$ ■

EXERCISES 5-1

In Exercises 1–8 what value(s) of the variable (if any) must be excluded?

1. $\frac{10-7y}{y+4}$
2. $\frac{3z+2}{5-z}$
3. $\frac{5x}{9}$
4. $\frac{7}{2x}$
5. $\frac{a^2+1}{a^2-25}$
6. $\frac{h^2+5}{h^2-h-6}$
7. $\frac{4c+3}{c^4-13c^2+36}$
8. $\frac{2x-5}{x^3-5x^2-9x+45}$

In Exercises 9–14 use the rule about the three signs of a fraction to find the missing term.

9. $-\frac{5}{8} = \frac{5}{?}$
10. $\frac{6}{-k} = \frac{?}{k}$
11. $\frac{8-y}{4y-7} = \frac{y-8}{?}$
12. $\frac{w-2}{5-w} = \frac{?}{w-5}$
13. $\frac{a-b}{(3a+2b)(a-5b)} = \frac{?}{(3a+2b)(5b-a)}$
14. $\frac{(e+4f)(7e-3f)}{2e-f} = \frac{(e+4f)(3f-7e)}{?}$

In Exercises 15–38 reduce each fraction to lowest terms.

15. $\dfrac{12m^3n}{4mn}$ **16.** $\dfrac{-6hk^4}{24hk}$ **17.** $\dfrac{15a^4b^3c^2}{-35ab^5c}$

18. $\dfrac{40e^2f^2g}{16e^4fg^3}$ **19.** $\dfrac{40x - 8x^2}{5x^2 + 10x}$ **20.** $\dfrac{16y^4 - 16y^3}{24y^2 - 24y}$

21. $\dfrac{c^2 - 4}{4}$ **22.** $\dfrac{9 + d^2}{9}$ **23.** $\dfrac{24w^2x^3 - 16wx^4}{18w^3x - 12w^2x^2}$

24. $\dfrac{18c^5d + 45c^4d^2}{12c^2d^3 + 30cd^4}$ **25.** $\dfrac{x^2 - 16}{x^2 - 9x + 20}$ **26.** $\dfrac{x^2 - 2x - 15}{x^2 - 9}$

27. $\dfrac{(2x + y)(x - y)}{(y - 3x)(y - x)}$ **28.** $\dfrac{(k - h)(2k + 5h)}{(h - k)(3h - 4k)}$ **29.** $\dfrac{2x^2 - 3x - 9}{12 - 7x + x^2}$

30. $\dfrac{15 + 7y - 2y^2}{4y^2 - 21y + 5}$ **31.** $\dfrac{2y^2 + xy - 6x^2}{3x^2 + xy - 2y^2}$ **32.** $\dfrac{10y^2 + 11xy - 6x^2}{4x^2 - 4xy - 15y^2}$

33. $\dfrac{a^3 - 1}{1 - a^2}$ **34.** $\dfrac{x^3 + y^3}{y^2 - x^2}$ **35.** $\dfrac{4x^4 + y^4}{6x^2 + 6xy + 3y^2}$

36. $\dfrac{x^4 + x^2y^2 + y^4}{2x^2 - 2xy + 2y^2}$ **37.** $\dfrac{x^3 + 8}{x^3 - 3x^2 + 6x - 4}$ **38.** $\dfrac{x^3 + 4x^2 + 12x + 9}{x^3 - 27}$

5-2 Multiplication and Division of Fractions

MULTIPLICATION OF FRACTIONS

We multiply algebraic fractions the same way we multiply fractions in arithmetic.

> **TO MULTIPLY FRACTIONS**
>
> **1.** Factor the numerator and denominator of the fractions.
>
> **2.** Divide the numerator and denominator by all factors common to both.
>
> **3.** The answer is the product of the factors remaining in the numerator divided by the product of the factors remaining in the denominator. A factor of 1 will always remain in both numerator and denominator [see Examples 1(a) and 2(c)].

Example 1 Multiply the fractions

(a) $\dfrac{a}{3} \cdot \dfrac{1}{a^2} = \dfrac{\overset{1}{\cancel{a}}}{3} \cdot \dfrac{1}{\underset{a}{\cancel{a^2}}} = \dfrac{1}{3a}$

(b) $$\frac{2y^3}{3x^2}\cdot\frac{12x}{5y^2} = \frac{2y^3\cdot \overset{4}{\cancel{12}}x}{\underset{1}{\cancel{3}}x^2\cdot 5y^2} = \frac{8y}{5x}$$

(c) $$\frac{10xy^3}{x^2-y^2}\cdot\frac{2x^2+xy-y^2}{15x^2y} = \frac{\overset{2}{\cancel{10}}xy^3}{\cancel{(x+y)}(x-y)}\cdot\frac{\cancel{(x+y)}(2x-y)}{\underset{3}{\cancel{15}}x^2y}$$

$$= \frac{2y^2(2x-y)}{3x(x-y)}$$

DIVISION OF FRACTIONS

In Arithmetic

$$\frac{3}{5}\div\frac{4}{7} = \frac{3}{5}\cdot\boxed{\frac{7}{4}} = \frac{3\cdot 7}{5\cdot 4} = \frac{21}{20}$$

Invert the second fraction and multiply

This method works because:

$$\frac{3}{5}\div\frac{4}{7} = \frac{\frac{3}{5}}{\frac{4}{7}} = \frac{\frac{3}{5}}{\frac{4}{7}}\cdot\left(\frac{\frac{7}{4}}{\frac{7}{4}}\right) = \frac{\frac{3}{5}\cdot\frac{7}{4}}{\frac{4}{7}\cdot\frac{7}{4}} = \frac{\frac{3}{5}\cdot\frac{7}{4}}{1} = \frac{3}{5}\cdot\frac{7}{4}$$

The value of this fraction is 1

Therefore $\frac{3}{5}\div\frac{4}{7} = \frac{3}{5}\cdot\frac{7}{4}$

In Algebra We divide algebraic fractions the same way we divide fractions in arithmetic.

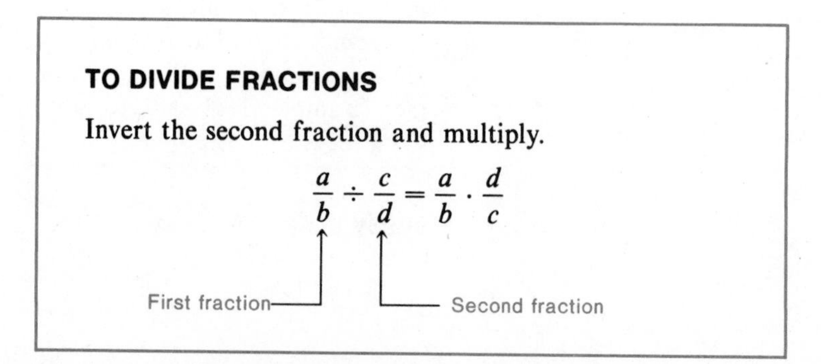

TO DIVIDE FRACTIONS

Invert the second fraction and multiply.

$$\frac{a}{b}\div\frac{c}{d} = \frac{a}{b}\cdot\frac{d}{c}$$

Example 2 Divide the fractions

(a) $$\frac{4r^3}{9s^2} \div \frac{8r^2s^4}{15rs} = \frac{\overset{1}{\cancel{4}}r^3}{\underset{3}{\cancel{9}}s^2} \cdot \frac{\overset{5}{\cancel{15}}rs}{\underset{2}{\cancel{8}}r^2s^4} = \frac{5r^2}{6s^5}$$

(b) $$\frac{3y^3 - 3y^2}{16y^5 + 8y^4} \div \frac{y^2 + 2y - 3}{4y + 12} = \frac{3y^2\cancel{(y-1)}}{\underset{2}{\cancel{8}}y^4(2y+1)} \cdot \frac{\overset{1}{\cancel{4}}\cancel{(y+3)}}{\cancel{(y-1)}\cancel{(y+3)}}$$

$$= \frac{3}{2y^2(2y+1)}$$

(c) $$\frac{x^2 + x - 2}{x + 2} \div \frac{x^2 + 2x - 3}{x + 3} = \frac{\overset{1}{\cancel{(x+2)}}\overset{1}{\cancel{(x-1)}}}{\underset{1}{\cancel{x+2}}} \cdot \frac{\overset{1}{\cancel{x+3}}}{\underset{1}{\cancel{(x+3)}}\underset{1}{\cancel{(x-1)}}} = 1$$

EXERCISES 5-2

Perform the indicated operations.

1. $\dfrac{27x^4y^3}{22x^5yz} \cdot \dfrac{55x^2z^2}{9y^3z}$

2. $\dfrac{13b^2c^4}{42a^4b^3} \cdot \dfrac{35a^3bc^2}{39ac^5}$

3. $\dfrac{mn^3}{18n^2} \div \dfrac{5m^4}{24m^3n}$

4. $\dfrac{27k^3}{h^5k} \div \dfrac{15hk^2}{-4h^4}$

5. $\dfrac{15u - 6u^2}{10u^2} \cdot \dfrac{15u^3}{35 - 14u}$

6. $\dfrac{-22v^2}{63v + 84} \cdot \dfrac{42v^3 + 56v^2}{55v^3}$

7. $\dfrac{-15c^4}{40c^3 - 24c^2} \div \dfrac{35c}{35c^2 - 21c}$

8. $\dfrac{40d - 30d^2}{d^2} \div \dfrac{24d^2 - 18d^3}{12d^3}$

9. $\dfrac{d^2e^2 - d^3e}{12e^2d} \div \dfrac{d^2e^2 - de^3}{3e^2d + 3e^3}$

10. $\dfrac{9m^2n + 3mn^2}{16mn^2} \div \dfrac{2mn^2 - m^2n}{10mn^2 - 20n^3}$

11. $\dfrac{w^2 - 2w - 8}{6w - 24} \cdot \dfrac{5w^2}{w^2 - 3w - 10}$

12. $\dfrac{-15k^3}{8k + 32} \div \dfrac{15k^2 - 5k}{k^2 + k - 12}$

13. $\dfrac{4a^2 + 8ab + 4b^2}{a^2 - b^2} \div \dfrac{6ab + 6b^2}{b - a}$

14. $\dfrac{u^2 - v^2}{7u^2 - 14uv + 7v^2} \div \dfrac{2u^2 + 2uv}{14v - 14u}$

15. $\dfrac{4 - 2a}{2a + 2} \div \dfrac{2a^3 - 16}{a^2 + 2a + 1}$

16. $\dfrac{18 + 6a}{4 - 2a} \div \dfrac{2a^3 + 54}{4a - 8}$

17. $\dfrac{x^3 + y^3}{2x - 2y} \div \dfrac{x^2 - xy + y^2}{x^2 - y^2}$

18. $\dfrac{x^3 - y^3}{3x + 3y} \div \dfrac{x^2 + xy + y^2}{x^2 - y^2}$

19. $\dfrac{m^4 + 4}{m^2 - 4} \cdot \dfrac{m + 2}{m^2 - 2m + 2}$

20. $\dfrac{R^4 + 5R^2 + 9}{R^2 - 9} \cdot \dfrac{R - 3}{R^2 - R + 3}$

21. $\dfrac{5h^2k - h^3}{9hk + 18k^2} \cdot \dfrac{3h^2 - 12k^2}{h^2 - 7hk + 10k^2}$

22. $\dfrac{32 + 4b^3}{b^2 - b - 6} \div \dfrac{b^2 - 2b + 4}{15b - 5b^2}$

23. $\dfrac{5y^2 + 15yz + 45z^2}{9z^2 + 3zy} \div \dfrac{2y^3 - 54z^3}{9z^2 - y^2}$

24. $\dfrac{4c^4 + 1}{36 - c^2} \cdot \dfrac{c - 6}{2c^2 - 2c + 1}$

25. $\dfrac{6uv - 6v^2 + 6v}{10uv} \cdot \dfrac{(u+v) - (u+v)^2}{(u-v)^2 + (u-v)} \cdot \dfrac{5uv - 5u^2}{3v^2 + 3vu}$

26. $\dfrac{10nm - 15n^2}{3n + 2m} \div \dfrac{9mn - 6m^2}{4m^2 + 12mn + 9n^2} \div \dfrac{4m^2 - 9n^2}{27n - 18m}$

27. $\dfrac{e^2 + 10ef + 25f^2}{e^2 - 25f^2} \cdot \dfrac{3e - 3f}{f - e} \div \dfrac{e + 5f}{5f - e}$

28. $\dfrac{3x - y}{y + x} \div \dfrac{y^2 + yx - 12x^2}{4x^2 + 5xy + y^2} \cdot \dfrac{10x^2 - 8xy}{8xy - 10x^2}$

29. $\dfrac{(x+y)^2 + x + y}{(x-y)^2 - x + y} \cdot \dfrac{x^2 - 2xy + y^2}{x^2 + 2xy + y^2} \cdot \dfrac{x + y}{x - y}$

30. $\dfrac{ac + bc + ad + bd}{ac - ad - bc + bd} \cdot \dfrac{c^2 - 2cd + d^2}{a^2 + 2ab + b^2} \cdot \dfrac{a - b}{c + d}$

5–3 Addition of Fractions

Like fractions are fractions that have the same denominator.

Like Fractions

Example 1 Like fractions

(a) $\dfrac{8}{x}, \quad \dfrac{x+1}{x}$ (b) $\dfrac{2}{a-b}, \quad \dfrac{5a}{a-b}, \quad \dfrac{a+b}{a-b}$

TO ADD LIKE FRACTIONS

1. Add the numerators.
2. Write the sum of the numerators over the denominator of the like fractions.

$$\frac{a}{c} + \frac{b}{c} = \frac{a+b}{c}$$

3. Reduce the resulting fraction to lowest terms.

Note: Any **subtraction of fractions** can be done as an addition problem. ■

$$\frac{a}{c} - \frac{b}{c} = \frac{a}{c} + \frac{-b}{c} = \frac{a + (-b)}{c} = \frac{a - b}{c}$$

Example 2 Adding like fractions

(a) $\underbrace{\dfrac{3}{4a} - \dfrac{5}{4a} = \dfrac{3}{4a} + \dfrac{-5}{4a}} = \dfrac{3-5}{4a} = \dfrac{-2}{4a} = -\dfrac{\overset{1}{\cancel{2}}}{\underset{2}{\cancel{4}}a} = -\dfrac{1}{2a}$

Changing the subtraction problem to an addition problem

(b) $\dfrac{4x}{2x-y} - \dfrac{2y}{2x-y} = \dfrac{4x - 2y}{2x - y} = \dfrac{2(2x-y)}{(2x-y)} = \dfrac{2\cancel{(2x-y)}}{\cancel{(2x-y)}} = 2$

(c) $\frac{15}{d-5} + \frac{-3d}{d-5} = \frac{15-3d}{d-5} = \frac{3(5-d)}{d-5}$

$= -\frac{3(d-5)}{(d-5)} = -\frac{3\cancel{(d-5)}}{\cancel{(d-5)}} = -3$

Changing sign of fraction and numerator

Unlike Fractions

Unlike fractions are fractions that have different denominators.

Example 3 Unlike fractions

(a) $\frac{2}{x}, \quad \frac{6}{x^2}, \quad \frac{x-1}{5x}$ (b) $\frac{5}{x+2}, \quad \frac{2x}{x-1}$

LCD

The **lowest common denominator (LCD)** is the smallest algebraic expression that is exactly divisible by each of the denominators.

We ordinarily use the lowest common denominator when adding or subtracting unlike fractions.

TO FIND THE LCD

1. Factor each denominator completely. Repeated factors should be expressed as powers.
2. Write each different factor that appears.
3. Raise each factor to the highest power it occurs in *any* denominator.
4. The LCD is the product of all the powers found in Step 3.

Example 4 Find the LCD for $\frac{7}{18x^2y} + \frac{5}{8xy^4}$

1. $2 \cdot 3^2 \cdot x^2 \cdot y, \quad 2^3 \cdot x \cdot y^4$ — Denominators in factored form
2. $2, \ 3, \ x, \ y$ — All the different factors
3. $2^3, \ 3^2, \ x^2, \ y^4$ — Highest powers of factors
4. $\text{LCD} = 2^3 \cdot 3^2 \cdot x^2 \cdot y^4 = 72x^2y^4$

Example 5 Find the LCD for $\frac{2}{x} + \frac{x}{x+2}$

1. The denominators are already factored.
2. $x, \quad (x+2)$

3. $x^1, \quad (x+2)^1$
4. $\text{LCD} = x(x+2)$

Example 6 Find the LCD for $\dfrac{16a}{a^2b} + \dfrac{a-2}{2a(a-b)^2} - \dfrac{b+1}{4b^3(a-b)}$

1. $a^2b, \quad 2a(a-b)^2, \quad 2^2b^3(a-b)$
2. $2, \quad a, \quad b, \quad (a-b)$ — All the different factors
3. $2^2, \quad a^2, \quad b^3, (a-b)^2$ — Highest powers of factors
4. $\text{LCD} = 2^2a^2b^3(a-b)^2 = 4a^2b^3(a-b)^2$

Equivalent Fractions

Equivalent fractions are fractions that have the same value. If a fraction is multiplied by 1, its value is unchanged.

$$\left.\begin{array}{r} \dfrac{2}{2} = 1 \\ \dfrac{x}{x} = 1 \\ \dfrac{x+2}{x+2} = 1 \end{array}\right\}$$

Multiplying a fraction by expressions like these will produce equivalent fractions

For example:

$$\frac{5}{6} = \frac{5}{6} \cdot \frac{2}{2} = \frac{10}{12}$$ A fraction equivalent to $\frac{5}{6}$

$$\frac{x}{x+2} = \frac{x}{x+2} \cdot \frac{x}{x} = \frac{x^2}{x(x+2)}$$ A fraction equivalent to $\frac{x}{x+2}$

$$\frac{2}{x} = \frac{2}{x} \cdot \frac{x+2}{x+2} = \frac{2x+4}{x(x+2)}$$ A fraction equivalent to $\frac{2}{x}$

TO ADD UNLIKE FRACTIONS

1. Find the LCD.
2. Convert all fractions to equivalent fractions that have the LCD as denominator.
3. Add the resulting like fractions.
4. Reduce the resulting fraction to lowest terms.

Example 7 Add: $\dfrac{2}{x} + \dfrac{5}{x^2}$

1. $\text{LCD} = x^2$

We multiply numerator and denominator by x in order to obtain an equivalent fraction whose denominator is the LCD x^2

2. $\dfrac{2}{x} = \dfrac{2}{x} \cdot \dfrac{x}{x} = \dfrac{2x}{x^2}$

$\dfrac{5}{x^2}$ The denominator of this fraction is already the LCD

3. $\dfrac{2}{x} + \dfrac{5}{x^2} = \dfrac{2x}{x^2} + \dfrac{5}{x^2} = \dfrac{2x + 5}{x^2}$

Example 8 Add: $\dfrac{7}{18x^2y} + \dfrac{5}{8xy^4}$

1. LCD $= 72x^2y^4$ (see Example 4)

We multiply numerator and denominator by $4y^3$ in order to obtain an equivalent fraction whose denominator is the LCD $72x^2y^4$

2. $\dfrac{7}{18x^2y} = \dfrac{7}{18x^2y} \cdot \dfrac{4y^3}{4y^3} = \dfrac{28y^3}{72x^2y^4}$

$\dfrac{5}{8xy^4} = \dfrac{5}{8xy^4} \cdot \dfrac{9x}{9x} = \dfrac{45x}{72x^2y^4}$

We multiply numerator and denominator by $9x$ in order to obtain an equivalent fraction whose denominator is the LCD $72x^2y^4$

3. $\dfrac{7}{18x^2y} + \dfrac{5}{8xy^4} = \dfrac{28y^3}{72x^2y^4} + \dfrac{45x}{72x^2y^4} = \dfrac{28y^3 + 45x}{72x^2y^4}$

Example 9 Add: $\dfrac{2}{x} + \dfrac{x}{x + 2}$

1. LCD $= x(x + 2)$

We multiply numerator and denominator by $(x + 2)$ in order to obtain an equivalent fraction whose denominator is the LCD $x(x + 2)$

2. $\dfrac{2}{x} = \dfrac{2}{x} \cdot \dfrac{x + 2}{x + 2} = \dfrac{2(x + 2)}{x(x + 2)}$

$\dfrac{x}{x + 2} = \dfrac{x}{x + 2} \cdot \dfrac{x}{x} = \dfrac{x^2}{x(x + 2)}$

We multiply numerator and denominator by x in order to obtain an equivalent fraction whose denominator is the LCD $x(x + 2)$

3. $\dfrac{2}{x} + \dfrac{x}{x+2} = \dfrac{2(x+2)}{x(x+2)} + \dfrac{x^2}{x(x+2)} = \dfrac{2x+4+x^2}{x(x+2)} = \dfrac{x^2+2x+4}{x(x+2)}$

Example 10 Add: $3 - \dfrac{2a}{a+2}$

1. $\text{LCD} = a + 2$
2. $3 = \dfrac{3}{1} \cdot \dfrac{a+2}{a+2} = \dfrac{3(a+2)}{a+2} = \dfrac{3a+6}{a+2}$

   $\dfrac{2a}{a+2}$ The denominator of this fraction is already the LCD
3. $3 - \dfrac{2a}{a+2} = \dfrac{3a+6}{a+2} - \dfrac{2a}{a+2} = \dfrac{3a+6-2a}{a+2} = \dfrac{a+6}{a+2}$

EXERCISES 5-3

In Exercises 1–10 find the LCD.

1. $\dfrac{9}{25a^3} + \dfrac{7}{15a}$
2. $\dfrac{13}{18b^2} + \dfrac{11}{12b^4}$
3. $\dfrac{11}{2w-10} - \dfrac{15}{4w}$
4. $\dfrac{27}{2m^2} + \dfrac{19}{8m-48}$
5. $\dfrac{15b}{9b^2-c^2} + \dfrac{12c}{(3b-c)^2}$
6. $\dfrac{14e}{(5f-2e)^2} + \dfrac{27f}{4e^2-25f^2}$
7. $\dfrac{2x-5}{2x^2-16x+32} + \dfrac{4x+7}{x^2+x-20}$
8. $\dfrac{8k-1}{5k^2-30k+45} + \dfrac{3k-4}{k^2+4k-21}$
9. $\dfrac{x^2+1}{12x^3+24x^2} - \dfrac{4x+3}{x^2-4x+4} - \dfrac{1}{x^2-4}$
10. $\dfrac{2y+5}{y^2+6y+9} + \dfrac{7y}{y^2-9} - \dfrac{11}{8y^2-24y}$

In Exercises 11–56 add the fractions.

11. $\dfrac{8m}{2m-3n} - \dfrac{12n}{2m-3n}$
12. $\dfrac{21k}{4h+3k} + \dfrac{28h}{4h+3k}$
13. $\dfrac{-15w}{1-5w} - \dfrac{3}{5w-1}$
14. $\dfrac{-35}{6w-7} - \dfrac{30w}{7-6w}$
15. $\dfrac{7z}{8z-4} + \dfrac{6-5z}{4-8z}$
16. $\dfrac{13-30w}{15-10w} - \dfrac{10w+17}{10w-15}$
17. $\dfrac{9}{25a^3} + \dfrac{7}{15a}$
18. $\dfrac{13}{18b^2} + \dfrac{11}{12b^4}$
19. $\dfrac{49}{60h^2k^2} - \dfrac{71}{90hk^4}$
20. $\dfrac{44}{42x^2y^2} - \dfrac{45}{49x^3y}$
21. $\dfrac{1}{2} + \dfrac{3}{x} - \dfrac{5}{x^2}$
22. $\dfrac{2}{3} - \dfrac{1}{y} + \dfrac{4}{y^2}$
23. $3x - \dfrac{3}{x}$
24. $\dfrac{5}{y} - 4y$

25. $a + b - \frac{2ab}{a + b}$

26. $\frac{9}{x + 3} + x - 3$

27. $x - 3 - \frac{6x - 18}{x + 3}$

28. $\frac{4hk - 8h^2}{2h + k} - k + 2h$

29. $\frac{5}{t} + \frac{2t}{t - 4}$

30. $\frac{6r}{r - 8} - \frac{11}{r}$

31. $\frac{3k}{8k - 4} - \frac{7}{6k}$

32. $\frac{2}{9j} + \frac{4j}{18j + 12}$

33. $x + \frac{2}{x} - \frac{3}{x - 2}$

34. $m - \frac{3}{m} + \frac{2}{m + 4}$

35. $\frac{a + b}{b} + \frac{b}{a - b}$

36. $\frac{x - y}{x} - \frac{x}{x + y}$

37. $\frac{2}{a + 3} - \frac{4}{a - 1}$

38. $\frac{5}{b - 2} - \frac{3}{b + 4}$

39. $\frac{x + 2}{x - 3} - \frac{x + 3}{x - 2}$

40. $\frac{x - 4}{x + 6} - \frac{x - 6}{x + 4}$

41. $\frac{y}{x^2 - xy} - \frac{x}{xy - y^2}$

42. $\frac{b}{ab - a^2} + \frac{a}{ab - b^2}$

43. $\frac{x + 2}{x^2 + x - 2} + \frac{3}{x^2 - 1}$

44. $\frac{5}{x^2 - 4} + \frac{x + 1}{x^2 - x - 2}$

45. $\frac{2x}{x - 3} - \frac{2x}{x + 3} + \frac{36}{x^2 - 9}$

46. $\frac{x}{x + 4} - \frac{x}{x - 4} - \frac{32}{x^2 - 16}$

47. $\frac{x - 2}{x^2 + 4x + 4} - \frac{x + 1}{x^2 - 4}$

48. $\frac{x - 2}{x^2 - 1} - \frac{x + 1}{x^2 - 2x + 1}$

49. $\frac{4}{x^2 + 2x + 4} + \frac{x - 2}{x + 2}$

50. $\frac{x + 3}{x - 9} + \frac{3}{x^2 - 3x + 9}$

51. $\frac{5}{2g^3} - \frac{3g - 9}{g^2 - 6g + 9} + \frac{12g}{4g^2 - 12g}$

52. $\frac{5y - 30}{y^2 - 12y + 36} + \frac{7}{9y^2} - \frac{15y}{3y^2 - 18y}$

53. $\frac{2x - 5}{2x^2 - 16x + 32} + \frac{4x + 7}{x^2 + x - 20}$

54. $\frac{8k - 1}{5k^2 - 30k + 45} + \frac{3k - 4}{k^2 + 4k - 21}$

55. $\frac{35}{3e^2} - \frac{2e}{e^2 - 9} - \frac{3}{4e - 12}$

56. $\frac{3}{8u^3} - \frac{5u - 1}{6u^2 + 18u} - \frac{6u + 7}{u^2 - 5u - 24}$

5-4 Complex Fractions

Simple Fractions

A **simple fraction** is a fraction that has only one fraction line.

Examples: $\frac{2}{x}$, $\frac{3 + y}{12}$, $\frac{7a - 7b}{ab^2}$, $\frac{5}{x + y}$

Complex Fractions

A **complex fraction** is a fraction that has more than one fraction line.

Examples: $\dfrac{\frac{2}{x}}{3}$, $\dfrac{a}{\frac{1}{c}}$, $\dfrac{\frac{3}{z}}{\frac{5}{z}}$, $\dfrac{\frac{3}{x}-\frac{2}{y}}{\frac{5}{x}+\frac{3}{y}}$

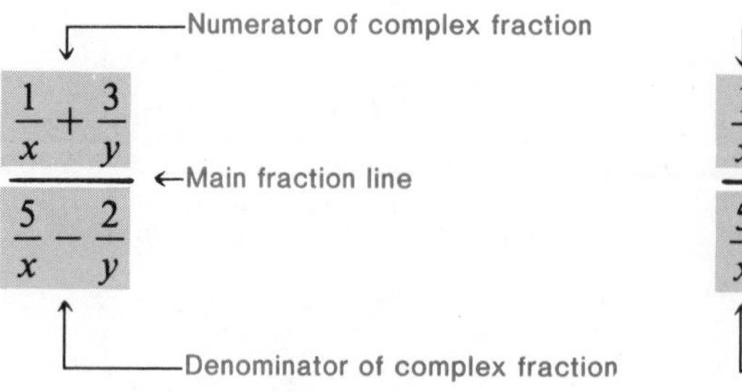

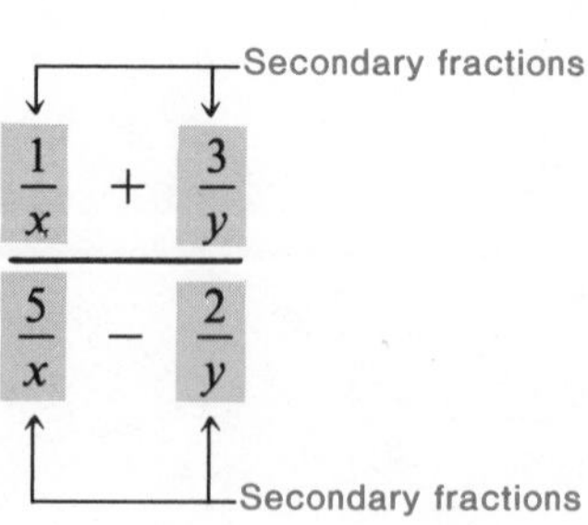

Methods for Simplifying Complex Fractions

> **TO SIMPLIFY COMPLEX FRACTIONS**
>
> *Method 1.* Multiply both numerator and denominator of the complex fraction by the LCD of the secondary fractions; then simplify the results.
>
> *Method 2.* First, simplify the numerator and denominator of the complex fraction; then divide the simplified numerator by the simplified denominator.

Note that in some of the examples below, the solution by Method 1 is easier than that by Method 2. In others, the opposite is true.

Example 1 Simplify: $\dfrac{\frac{4b^2}{9a^2}}{\frac{8b}{3a^3}}$ ← Main fraction line

Method 1. The LCD of the secondary denominators $9a^2$ and $3a^3$ is $9a^3$.

$$\frac{9a^3}{9a^3}\left(\frac{\frac{4b^2}{9a^2}}{\frac{8b}{3a^3}}\right) = \frac{\frac{9a^3}{1}\left(\frac{4b^2}{9a^2}\right)}{\frac{9a^3}{1}\left(\frac{8b}{3a^3}\right)} = \frac{4ab^2}{24b} = \frac{ab}{6}$$

The value of this fraction is 1

Method 2. $\dfrac{\dfrac{4b^2}{9a^2}}{\dfrac{8b}{3a^3}} = \dfrac{4b^2}{9a^2} \div \dfrac{8b}{3a^3} = \dfrac{\overset{1}{\cancel{4}}b^2}{\underset{3}{\cancel{9}}a^2} \cdot \dfrac{\overset{1}{\cancel{3}}a^3}{\underset{2}{\cancel{8}}b} = \dfrac{ab}{6}$

Example 2 Simplify: $\dfrac{\dfrac{2}{x} - \dfrac{3}{x^2}}{5 + \dfrac{1}{x}}$

Method 1. The LCD of the secondary denominators x and x^2 is x^2.

$$\frac{x^2}{x^2}\left(\frac{\dfrac{2}{x} - \dfrac{3}{x^2}}{5 + \dfrac{1}{x}}\right) = \frac{\dfrac{x^2}{1}\left(\dfrac{2}{x}\right) - \dfrac{x^2}{1}\left(\dfrac{3}{x^2}\right)}{\dfrac{x^2}{1}\left(\dfrac{5}{1}\right) + \dfrac{x^2}{1}\left(\dfrac{1}{x}\right)} = \frac{2x - 3}{5x^2 + x}$$

Method 2.
$$\frac{\dfrac{2}{x} - \dfrac{3}{x^2}}{5 + \dfrac{1}{x}} = \left(\frac{2}{x} - \frac{3}{x^2}\right) \div \left(5 + \frac{1}{x}\right)$$

$$= \left(\frac{2x}{x^2} - \frac{3}{x^2}\right) \div \left(\frac{5x}{x} + \frac{1}{x}\right)$$

$$= \frac{2x - 3}{x^2} \div \frac{5x + 1}{x}$$

$$= \frac{2x - 3}{x^2} \cdot \frac{x}{5x + 1} = \frac{2x - 3}{5x^2 + x}$$

Example 3 Simplify: $\dfrac{x - \dfrac{4}{x}}{x - \dfrac{2}{x + 1}}$

Method 1. The LCD of the secondary denominators x and $(x + 1)$ is $x(x + 1)$.

$$\frac{x(x + 1)}{x(x + 1)}\left(\frac{x - \dfrac{4}{x}}{x - \dfrac{2}{x + 1}}\right) = \frac{\dfrac{x(x + 1)}{1} \cdot \dfrac{x}{1} - \dfrac{x(x + 1)}{1} \cdot \dfrac{4}{x}}{\dfrac{x(x + 1)}{1} \cdot \dfrac{x}{1} - \dfrac{x(x + 1)}{1} \cdot \dfrac{2}{(x + 1)}}$$

$$= \frac{x^2(x + 1) - 4(x + 1)}{x^2(x + 1) - 2x} = \frac{(x + 1)(x^2 - 4)}{x^3 + x^2 - 2x}$$

$$= \frac{(x + 1)\cancel{(x + 2)}(x - 2)}{x(x - 1)\cancel{(x + 2)}} = \frac{(x + 1)(x - 2)}{x(x - 1)}$$

Method 2. This method is left to the student.

EXERCISES 5-4

Simplify each complex fraction.

1. $\dfrac{\frac{21m^3n}{14mn^2}}{\frac{20m^2n^2}{8mn^3}}$

2. $\dfrac{\frac{10a^2b}{12a^4b^3}}{\frac{5ab^2}{16a^2b^3}}$

3. $\dfrac{\frac{15h-6}{18h}}{\frac{30h^2-12h}{8h}}$

4. $\dfrac{\frac{9k^4}{20k^2-35k^3}}{\frac{12k}{16-28k}}$

5. $\dfrac{\frac{c}{d}+2}{\frac{c^2}{d^2}-4}$

6. $\dfrac{\frac{x^2}{y^2}-1}{\frac{x}{y}-1}$

7. $\dfrac{a+2-\frac{9}{a+2}}{a+1+\frac{a-7}{a+2}}$

8. $\dfrac{x-3+\frac{x-3}{x+2}}{x+4-\frac{4x+23}{x+2}}$

9. $\dfrac{\frac{x+y}{y}+\frac{y}{x-y}}{\frac{y}{x-y}}$

10. $\dfrac{\frac{a-b}{a}-\frac{a}{a+b}}{\frac{b^2}{a+b}}$

11. $\dfrac{\frac{x}{x+1}+\frac{4}{x}}{\frac{x}{x+1}-2}$

12. $\dfrac{\frac{4x}{4x+1}+\frac{1}{x}}{\frac{2}{4x+1}+2}$

13. $\dfrac{\frac{x+4}{x}-\frac{3}{x-1}}{x+1+\frac{2x+1}{x-1}}$

14. $\dfrac{\frac{1}{x}+\frac{x}{x-6}}{x-1-\frac{12}{x}}$

15. $\dfrac{4x^{-2}-y^{-2}}{2x^{-1}+y^{-1}}$

16. $\dfrac{x^{-2}-9y^{-2}}{x^{-1}-3y^{-1}}$

17. $\dfrac{5c^{-1}+2d^{-1}}{25c^{-2}-4d^{-2}}$

18. $\dfrac{2x-5-3x^{-1}}{x^{-1}+x(x-12)^{-1}}$

19. $\dfrac{\frac{x-2}{x+2}-\frac{x+2}{x-2}}{\frac{x-2}{x+2}+\frac{x+2}{x-2}}$

20. $\dfrac{\frac{m+3}{m-3}+\frac{m-3}{m+3}}{\frac{m+3}{m-3}-\frac{m-3}{m+3}}$

21. $\dfrac{\frac{2x+y}{x}+\frac{3x+y}{y-x}}{\frac{x+y}{y}+\frac{2(x+y)}{x-y}}$

22. $\dfrac{\frac{a+b}{b}+\frac{2(a+b)}{a-b}}{\frac{2a-b}{a}-\frac{5b-a}{b-a}}$

23. $\dfrac{1}{x+\frac{1}{x+\frac{1}{x+x}}}$

24. $\dfrac{2}{y+\frac{2}{y+\frac{2}{y+y}}}$

5–5 Solving Equations that Have Fractions

In Section 2–1 we removed fractions from simple equations by multiplying both sides by the same *number*. In this section we sometimes multiply both sides of an equation by an *expression* containing the unknown *letter*. This procedure *can* lead to an equation that is *not equivalent* to the one given to be solved.

Equivalent Equations

Equivalent equations are equations that have identical solution sets.

Equivalent Equations Are Obtained When:

1. The same number or expression is added to or subtracted from both sides of the given equation (see Example 1).
2. Both sides of the given equation are multiplied or divided by the same nonzero number (see Examples 1 and 2).

Nonequivalent Equations May Be Obtained When:

1. Both sides of the given equation are multiplied by an expression containing the unknown letter. In this case some roots may be introduced that are not roots of the given equation (see Example 3). This makes it necessary to *check all roots in the given equation.*
2. Both sides of the given equation are divided by an expression containing the unknown letter. In this case some roots of the given equation may be lost (see Example 4).

Example 1 Multiplying both sides by a *number* gives an equivalent equation

(1) $x - 1 = 0$ Solution set = $\{1\}$

$5(x - 1) = 5(0)$ Multiplied both sides by 5

(2) $5x - 5 = 0$

$5x = 5$ The result of adding 5 to both sides

$x = 1$ Solution set = $\{1\}$; the result of dividing both sides by 5

Equations (1) and (2) are equivalent because they have the same solution set.

Example 2 Dividing both sides by a *number* gives an equivalent equation

(1) $2(x + 3) = 0$ Solution set = $\{-3\}$

$\frac{2(x + 3)}{2} = \frac{0}{2}$ Divided both sides by 2

(2) $x + 3 = 0$ Solution set = $\{-3\}$

Equations (1) and (2) are equivalent because they have the same solution set.

Extraneous Roots

An **extraneous root** is a number obtained in the process of solving an equation that is not a root of the original equation (see Example 3).

Example 3 Multiplying both sides by an expression containing the unknown *may* give a non-equivalent equation

(1) $x - 1 = 0$ Solution set = $\{1\}$

$(x + 2)(x - 1) = (x + 2) \cdot 0$ Multiplied both sides by $(x + 2)$

(2) $(x + 2)(x - 1) = 0$ Solution set = $\{-2, 1\}$ (Extraneous root: -2)

Equations (1) and (2) are *not equivalent* because they have different solution sets.

Example 4 Dividing both sides by an expression containing the unknown may give a non-equivalent equation

(1) $(x - 5)(x + 4) = 0$ Solution set = $\{5, -4\}$

$$\frac{(x-5)(x+4)}{(x-5)} = \frac{0}{(x-5)}$$ Divided both sides by $(x - 5)$

(2) $x + 4 = 0$ Solution set = $\{-4\}$

Equations (1) and (2) are *not equivalent* because they have different solution sets. For this reason *do not divide both sides of an equation by an expression containing the unknown.*

TO SOLVE AN EQUATION THAT HAS FRACTIONS

1. **Remove fractions** by multiplying each term by the LCD.
2. **Remove grouping symbols** (see Section 1–9C).
3. **Combine like terms.**

If a *first-degree equation* is obtained in Step 3:	If a *second-degree equation* (quadratic) is obtained in Step 3:
4. Get all terms with the unknown on one side and the remaining terms on the other side.	**4.** Get all nonzero terms on one side and arrange them in descending powers. *Only zero must remain on the other side.*
5. Divide both sides by the coefficient of the unknown.	**5.** Factor the polynomial.
	6. Set each factor equal to zero and the solve for the unknown.

Check apparent solutions in the original equation. Any value of a letter that makes any denominator in the equation zero is not a solution.

Example 5 Solve: $\dfrac{x-4}{2} - \dfrac{x}{5} = \dfrac{1}{10}$

Multiply both sides by the LCD 10.

$$\frac{\overset{5}{\cancel{10}}}{1}\left(\frac{x-4}{\underset{1}{\cancel{2}}}\right) - \frac{\overset{2}{\cancel{10}}}{1}\left(\frac{x}{\underset{1}{\cancel{5}}}\right) = \frac{\cancel{10}}{1}\left(\frac{1}{\cancel{10}}\right)$$

$$5(x-4) - 2x = 1$$

$$5x - 20 - 2x = 1$$

$$3x = 21$$

$$x = 7$$

Check for $x = 7$:

$$\frac{x-4}{2} - \frac{x}{5} = \frac{1}{10}$$

$$\frac{7-4}{2} - \frac{7}{5} \stackrel{?}{=} \frac{1}{10}$$

$$\frac{3}{2} - \frac{7}{5} \stackrel{?}{=} \frac{1}{10}$$

$$\frac{15}{10} - \frac{14}{10} = \frac{1}{10}$$

Example 6 Solve: $\dfrac{3}{x+1} - \dfrac{2}{x} = \dfrac{5}{2x}$

LCD $= 2x(x+1)$

$$\frac{2x\cancel{(x+1)}}{1}\cdot\frac{3}{\cancel{(x+1)}} - \frac{2\cancel{x}(x+1)}{1}\cdot\frac{2}{\cancel{x}} = \frac{\cancel{2x}(x+1)}{1}\cdot\frac{5}{\cancel{2x}}$$

$$6x - 4(x+1) = 5(x+1)$$

$$6x - 4x - 4 = 5x + 5$$

$$2x - 4 = 5x + 5$$

$$-9 = 3x$$

$$-3 = x$$

Check for $x = -3$:

$$\frac{3}{x+1} - \frac{2}{x} = \frac{5}{2x}$$

$$\frac{3}{-3+1} - \frac{2}{-3} \stackrel{?}{=} \frac{5}{2(-3)}$$

$$\frac{3}{-2} - \frac{2}{-3} \stackrel{?}{=} \frac{5}{-6}$$

$$-\frac{9}{6} + \frac{4}{6} = -\frac{5}{6}$$

Example 7 Solve: $\dfrac{x}{x-3} = \dfrac{3}{x-3} + 4$

Note: Before you try to solve an equation, note all excluded values of the variable by inspection. In this example, 3 is an excluded value because it makes the denominators $(x - 3)$ equal to zero. Therefore 3 cannot be a solution of this equation. ■

LCD $= x - 3$

$$\frac{\cancel{(x-3)}}{1} \cdot \frac{x}{\cancel{(x-3)}} = \frac{\cancel{(x-3)}}{1} \cdot \frac{3}{\cancel{(x-3)}} + \frac{(x-3)}{1} \cdot \frac{4}{1}$$

$$x = 3 + 4(x-3)$$

$$x = 3 + 4x - 12$$

$$x = 4x - 9$$

$$9 = 3x$$

$$3 = x$$

Since 3 is an excluded value, this equation has *no solution*

Extraneous root

Check: If we try to check the value 3 in the equation, we obtain numbers that are not real numbers.

$$\frac{x}{x-3} = \frac{3}{x-3} + 4$$

$$\frac{3}{3-3} \stackrel{?}{=} \frac{3}{3-3} + 4$$

$$\frac{3}{0} \stackrel{?}{=} \frac{3}{0} + 4$$

Not a real number

Therefore 3 is not a root of the given equation. This equation has no real roots.

When an equation has only one fraction on each side of the equal sign, it is a *proportion.*

Example 8 Solve: $\dfrac{x+1}{x-2} = \dfrac{7}{4}$

$$\frac{x+1}{x-2} = \frac{7}{4}$$ This is a proportion

$$4(x+1) = 7(x-2)$$ Product of means = Product of extremes

$$4x + 4 = 7x - 14$$ Removed ( )

$$18 = 3x$$

$$6 = x$$ Divided both sides by 3

After removing fractions and grouping symbols, there may be second-degree terms. When this is the case, the equation can sometimes be solved by factoring.

Example 9 Solve: $\frac{2}{x}+\frac{3}{x^2}=1$

LCD $= x^2$

$$\frac{x^2}{1}\left(\frac{2}{x}\right)+\frac{x^2}{1}\left(\frac{3}{x^2}\right)=\frac{x^2}{1}\left(\frac{1}{1}\right)$$

Second-degree term

$$2x+3=x^2$$
$$0=x^2-2x-3$$
$$0=(x-3)(x+1)$$
$$x-3=0 \quad\mid\quad x+1=0$$
$$x=3 \quad\mid\quad x=-1$$

Check for $x = 3$:

$$\frac{2}{x}+\frac{3}{x^2}=1$$
$$\frac{2}{3}+\frac{3}{3^2}\stackrel{?}{=}1$$
$$\frac{2}{3}+\frac{1}{3}=1$$

Check for $x = -1$:

$$\frac{2}{x}+\frac{3}{x^2}=1$$
$$\frac{2}{-1}+\frac{3}{(-1)^2}\stackrel{?}{=}1$$
$$-2+3=1$$

A Word of Caution. A common mistake students make is to confuse *an equation* such as $\frac{2}{x}+\frac{3}{x^2}=1$ with *a sum* such as $\frac{2}{x}+\frac{3}{x^2}$.

The Equation

Both sides are multiplied by the LCD to remove fractions.

$$\frac{2}{x}+\frac{3}{x^2}=1 \qquad \text{LCD} = x^2$$
$$\frac{x^2}{1}\cdot\frac{2}{x}+\frac{x^2}{1}\cdot\frac{3}{x^2}=\frac{x^2}{1}\cdot\frac{1}{1}$$
$$2x \quad + \quad 3 \quad = x^2$$

This equation is then solved by factoring (see Example 9). Here, the result is two numbers (-1 and 3) that make both sides of the given equation equal.

The Sum

Each fraction is changed into an equivalent fraction that has the LCD for a denominator.

$$\frac{2}{x}+\frac{3}{x^2} \qquad \text{LCD} = x^2$$

This fraction is 1

$$=\frac{2}{x}\cdot\frac{x}{x}+\frac{3}{x^2}$$
$$=\frac{2x}{x^2}+\frac{3}{x^2}=\frac{2x+3}{x^2}$$

Here, the result is a fraction that represents the sum of the given fractions.

The usual mistake made is to multiply both terms of *the sum* by the LCD.

$$\frac{x^2}{1}\cdot\frac{2}{x}+\frac{x^2}{1}\cdot\frac{3}{x^2}$$
$$= 2x + 3 \neq \frac{2}{x}+\frac{3}{x^2}$$

The sum has been multiplied by x^2, and therefore is no longer equal to its original value. ■

EXERCISES 5-5

Solve the equations.

1. $\frac{a-4}{5}+\frac{2a}{3}=\frac{4}{5}$
2. $\frac{5p}{6}+\frac{p+2}{3}=\frac{7}{9}$
3. $\frac{3u}{7}+\frac{6-u}{14}=\frac{1}{4}$
4. $\frac{9f-5}{6}-\frac{3f-11}{21}=\frac{1}{7}$
5. $\frac{2c+5}{9}-\frac{4c-7}{4}=\frac{11}{12}$
6. $\frac{3d-4}{8}-\frac{d+2}{3}=-\frac{3}{2}$
7. $\frac{2}{k-5}-\frac{5}{k}=\frac{3}{4k}$
8. $\frac{4}{h}-\frac{6}{h-7}=\frac{2}{3h}$
9. $\frac{7}{2t-3}-\frac{4}{5t}=\frac{3}{t}$
10. $\frac{8}{6+5e}=3-\frac{15e}{6+5e}$
11. $\frac{x}{x-2}=\frac{2}{x-2}+5$
12. $\frac{x}{x+5}=4-\frac{5}{x+5}$
13. $\frac{12m}{2m-3}=6+\frac{18}{2m-3}$
14. $\frac{40w}{5w+6}=8-\frac{48}{5w+6}$
15. $\frac{36}{4-3d}=9+\frac{27d}{4-3d}$
16. $\frac{8}{7m}=\frac{6}{10m-3}$
17. $\frac{2w}{3(w-4)}=\frac{6}{11}$
18. $\frac{4a}{33-5a}=\frac{2}{a}$
19. $\frac{9}{3g-7}=\frac{2}{5g}$
20. $\frac{7}{9+2q}=\frac{5}{4q}$
21. $\frac{5z}{2(z-6)}=\frac{5}{12}$
22. $\frac{3x}{4(x-1)}=\frac{9}{8}$
23. $\frac{2y}{7y+5}=\frac{1}{3y}$
24. $\frac{2b}{3-4b}=\frac{1}{2b}$
25. $\frac{3e-5}{4e}=\frac{e}{2e+3}$
26. $\frac{2x-1}{3x}=\frac{3}{2x+7}$
27. $\frac{4y-3}{2}=\frac{5y}{y+2}$
28. $\frac{1}{2}-\frac{1}{6x}=\frac{7}{3x^2}$
29. $\frac{2}{x}-\frac{2}{x^2}=\frac{1}{2}$
30. $\frac{7}{4}-\frac{17}{4x}=\frac{3}{x^2}$
31. $\frac{1}{x-1}+\frac{2}{x+1}=\frac{5}{3}$
32. $\frac{2}{3x+1}+\frac{1}{x-1}=\frac{7}{10}$
33. $\frac{4}{x+1}=\frac{3}{x}+\frac{1}{15}$
34. $\frac{1}{x+1}=\frac{3}{x}+\frac{1}{2}$
35. $\frac{6}{x+4}=\frac{5}{x+3}+\frac{4}{x}$
36. $\frac{7}{x+5}=\frac{3}{x-1}-\frac{4}{x}$
37. $\frac{3}{1-2x}+\frac{5}{2-x}=2$
38. $\frac{4}{3x-1}=\frac{2}{x}+1$
39. $\frac{11}{2-3x}=\frac{6}{x+6}+\frac{3}{x}$
40. $\frac{x+25}{x^2-25}=1-\frac{12x}{x-5}$
41. $\frac{6}{x^2-9}+\frac{1}{5}=\frac{1}{x-3}$
42. $\frac{6-x}{x^2-4}-2=\frac{x}{x+2}$

5–6 Literal Equations

Literal equations are equations that have more than one letter.

Formulas

Formulas are literal equations that have meaning in real-life situations.

Example 1 Literal equations

Literal equations that *are not formulas:*

(a) $3x - 4y = 7$ (b) $\dfrac{4ab}{d} = 15$

Literal equations that *are formulas:*

(c) $A = P(1 + rt)$ Formula from business

(d) $I = \dfrac{nE}{R + nr}$ Formula from physics

Example 2 When we solve a literal equation for one of its letters, the solution will contain the other letters as well as numbers.

(a) Solve $x + y = 5$ for x

$$x + y = 5$$

$$x = 5 - y \quad \text{Solution for } x$$

(b) Solve $x + y = 5$ for y

$$x + y = 5$$

$$y = 5 - x \quad \text{Solution for } y$$

TO SOLVE A LITERAL EQUATION

1. *Remove fractions* (if there are any) by multiplying both sides by the LCD.
2. *Remove grouping symbols* (if there are any).
3. *Collect like terms:* all terms containing the letter you are solving for on one side, all other terms on the other side.
4. *Factor out the letter you are solving for* (if it appears in more than one term).
5. *Divide both sides by the coefficient of the letter you are solving for.*

Example 3 Solve $A = P(1 + rt)$ for t

$$A = P(1 + rt)$$

$$A = P + Prt$$ Removed ( ) by using the distributive rule

$$A - P = Prt$$ Collected terms with the letter being solved for (t) on one side and all other terms on the other side

$$\frac{A - P}{Pr} = \frac{\cancel{Pr}t}{\cancel{Pr}}$$ Divided both sides by Pr

$$\frac{A - P}{Pr} = t$$ Solution

Example 4 Solve $\frac{1}{F} = \frac{1}{u} + \frac{1}{v}$ for u

LCD $= Fuv$

$$\frac{Fuv}{1} \cdot \frac{1}{\cancel{F}} = \frac{F\cancel{u}v}{1} \cdot \frac{1}{\cancel{u}} + \frac{Fu\cancel{v}}{1} \cdot \frac{1}{\cancel{v}}$$ Removed fractions by multiplying both sides by the LCD

$$uv = Fv - Fu$$

$$uv - Fu = Fv$$ Collected terms with the letter being solved for (u) on one side and all other terms on the other side

$$u(v - F) = Fv$$ Removed u as a common factor

$$\frac{u\cancel{(v - F)}}{\cancel{(v - F)}} = \frac{Fv}{(v - F)}$$ Divided both sides by $(v - F)$

$$u = \frac{Fv}{v - F}$$ Solution

Example 5 Solve $I = \frac{nE}{R + nr}$ for n Sometimes a literal equation can be solved as a proportion

$$\frac{I}{1} = \frac{nE}{R + nr}$$ This is a proportion

$$1(nE) = I(R + nr)$$ Product of means = Product of extremes

$$nE = IR + Inr$$ Removed ( ) using distributive rule

$$nE - Inr = IR$$ Collected terms with the letter being solved for (n) on one side and all other terms on the other side

$$n(E - Ir) = IR$$ Removed n as a common factor

$$\frac{n\cancel{(E - Ir)}}{\cancel{(E - Ir)}} = \frac{IR}{(E - Ir)}$$ Divided both sides by $(E - Ir)$

$$n = \frac{IR}{E - Ir}$$ Solution

EXERCISES 5-6

Solve for the letter listed after each equation.

1. $2(3x - y) = xy - 12;\ y$

2. $2(3x - y) = xy - 12;\ x$

3. $2x - y = \dfrac{2xy + 25}{5};\ x$

4. $I = \dfrac{nE}{R + nr};\ n$

5. $z = \dfrac{x - m}{s};\ m$

6. $s^2 = \dfrac{N - n}{N - 1};\ N$

7. $C = \dfrac{5}{9}(F - 32);\ F$

8. $A = \dfrac{h}{2}(B + b);\ B$

9. $s = c\left(1 + \dfrac{a}{c}\right);\ c$

10. $Z = \dfrac{Rr}{R + r};\ R$

11. $A = P(1 + rt);\ r$

12. $S = \dfrac{1}{2}g(2t - 1);\ t$

13. $\dfrac{2}{15ab - 8a^2} = \dfrac{7}{20a^2 - 12ab};\ b$

14. $\dfrac{1}{a} = \dfrac{2}{r} - v^2;\ r$

15. $v^2 = \dfrac{2}{r} - \dfrac{1}{a};\ a$

16. $\dfrac{1}{p} = 1 + \dfrac{1}{s};\ s$

17. $\dfrac{1}{R} = \dfrac{1}{r_1} + \dfrac{1}{r_2};\ r_1$

18. $\dfrac{1}{R} = \dfrac{1}{r_1} + \dfrac{1}{r_2} + \dfrac{1}{r_3};\ R$

19. $C = \dfrac{a}{1 + \dfrac{a}{\pi A}};\ A$

20. $E = \dfrac{k(1 - A)\pi R^2}{r^2};\ A$

21. $C = \dfrac{a}{1 + \dfrac{a}{\pi A}};\ a$

22. $R = \dfrac{V + v}{1 + \dfrac{vV}{c^2}};\ v$

5-7 Word Problems Leading to Equations that Have Fractions

Work Problems

In this section we introduce a new type of word problem called **work problems.** One basic idea used to solve work problems is:

Rate × Time = Amount of work

For example, suppose Jim can assemble a radio in 3 days.

$$\text{Jim's } rate = \frac{1 \text{ radio}}{3 \text{ days}} = \frac{1}{3} \text{ radio per day}$$

If his working *time* is 5 days; then

$$\text{Rate} \cdot \text{Time} = \text{Amount of work}$$

$$\frac{1}{3} \cdot 5 = \frac{5}{3} \text{ radios}$$

The other basic idea used to solve work problems is:

> If two people, A and B, work together,
>
> then $\begin{pmatrix}\text{Amount A}\\ \text{does}\end{pmatrix} + \begin{pmatrix}\text{Amount B}\\ \text{does}\end{pmatrix} = \begin{pmatrix}\text{Amount done}\\ \text{together}\end{pmatrix}$

Example 1 Albert can paint a house in 6 days. Ben can do the same job in 8 days. How long would it take them to paint the same house if they work together?

$$\text{Albert's rate} = \frac{1 \text{ house}}{6 \text{ days}} = \frac{1}{6} \text{ house per day}$$

$$\text{Ben's rate} = \frac{1 \text{ house}}{8 \text{ days}} = \frac{1}{8} \text{ house per day}$$

Let x = Number of days to paint the house working together.

Then

$$\text{Albert's rate} \cdot \text{Albert's time} = \text{Amount Albert paints}$$

$$\frac{1}{6} \cdot x = \frac{x}{6}$$

$$\text{Ben's rate} \cdot \text{Ben's time} = \text{Amount Ben paints}$$

$$\frac{1}{8} \cdot x = \frac{x}{8}$$

Therefore

$$\text{Amount Albert paints} + \text{Amount Ben paints} = \text{Amount painted together}$$

$$\frac{x}{6} + \frac{x}{8} = 1$$

(One house painted)

$$\text{LCD} = 24 \qquad \frac{24}{1}\cdot\frac{x}{\not{6}} + \frac{24}{1}\cdot\frac{x}{\not{8}} = \frac{24}{1}\cdot\frac{1}{1}$$

$$4x + 3x = 24$$

$$7x = 24$$

$$x = \frac{24}{7} = 3\frac{3}{7}\text{ days}$$

Therefore it takes them $3\frac{3}{7}$ days to paint the house when they work together.

Example 2 It takes pipe 1 twelve minutes to fill a particular tank. It takes pipe 2 only eight minutes to fill the same tank. Pipe 3 takes six minutes to *empty* the same tank. How long does it take to fill the tank when all three pipes are open?

$$\text{Pipe 1's rate} = \frac{1\text{ tank}}{12\text{ min}} = \frac{1}{12}\text{ tank per minute}$$

$$\text{Pipe 2's rate} = \frac{1\text{ tank}}{8\text{ min}} = \frac{1}{8}\text{ tank per minute}$$

$$\text{Pipe 3's rate} = \frac{1\text{ tank}}{6\text{ min}} = \frac{1}{6}\text{ tank per minute}$$

Let x = Time for all three together to fill tank

Amount 1 does	+	Amount 2 does	−	Amount 3 does	=	1 full tank

↑ Because pipe 3 *empties* the tank

$$\frac{1}{12}x + \frac{1}{8}x - \frac{1}{6}x = 1$$

LCD = 24

$$\frac{\overset{2}{24}}{1}\cdot\frac{x}{\not{12}} + \frac{\overset{3}{24}}{1}\cdot\frac{x}{\not{8}} - \frac{\overset{4}{24}}{1}\cdot\frac{x}{\not{6}} = \frac{24}{1}\cdot\frac{1}{1}$$

$$2x + 3x - 4x = 24$$

$$x = 24\text{ min}$$

Therefore it takes 24 min to fill the tank when all three pipes are open.

EXERCISES 5-7

In this problem set, in addition to work problems, we have included word problems like those discussed in Chapter 3 whose equations contain fractions.

1. Henry can paint a house in 5 days and Jean can do the same work in 4 days. If they work together, how many days will it take them to paint the house?

2. Juan can paint a house in 6 days and George can do it in 8 days. How many days will it take the two men working together to paint the house?

3. Merwin can wax all his floors in 4 hr. His wife can do that same job in 3 hr. How long will it take them working together?

4. It takes Bert 8 hr to clean the yard and mow the lawn. He and his brother Howard can do that same job in 3 hr working together. How long does it take Howard working alone?

5. Abby can type 100 pages of manuscript in 3 hr. How long does it take David to type 80 pages if he and Abby working together can type a 500-page manuscript in 10 hr?

6. Carlos can wrap 200 newspapers in 55 min. How long does it take Bernice to wrap 150 papers if she and Carlos working together can wrap 400 papers in 1 hr?

7. Machine A takes 36 hr to do a job that machine B does in 24 hr. If machine A runs for 12 hr before machine B is turned on, how long will it take both machines running together to finish the job?

8. Machine A takes 18 hr to do a job that machine B does in 15 hr. If machine A runs for 3 hr before machine B is turned on, how long will it take both machines running together to finish the job?

9. Two numbers differ by 8. One-fourth the larger is 1 more than one-third the smaller. Find the numbers.

10. Two numbers differ by 6. One-fifth the smaller exceeds one-half the larger by 3. Find the numbers.

11. It takes one plane one-half hour longer than another to fly a certain distance. Find the distance if one plane travels 500 mph and the other 400 mph.

12. It takes Barbara 10 min longer than Ed to ride her bicycle a certain distance. Find the distance if Barbara's speed is 9 mph and Ed's is 12 mph.

13. An automobile radiator contains 14 qt of a 45% solution of antifreeze. How much must be drained out and replaced by pure antifreeze to make a 50% solution?

14. An automobile radiator contains 12 qt of a 30% solution of antifreeze. How much must be drained out and replaced with pure antifreeze to make a 50% solution?

15. One-fifth of the 10% antifreeze solution in a car radiator is drained and replaced by an antifreeze solution of unknown concentration. If the resulting mixture is 25% antifreeze, what was the unknown concentration?

16. The units digit of a two-digit number is 1 more than the tens digit. The product of the digits divided by the sum of the digits is $12/7$. Find the number.

17. The tens digit of a two-digit number is 1 more than its units digit. The product of the digits divided by the sum of the digits is $6/5$. Find the number.

18. The tens digit of a two-digit number is 2 more than its units digit. The ratio of the product of the digits to the sum of the digits is $12:5$. Find the number.

19. Pipe 2 takes 1 hr longer to fill a particular tank than pipe 1. Pipe 3 can drain the tank in 2 hr. If it takes 3 hr to fill the tank when all three pipes are open, how long does it take pipe 1 to fill the tank alone?

20. Pipe 2 takes 2 hr more time to fill a particular tank than pipe 1. Pipe 3 can drain the tank in 4 hr. If it takes 2 hr to fill the tank when all three pipes are open, how long does it take pipe 1 to fill the tank alone?

21. Three times a fraction plus one-third that fraction, times that fraction, equals that fraction. What is the fraction?

Chapter Five Summary

To Reduce a Fraction to Lowest Terms. (Section 5–1)

1. Factor the numerator and denominator completely.
2. Divide the numerator and denominator by all factors common to both.

To Multiply Fractions. (Section 5–2)

1. Factor the numerator and denominator of the fractions.
2. Divide the numerator and denominator by all factors common to both.
3. The answer is the product of factors remaining in the numerator divided by the product of factors remaining in the denominator. A factor of 1 will always remain in both numerator and denominator.

To Divide Fractions. (Section 5–2) Invert the second fraction and multiply.

$$\frac{a}{b} \div \frac{c}{d} = \frac{a}{b} \cdot \frac{d}{c}$$

First fraction ($\frac{a}{b}$) — Second fraction ($\frac{c}{d}$)

To Find the LCD. (Section 5–3)

1. Factor each denominator completely. Repeated factors should be expressed as powers.
2. Write down each different factor that appears.
3. Raise each factor to the highest power it occurs in *any* denominator.
4. The LCD is the product of all the powers found in Step 3.

To Add Like Fractions. (Section 5–3)

1. Add the numerators.
2. Write the sum of the numerators over the denominator of the like fractions.
3. Reduce the resulting fraction to lowest terms.

To Add Unlike Fractions. (Section 5–3)

1. Find the LCD.
2. Convert all fractions to equivalent fractions that have the LCD as denominator.
3. Add the resulting like fractions.
4. Reduce the resulting fraction to lowest terms.

To Simplify Complex Fractions. (Section 5–4)

Method 1. Multiply both numerator and denominator of the complex fraction by the LCD of the secondary fractions; then simplify the results.

Method 2. First, simplify the numerator and denominator of the complex fraction; then divide the simplified numerator by the simplified denominator.

To Solve an Equation that Has Fractions. (Section 5–5)

1. Remove fractions by multiplying each term by the LCD.
2. Remove grouping symbols (Section 1–9C).
3. Combine like terms.

If a *first-degree equation* is obtained in Step 3:

4. Get all terms with the unknown on one side and the remaining terms on the other side.
5. Divide both sides by the coefficient of the unknown.

If a *second-degree equation* (quadratic) is obtained in Step 3:

4. Get all nonzero terms on one side and arrange them in descending powers. *Only zero must remain on the other side.*
5. Factor the polynomial.
6. Set each factor equal to zero and then solve for unknown.

Literal Equations are equations that have more than one letter (Section 5–6).

To Solve a Literal Equation. (Section 5–6) Proceed in the same way used to solve an equation with a single letter. The solution will be expressed in terms of the other letters given in the literal equation, as well as numbers.

Chapter Five Diagnostic Test or Review

Allow yourself about one hour to do these problems. Complete solutions for every problem, together with section references, are given in the answer section at the end of the book.

---

1. What values of the variable must be excluded, if any, in each expression?

(a) $\dfrac{2x + 3}{x^2 - 4x}$ (b) $\dfrac{y + 2}{3y^2 - y - 10}$

2. Use the rule about the three signs of a fraction to find the missing term in each expression.

(a) $\dfrac{4}{-h} = \dfrac{-4}{?}$ (b) $\dfrac{-3}{k - 2} = \dfrac{?}{2 - k}$

3. Reduce each fraction to lowest terms.

(a) $\dfrac{4a^4b}{10a^2b^3}$ (b) $\dfrac{f^2 + 5f + 6}{f^2 - 9}$ (c) $\dfrac{x^4 - 2x^3 + 5x^2 - 10x}{x^3 - 8}$

In Problems 4–9 perform the indicated operations.

4. $\dfrac{z}{z - 2} \cdot \dfrac{3z - 6}{6z^2}$

5. $\dfrac{3x^2}{x + 2} \div \dfrac{6x}{x^2 - 4}$

6. $\dfrac{3m + 3n}{m^3 - n^3} \div \dfrac{m^2 - n^2}{m^2 + mn + n^2}$

7. $\dfrac{20a + 27b}{12a - 20b} + \dfrac{44a - 13b}{20b - 12a}$

8. $\dfrac{y-2}{y} - \dfrac{y}{y+2}$

9. $\dfrac{x}{x+1} - \dfrac{x-1}{2x-2}$

In Problems 10 and 11 simplify the complex fractions.

10. $\dfrac{\dfrac{8h^4}{5k}}{\dfrac{4h^2}{15k^3}}$

11. $\dfrac{6 - \dfrac{4}{w}}{\dfrac{3w}{w-2} + \dfrac{1}{w}}$

12. Solve: $\dfrac{6}{2c-7} = \dfrac{9}{5-3c}$

13. Solve: $\dfrac{y+6}{3} = \dfrac{y+4}{2} + \dfrac{1}{3}$

14. Solve: $\dfrac{2}{3a+5} - \dfrac{6}{a-2} = 3$

15. Solve: $\dfrac{3}{2z} + \dfrac{3}{z^2} = -\dfrac{1}{6}$

16. Solve for r in terms of the other letters: $I = \dfrac{E}{R+r}$

17. Arlene can make 24 bushings on a lathe in 8 hr. How long does it take Ruben to make 8 bushings if he and Arlene can make 14 bushings in 4 hr when working together?

6 RATIONAL EXPONENTS AND RADICALS

Integral exponents were discussed in Section 1–6. In this chapter the rules of exponents are extended to include rational exponents. The relation between exponents and radicals is also discussed.

Roots of *numbers* were introduced in Section 1–5. Square roots of *algebraic terms* were discussed in Section 4–3. Both of these are examples of *radicals,* which will be more fully discussed in this chapter.

6–1 Rational Exponents

In Section 1–6 the following basic rules for exponents were introduced. These rules were used only for exponents that were *integers*.

THE RULES OF EXPONENTS

Rule 1 $\quad x^a x^b = x^{a+b}$

Rule 2 $\quad (x^a)^b = x^{ab}$

Rule 2A $\quad (xy)^a = x^a y^a$

Rule 3 $\quad \dfrac{x^a}{x^b} = x^{a-b} \qquad (x \neq 0)$

Rule 3A $\quad \left(\dfrac{x}{y}\right)^a = \dfrac{x^a}{y^a} \qquad (y \neq 0)$

Rule 4 $\quad x^{-n} = \dfrac{1}{x^n} \quad (x \neq 0)$

Rule 5 $\quad x^0 = 1 \quad (x \neq 0)$

Rule 6 $\quad \left(\dfrac{x^a y^b}{z^c}\right)^n = \dfrac{x^{an} y^{bn}}{z^{cn}}$ None of the letters can have a value that makes the denominator zero

The rules of exponents can be used when the exponents are *any* rational numbers.

Example 1 Using the rules of exponents with rational exponents

(a) $a^{1/2}a^{-1/4} = a^{1/2+(-1/4)} = a^{1/4}$ Rule 1

(b) $(8)^{2/3} = (2^3)^{2/3} = 2^{3(2/3)}$ Rule 2

$= 2^2 = 4$

(c) $x^{-1/3} = \dfrac{1}{x^{1/3}}$ Rule 4

(d) $(z^{-1/2})^4 = z^{(-1/2)4} = z^{-2} = \dfrac{1}{z^2}$ Rules 2 and 4

(e) $\dfrac{y^{2/3}}{y^{1/2}} = y^{2/3-1/2} = y^{1/6}$ Rule 3

(f) $x^{-1/2}xx^{2/5} = x^{-1/2+1+2/5}$

$= x^{-5/10+10/10+4/10} = x^{9/10}$ Rule 1

(g) $\dfrac{b^{3/2}b^{-2/3}}{b^{5/6}} = b^{3/2-2/3-5/6}$

$= b^{9/6-4/6-5/6} = b^0 = 1$ Rules 1, 3, and 5

(h) $(9h^{-2/5}k^{4/3})^{-3/2} = (3^2h^{-2/5}k^{4/3})^{-3/2}$

$= 3^{2(-3/2)}h^{-2/5(-3/2)}k^{4/3(-3/2)}$

$= 3^{-3}h^{3/5}k^{-2} = \dfrac{h^{3/5}}{27k^2}$ Rule 6

(i) $\left(\dfrac{xy^{-2/3}}{16z^{-4}}\right)^{-3/4} = \dfrac{x^{(1)(-3/4)}y^{(-2/3)(-3/4)}}{(2^{+4})^{-3/4}z^{-4(-3/4)}}$

$= \dfrac{x^{-3/4}y^{1/2}}{2^{-3}z^3} = \dfrac{8y^{1/2}}{x^{3/4}z^3}$ Rules 4 and 6

(j) $\left(\dfrac{24c^{5/2}d^{-3}}{3c^{-1/2}}\right)^{2/3} = (8c^{5/2+1/2}d^{-3})^{2/3} = \left(\dfrac{2^3c^3}{d^3}\right)^{2/3}$

$= \dfrac{2^{3(2/3)}c^{3(2/3)}}{d^{3(2/3)}} = \dfrac{2^2c^2}{d^2} = \dfrac{4c^2}{d^2}$ Rules 3 and 6

EXERCISES 6-1

In Exercises 1–46 perform the indicated operations. Express answers in exponential form using positive exponents.

1. $x^{1/2}x^{3/2}$
2. $y^{5/4}y^{3/4}$
3. $a^{1/4}a^{3/4}$
4. $x^{5/6}x^{-1/3}$
5. $a^{3/4}a^{-1/2}$
6. $b^{5/6}b^{-1/3}$
7. $z^{-1/2}z^{2/3}$
8. $N^{-1/3}N^{3/4}$
9. $w^{1/3}w^{-1/2}$
10. $u^{-3/5}u^{1/2}$
11. $(H^{3/4})^2$
12. $(s^{5/6})^3$
13. $(x^{-3/4})^{1/3}$
14. $(y^{-2/3})^{1/2}$
15. $(e^{-2/5})^{-1/2}$
16. $(f^{-3/4})^{-1/3}$
17. $\dfrac{a^{3/4}}{a^{1/2}}$
18. $\dfrac{b^{2/3}}{b^{1/6}}$
19. $\dfrac{P^{3/5}}{P^{-1/2}}$
20. $\dfrac{R^{3/4}}{R^{-1/3}}$
21. $\dfrac{z^{-1/4}}{z^{1/2}}$
22. $\dfrac{s^{1/3}}{s^{-1/6}}$
23. $\dfrac{h^{-1/3}}{h^{-1/2}}$
24. $\dfrac{k^{-1/2}}{k^{-1/4}}$
25. $x^{2/3}xx^{-1/2}$
26. $x^{3/4}xx^{-1/3}$
27. $z^{2/5}z^{-3/5}z^{7/10}$
28. $m^{-3/4}m^{1/2}m^{7/8}$
29. $(x^{-1/2})^3(x^{2/3})^2$
30. $(x^{3/4})^2(x^{-2/3})^3$
31. $(2u^{-1/3})^2$
32. $(3v^{-3/5})^2$
33. $(16x^{-2/5}y^{4/9})^{3/2}$
34. $(8x^{9/8}y^{-3/4})^{2/3}$
35. $(36R^{-4/3}S^{2/5})^{-1/2}$
36. $(16a^{2/3}b^{-4/5})^{-1/4}$
37. $\dfrac{u^{1/2}v^{-2/3}}{u^{-1/4}v^{-1}}$
38. $\dfrac{u^{2/3}v^{-3/5}}{u^{-1/3}v^{-1}}$
39. $\dfrac{18a^{1/2}b^{-1/3}}{12a^{2/3}b^{-1/2}}$
40. $\dfrac{15c^{-1/2}d^{2/3}}{24c^{-3/4}d}$
41. $\left(\dfrac{R^{-3/2}S^{1/2}}{8R^{-2}}\right)^{2/3}$
42. $\left(\dfrac{y^{2/3}z^{-2/3}}{4z^{-3}}\right)^{3/2}$
43. $\left(\dfrac{x^{-1}y^{2/3}}{z^{-5}}\right)^{-3/5}$
44. $\left(\dfrac{a^{3/4}b^{-1}}{c^{-2}}\right)^{-2/3}$
45. $\left(\dfrac{9x^{-2/3}y^{2/9}}{x^{-2}}\right)^{-3/2}$
46. $\left(\dfrac{27R^{3/5}S^{-5/2}}{S^{-4}}\right)^{-2/3}$

In Exercises 47–56 evaluate each expression.

47. $(8)^{-2/3}$
48. $(27)^{-2/3}$
49. $(4)^{1/2}(9)^{-3/2}$
50. $(8)^{1/3}(16)^{-3/2}$
51. $(100)^{-1/2}(-8)^{1/3}$
52. $(10^4)^{-1/2}(-27)^{1/3}$
53. $(10^6)^{-1/2}(125)^{1/3}$
54. $\dfrac{(10^4)^{1/2}(10^{-2})^{3/2}}{(10^5)^{-1/5}}$
55. $\left(\dfrac{10^{-3/2}10^{1/6}}{10^{-4/3}}\right)^{1/2}$
56. $\left(\dfrac{10^{3/4}10^{-1}}{10^{-1/2}10^{3/2}}\right)^{-4}$

6-2 The Relation between Rational Exponents and Radicals

Radicals

A **radical** is an indicated root of a number.

Some Symbols Used to Indicate Roots

$\sqrt[3]{\ }$ indicates cube root. Therefore $\sqrt[3]{8} = 2$, because $2^3 = 8$.
$\sqrt[4]{\ }$ indicates fourth root. Therefore $\sqrt[4]{16} = 2$, because $2^4 = 16$.
$\sqrt[5]{\ }$ indicates fifth root. Therefore $\sqrt[5]{32} = 2$, because $2^5 = 32$.

The entire expression $\sqrt[n]{p}$ is called a **radical expression,** or more simply, a **radical.**

The Parts of a Radical

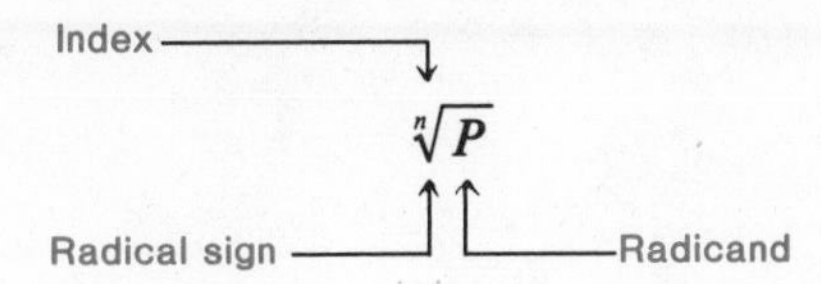

In this book the index n will always represent a natural number.

Example 1 The parts of radicals

(a) $\sqrt{7}$ Read "the square root of 7"
7 is the radicand
2 is the index (the index is *not written for square roots*)

(b) $\sqrt[3]{2x}$ Read "the cube root of $2x$"
$2x$ is the radicand
3 is the index

(c) $\sqrt[5]{25a^4b^2}$ Read "the fifth root of $25a^4b^2$"
$25a^4b^2$ is the radicand
5 is the index

Principal Roots

Every even index radical has both a positive and a negative real root; the positive root is called the **principal root.**

Every odd index radical has a single real root called the principal root, which is positive when the radicand is positive, and negative when the radicand is negative.

Whenever the radical symbol is used, we agree that it is to stand for the principal root. We summarize this information about principal roots in the following box.

> **PRINCIPAL ROOTS**
>
> The symbol $\sqrt[n]{p}$ always represents the *principal* nth root of p.
>
> *Positive radicand:* The principal root is positive.
>
> *Negative radicand:*
>
> Odd index: The principal root is negative.
>
> Even index: Roots are not real numbers. These are discussed in Section 6–8.

Example 2 Principal roots

(a) $\sqrt{9} = 3$ *Even* index, *positive* radicand
9 has two square roots, 3 and -3
3 is the principal root

(b) $\sqrt[4]{16} = 2$ — *Even* index, *positive* radicand
16 has two real fourth roots, 2 and -2
2 is the princial root

(c) $\sqrt{-4}$ — *Even* index, *negative* radicand
Roots are not real numbers
(Discussed in Section 6-8)

(d) $\sqrt[3]{8} = 2$ — *Odd* index, *positive* radicand
8 has one real cube root, 2
2 is the principal root

(e) $\sqrt[3]{-8} = -2$ — *Odd* index, *negative* radicand
-8 has one real cube root, -2
-2 is the principal root

(f) $\sqrt{x^2} = |x|$ — *Even* index, *positive* radicand (or zero)
x^2 has two square roots, x and $-x$
$|x|$ is the principal root

Since x can be a positive number or a negative number, we don't know whether x or $-x$ is the principal root. However, $|x|$ must be positive (provided $x \neq 0$); therefore, $|x|$ is the principal square root of x^2.

RELATING RATIONAL EXPONENTS TO RADICALS

We know that $\sqrt{x}$ is a number such that $\sqrt{x}\sqrt{x} = x$. If we represent $\sqrt{x}$ as x^p, a power of x, then

$$\sqrt{x} \cdot \sqrt{x} = x$$
$$x^p \cdot x^p = x^1$$
$$x^{p+p} = x^1 \quad \text{Rule 1}$$
$$x^{2p} = x^1$$
$$2p = 1$$
$$p = \frac{1}{2} \quad \text{That is, } x^p = x^{1/2}$$

Therefore $x^{1/2} = \sqrt{x}$.

In the same way, if we represent $\sqrt[3]{x}$ as x^r, then

$$\sqrt[3]{x} \cdot \sqrt[3]{x} \cdot \sqrt[3]{x} = x$$
$$x^r \cdot x^r \cdot x^r = x$$
$$x^{r+r+r} = x^1 \quad \text{Rule 1}$$
$$x^{3r} = x^1$$
$$3r = 1$$
$$r = \frac{1}{3} \quad \text{That is, } x^r = x^{1/3}$$

Therefore, $x^{1/3} = \sqrt[3]{x}$

In general, we define

RULE 7

$$x^{1/n} = \sqrt[n]{x}$$

When n is even and x is negative, $\sqrt[n]{x}$ does not represent a real number. This case is discussed in Section 6–8.

Example 3 Using Rule 7

(a) $a^{1/4} = \sqrt[4]{a}$ (b) $\sqrt[5]{y} = y^{1/5}$ (c) $-10^{1/3} = -(10^{1/3}) = -\sqrt[3]{10}$

(d) $27^{1/3} = \sqrt[3]{27} = \sqrt[3]{3^3} = 3$

or $27^{1/3} = (3^3)^{1/3} = 3^1 = 3$

(e) $x^{2/3} = (x^2)^{1/3} = \sqrt[3]{x^2}$

(f) $x^{2/3} = (x^{1/3})^2 = (\sqrt[3]{x})^2$

Therefore, $x^{2/3} = \sqrt[3]{x^2} = (\sqrt[3]{x})^2$

Examples 3(e) and 3(f) lead to the following rule:

RULE 8

$$x^{a/b} = \sqrt[b]{x^a} = (\sqrt[b]{x})^a$$

Example 4 Using Rule 8

(a) $m^{3/5} = \sqrt[5]{m^3}$

(b) $d^{4/5} = \sqrt[5]{d^4}$ or $(\sqrt[5]{d})^4$

(c) $8^{2/3} = (\sqrt[3]{8})^2 = (\sqrt[3]{2^3})^2 = (2)^2 = 4$

or $8^{2/3} = (2^3)^{2/3} = 2^2 = 4$

EXERCISES 6-2

In Exercises 1–18 find the principal root if it exists.

1. $\sqrt{16}$
2. $\sqrt{25}$
3. $\sqrt[3]{27}$
4. $\sqrt[3]{64}$
5. $\sqrt[3]{-64}$
6. $\sqrt[3]{-27}$
7. $\sqrt{-25}$
8. $\sqrt{-16}$
9. $\sqrt[4]{81}$
10. $\sqrt[4]{16}$
11. $\sqrt[5]{32}$
12. $\sqrt[5]{-32}$
13. $\sqrt[4]{-1}$
14. $\sqrt[5]{-1}$
15. $\sqrt[6]{64}$
16. $\sqrt[6]{1}$
17. $\sqrt[3]{-1000}$
18. $\sqrt[4]{10{,}000}$

In Exercises 19–26 rewrite each expression in radical form.

19. $a^{3/5}$ **20.** $b^{2/3}$ **21.** $7^{1/2}$ **22.** $5^{1/3}$

23. $x^{-1/2}$ **24.** $y^{-3/4}$ **25.** $7^{m/n}$ **26.** $x^{(a+b)/a}$

In Exercises 27–38 replace each radical by an equivalent exponential expression.

27. $\sqrt{5}$ **28.** $\sqrt{7}$ **29.** $\sqrt[3]{z}$ **30.** $\sqrt[4]{x}$

31. $\sqrt[4]{x^3}$ **32.** $\sqrt[5]{y^2}$ **33.** $\sqrt[3]{x^2}$ **34.** $\sqrt{x^n}$

35. $(\sqrt[3]{x^2})^2$ **36.** $(\sqrt[4]{x^3})^2$ **37.** $\sqrt[n]{x^{2n}}$ **38.** $\sqrt[n]{x^{n^2+n}}$

In Exercises 39–50 evaluate each expression.

39. $8^{1/3}$ **40.** $27^{1/3}$ **41.** $(-27)^{2/3}$ **42.** $(-8)^{2/3}$

43. $4^{3/2}$ **44.** $9^{3/2}$ **45.** $(-16)^{3/4}$ **46.** $(-1)^{1/4}$

47. $64^{2/3}$ **48.** $16^{3/4}$ **49.** $(-1)^{1/5}$ **50.** $27^{4/3}$

6–3 Simplifying Radicals

We will consider radicals of two kinds:

1. Radicals *without* fractions in the radicand (6–3A).
2. Radicals *with* fractions in the radicand (6–3B).

6-3A RADICALS NOT HAVING FRACTIONS IN THE RADICAND

Perfect Power

A factor of a radicand raised to an exponent exactly divisible by the index of the radical is called a **perfect power.**

If a^p is a perfect power, then

$$\sqrt[n]{a^p} = a^{p/n} = a^{p \div n}$$

Exponent exactly divisible by index. The quotient is an integer. Rule 8, Section 6-2

If a^p is not a perfect power, then

$$\sqrt[n]{a^p} = a^{p \div n}$$

The quotient is not an integer

TO FIND THE PRINCIPAL ROOT OF A PERFECT POWER (Exponent Exactly Divisible by Index)

Divide the exponent by the index.

$$\sqrt[n]{a^p} = a^{p \div n}$$

where n is a positive integer.

Example 1 Principal root of perfect powers

(a) $\sqrt{x^6} = x^{6 \div 2} = x^3$ To find the square root of a perfect power, divide its exponent by 2

(b) $\sqrt[3]{2^3} = 2^{3 \div 3} = 2^1$ To find the cube root of a perfect power, divide its exponent by 3

(c) $\sqrt[4]{z^{12}} = z^{12 \div 4} = z^3$ To find the fourth root of a perfect power, divide its exponent by 4

To find the nth root of a product, proceed as follows:

$$\sqrt[n]{ab} = (ab)^{1/n} = a^{1/n}b^{1/n} = \sqrt[n]{a}\sqrt[n]{b}$$

nth Root of a Product

TO FIND THE PRINCIPAL ROOT OF SEVERAL FACTORS

Rule 1 (for radicals) $\sqrt[n]{ab} = \sqrt[n]{a}\sqrt[n]{b}$

There are restrictions on the use of Rule 1 which will be discussed in Section 6–8.

Example 2 Principal root of several factors

(a) $\sqrt{4x^2} = \sqrt{4}\sqrt{x^2}$ Rule 1

$= 2 \cdot x = 2x$

(b) $\sqrt[3]{8a^3b^6} = \sqrt[3]{8}\sqrt[3]{a^3}\sqrt[3]{b^6}$ Rule 1

$= 2 \cdot a \cdot b^2 = 2ab^2$

TO FIND THE PRINCIPAL ROOT OF A NONPERFECT POWER (Exponent Not Exactly Divisible by Index)

A. *If exponent is less than index,*
radical cannot be simplified

B. *If exponent is greater than index,*

1. Factor power into product of perfect power and nonperfect power.
2. Use Rule 1.

A radical with no fractions in its radicand is considered **simplified** if there are no prime factors of the radicand that have an exponent equal to or greater than the index.

Example 3 Principal root of nonperfect powers

(a) $\sqrt{x^3}$ — x^3 is not a perfect power

$\sqrt{x^3} = \sqrt{x^2 \cdot x}$ — x^3 is factored into the product of a perfect power, x^2, and x

$= \sqrt{x^2}\sqrt{x}$ — Rule 1

$= x\sqrt{x}$

(b) $\sqrt[3]{16} = \sqrt[3]{2^4}$ — 2^4 is not a perfect power

$= \sqrt[3]{2^3 \cdot 2}$ — 2^4 is factored into the product of a perfect power, 2^3, and 2

$= \sqrt[3]{2^3}\sqrt[3]{2}$ — Rule 1

$= 2\sqrt[3]{2}$

(c) $\sqrt[5]{y^7}$ — y^7 is not a perfect power

$\sqrt[5]{y^7} = \sqrt[5]{y^5 \cdot y^2}$ — y^7 is factored into the product of a perfect power, y^5, and y^2.

$= \sqrt[5]{y^5}\sqrt[5]{y^2}$ — Rule 1

$= y\sqrt[5]{y^2}$

(d) $\sqrt{360} = \sqrt{2^3 \cdot 3^2 \cdot 5}$ — Expressed the radicand in prime-factored form

$= \sqrt{2^2 \cdot 2 \cdot 3^2 \cdot 5}$

$= 2 \cdot 3 \cdot \sqrt{2 \cdot 5} = 6\sqrt{10}$ — Simplified form

(e) $\sqrt{12x^4y^3} = \sqrt{2^2 \cdot 3 \cdot x^4 \cdot y^2 \cdot y}$ — Expressed the radicand in prime-factored form

$= 2 \cdot x^2 \cdot y\sqrt{3y} = 2x^2y\sqrt{3y}$ — Simplified form

(f) $\sqrt[3]{-16a^5b^7} = \sqrt[3]{-2^4a^5b^7}$ — Expressed the radicand in prime-factored form

$= \sqrt[3]{-2^3 \cdot 2 \cdot a^3 \cdot a^2 \cdot b^6 \cdot b}$

$= -2 \cdot a \cdot b^2\sqrt[3]{2 \cdot a^2 \cdot b} = -2ab^2\sqrt[3]{2a^2b}$

(g) $\sqrt[4]{96h^{11}k^3m^8} = \sqrt[4]{2^5 \cdot 3h^{11}k^3m^8}$

$= \sqrt[4]{2^4 \cdot 2 \cdot 3 \cdot h^8 \cdot h^3 \cdot k^3 \cdot m^8}$

$= 2 \cdot h^2 \cdot m^2\sqrt[4]{2 \cdot 3 \cdot h^3 \cdot k^3}$

$= 2h^2m^2\sqrt[4]{6h^3k^3}$

(h) $5xy^2\sqrt{28x^4y^9z^3} = 5xy^2\sqrt{2^2 \cdot 7 \cdot x^4 \cdot y^8 \cdot y \cdot z^2 \cdot z}$

$= 5xy^2 \cdot 2 \cdot x^2 \cdot y^4 \cdot z\sqrt{7 \cdot y \cdot z}$

$= 10x^3y^6z\sqrt{7yz}$

Example 4 Sometimes finding the root of an integer can be simplified by noticing that the number has a factor which is a perfect power.

(a) $\sqrt{50} = \sqrt{25 \cdot 2} = \sqrt{25}\sqrt{2} = 5\sqrt{2}$

(b) $\sqrt[3]{24} = \sqrt[3]{8 \cdot 3} = \sqrt[3]{8}\sqrt[3]{3} = 2\sqrt[3]{3}$

(c) $\sqrt[4]{32} = \sqrt[4]{16 \cdot 2} = \sqrt[4]{16}\sqrt[4]{2} = 2\sqrt[4]{2}$

When simplifying radicals, we assume that all letters represent positive numbers (unless otherwise stated). Since the letters are assumed to be positive, the absolute value symbol need not be used.

If x is assumed positive, $\sqrt{x^2} = x$

If x can be negative, $\sqrt{x^2} = |x|$

EXERCISES 6-3A

Simplify each radical.

1. $\sqrt{4x^2}$ **2.** $\sqrt{9y^2}$ **3.** $\sqrt[3]{25x^2}$ **4.** $\sqrt{64a^4}$

5. $\sqrt[3]{8x^3}$ **6.** $\sqrt[3]{27y^3}$ **7.** $\sqrt[4]{16x^4y^8}$ **8.** $\sqrt[4]{81u^8v^4}$

9. $\sqrt{(2x-3)^4}$ **10.** $\sqrt[3]{(a+b)^6}$ **11.** $\sqrt{(-2)^2}$ **12.** $\sqrt{(-3)^2}$

13. $\sqrt{2^5}$ **14.** $\sqrt{3^3}$ **15.** $\sqrt[3]{-3^5}$ **16.** $\sqrt[3]{-2^8}$

17. $\sqrt{2^7}$ **18.** $\sqrt[3]{-2^5}$ **19.** $\sqrt{m^5}$ **20.** $\sqrt[3]{x^7}$

21. $\sqrt[5]{-x^7}$ **22.** $\sqrt[5]{-z^8}$ **23.** $\sqrt[4]{32}$ **24.** $\sqrt[3]{z^8}$

25. $\sqrt[4]{48}$ **26.** $\sqrt{75x^5y^4}$ **27.** $\sqrt{8a^4b^2}$ **28.** $\sqrt{20m^8u^2}$

29. $\sqrt{18m^3n^5}$ **30.** $\sqrt{50h^5k^3}$ **31.** $5\sqrt[3]{-24a^5b^2}$ **32.** $6\sqrt[3]{-54c^4d}$

33. $\sqrt[5]{64m^{11}p^{15}u}$ **34.** $\sqrt[5]{128u^4v^{10}w^{16}}$ **35.** $\frac{2}{3}\sqrt{27xy^8z^5}$ **36.** $\sqrt[7]{21x^4y^7z^{10}}$

37. $\frac{3}{2abc}\sqrt[4]{2^5a^8b^9c^{10}}$ **38.** $\frac{7}{2bc}\sqrt[4]{2^6a^5b^6c^7}$ **39.** $\frac{1}{2}\sqrt[4]{32x^5y^9}$ **40.** $\frac{3}{4xy}\sqrt[4]{2^6x^5y^6z^5}$

41. $\sqrt[3]{8(a+b)^3}$ **42.** $\sqrt[3]{27(x-y)^6}$ **43.** $\sqrt{4x^2+4x+1}$ **44.** $\sqrt{a^3+6a^2+9a}$

In Exercises 45 and 46 make no assumption about the value of the letters.

45. $\sqrt{4-4y+y^2}$ **46.** $\sqrt[4]{32(2x-y)^4}$

6-3B RADICALS HAVING FRACTIONS IN THE RADICAND

To find the nth root of a quotient, proceed as follows:

$$\sqrt[n]{\frac{a}{b}} = \left(\frac{a}{b}\right)^{1/n} = \frac{a^{1/n}}{b^{1/n}} = \frac{\sqrt[n]{a}}{\sqrt[n]{b}}$$

nth Root of a Quotient

TO FIND THE PRINCIPAL ROOT OF A QUOTIENT

Rule 2 (for radicals) $$\sqrt[n]{\frac{a}{b}} = \frac{\sqrt[n]{a}}{\sqrt[n]{b}} \qquad (b \neq 0)$$

Example 5 Principal root of quotients

(a) $\sqrt{\frac{25}{36}} = \frac{\sqrt{25}}{\sqrt{36}} = \frac{5}{6}$ (b) $\sqrt[3]{\frac{8}{27}} = \frac{\sqrt[3]{8}}{\sqrt[3]{27}} = \frac{2}{3}$

(c) $\sqrt{\frac{x^4}{y^6}} = \frac{\sqrt{x^4}}{\sqrt{y^6}} = \frac{x^2}{y^3}$ (d) $\sqrt{\frac{50h^2}{2k^4}} = \sqrt{\frac{\overset{25}{\cancel{50}}h^2}{\cancel{2}k^4}} = \frac{\sqrt{25h^2}}{\sqrt{k^4}} = \frac{5h}{k^2}$

(e) $\sqrt[5]{\frac{32x^{10}}{y^{15}}} = \frac{\sqrt[5]{2^5x^{10}}}{\sqrt[5]{y^{15}}} = \frac{2x^2}{y^3}$ (f) $\sqrt{\frac{3x^2}{4}} = \frac{\sqrt{3x^2}}{\sqrt{4}} = \frac{x\sqrt{3}}{2}$

(g) $\sqrt{\frac{4}{5}} = \frac{\sqrt{4}}{\sqrt{5}} = \frac{2}{\sqrt{5}}$

↑ This denominator is an irrational number

An algebraic fraction is not considered simplified if a radical appears in the denominator.

Rationalizing the Denominator

When the denominator of a fraction is not a rational number, the procedure for changing it into a rational number is called **rationalizing the denominator.**

Example 6 Rationalizing the denominator.

(a)

$$\frac{2}{\sqrt{5}} = \frac{2}{\sqrt{5}} \cdot \frac{\sqrt{5}}{\sqrt{5}} = \frac{2\sqrt{5}}{\sqrt{5}\sqrt{5}} = \frac{2\sqrt{5}}{5}$$

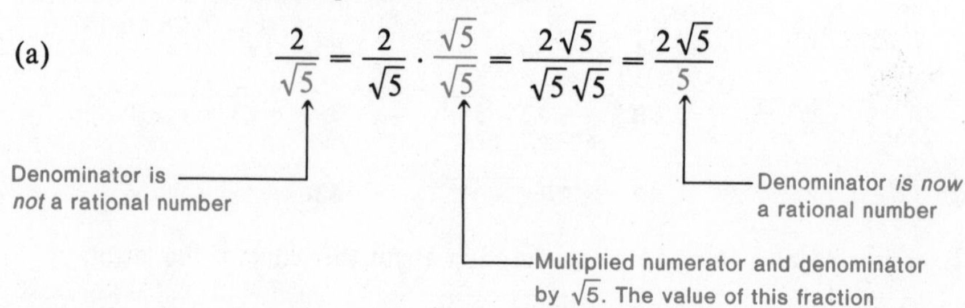

(b) $\dfrac{6}{\sqrt[3]{2}} = \dfrac{6}{\sqrt[3]{2}} \cdot \dfrac{\sqrt[3]{2^2}}{\sqrt[3]{2^2}} = \dfrac{6\sqrt[3]{4}}{\sqrt[3]{2^3}} = \dfrac{6\sqrt[3]{4}}{2} = 3\sqrt[3]{4}$

Since we have *cube* roots, the numerator and denominator are multiplied by $\sqrt[3]{2^2}$, which makes the radicand in the denominator a perfect cube, $\sqrt[3]{2^3}$

(c) $\dfrac{3xy}{\sqrt[4]{y}} = \dfrac{3xy}{\sqrt[4]{y}} \cdot \dfrac{\sqrt[4]{y^3}}{\sqrt[4]{y^3}} = \dfrac{3xy\sqrt[4]{y^3}}{\sqrt[4]{y^4}} = \dfrac{3xy\sqrt[4]{y^3}}{y} = 3x\sqrt[4]{y^3}$

Multiplied numerator and denominator by $\sqrt[4]{y^3}$ to make the radicand in the denominator a perfect fourth power, $\sqrt[4]{y^4}$

Simplified Form of a Radical

THE SIMPLIFIED FORM OF A RADICAL

1. No prime factor of a radicand has an exponent equal to or greater than the index.
2. No radicand contains a fraction.
3. No denominator contains a radical.

Rationalizing the denominator can often be combined with simplifying a radical (see Example 7).

Example 7 Simplifying a radical that has a fractional radicand

(a) $\sqrt{\dfrac{x^3}{8}} = \sqrt{\dfrac{x^3}{8} \cdot \dfrac{2}{2}} = \sqrt{\dfrac{2x^3}{16}} = \sqrt{\dfrac{2 \cdot x^2 \cdot x}{16}} = \dfrac{x}{4}\sqrt{2x}$ Simplified form

Multiplied by $\frac{2}{2}$ in order to make the denominator a perfect square, 16

(b) $\sqrt[3]{\dfrac{a^4b^3}{-2c}} = \sqrt[3]{\dfrac{a^4b^3}{-2c} \cdot \dfrac{2^2c^2}{2^2c^2}} = \dfrac{\sqrt[3]{a^3ab^3 2^2c^2}}{\sqrt[3]{-2^3c^3}} = \dfrac{ab\sqrt[3]{4ac^2}}{-2c} = -\dfrac{ab\sqrt[3]{4ac^2}}{2c}$

Multiplied by $\dfrac{2^2c^2}{2^2c^2}$ in order to make the denominator a perfect cube, -2^3c^3

(c) $$\sqrt[5]{\frac{24x^2y^6}{64x^4y}} = \sqrt[5]{\frac{\overset{3}{24}x^2y^6}{\underset{8}{64}x^4y}} = \sqrt[5]{\frac{3y^5}{2^3x^2} \cdot \frac{2^2x^3}{2^2x^3}} = \frac{\sqrt[5]{12y^5x^3}}{\sqrt[5]{2^5x^5}} = \frac{y\sqrt[5]{12x^3}}{2x}$$

Multiplied by $\frac{2^2x^3}{2^2x^3}$ in order to make the denominator a perfect fifth power, 2^5x^5

(d) $$\frac{4x^2}{5y^3}\sqrt{\frac{3x^2y^3}{8x^5y}} = \frac{4x^2}{5y^3}\sqrt{\frac{3y^2}{8x^3}} = \frac{4x^2}{5y^3}\sqrt{\frac{3 \cdot y^2}{2^2 \cdot 2 \cdot x^2 \cdot x}}$$

$$= \frac{4x^2}{5y^3} \cdot \frac{y}{2x}\sqrt{\frac{3}{2x} \cdot \frac{2x}{2x}} = \frac{\cancel{4x^2}}{\underset{y^2}{5\cancel{y^3}}} \cdot \frac{\cancel{y}}{\cancel{2x} \cdot \cancel{2x}}\sqrt{6x} = \frac{1}{5y^2}\sqrt{6x}$$

TO SIMPLIFY A RADICAL THAT HAS A FRACTION AS ITS RADICAND

1. Reduce the fraction of the radicand to lowest terms.
2. Write the numerator and denominator in prime-factored form.
3. Remove perfect powers by dividing their exponents by the index. (The root of a perfect power in the numerator of the radicand becomes a factor of the numerator outside the radical. The root of a perfect power in the denominator becomes a factor of the denominator outside the radical.)
4. Find the expression which, when multiplied by the denominator, makes the denominator a perfect power. Multiply the numerator and denominator of the remaining radicand by this expression.
5. Remove the perfect power in the denominator formed in Step 4.
6. The simplified radical is the fraction formed by those factors that were removed (reduced to lowest terms) multiplied by the radical whose radicand is the product of all factors that were not removed (see Example 7).

EXERCISES 6-3B

Simplify each radical.

1. $\sqrt{\frac{16}{25}}$ 2. $\sqrt{\frac{64}{100}}$ 3. $\sqrt[3]{\frac{-27}{64}}$ 4. $\sqrt[3]{\frac{-8}{125}}$

5. $\sqrt{\frac{36}{81}}$ 6. $\sqrt{\frac{x^6}{v^8}}$ 7. $\sqrt[4]{\frac{a^4b^8}{16}}$ 8. $\sqrt[4]{\frac{c^8d^{12}}{81}}$

9. $\sqrt{\frac{4x^3y}{xy^3}}$ 10. $\sqrt{\frac{x^5y}{9xy^3}}$ 11. $\sqrt[5]{\frac{x^{15}y^{10}}{-243}}$ 12. $\frac{3ab}{\sqrt[4]{a}}$

13. $\frac{10}{\sqrt{5}}$ 14. $\frac{14}{\sqrt{2}}$ 15. $\frac{9}{\sqrt[3]{3}}$ 16. $\frac{10}{\sqrt[3]{5}}$

17. $\frac{3c^2d}{\sqrt[5]{-d}}$ 18. $\frac{6x}{\sqrt[4]{2x^3}}$ 19. $\sqrt[3]{\frac{m^5}{-3}}$ 20. $\sqrt[3]{\frac{k^4}{-5}}$

21. $\frac{n}{2m}\sqrt{\frac{8m^2n}{2n^3}}$ 22. $\frac{x}{3y}\sqrt{\frac{18xy^2}{2x^3}}$ 23. $\frac{4}{a}\sqrt{\frac{3a^2b}{4b^3}}$ 24. $\frac{6}{v}\sqrt{\frac{10uv^2}{8u}}$

25. $\sqrt[4]{\frac{3m^7}{4m^3p^2}}$ 26. $\sqrt[4]{\frac{5a^9}{2a^5b^3}}$ 27. $\sqrt[5]{\frac{15x^4y^7}{24x^6y^2}}$ 28. $\sqrt[5]{\frac{30mp^8}{48m^4p^3}}$

29. $9x^2\sqrt[3]{\frac{x^2y^5}{9x^3y^2}}$ 30. $\frac{5x}{2y^4}\sqrt{\frac{8xy^6}{5x^3y}}$ 31. $\frac{4x^2}{5y^3}\sqrt{\frac{3x^2y^3}{8x^5y}}$ 32. $\frac{3x}{4y^2}\sqrt{\frac{2xy^5}{3x^3y}}$

33. $\sqrt{\frac{a^2+2a-3}{a^2+4a+3}}$, $a > 1$ 34. $\sqrt{\frac{m^2-m-2}{m^2-3m+2}}$, $m > 2$ 35. $\sqrt{\frac{a^2+2ab+b^2}{a^2-b^2}}$, $a > b$

36. $\sqrt{\frac{x^2-y^2}{x^2-2xy+y^2}}$, $x > y$

In Exercises 37 and 38, make no assumption about the value of the letters.

37. $\sqrt{\frac{4x^2-16}{4-4x+x^2}}$ 38. $\sqrt[4]{\frac{16x^3}{9(a-2b)^4}}$

6–4 Multiplying and Dividing Radicals

MULTIPLYING RADICALS

To find the product of two nth roots, proceed as follows:

$$\sqrt[n]{a}\sqrt[n]{b} = a^{1/n}b^{1/n} = (ab)^{1/n} = \sqrt[n]{ab}$$

Product of Two *n*th Roots

TO MULTIPLY RADICALS

1. Replace $\sqrt[n]{a}\sqrt[n]{b}$ by $\sqrt[n]{a \cdot b}$.
2. Simplify the resulting radical.

Example 1 Multiplying radicals

(a) $\sqrt{3y}\sqrt{12y^3} = \sqrt{3y \cdot 12y^3} = \sqrt{36y^4} = 6y^2$

(b) $\sqrt[3]{2x^2}\sqrt[3]{4x} = \sqrt[3]{2x^2 \cdot 4x} = \sqrt[3]{2^3x^3} = 2x$

(c) $\sqrt[4]{8ab^3}\sqrt[4]{4a^3b^2} = \sqrt[4]{2^3ab^3 \cdot 2^2a^3b^2} = \sqrt[4]{2^5a^4b^5}$
$= \sqrt[4]{2^4 \cdot 2 \cdot a^4 \cdot b^4 \cdot b} = 2ab\sqrt[4]{2b}$

(d) $2\sqrt[5]{4x} \cdot 4\sqrt[5]{8x^3}\sqrt[5]{x} = 2 \cdot 4\sqrt[5]{2^2x \cdot 2^3x^3 \cdot x} = 8\sqrt[5]{2^5x^5} = 8 \cdot 2x = 16x$

(e) $(4\sqrt{3})^2 = 4\sqrt{3} \cdot 4\sqrt{3} = 16\sqrt{3 \cdot 3} = 16 \cdot 3 = 48$

(f) $(2\sqrt{5x})^3 = (2\sqrt{5x})(2\sqrt{5x})(2\sqrt{5x}) = 2 \cdot 2 \cdot 2\sqrt{5x \cdot 5x \cdot 5x}$
$= 8\sqrt{(5x)^2 \cdot 5x} = 8 \cdot 5x\sqrt{5x} = 40x\sqrt{5x}$

DIVIDING RADICALS

To find the quotient of two nth roots, proceed as follows:

$$\frac{\sqrt[n]{a}}{\sqrt[n]{b}} = \frac{a^{1/n}}{b^{1/n}} = \left(\frac{a}{b}\right)^{1/n} = \sqrt[n]{\frac{a}{b}}$$

TO DIVIDE RADICALS

1. Replace $\dfrac{\sqrt[n]{a}}{\sqrt[n]{b}}$ by $\sqrt[n]{\dfrac{a}{b}}$.
2. Simplify the resulting radical.

Example 2 Dividing radicals

(a) $\dfrac{12\sqrt[3]{a^5}}{2\sqrt[3]{a^2}} = \dfrac{6}{1}\sqrt[3]{\dfrac{a^5}{a^2}} = 6\sqrt[3]{a^3} = 6a$

(b) $\dfrac{\sqrt{27x}}{\sqrt{3x^3}} = \sqrt{\dfrac{27x}{3x^3}} = \sqrt{\dfrac{9}{x^2}} = \dfrac{3}{x}$

(c) $\dfrac{5\sqrt[4]{28xy^6}}{10\sqrt[4]{7xy}} = \dfrac{1}{2}\sqrt[4]{\dfrac{28xy^6}{7xy}} = \dfrac{1}{2}\sqrt[4]{4y^5} = \dfrac{1}{2}\sqrt[4]{4y^4y} = \dfrac{y}{2}\sqrt[4]{4y}$

(d) $\dfrac{\sqrt{5x}}{\sqrt{10x^2}} = \sqrt{\dfrac{5x}{10x^2}} = \sqrt{\dfrac{1}{2x}} = \sqrt{\dfrac{1}{2x}\cdot\dfrac{2x}{2x}} = \dfrac{1}{2x}\sqrt{2x}$

USING RATIONAL EXPONENTS TO WORK PROBLEMS THAT HAVE RADICALS

Example 3 Using the rules of exponents to simplify the work with radicals

(a) $\sqrt[4]{x^2} = x^{2/4} = x^{1/2} = \sqrt{x}$

(b) $\sqrt[6]{8b^3} = (2^3b^3)^{1/6} = 2^{3/6}b^{3/6} = 2^{1/2}b^{1/2} = (2b)^{1/2} = \sqrt{2b}$

(c) $\sqrt{x}\sqrt[3]{x} = x^{1/2}x^{1/3} = x^{3/6}x^{2/6} = x^{5/6} = \sqrt[6]{x^5}$

(d) $\sqrt{2}\sqrt[3]{-32} = 2^{1/2}(-2^5)^{1/3} = -(2^{1/2+5/3}) = -2^{3/6+10/6} = -2^{13/6}$
$= -2^{2+1/6} = -2^2\cdot 2^{1/6} = -4\sqrt[6]{2}$

(e) $\sqrt[4]{a^3}\sqrt{a}\sqrt[3]{a^2} = a^{3/4}a^{1/2}a^{2/3} = a^{9/12}a^{6/12}a^{8/12} = a^{23/12} = a^{1+11/12}$
$= a\cdot a^{11/12} = a\sqrt[12]{a^{11}}$

(f) $\dfrac{\sqrt[3]{-d^2}}{\sqrt[4]{d^3}} = \dfrac{-d^{2/3}}{d^{3/4}} = -d^{2/3-3/4} = -d^{8/12-9/12} = -d^{-1/12} = -\dfrac{1}{\sqrt[12]{d}}$

Rationalizing Denominator

You can use fractional exponents to rationalize the denominator as follows:

Multiply by an expression that makes the exponent in the denominator a whole number
In this case,
$d^{1/12}\cdot d^{11/12} = d^{12/12} = d^1$

$$-\frac{1}{\sqrt[12]{d}} = -\frac{1}{d^{1/12}}\cdot\frac{d^{11/12}}{d^{11/12}} = -\frac{\sqrt[12]{d^{11}}}{d}$$

EXERCISES 6-4

In Exercises 1–30 multiply the radicals and simplify.

1. $\sqrt{3}\sqrt{3}$
2. $\sqrt{7}\sqrt{7}$
3. $\sqrt{2}\sqrt{32}$
4. $\sqrt[3]{4}\sqrt[3]{2}$
5. $\sqrt[3]{3}\sqrt[3]{9}$
6. $\sqrt[3]{4}\sqrt[3]{16}$
7. $\sqrt[4]{3}\sqrt[4]{27}$
8. $\sqrt[4]{125}\sqrt[4]{5}$
9. $\sqrt{9x}\sqrt{x}$
10. $\sqrt[4]{2}\sqrt[4]{8}$
11. $\sqrt{5ab^2}\sqrt{20ab}$
12. $\sqrt{3x^2y}\sqrt{27xy}$
13. $\sqrt[5]{2a^3b}\sqrt[5]{16a^2b}$
14. $\sqrt[5]{4cb^4}\sqrt[5]{8c^2b}$
15. $\sqrt[5]{4x^2y^3}\sqrt[5]{8x^3y^2}$
16. $\sqrt[3]{4x^2y^2}\sqrt[3]{2x^2y}$
17. $(5\sqrt{7})^2$
18. $(4\sqrt{6})^2$

19. $(2\sqrt[3]{4x^2})^2$
20. $(4\sqrt[3]{3x^2})^2$
21. $2\sqrt{7x^3y^3} \cdot 5\sqrt{3xy} \cdot 2\sqrt{7x^3y}$
22. $3\sqrt{5xy} \cdot 4\sqrt{2x^5y^3} \cdot 5\sqrt{5x^3y^5}$
23. $5\sqrt{3x^3}\sqrt{6x} \cdot 4\sqrt{2x}$
24. $\sqrt{2}(\sqrt{2}+1)$
25. $(2\sqrt[3]{4x})^3$
26. $(5\sqrt{2x^2})^3$
27. $(3\sqrt{2x}+5)^2$
28. $(4\sqrt{3x}-2)^2$
29. $(5x\sqrt{3x}-4)^2$
30. $(\sqrt{a+b}\sqrt{a-b})^2$

In Exercises 31–48 divide the radicals and simplify.

31. $\dfrac{\sqrt{32}}{\sqrt{2}}$
32. $\dfrac{\sqrt{98}}{\sqrt{2}}$
33. $\dfrac{\sqrt[3]{5}}{\sqrt[3]{4}}$
34. $\dfrac{\sqrt[3]{7}}{\sqrt[3]{2}}$
35. $\dfrac{\sqrt{72}}{\sqrt{2}}$
36. $\dfrac{30\sqrt[4]{64x}}{6\sqrt[4]{4x}}$
37. $\dfrac{\sqrt{x^4y}}{\sqrt{5y}}$
38. $\dfrac{\sqrt{m^6n}}{\sqrt{3n}}$
39. $\dfrac{12\sqrt[4]{15x}}{4\sqrt[4]{5x}}$
40. $\dfrac{15\sqrt[4]{18y}}{5\sqrt[4]{3y}}$
41. $\dfrac{\sqrt[5]{128z^7}}{\sqrt[5]{2z}}$
42. $\dfrac{\sqrt[5]{3^7b^8}}{\sqrt[5]{3b^2}}$
43. $\dfrac{\sqrt{72x^3y^2}}{\sqrt{2xy^2}}$
44. $\dfrac{\sqrt{27x^2y^3}}{\sqrt{3x^2y}}$
45. $\dfrac{6\sqrt[4]{2^5m^2}}{2\sqrt[4]{2m^3}}$
46. $\dfrac{7\sqrt[4]{3^5H^3}}{14\sqrt[4]{3H^4}}$
47. $\dfrac{6\sqrt[3]{4B^2}}{9\sqrt[3]{8B^4}}$
48. $\dfrac{5\sqrt[3]{3K^3}}{15\sqrt[3]{9K^5}}$

In Exercises 49–78 first change the radical expression into an equivalent expression that has exponents. Then simplify it by using the rules of exponents. Finally, convert back to a radical expression.

49. $\sqrt[6]{x^3}$
50. $\sqrt[6]{x^2}$
51. $\sqrt[8]{a^6}$
52. $\sqrt[4]{a^2}$
53. $\sqrt[6]{x^4}$
54. $\sqrt[8]{16a^4}$
55. $\sqrt[6]{27b^3}$
56. $\sqrt[6]{4b^4}$
57. $(\sqrt[6]{m^3})^2$
58. $(\sqrt[6]{x^2})^2$
59. $(\sqrt[4]{m^2})^2$
60. $\sqrt[3]{a^2}\sqrt[6]{a^2}$
61. $\sqrt{a}\sqrt[4]{a}$
62. $\sqrt{b}\sqrt[3]{b}$
63. $\sqrt{8}\sqrt[3]{16}$
64. $\sqrt{27}\sqrt[3]{81}$
65. $\sqrt{2}\sqrt[6]{8}$
66. $\sqrt{3}\sqrt[3]{3}\sqrt[6]{3}$
67. $\sqrt[3]{x^2}\sqrt[4]{x^3}\sqrt{x}$
68. $\sqrt{y}\sqrt[3]{y}\sqrt[4]{y^3}$
69. $\sqrt[3]{-8z^2}\sqrt[3]{-z}\sqrt[4]{16z^3}$
70. $\sqrt[3]{-27w}\sqrt[3]{-w^2}\sqrt[4]{16w^3}$
71. $\sqrt{x}\sqrt[4]{x^3}\sqrt[8]{x^6}$
72. $\dfrac{\sqrt[4]{32x^3}}{\sqrt{2x}}$
73. $\dfrac{\sqrt[4]{G^3}}{\sqrt[3]{G^2}}$
74. $\dfrac{\sqrt[5]{H^4}}{\sqrt{H}}$
75. $\dfrac{\sqrt[3]{-x^2}}{\sqrt[6]{x^5}}$
76. $\dfrac{\sqrt[6]{y^3}}{\sqrt[3]{-y^2}}$
77. $\dfrac{\sqrt[3]{x^{-1}}}{\sqrt{x}}$
78. $\dfrac{\sqrt[n]{x^3}}{\sqrt[n]{x}}$

In Exercises 79 and 80 make no assumption about the value of the letters.

79. $\sqrt{4-k}\sqrt[4]{16-8k+k^2}$

80. $\dfrac{10\sqrt[6]{(S-2t)^8}}{\sqrt[6]{64(S-2t)^2}}$

6–5 Adding Radicals

Like Radicals

Like radicals are radicals that have the *same index and the same radicand.*

Example 1 Like radicals

(a) $2\sqrt[3]{x},\quad -9\sqrt[3]{x},\quad 11\sqrt[3]{x}$ Index = 3; radicand = x

(b) $\dfrac{2}{3}\sqrt[4]{5ab},\quad -\dfrac{1}{2}\sqrt[4]{5ab}$ Index = 4; radicand = $5ab$

Unlike Radicals

Unlike radicals are radicals that have *different indices or different radicands.*

Example 2 Unlike radicals

(a) $\sqrt{7},\quad \sqrt{5}$ Different radicands

(b) $\sqrt[3]{x},\quad \sqrt{x}$ Different indices

(c) $\sqrt[5]{2y},\quad \sqrt[3]{2}$ Different indices and radicands

ADDING LIKE RADICALS

Like radicals are added in the same way as any other like things—by adding the coefficients.

$$\begin{aligned} 3\text{ cars} + 2\text{ cars} &= (3+2)\text{ cars} &&= 5\text{ cars}\\ 3c + 2c &= (3+2)c &&= 5c\\ 3\sqrt{7} + 2\sqrt{7} &= (3+2)\sqrt{7} &&= 5\sqrt{7} \end{aligned}$$

Example 3 Adding like radicals

(a) $6\sqrt[3]{x} - 4\sqrt[3]{x} = (6-4)\sqrt[3]{x} = 2\sqrt[3]{x}$

(b) $2x\sqrt[4]{5x} - x\sqrt[4]{5x} = (2x-x)\sqrt[4]{5x} = x\sqrt[4]{5x}$

(c) $\dfrac{2}{3}\sqrt[5]{4xy^2} - 6\sqrt[5]{4xy^2} = \left(\dfrac{2}{3}-6\right)\sqrt[5]{4xy^2} = -\dfrac{16}{3}\sqrt[5]{4xy^2}$

ADDING UNLIKE RADICALS

Not all unlike radicals can be added. Unlike radicals can be added *only if they become like radicals when simplified.*

TO ADD UNLIKE RADICALS

1. Simplify each radical in the sum.
2. Then combine any terms that have like radicals by adding their coefficients and multiplying that sum by the like radical.

Example 4 $\sqrt{8} + \sqrt{18}$

$$= \sqrt{4 \cdot 2} + \sqrt{9 \cdot 2}$$

$$= 2\sqrt{2} + 3\sqrt{2} = 5\sqrt{2}$$

Example 5 $\sqrt{12x} - \sqrt{27x} + 5\sqrt{3x}$

$$= \sqrt{4 \cdot 3 \cdot x} - \sqrt{9 \cdot 3 \cdot x} + 5\sqrt{3x}$$

$$= 2\sqrt{3x} - 3\sqrt{3x} + 5\sqrt{3x}$$

$$= (2 - 3 + 5)\sqrt{3x} = 4\sqrt{3x}$$

Example 6 $\sqrt[3]{24a^2} - 5\sqrt[3]{3a^5} + 2\sqrt[3]{81a^2}$

$$= \sqrt[3]{2^3 \cdot 3 \cdot a^2} - 5\sqrt[3]{3 \cdot a^3 \cdot a^2} + 2\sqrt[3]{3^3 \cdot 3 \cdot a^2}$$

$$= 2\sqrt[3]{3a^2} - 5a\sqrt[3]{3a^2} + 2 \cdot 3\sqrt[3]{3a^2}$$

$$= (2 - 5a + 6)\sqrt[3]{3a^2} = (8 - 5a)\sqrt[3]{3a^2}$$

Example 7 $2\sqrt{\frac{1}{2}} - 6\sqrt{\frac{1}{8}} - 10\sqrt{\frac{4}{5}}$

$$= \frac{2}{1}\sqrt{\frac{1}{2} \cdot \frac{2}{2}} - \frac{6}{1}\sqrt{\frac{1}{2^3} \cdot \frac{2}{2}} - \frac{10}{1}\sqrt{\frac{4}{5} \cdot \frac{5}{5}}$$

$$= \frac{2}{1} \cdot \frac{1}{2}\sqrt{2} - \frac{6}{1}\ \frac{1}{4}\sqrt{2} - \frac{10}{1} \cdot \frac{2}{5}\sqrt{5}$$

$$= \sqrt{2} - \frac{3}{2}\sqrt{2} - 4\sqrt{5}$$

$$= \left(1 - \frac{3}{2}\right)\sqrt{2} - 4\sqrt{5} = -\frac{1}{2}\sqrt{2} - 4\sqrt{5}$$

Example 8 $5\sqrt[4]{\dfrac{3x}{8}} + \dfrac{2x}{3}\sqrt[4]{\dfrac{1}{6^3x^3}}$

$$= \frac{5}{1}\sqrt[4]{\frac{3x}{2^3}\cdot\frac{2}{2}} + \frac{2x}{3}\sqrt[4]{\frac{1}{6^3x^3}\cdot\frac{6x}{6x}}$$

$$= \frac{5}{2}\sqrt[4]{6x} + \frac{2x}{3}\cdot\frac{1}{6x}\sqrt[4]{6x}$$

$$= \frac{5}{2}\sqrt[4]{6x} + \frac{1}{9}\sqrt[4]{6x}$$

$$= \left(\frac{5}{2}+\frac{1}{9}\right)\sqrt[4]{6x} = \left(\frac{45}{18}+\frac{2}{18}\right)\sqrt[4]{6x} = \frac{47}{18}\sqrt[4]{6x}$$

Examples 9–11 involve both multiplication *and* addition of radicals.

Example 9 $\sqrt[3]{x}(4\sqrt[3]{x^2} - 5) = \sqrt[3]{x}\cdot 4\sqrt[3]{x^2} - \sqrt[3]{x}\cdot 5$ Distributive rule

$$= 4\sqrt[3]{x\cdot x^2} - 5\sqrt[3]{x}$$

$$= 4\sqrt[3]{x^3} - 5\sqrt[3]{x}$$

$$= 4x - 5\sqrt[3]{x}$$

Example 10 $(2\sqrt{3} - 5)(4\sqrt{3} - 6)$ Product of two binomials

$2\sqrt{3}\cdot 4\sqrt{3} = 8\cdot 3 = 24$

$(-5)(-6) = 30$

$(2\sqrt{3} - 5)\ (4\sqrt{3} - 6)$

$-20\sqrt{3}$

$\underline{-12\sqrt{3}}$

$$= 24 - 32\sqrt{3} + 30$$

$$= 54 - 32\sqrt{3}$$

Example 11 Examples involving both division *and* addition of radicals

(a) $\dfrac{3\sqrt{14} - \sqrt{8}}{\sqrt{2}} = \dfrac{3\sqrt{14}}{\sqrt{2}} - \dfrac{\sqrt{8}}{\sqrt{2}} = 3\sqrt{\dfrac{14}{2}} - \sqrt{\dfrac{8}{2}}$

$$= 3\sqrt{7} - \sqrt{4} = 3\sqrt{7} - 2$$

(b) $\dfrac{3\sqrt[5]{64a} - 9\sqrt[5]{4a^3}}{3\sqrt[5]{2a}} = \dfrac{3\sqrt[5]{64a}}{3\sqrt[5]{2a}} - \dfrac{9\sqrt[5]{4a^3}}{3\sqrt[5]{2a}} = \sqrt[5]{\dfrac{64a}{2a}} - 3\sqrt[5]{\dfrac{4a^3}{2a}}$

$$= \sqrt[5]{32} - 3\sqrt[5]{2a^2} = 2 - 3\sqrt[5]{2a^2}$$

RATIONALIZING A BINOMIAL DENOMINATOR THAT CONTAINS SQUARE ROOTS

Conjugate

*The **conjugate** of a binomial containing square roots* is a binomial that has the same two terms with the sign of the second term changed.

Example 12 (a) The conjugate of $1 - \sqrt{2}$ is $1 + \sqrt{2}$.

(b) The conjugate of $-2\sqrt{3} + 5$ is $-2\sqrt{3} - 5$.

(c) The conjugate of $\sqrt{x} - \sqrt{y}$ is $\sqrt{x} + \sqrt{y}$.

The product of a binomial containing square roots, and its conjugate, is a rational number. For example:

$$(1 - \sqrt{2})(1 + \sqrt{2}) = 1 - 2 = -1 \quad \text{Rational number}$$

because $(a - b)(a + b) = a^2 - b^2$

Because of this fact, the following procedure should be used when a binomial denominator contains a square root:

Rationalizing Binomial Denominators

TO RATIONALIZE A BINOMIAL DENOMINATOR THAT CONTAINS SQUARE ROOTS

Multiply the numerator and the denominator by the conjugate of the denominator:

$$\frac{a}{b + \sqrt{c}} \cdot \frac{b - \sqrt{c}}{b - \sqrt{c}} = \frac{a(b - \sqrt{c})}{b^2 - c}$$

Example 13 Rationalizing binomial denominators containing square roots

(a) $$\frac{2}{1 + \sqrt{3}} = \frac{2}{1 + \sqrt{3}} \cdot \frac{1 - \sqrt{3}}{1 - \sqrt{3}} = \frac{2(1 - \sqrt{3})}{1 - 3} = \frac{2(1 - \sqrt{3})}{-2} = \sqrt{3} - 1$$

Multiplied numerator and denominator by $1 - \sqrt{3}$ (the conjugate of the denominator $1 + \sqrt{3}$) — since the value of this fraction is 1, multiplying $\frac{2}{1 + \sqrt{3}}$ by 1 does not change its value

(b) $$\frac{6}{\sqrt{5} - \sqrt{3}} = \frac{6}{\sqrt{5} - \sqrt{3}} \cdot \frac{\sqrt{5} + \sqrt{3}}{\sqrt{5} + \sqrt{3}} = \frac{6(\sqrt{5} + \sqrt{3})}{5 - 3} = \frac{\overset{3}{\cancel{6}}(\sqrt{5} + \sqrt{3})}{\cancel{2}}$$

$$= 3\sqrt{5} + 3\sqrt{3}$$

Multiplied numerator and denominator by $\sqrt{5} + \sqrt{3}$ (the conjugate of the denominator $\sqrt{5} - \sqrt{3}$)

EXERCISES 6-5

In Exercises 1–35 find each sum.

1. $3\sqrt{5} - \sqrt{5}$
2. $4\sqrt{3} - \sqrt{3}$
3. $5\sqrt[4]{xy} + 2\sqrt[4]{xy}$
4. $7\sqrt[4]{ab} + 3\sqrt[4]{ab}$
5. $\frac{3\sqrt[3]{2}}{2} - \frac{\sqrt[3]{2}}{2} + \frac{1}{2}\sqrt[3]{2}$
6. $\frac{4\sqrt[5]{3}}{3} - \frac{\sqrt[5]{3}}{3} + \frac{1}{3}\sqrt[5]{3}$
7. $2\sqrt{50} - \sqrt{32}$
8. $3\sqrt{24} - \sqrt{54}$
9. $3\sqrt{32x} - \sqrt{8x}$
10. $4\sqrt{27y} - 3\sqrt{12y}$
11. $\sqrt{125M} + \sqrt{20M} - \sqrt{45M}$
12. $\sqrt{75P} - \sqrt{48P} + \sqrt{27P}$
13. $\sqrt[3]{27x} + \frac{1}{2}\sqrt[3]{8x}$
14. $\frac{3}{4}\sqrt[3]{64a} + \sqrt[3]{27a}$
15. $\sqrt[3]{a^4} + 2a\sqrt[3]{8a}$
16. $H\sqrt[3]{8H^2} + 3\sqrt[3]{H^5}$
17. $\sqrt[5]{x^2y^6} + \sqrt[5]{x^7y}$
18. $\sqrt[5]{a^3b^8} + \sqrt[5]{a^8b^3}$
19. $b\sqrt[4]{4a^4b} + ab\sqrt[4]{64b} - \sqrt[4]{4a^4b^5}$
20. $\sqrt[4]{16a^7b^4} - a\sqrt[4]{a^3b^4} + 3ab\sqrt[4]{a^3}$
21. $3\sqrt{\frac{1}{6}} + \sqrt{12} - 5\sqrt{\frac{3}{2}}$
22. $3\sqrt{\frac{5}{2}} + \sqrt{20} - 5\sqrt{\frac{1}{10}}$
23. $5\sqrt{\frac{1}{8}} - 7\sqrt{\frac{1}{18}} - 5\sqrt{\frac{1}{50}}$
24. $12\sqrt[3]{\frac{x^3}{16}} + x\sqrt[3]{\frac{1}{2}}$
25. $8\sqrt[3]{\frac{a^3}{32}} + a\sqrt[3]{54}$
26. $12\sqrt{\frac{7z}{9}} - \frac{4z}{5}\sqrt{\frac{25}{7z}}$
27. $10\sqrt{\frac{5b}{4}} - \frac{3b}{2}\sqrt{\frac{4}{5b}}$
28. $6\sqrt[3]{\frac{a^4}{54}} + 2a\sqrt[3]{\frac{a}{2}}$
29. $12\sqrt[3]{\frac{x^5}{24}} + 6x\sqrt[3]{\frac{x^2}{3}}$
30. $3h^2\sqrt[4]{\frac{4}{27h}} + 4h^2\sqrt[4]{\frac{5}{8}} - 2h^2\sqrt[4]{\frac{12}{h}}$
31. $2k\sqrt[4]{\frac{3}{8k}} - \frac{1}{k}\sqrt[4]{\frac{2k^3}{27}} + 5k^2\sqrt[4]{\frac{6}{k^2}}$
32. $\sqrt{\frac{x^3}{3y^3}} + xy\sqrt{\frac{x}{3y^5}} - \frac{y^2}{3}\sqrt{\frac{3x^3}{y^7}}$
33. $\sqrt[4]{x}\sqrt[4]{x^3} + 3\sqrt{x^2}$
34. $\sqrt[3]{\frac{3}{5}}\sqrt[3]{\frac{40}{81}} - 1$
35. $\sqrt{\frac{3}{5}}\sqrt{\frac{2}{9}}\sqrt{\frac{5}{6}} - \sqrt{\frac{1}{3}}$

In Exercises 36–54 find each product.

36. $\sqrt{3}(\sqrt{3}+1)$
37. $\sqrt{2}(\sqrt{2}+1)$
38. $\sqrt{y}(4-\sqrt{y})$
39. $\sqrt{3}(2\sqrt{3}+1)$
40. $\sqrt{5}(3\sqrt{5}+1)$
41. $\sqrt{x}(\sqrt{x}-3)$
42. $4\sqrt{3x}(5\sqrt{4x}+\sqrt{3x})$
43. $3\sqrt{2x}(5\sqrt{8x}-7\sqrt{2x})$
44. $(\sqrt{2x})^3(\sqrt{2x}+1)$
45. $\sqrt[3]{x}(2\sqrt[3]{x}-3)$
46. $\sqrt[3]{y}(3\sqrt[3]{y}-5)$
47. $(\sqrt{7}+2)(\sqrt{7}+3)$
48. $(\sqrt{3}+2)(\sqrt{3}+4)$
49. $(\sqrt{7x}-4)(\sqrt{7x}+4)$
50. $(\sqrt{2x}+3)(\sqrt{2x}-3)$
51. $(2\sqrt{3}-5x)^2$
52. $(5\sqrt{2}-3x)^2$
53. $(\sqrt{ab}+2\sqrt{a})^2$
54. $(\sqrt{cd}-3\sqrt{c})^2$

In Exercises 55–60 find each quotient.

55. $\dfrac{\sqrt{20}+5\sqrt{10}}{\sqrt{5}}$
56. $\dfrac{2\sqrt{6}+\sqrt{18}}{\sqrt{6}}$
57. $\dfrac{4\sqrt[3]{8x}+6\sqrt[3]{32x^4}}{2\sqrt[3]{4x}}$
58. $\dfrac{10\sqrt[3]{81a^7}+15\sqrt[3]{6a}}{5\sqrt[3]{3a}}$
59. $\dfrac{30\sqrt[4]{32a^6}-6\sqrt[4]{24a^2}}{3\sqrt[4]{2a^2}}$
60. $\dfrac{15\sqrt[4]{6x^3}-50\sqrt[4]{32x^7}}{5\sqrt[4]{2x^3}}$

In Exercises 61–72 rationalize each denominator.

61. $\dfrac{14}{\sqrt{7}}$
62. $\dfrac{5}{\sqrt{10}}$
63. $\dfrac{a^2}{2\sqrt[3]{a}}$
64. $\dfrac{b^2}{3\sqrt[3]{b^2}}$
65. $\dfrac{3}{\sqrt{2}-1}$
66. $\dfrac{5}{\sqrt{2}+1}$
67. $\dfrac{6}{\sqrt{5}+\sqrt{2}}$
68. $\dfrac{4}{\sqrt{7}+\sqrt{5}}$
69. $\dfrac{4\sqrt{3}-\sqrt{2}}{4\sqrt{3}+\sqrt{2}}$
70. $\dfrac{y-9}{\sqrt{y}-3}$
71. $\dfrac{\sqrt{x+1}-\sqrt{x}}{\sqrt{x+1}+\sqrt{x}}$
72. $\sqrt{\dfrac{\sqrt{x+4}-\sqrt{x}}{\sqrt{x+4}+\sqrt{x}}}$

73. Find the value of y^2-2y-5 if $y=1-\sqrt{6}$.
74. Find the value of $x^2-10x+22$ if $x=\sqrt{3}+5$.

In Exercises 75 and 76 add using rational exponents.

75. $\sqrt{2}+\sqrt[4]{4}$
76. $\sqrt{3}+\sqrt[4]{9}$

6–6 Radical Equations

A **radical equation** is an equation in which the unknown letter appears in a radicand.

Example 1 Radical equations

(a) $\sqrt{x+2}=3$
(b) $-3=\sqrt[3]{2x+5}$
(c) $\sqrt{2x-3}=\sqrt{x}+5$
(d) $\sqrt[4]{x-6}=2$

Solving Radical Equations

Example 2 Solve: $\sqrt{x-1} = 6$

Therefore $(\sqrt{x-1})^2 = (6)^2$

$x - 1 = 36$

$x = 37$

Check for x = 37:

$\sqrt{x-1} = 6$

$\sqrt{(37)-1} \stackrel{?}{=} 6$

$\sqrt{36} = 6$

A similar procedure can be used for radicals that are not square roots.

TO SOLVE A RADICAL EQUATION

1. Arrange the terms so that one term with a radical is by itself on one side of the equation.
2. Raise each side of the equation to a power equal to the index of the radical.
3. Collect like terms.
4. If a radical still remains, repeat Steps 1, 2, and 3.
5. Solve the resulting equation for the unknown letter.

Check apparent solutions in the *original* equation.

Example 3 Solve: $\sqrt[3]{x-1} + 4 = 0$

$\sqrt[3]{x-1} = -4$

$(\sqrt[3]{x-1})^3 = (-4)^3$

$x - 1 = -64$

$x = -63$

Check for x = −63:

$\sqrt[3]{x-1} + 4 = 0$

$\sqrt[3]{(-63)-1} + 4 \stackrel{?}{=} 0$

$\sqrt[3]{-64} + 4 \stackrel{?}{=} 0$

$-4 + 4 = 0$

Therefore, the solution set $= \{-63\}$

A Word of Caution. In solving radical equations, we raise both sides to an integral power. This amounts to multiplying both sides by an expression containing the unknown. This procedure may introduce extraneous roots (Example 3, Section 5–5). For this reason it is necessary to *check all roots in the given equation.* ■

Example 4 Solve: $\sqrt[4]{2x+1} = -2$

We know the principal root of an even index radical must be *positive*. Therefore this equation does not have a solution that is a real number. If you did not notice this and followed the procedure we outlined for solving radical equations, the same result would be obtained.

	Check for $x = \frac{15}{2}$:
$\sqrt[4]{2x+1} = -2$	$\sqrt[4]{2x+1} = -2$
$(\sqrt[4]{2x+1})^4 = (-2)^4$	$\sqrt[4]{2\left(\frac{15}{2}\right)+1} \stackrel{?}{=} -2$
$2x + 1 = 16$	$\sqrt[4]{15+1} \stackrel{?}{=} -2$
$2x = 15$	$\sqrt[4]{16} \stackrel{?}{=} -2$
$x = \frac{15}{2}$	$2 \neq -2$

Principal root

Since $\frac{15}{2}$ does not check, this equation has no real root.

The solution set = { }

Example 5 Solve: $\sqrt{2x+1} = x - 1$

$$(\sqrt{2x+1})^2 = (x-1)^2$$

When squaring $(x - 1)$, do not forget this middle term

$$2x + 1 = x^2 - 2x + 1$$

$$0 = x^2 - 4x$$

$$0 = x(x - 4)$$

$$x = 0 \quad \Big| \quad x - 4 = 0$$

$$x = 4$$

Check for $x = 0$:

$$\sqrt{2x+1} = x - 1$$

$$\sqrt{2(0)+1} \stackrel{?}{=} (0) - 1$$

$$\sqrt{1} \neq -1$$

$$1 \neq -1$$

The symbol $\sqrt{1}$ *always* stands for the *principal* square root of 1, which is 1 (*not* -1)

Therefore 0 *is not a solution* of $\sqrt{2x+1} = x - 1$ because it does not satisfy the equation.

Check for $x = 4$:

$$\sqrt{2x+1} = x - 1$$

$$\sqrt{2(4)+1} \stackrel{?}{=} (4) - 1$$

$$\sqrt{9} \stackrel{?}{=} 3$$

$$3 = 3$$

Therefore 4 *is a solution* because it does satisfy the equation.

The solution set = {4}

Example 6

Solve: $\sqrt{4x+5} - \sqrt{x-1} = 3$

$$\sqrt{4x+5} = \sqrt{x-1} + 3$$

Get one radical by itself on one side

$$(\sqrt{4x+5})^2 = (\sqrt{x-1} + 3)^2$$

When squaring, don't forget this term

$$4x + 5 = x - 1 + 6\sqrt{x-1} + 9$$

$$3x - 3 = 6\sqrt{x-1}$$

$$x - 1 = 2\sqrt{x-1}$$

Divide both sides by 3

$$(x-1)^2 = (2\sqrt{x-1})^2$$

$$x^2 - 2x + 1 = 4(x-1)$$

$$x^2 - 6x + 5 = 0$$

$$(x-1)(x-5) = 0$$

$$x - 1 = 0 \quad \Big| \quad x - 5 = 0$$

$$x = 1 \quad \Big| \quad x = 5$$

You should verify that both 1 and 5 are solutions.

The same method used to solve radical equations can be used to solve some equations that have rational exponents.

Example 7 Solve: $x^{-2/3} = 4$

$$x^{-2/3} = 2^2$$

$$(x^{-2/3})^{-3/2} = (2^2)^{-3/2}$$

$$x^{(-2/3)(-3/2)} = 2^{2(-3/2)}$$

$$x = 2^{-3} = \frac{1}{8}$$

We raise both sides to the $-\frac{3}{2}$ power in order to make the exponent of x equal to 1: $(-\frac{2}{3})(-\frac{3}{2}) = 1$

Check for $x = \frac{1}{8}$: $x^{-2/3} = 4$

$$\left(\frac{1}{8}\right)^{-2/3} \stackrel{?}{=} 4$$

$$(2^{-3})^{-2/3} \stackrel{?}{=} 4$$

$$2^2 = 4$$

The solution set $= \left\{\frac{1}{8}\right\}$

EXERCISES 6-6

Solve and check each equation.

1. $\sqrt{3x+1} = 5$
2. $\sqrt{7x+8} = 6$
3. $\sqrt{3x-2} = x$
4. $\sqrt{5x-6} = x$
5. $\sqrt{x+1} = \sqrt{2x-7}$
6. $\sqrt{3x-2} = \sqrt{x+4}$
7. $\sqrt[4]{4x-11} - 1 = 0$
8. $\sqrt[4]{3x+1} - 2 = 0$
9. $\sqrt{4x-1} = 2x$
10. $\sqrt{6x-1} = 3x$
11. $\sqrt{x-3} + 5 = x$
12. $\sqrt{4x+5} + 5 = 2x$
13. $\sqrt{x+7} = 2x - 1$
14. $\sqrt{2x} + 1 = \sqrt{2x-3}$
15. $\sqrt{v+7} = 5 + \sqrt{v-2}$
16. $\sqrt{x} = \sqrt{x+16} - 2$
17. $\sqrt{3x+4} - \sqrt{2x-4} = 2$
18. $\sqrt{2x-1} + 2 = \sqrt{3x+10}$
19. $\sqrt[3]{2x+3} - 2 = 0$
20. $\sqrt[3]{4x-3} - 3 = 0$
21. $\sqrt[5]{7x+4} - 2 = 0$
22. $\sqrt[5]{3x-14} - 1 = 0$
23. $x^{1/2} = 5$
24. $x^{1/2} = 7$
25. $x^{3/5} = 8$
26. $x^{2/5} = 4$
27. $3x^{-2/3} = 48$
28. $5x^{-3/2} = 40$
29. $\sqrt{2x-9} - \sqrt{4x} + 3 = 0$
30. $\sqrt{2x-3} - \sqrt{8x} + 3 = 0$
31. $\sqrt{4u+1} - \sqrt{u-2} = \sqrt{u+3}$
32. $\sqrt{2-v} + \sqrt{v+3} = \sqrt{7+2v}$
33. $\sqrt{x+4} - \sqrt{2} = \sqrt{x-6}$
34. $\sqrt{4v+1} = \sqrt{v+4} + \sqrt{v-3}$

6–7 The Pythagorean Theorem

Right Triangles

A triangle that has a **right angle** (90°) is called a **right triangle.** The **diagonal** of a rectangle divides the rectangle into two right triangles. The parts of a right triangle are shown in Figure 6–7A.

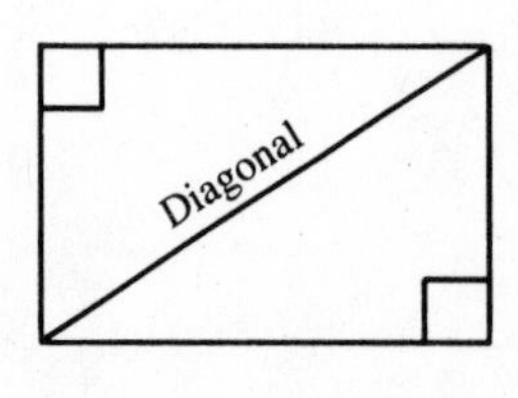

Rectangle

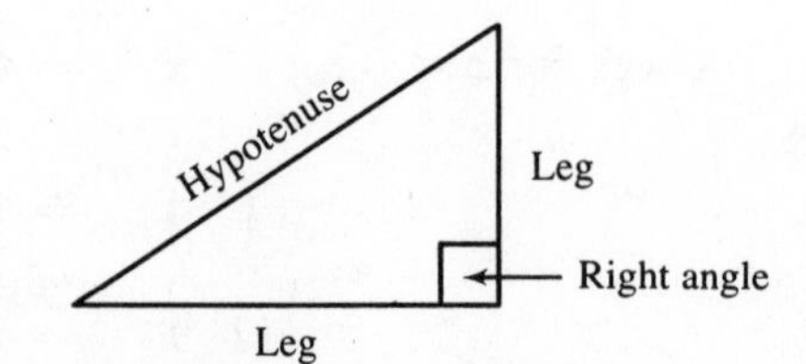

Right triangle

Figure 6–7A

THE PYTHAGOREAN THEOREM

The square of the hypotenuse of a right triangle is equal to the sum of the squares of the other sides.

$$c^2 = a^2 + b^2$$

Note: The Pythagorean theorem applies only to *right triangles*. ■

The Pythagorean theorem can be verified by comparing the areas shown in Figure 6–7B. The area of the large square $(a + b)^2$ is equal to the area of the inside square c^2 plus the area of four equal triangles, $4\left(\frac{1}{2}ab\right)$.

$$(a + b)^2 = c^2 + \frac{\overset{2}{\cancel{4}}}{1}\left(\frac{1}{\cancel{2}}ab\right)$$

$$\begin{array}{rcl} a^2 + 2ab + b^2 & = & c^2 + 2ab \\ -\ 2ab & & -\ 2ab \\ \hline a^2 + b^2 & = & c^2 \end{array}$$

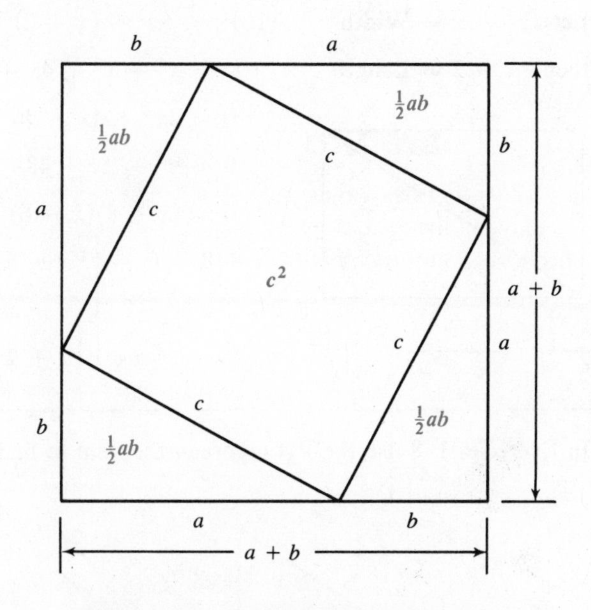

Figure 6–7B Pythagorean Theorem

Example 1 Find the hypotenuse of a right triangle with legs that are 8 and 6 units long

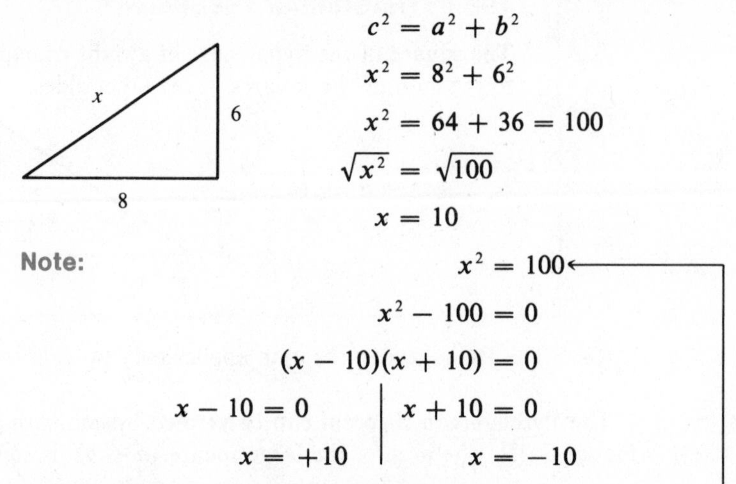

$$c^2 = a^2 + b^2$$
$$x^2 = 8^2 + 6^2$$
$$x^2 = 64 + 36 = 100$$
$$\sqrt{x^2} = \sqrt{100}$$
$$x = 10$$

Note:

$$x^2 = 100$$
$$x^2 - 100 = 0$$
$$(x - 10)(x + 10) = 0$$
$$x - 10 = 0 \quad\Big|\quad x + 10 = 0$$
$$x = +10 \quad\Big|\quad x = -10$$

$x = 10, -10$ are solutions of this *equation*

However, -10 cannot be a solution of this *geometric problem* because we usually consider lengths in geometric figures as positive numbers. For this reason in the problems of this section we will take only the positive *(principal)* square root. (For more information on principal square roots, see Sections 6–2 and 6–3.) ■

Example 2 The length of a rectangle is 2 more than its width. If the length of its diagonal is 10, find the dimensions of the rectangle.

Let $x = \text{Width}$

then $x + 2 = \text{Length}$

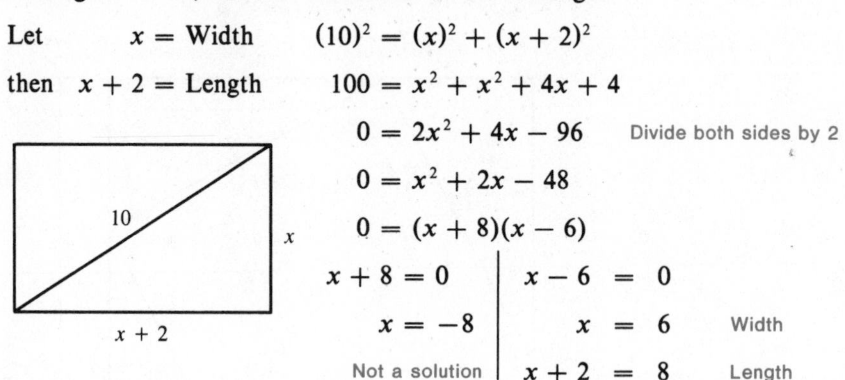

$$(10)^2 = (x)^2 + (x + 2)^2$$
$$100 = x^2 + x^2 + 4x + 4$$
$$0 = 2x^2 + 4x - 96 \qquad \text{Divide both sides by 2}$$
$$0 = x^2 + 2x - 48$$
$$0 = (x + 8)(x - 6)$$
$$x + 8 = 0 \quad\Big|\quad x - 6 = 0$$
$$x = -8 \quad\Big|\quad x = 6 \qquad \text{Width}$$
$$\text{Not a solution} \quad\Big|\quad x + 2 = 8 \qquad \text{Length}$$

EXERCISES 6-7

In Exercises 1–8 use the Pythagorean theorem to find x in each figure.

1.

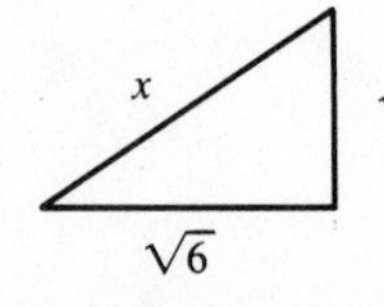

2.

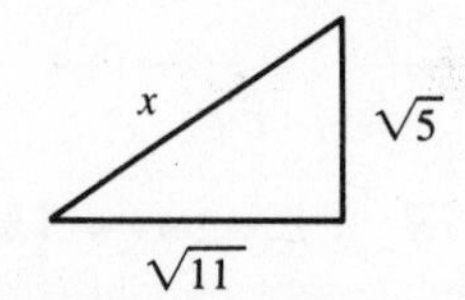

3.

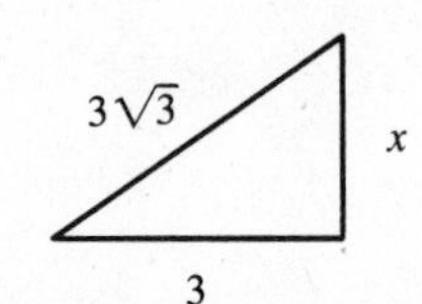

4.
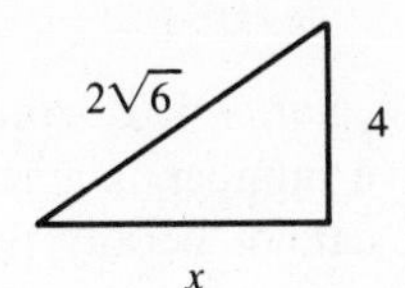

5.
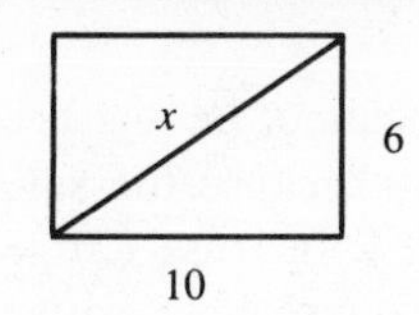

6.
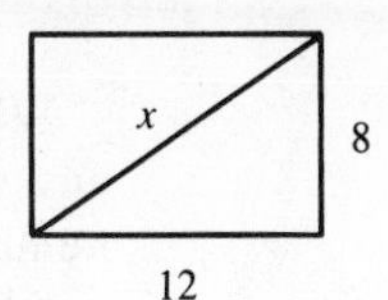

7.
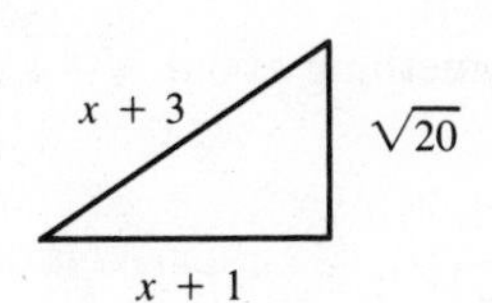

8.
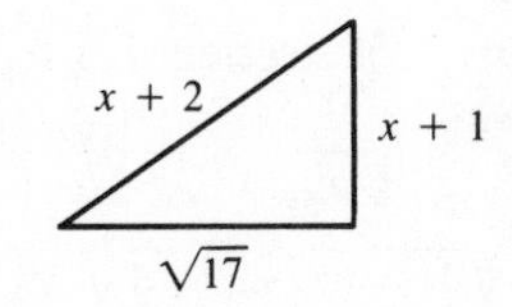

9. Find the diagonal of a square with side equal to 4.
10. Find the diagonal of a square with side equal to 3.
11. Find the width of a rectangle that has a diagonal of 25 and a length of 24.
12. Find the width of a rectangle that has a diagonal of 41 and a length of 40.
13. One leg of a right triangle is 4 less than twice the other leg. If its hypotenuse is 10, how long are the two legs?
14. One leg of a right triangle is 4 more than twice the other leg. If its hypotenuse is $\sqrt{61}$, how long are the two legs?
15. One side of a triangle is 2 more than twice the other side. If the hypotenuse is 2 less than three times the shorter side, find the length of the sides and the hypotenuse of the triangle.
16. The length of a rectangle is 3 more than its width. If the length of its diagonal is 15, find the dimensions of the rectangle.
17. *Brainteaser.* Rectangle $OABC$ is inscribed in a quarter of a circle as shown in the figure. Given the lengths indicated, find:
(a) The length of diagonal AC.
(b) The length of side AB.

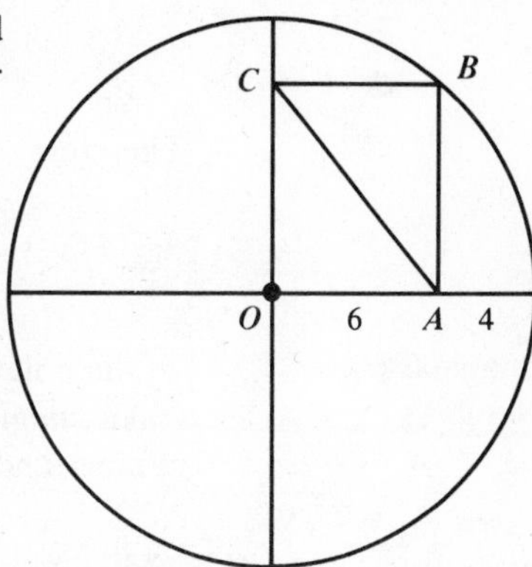

6–8 Complex Numbers

All numbers discussed up to this point have been real numbers. In this section we discuss a new kind of number which is *not* a real number, but which is essential in many applications of mathematics.

6-8A BASIC DEFINITIONS

What is $\sqrt{-4}$? It cannot be -2 because $(-2)^2 = 4$, *not* -4. It cannot be any negative number, because the square of a negative number is a positive number. It cannot be 2 because $(2)^2 = 4$, *not* -4. It cannot be any positive number, because the square of a positive number is positive. It cannot be zero—the only remaining real number—because $(0)^2 = 0$, *not* -4. Therefore $\sqrt{-4}$ is *not* a real number.

We can use Rule 1 of Section 6–3 to discover something about $\sqrt{-4}$.

$$\sqrt[n]{ab} = \sqrt[n]{a}\sqrt[n]{b} \qquad \text{Rule 1}$$

$$\sqrt{-4} = \sqrt{4(-1)} = \sqrt{4}\sqrt{-1} = 2\sqrt{-1}$$

Also, $\sqrt{-9} = \sqrt{9(-1)} = \sqrt{9}\sqrt{-1} = 3\sqrt{-1}$

and $\sqrt{-17} = \sqrt{17(-1)} = \sqrt{17}\sqrt{-1} = \sqrt{17}\sqrt{-1}$

Each of these numbers has been written as the product of a real number and $\sqrt{-1}$

2, 3, $\sqrt{17}$ are real numbers

The square root of any negative number can be written as the product of a real number and $\sqrt{-1}$. Since $\sqrt{-1}$ enters as a factor in *every* number of this type, it is given a special symbol, i.

DEFINITIONS

$$i = \sqrt{-1}$$

$$i^2 = -1$$

Therefore

$$\sqrt{-4} = \sqrt{4(-1)} = \sqrt{4}\sqrt{-1} = 2i$$

$$\sqrt{-9} = \sqrt{9(-1)} = \sqrt{9}\sqrt{-1} = 3i$$

$$\sqrt{-17} = \sqrt{17(-1)} = \sqrt{17}\sqrt{-1} = \sqrt{17}i$$

Imaginary Numbers

These new numbers, $2i$, $3i$, and $\sqrt{17}i$, are called **imaginary numbers.**

An *imaginary number* is a number that can be written as bi, where b is a real number and $i = \sqrt{-1}$.

Example 1 Imaginary numbers

$$-5i, \qquad 2\sqrt{3}i, \qquad -\frac{2}{5}i, \qquad 0.612i$$

Complex Numbers

A **complex number** is a number of the form $a + bi$, where a and b are real numbers and $i = \sqrt{-1}$.

Real part → a, Imaginary part → bi

$$a + bi$$

Example 2 Complex numbers

$$2 - 3i, \qquad \sqrt{5} - \frac{1}{2}i, \qquad -0.7 + 0.4i$$

When the real part is zero, the complex number is a *pure imaginary number.*

$$0 + 3i = 3i$$

a is zero (points to 0) — Pure imaginary number (points to $3i$)

Therefore the set of pure imaginary numbers is a subset of the set of complex numbers.

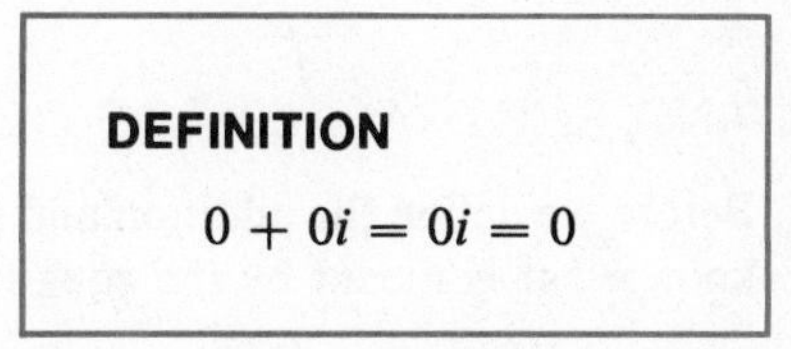

DEFINITION

$$0 + 0i = 0i = 0$$

This means that 0 is a pure imaginary number as well as a real number.

When the imaginary part is zero, the complex number is a *real number.*

$$2 + 0i = 2$$

b is zero (points to $0i$) — Real number (points to 2)

Therefore the set of real numbers is a subset of the set of complex numbers.

Example 3 Writing complex numbers in $a + bi$ form

(a) $2 - \sqrt{-25} = 2 - \sqrt{25(-1)} = 2 - \sqrt{25}\sqrt{-1} = 2 - 5i$

(b) $\sqrt{-17} = \sqrt{17(-1)} = \sqrt{17}\sqrt{-1} = \sqrt{17}i = 0 + \sqrt{17}i$

A Word of Caution. In writing complex numbers in $a + bi$ form, we have used Rule 1, Section 6–3:

$$\sqrt[n]{a}\sqrt[n]{b} = \sqrt[n]{ab}$$

This rule does *not* apply when both *a* and *b* are negative.

Incorrect	*Correct*
~~$\sqrt{-4}\sqrt{-9} = \sqrt{(-4)(-9)}$~~	$\sqrt{-4}\sqrt{-9}$ First, convert radicals to *bi* form
~~$= \sqrt{36} = 6$~~	$= 2i \cdot 3i = 6i^2$
	$= 6(-1) = -6$

Therefore $\sqrt{(-4)(-9)} \neq \sqrt{-4}\sqrt{-9}$. Rule 1 does not apply in this case. If it did, 6 would equal -6! ■

EXERCISES 6-8A

Convert each expression to the $a + bi$ form.

1. $3 + \sqrt{-16}$
2. $4 - \sqrt{-25}$
3. $5 + \sqrt{-49}$
4. $\sqrt{-4} - 6$
5. $5 + \sqrt{-32}$
6. $6 + \sqrt{-18}$
7. $\sqrt{-36} + \sqrt{4}$
8. $\sqrt{9} - \sqrt{-25}$
9. $3 + \sqrt{-8}$
10. $\sqrt{-81}$
11. $\sqrt{-64}$
12. $\sqrt{-100}$
13. $2i - \sqrt{9}$
14. $3i - \sqrt{16}$
15. 14
16. -7

6-8B ADDITION AND SUBTRACTION OF COMPLEX NUMBERS

Before we define the addition and subtraction of complex numbers, you need to know what is meant by the equality of complex numbers.

Equality of Complex Numbers

TWO COMPLEX NUMBERS ARE EQUAL IF

1. their real parts are equal, and
2. their imaginary parts are equal.

If $(a + bi) = (c + di)$,

then $a = c$ Real parts equal

and $b = d$ Imaginary parts equal

Example 4 If $(3x + 7yi) = (10 - 2i)$, find x and y

$$3x = 10 \quad \text{Real parts equal}$$

$$x = \frac{10}{3}$$

and $$7y = -2 \quad \text{Imaginary parts equal}$$

$$y = -\frac{2}{7}$$

Addition of Complex Numbers

To add two complex numbers, (1) add their real parts, and (2) add their imaginary parts as follows:

$$(a + bi) + (c + di) = (a + c) + (b + d)i$$

Subtraction of Complex Numbers

To subtract one complex number from another, (1) subtract their real parts, and (2) subtract their imaginary parts as follows:

$$(a + bi) - (c + di) = (a - c) + (b - d)i$$

These definitions lead to the following procedure:

> **TO SIMPLIFY AN EXPRESSION CONTAINING ADDITION OR SUBTRACTION OF COMPLEX NUMBERS**
>
> **1.** Remove grouping symbols
>
> **2.** Combine like terms
>
> **3.** Write the result in $a + bi$ form

Example 5 Addition and subtraction of complex numbers.

(a) $(-7 + 4i) + (6 - 3i) = \underline{-7} + \underline{\underline{4i}} + \underline{6} - \underline{\underline{3i}} = -1 + i$

(b) $(-13 + 8i) - (7 - 11i) = \underline{-13} + \underline{\underline{8i}} - \underline{7} + \underline{\underline{\underline{11i}}} = -20 + 19i$

(c) $(7 - i) - (6 - 10i) - (-4) = \underline{7} - \underline{\underline{i}} - \underline{6} + \underline{\underline{\underline{10i}}} + \underline{4} = 5 + 9i$

(d) $(-9 + 5i) - (4 - 2i) + (-7 - 3i)$
$= \underline{-9} + \underline{\underline{5i}} - \underline{4} + \underline{\underline{2i}} - \underline{7} - \underline{\underline{3i}} = -20 + 4i$

When complex numbers are not written in $a + bi$ form, they must be changed into $a + bi$ form before performing any operation with them.

Example 6 Add: $(2 + \sqrt{-4}) + (3 - \sqrt{-9}) + (\sqrt{-16})$

$$(2 + 2i) + (3 - 3i) + (4i) = 5 + 3i$$

EXERCISES 6-8B

In Exercises 1–6 solve for x and y.

1. $3 - 4i = x + 2yi$

2. $3x + 5i = 6 + yi$

3. $(4 + 3i) - (5 - i) = (3x + 2yi) + (2x - 3yi)$

4. $(3 - 2i) - (x + i) = (2x - yi) - (x + 2yi)$

5. $(3 - 2i) - (x + yi) = (2x + 5i) - (4 + i)$

6. $(5 + 2i) - (3i) - (x - 3i) = (x + yi) - (3 + 2yi)$

In Exercises 7–20 simplify and write answers in $a + bi$ form.

7. $(4 + 3i) + (5 - i)$

8. $(6 + 2i) + (-3 + 5i)$

9. $(7 - 4i) - (5 + 2i)$

10. $(8 - 3i) - (4 + i)$

11. $(-7 + 9i) + (3 - 4i) + (-9 - 5i)$

12. $(11 - 14i) - (-15 - 6i) + (8 + 12i)$

13. $(-13 - 8i) + (17 - 26i) - (19 + 14i)$

14. $(21 + 12i) - (15 - 6i) - (-26 + 16i)$

15. $(2 + i) + (3i) - (2 - 4i)$

16. $(3 - i) + (2i) - (-3 + 5i)$

17. $(5 - 2i) - (4) - (3 + 4i)$

18. $(7 + 3i) - (2i) - (2 - 5i)$

19. $(2 + 3i) - (x + yi)$

20. $(x - i) - (7 + yi)$

21. $\left(\frac{\sqrt{3}}{3} + 2i\right) - (\sqrt{3} - 4i)$

22. $\left(\frac{\sqrt{5}}{5} - 3i\right) - (\sqrt{5} + 7i)$

23. $(9 + \sqrt{-16}) + (2 + \sqrt{-25}) + (6 - \sqrt{-64})$

24. $(13 - \sqrt{-36}) - (10 - \sqrt{-49}) + (8 + \sqrt{-4})$

25. $3 + 5i - \sqrt{-75} - 2i + \sqrt{-27} - 15$

26. $-4i + 7 - \sqrt{-12} - 5 + \sqrt{-48} + 9i$

6-8C MULTIPLICATION OF COMPLEX NUMBERS

TO MULTIPLY TWO COMPLEX NUMBERS

1. Multiply them as you would two binomials.
2. Replace i^2 by -1.
3. Combine like terms and write the result in $a + bi$ form.

Example 7 Multiply: $(4 + 3i)\ (-5 + 2i)$

$$(4 + 3i)\ (-5 + 2i)$$

$$-15i$$

$$+\ 8i$$

$$= -20 - 7i + 6i^2$$

$$= -20 - 7i + 6(-1)$$

$$= -26 - 7i \quad \text{Product in } a + bi \text{ form}$$

Example 8 Multiply: $(2 - \sqrt{-25})(-3 + \sqrt{-9})$

$$\left.\begin{aligned} 2 - \sqrt{-25} &= 2 - 5i \\ -3 + \sqrt{-9} &= -3 + 3i \end{aligned}\right\} \text{First convert to } a + bi \text{ form}$$

Therefore $(2 - \sqrt{-25})(-3 + \sqrt{-9}) = (2 - 5i)(-3 + 3i)$

$$= -6 + 21i - 15i^2$$

$$= -6 + 21i + 15$$

$$= 9 + 21i$$

Simplifying Powers of i

Example 9 Powers of i

(a) $i = i$

(b) $i^2 = -1$

(c) $i^3 = i^2 \cdot i = (-1)i = -i$

(d) $i^4 = i^2 \cdot i^2 = (-1)(-1) = 1$

Each integral power of i must be one of these four values: $i, -1, -i, 1$

(e) $i^{13} = (i^4)^3 \cdot i = (1)^3 \cdot i = i$

(f) $i^{51} = (i^2)^{25} \cdot i = (-1)^{25} \cdot i = (-1)i = -i$

(g) $i^{100} = (i^4)^{25} = (1)^{25} = 1$

EXERCISES 6-8C

Perform the indicated operations. Express your answers in $a + bi$ form.

1. $(1 + i)(1 - i)$
2. $(3 + 2i)(3 - 2i)$
3. $(8 + 5i)(3 - 4i)$
4. $(6 - 3i)(4 + 5i)$
5. $(4 - i)(3 + 2i)$
6. $(5 + 2i)(2 - 3i)$
7. $(\sqrt{5} + 2i)(\sqrt{5} - 2i)$
8. $(\sqrt{7} - 3i)(\sqrt{7} + 3i)$
9. $i(3 + i)$
10. $i(4 - i)$
11. $5i(i - 2)$
12. $6i(2i - 1)$
13. $(\sqrt{11}\,i)^2$
14. $(\sqrt{17}\,i)^2$
15. $(2 + 5i)^2$
16. $(3 - 4i)^2$
17. $(-4 + 3i)^2$
18. $(-7 - 2i)^2$
19. $(5 + \sqrt{-36})(3 - \sqrt{-100})$
20. $(2 - \sqrt{-49})(4 + \sqrt{-81})$
21. $(3i)^3$
22. $(2i)^3$
23. $(2i)^4$
24. $(3i)^4$
25. $(2 - \sqrt{-1})^2$
26. $(3 + \sqrt{-1})^2$
27. $(4 - \sqrt{-1})^2$
28. $[3 + (-i)^7]^2$
29. $[3 + (-i)^6]^2$
30. $[4 - (-i)^{10}]^2$

6-8D DIVISION OF COMPLEX NUMBERS

Conjugate of a Complex Number

The **conjugate of a complex number** is obtained by changing the sign of its imaginary part.

The conjugate of $a + bi$ is $a - bi$

The conjugate of $a - bi$ is $a + bi$

Example 10 Conjugate complex numbers

(a) The conjugate of $3 - 2i$ is $3 + 2i$.

(b) The conjugate of $-5 + 4i$ is $-5 - 4i$.

(c) The conjugate of $7i$ is $-7i$, because $7i = 0 + 7i$ with conjugate $0 - 7i = -7i$.

(d) The conjugate of 5 is 5, because $5 = 5 + 0i$ with conjugate $5 - 0i = 5$.

The product of a complex number and its conjugate is a real number.

$$(a + bi)(a - bi) = a^2 - b^2i^2 = a^2 + b^2 \quad \text{Real number}$$

TO DIVIDE ONE COMPLEX NUMBER BY ANOTHER

1. Write the division as a fraction.
2. Multiply the numerator and denominator by the conjugate of the denominator.

$$\frac{a + bi}{c + di} \cdot \frac{c - di}{c - di} = \frac{(a + bi)(c - di)}{c^2 + d^2}$$

3. Simplify and write the result in $a + bi$ form.

Example 11 Division of complex numbers

(a) $$\frac{10i}{1 - 3i} = \frac{10i}{1 - 3i} \cdot \frac{1 + 3i}{1 + 3i} = \frac{10i + 30i^2}{1 - 9i^2} = \frac{10i - 30}{1 + 9}$$

$$= \frac{10(i - 3)}{10} = i - 3 = -3 + i$$

The denominator has been converted into a rational number

(b) $$(2 + i) \div (3 - 2i) = \frac{2 + i}{3 - 2i} = \frac{2 + i}{3 - 2i} \cdot \frac{3 + 2i}{3 + 2i}$$

$$= \frac{6 + 7i + 2i^2}{9 - 4i^2} = \frac{6 + 7i - 2}{9 + 4} = \frac{4 + 7i}{13}$$

$$= \frac{4}{13} + \frac{7}{13}i \quad \text{Quotient in } a + bi \text{ form}$$

(c) $$5 \div (-2i) = \frac{5}{-2i} = \frac{5}{-2i} \cdot \frac{2i}{2i} = \frac{10i}{-4i^2} = \frac{10i}{4} = \frac{5}{2}i = 0 + \frac{5}{2}i$$

EXERCISES 6-8D

In Exercises 1–8, write the conjugate of each complex number.

1. $3 - 2i$ **2.** $5 + 4i$ **3.** $-5 - 3i$ **4.** $-2 + 3i$

5. $5i$ **6.** $7i$ **7.** 10 **8.** -8

In Exercises 9–26 perform the indicated operations and write each answer in $a + bi$ form.

9. $\frac{10}{1 + 3i}$ **10.** $\frac{5}{1 + 2i}$ **11.** $\frac{1 + i}{1 - i}$ **12.** $\frac{1 - i}{1 + i}$

13. $\frac{8+i}{i}$ **14.** $\frac{4-i}{i}$ **15.** $\frac{3}{2i}$ **16.** $\frac{4}{5i}$

17. $\frac{3+i}{3-i}$ **18.** $\frac{5-i}{5+i}$ **19.** $\frac{15i}{1-2i}$ **20.** $\frac{20i}{1-3i}$

21. $\frac{10i}{3+\sqrt{-1}}$ **22.** $\frac{3\sqrt{-1}}{2-\sqrt{-1}}$ **23.** $\frac{2i+5}{3+2i}$ **24.** $\frac{3i+2}{4+5i}$

25. $\frac{i(2-\sqrt{-50})}{\sqrt{2}(2\sqrt{-1}+5\sqrt{2})}$ **26.** $\frac{-3\sqrt{-1}(4+5\sqrt{-1})}{2(12i-15)}$

Chapter Six Summary

Rules of Exponents. (Sections 6–1 and 6–2). We list all the rules of exponents for your reference.

1. $x^a x^b = x^{a+b}$
2. $(x^a)^b = x^{ab}$

2A. $(xy)^a = x^a y^a$

3. $\frac{x^a}{x^b} = x^{a-b}$ $(x \neq 0)$

3A. $\left(\frac{x}{y}\right)^a = \frac{x^a}{y^a}$ $(y \neq 0)$

4. $x^{-n} = \frac{1}{x^n}$ $(x \neq 0)$
5. $x^0 = 1$ $(x \neq 0)$
6. $\left(\frac{x^a y^b}{z^c}\right)^n = \frac{x^{an}y^{bn}}{z^{cn}}$ None of the letters can have a value that makes the denominator zero
7. $x^{1/n} = \sqrt[n]{x}$
8. $x^{a/b} = \sqrt[b]{x^a} = (\sqrt[b]{x})^a$

Rules 7 and 8 make it possible to convert a problem involving radicals into one involving rational exponents. Use the rules for exponents to simplify the expression; then convert your answer back into radical form.

Radicals. (Section 6–2). A radical is an indicated root of a number.

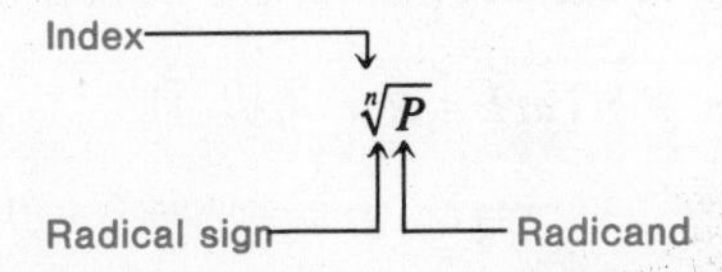

Principal Roots. (Section 6–2). The symbol $\sqrt[n]{p}$ always represents the *principal* nth root of p.

Positive radicand: The principal root is positive.

Negative radicand:

1. *Odd index:* The principal root is negative.
2. *Even index:* The principal root is a complex number.

Simplifying Radicals. (Section 6–3). Factor the radicand and use the formulas:

$$\sqrt[n]{ab} = \sqrt[n]{a}\sqrt[n]{b} \quad \text{or} \quad \sqrt[n]{\frac{a}{b}} = \frac{\sqrt[n]{a}}{\sqrt[n]{b}}$$

Simplified Form of a Radical Expression. (Section 6–3A)

1. No prime factor of a radicand has an exponent equal to or greater than the index.
2. No radicand contains a fraction.
3. No denominator contains a radical.
4. The index cannot be reduced. [See Examples 3(a) and 3(b), Section 6–4.]

Operations with Radicals.

Multiplication (Section 6–4): Use the formula $\sqrt[n]{a}\sqrt[n]{b} = \sqrt[n]{ab}$; then simplify.

Division (Section 6–4): Use the formula $\dfrac{\sqrt[n]{a}}{\sqrt[n]{b}} = \sqrt[n]{\dfrac{a}{b}}$; then simplify.

Addition (Section 6–5):

Like radicals: Add their coefficients.

Unlike radicals:

1. Simplify each radical.
2. Then combine any terms that have like radicals by adding their coefficients and multiplying their sum by the like radical.

Rationalizing the Denominator. (Section 6–3B)

Monomial denominator (Section 6–3B): Multiply the numerator and denominator by an expression that will make the exponent of any factor in the radicand of the denominator exactly divisible by the index.

Binomial denominator (Section 6–5): Multiply the numerator and denominator by the *conjugate* of the denominator.

Radical Equations. (Section 6–6)

The Pythagorean Theorem. (Section 6–7). The square of the hypotenuse of a right triangle is equal to the sum of the squares of the other sides.

Complex Numbers. (Section 6–8). A complex number is a number of the form $a + bi$, where a and b are real numbers and $i = \sqrt{-1}$.

The set of complex numbers: $C = \{a + bi \mid a, b \in R, i = \sqrt{-1}\}$.

Real part → a, Imaginary part → bi

$$a + bi$$

When the real part is zero, the complex number is a *pure imaginary number.*
When the imaginary part is zero, the complex number is a *real number.*

Equality of Complex Numbers. (Section 6–8B). Two complex numbers are equal if (1) their real parts are equal, and (2) their imaginary parts are equal.

Operations with Complex Numbers.

Addition (Section 6–8B)

1. Add the real parts.
2. Add the imaginary parts.
3. Write the result in $a + bi$ form.

Subtraction (Section 6–8B)

1. Subtract the real parts.
2. Subtract the imaginary parts.
3. Write the result in $a + bi$ form.

Multiplication (Section 6–8C)

1. Multiply them as you would two binomials.
2. Replace i^2 by -1.
3. Combine like terms and write the result in $a + bi$ form.

Division (Section 6–8D)

1. Write the division as a fraction.
2. Multiply the numerator and denominator by the conjugate of the denominator.
3. Simplify and write the result in $a + bi$ form.

Chapter Six Diagnostic Test or Review

Allow yourself about one hour to do these problems. Complete solutions for every problem, together with section references, are given at the end of the book.

In Problems 1–6 simplify each expression. Write the answers in radical form.

1. $\sqrt[3]{54x^6y^7}$ **2.** $\dfrac{4xy}{\sqrt{2x}}$ **3.** $\sqrt[6]{a^3}$

4. $(\sqrt[4]{z^2})^2$ **5.** $\dfrac{y}{2x}\sqrt[4]{\dfrac{16x^8y}{y^5}}$ **6.** $\sqrt{x}\,\sqrt[3]{x}$

In Problems 7 and 8 evaluate each expression.

7. $16^{3/2}$ **8.** $(-27)^{2/3}$

In Problems 9–12 perform the indicated operations. Express each answer in exponential form with positive exponents.

9. $x^{1/2}x^{-1/4}$ **10.** $(R^{-4/3})^3$

11. $\dfrac{a^{5/6}}{a^{1/3}}$ **12.** $\left(\dfrac{x^{-2/3}y^{3/5}}{x^{1/3}y}\right)^{-5/2}$

In Problems 13–18 perform the indicated operations. Leave your answers in simplified radical form.

13. $4\sqrt{8y} + 3\sqrt{32y}$ **14.** $3\sqrt{\dfrac{5x^2}{2}} - 5\sqrt{\dfrac{x^2}{10}}$ **15.** $\sqrt{2x^4}\,\sqrt{8x^3}$

16. $\sqrt{2x}(\sqrt{8x} - 5\sqrt{2})$ **17.** $\dfrac{\sqrt{10x} + \sqrt{5x}}{\sqrt{5x}}$ **18.** $\dfrac{5}{\sqrt{7} + \sqrt{2}}$

In Problems 19–21 perform the indicated operations and write each answer in $a + bi$ form.

19. $(5 - \sqrt{-8}) - (3 - \sqrt{-18})$ **20.** $(3 + i)(2 - 5i)$

21. $\dfrac{10}{1 - 3i}$ **22.** Solve and check: $x^{3/2} = 8$

23. Solve and check: $\sqrt{x - 3} + 5 = x$

24. Solve for x and y:

$$(2 + 5i) - (x + i) = (4 + 2yi) - (3 + yi).$$

25. Find the value of x shown in the right triangle.

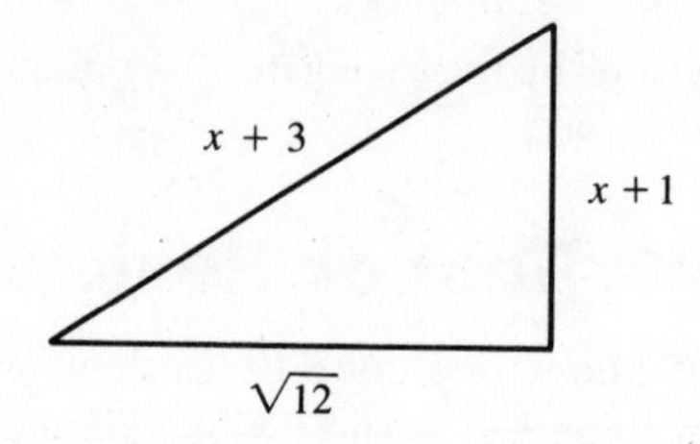

7 GRAPHS, FUNCTIONS, AND RELATIONS

Often a picture is a simple way to explain a complicated relationship. Graphs are mathematical pictures that help explain relationships between variables. The following are a few graphs that illustrate the widespread use of graphing.

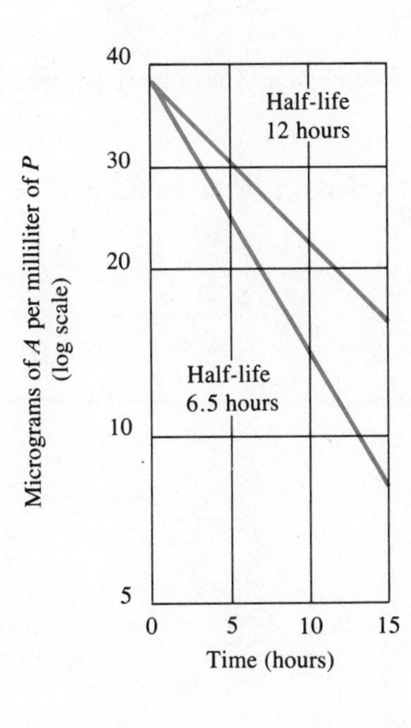

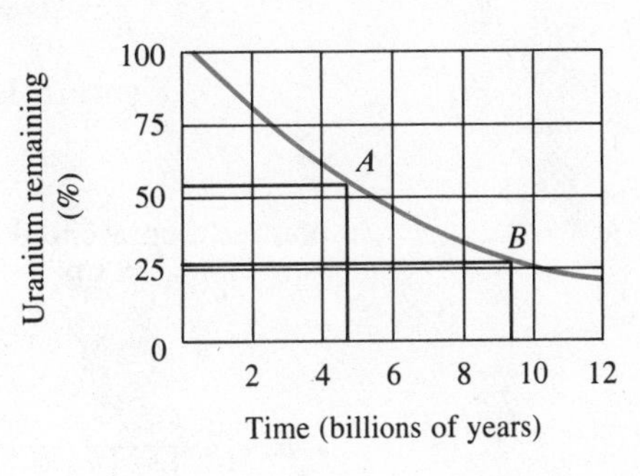

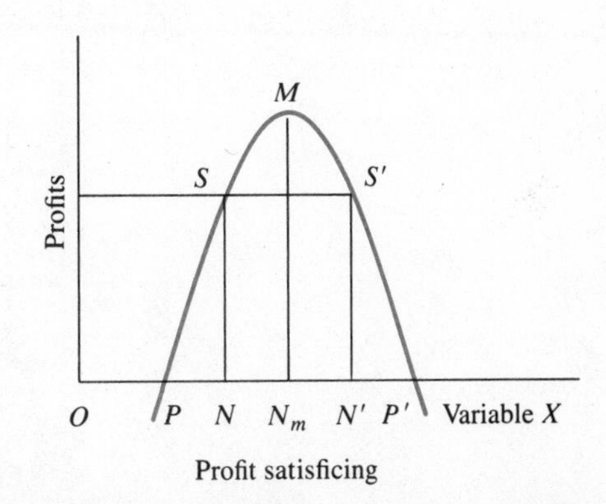

Profit satisficing

In this chapter we discuss how to draw graphs of points, lines, inequalities, relations, and functions.

7–1 The Rectangular Coordinate System

Origin

The **rectangular** (cartesian) **coordinate system** consists of two perpendicular number lines that intersect at the point representing zero on both lines called the **origin** (Figure 7–1A).

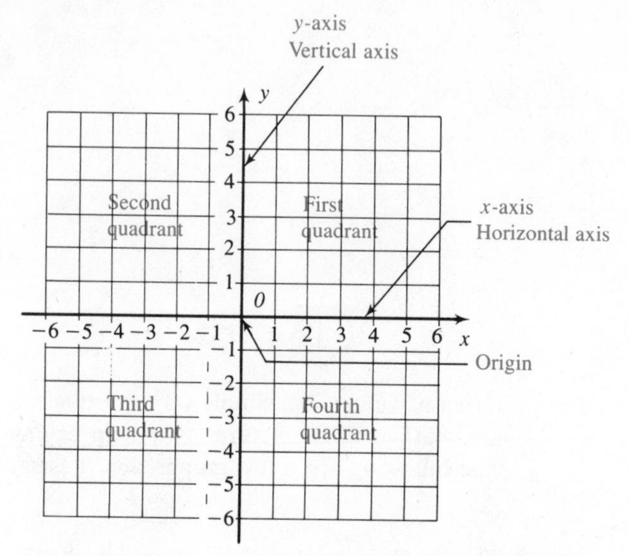

Figure 7–1A Rectangular Coordinate System

Graph of a Point

A point is represented by an **ordered pair** of numbers. The point (3,2) is shown in Figure 7–1B.

Abscissa and Ordinate

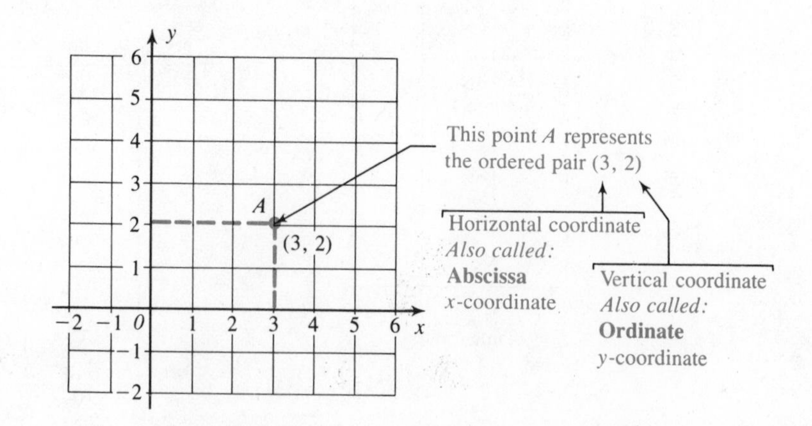

Figure 7–1B Graph of an Ordered Pair

The phrase "plot the points" means the same as "graph the points."

The Distance Between Two Points

To find the distance d between two points P_1 (x_1,y_1) and P_2 (x_2,y_2), draw a horizontal line through P_1 and a vertical line through P_2 that meet at a point C (Figure 7–1C). The triangle formed by joining points P_1, C, and P_2 is a right triangle having P_1P_2 as its hypotenuse. The points P_1, P_2, and C are called the **vertices** of the triangle.

Vertices

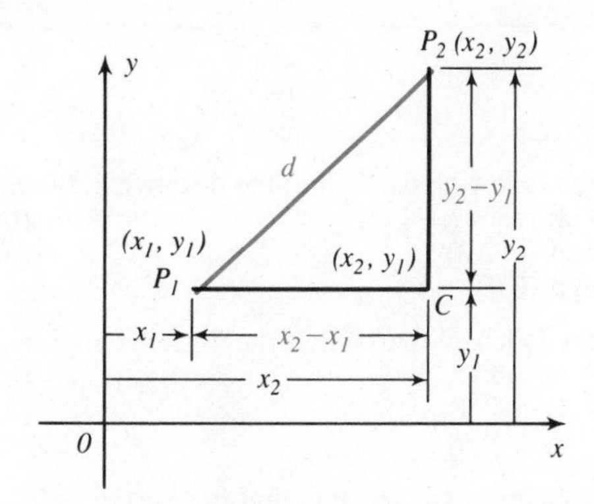

Figure 7–1C

We use the Pythagorean theorem to find the distance $d = P_1P_2$.

$$(d)^2 = (x_2 - x_1)^2 + (y_2 - y_1)^2$$

$$\sqrt{d^2} = \sqrt{(x_2 - x_1)^2 + (y_2 - y_1)^2}$$

$$d = \sqrt{(x_2 - x_1)^2 + (y_2 - y_1)^2}$$

P_2
d
y_2-y_1
P_1
x_2-x_1
C

This formula is true if points P_1 and P_2 are located *anywhere* in the plane.

TO FIND THE DISTANCE BETWEEN TWO POINTS P_1 (x_1,y_1) AND P_2 (x_2,y_2)

$$d = \sqrt{(x_2 - x_1)^2 + (y_2 - y_1)^2}$$

Note: This distance between two points is *always* positive, since the indicated square root is always the principal square root, which is positive. ■

Example 1 Find the distance between the points $(-6,5)$ and $(6,-4)$

Let $P_1 = (-6,5)$ and $P_2 = (6,-4)$. Then

$$d = \sqrt{(x_2 - x_1)^2 + (y_2 - y_1)^2}$$

$$= \sqrt{[6 - (-6)]^2 + [-4 - 5]^2}$$

$$= \sqrt{(12)^2 + (-9)^2} = \sqrt{144 + 81} = \sqrt{225} = 15$$

The distance is not changed if the points P_1 and P_2 are interchanged.

$$\left.\begin{aligned} \text{Let } P_1 &= (6,-4) \\ P_2 &= (-6,5) \end{aligned}\right\} \text{Then } d = \sqrt{[-6-6]^2 + [5-(-4)]^2}$$
$$= \sqrt{(-12)^2 + (9)^2} = \sqrt{144 + 81} = \sqrt{225} = 15$$

EXERCISES
7-1

1. In each of the following parts, find the distance between the two given points:
   (a) $(-2,-2)$ and $(2,1)$
   (b) $(-3,3)$ and $(3,-1)$
   (c) $(5,3)$ and $(-2,3)$
   (d) $(2,-2)$ and $(2,-5)$
   (e) $(4,6)$ and $(0,0)$
2. In each of the following parts, find the distance between the two given points:
   (a) $(-4,-3)$ and $(8,2)$
   (b) $(-3,2)$ and $(4,-3)$
   (c) $(-3,-4)$ and $(-3,2)$
   (d) $(-1,-2)$ and $(5,-2)$
   (e) $(-6,9)$ and $(0,0)$
3. Find the perimeter of the triangle that has vertices at $A(-2,2)$, $B(4,2)$, and $C(6,8)$.
4. Find the perimeter of the triangle that has vertices at $A(0,2)$, $B(11,2)$, and $C(8,6)$.
5. Use the distance formula to discover whether the triangle with vertices $A(-3,-2)$, $B(5,-1)$, and $C(3,2)$ is, or is not, a right triangle.
6. Use the distance formula to discover whether the triangle with vertices $A(3,2)$, $B(-5,1)$, and $C(-3,-2)$ is, or is not, a right triangle.
7. The vertices of a four-sided figure are $A(0,1)$, $B(-2,-2)$, $C(2,-3)$, and $D(4,0)$. Find:
   (a) the length of diagonal BD.
   (b) the length of diagonal AC.
8. The vertices of a four-sided figure are $A(-6,-2)$, $B(4,-7)$, $C(5,0)$, and $D(-5,5)$. Find:
   (a) the length of diagonal BD.
   (b) the length of diagonal AC.
9. Do the points $P(10,-6)$, $Q(-5,3)$, and $R(-15,9)$ lie in a straight line? Why?
10. Do the points $U(-9,3)$, $V(-3,-4)$, and $W(6,-12)$ lie in a straight line? Why?

7–2 Graphing Straight Lines

In Section 7–1 we showed how to graph points. In this section we show how to graph straight lines.

Linear Equations

Any first-degree equation (in no more than two variables) has a graph that is a straight line. Such equations are called linear equations. The general form of a linear equation is

$$Ax + By + C = 0$$

where A, B, and C are real numbers and A and B are not both 0.

Intercepts

A straight line can be drawn if we know two points that lie on that line. It is convenient to choose the two points where the line crosses the coordinate axes. These two points are called the **intercepts** of the line. The *x-intercept* is the point where the line crosses the x-axis. The *y-intercept* is the point where the line crosses the y-axis (Figure 7–2A).

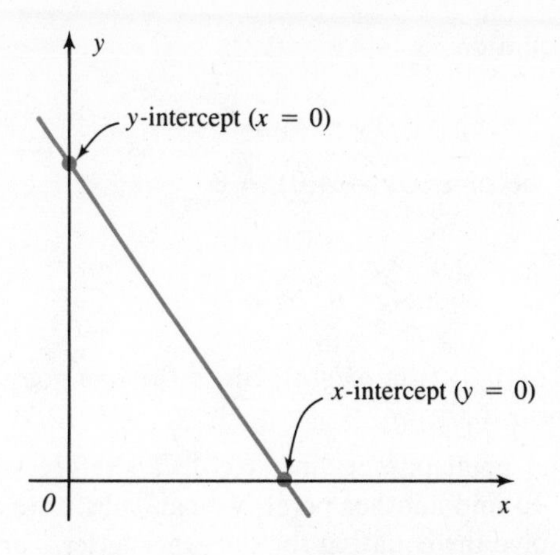

Figure 7–2A x- and y-intercepts

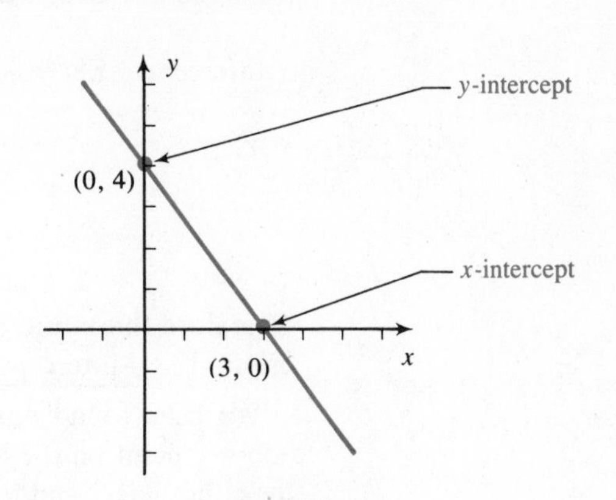

Figure 7–2B

THE INTERCEPT METHOD OF GRAPHING A STRAIGHT LINE

Example 1 Graph the equation $4x + 3y = 12$

x-intercept: Set $y = 0$.

Then $4x + 3y = 12$

becomes $4x + 3(0) = 12$

$4x = 12$

$x = 3$

Therefore the x-intercept is (3,0).

Table of Values

x	y
3	0

Sometimes we say: "The x-intercept is 3" (the *x-coordinate* of the point where the line crosses the x-axis, instead of the *point* itself).

y-intercept: Set $x = 0$.

Then $4x + 3y = 12$

becomes $4(0) + 3y = 12$

$3y = 12$

$y = 4$

Therefore the y-intercept is (0,4).

Table of Values

x	y
3	0
0	4

Sometimes we say: "The y-intercept is 4" (the *y-coordinate* of the point where the line crosses the y-axis, instead of the *point* itself).

Plot the x- and y-intercepts (3,0) and (0,4). Then draw the straight line through them (Figure 7–2B).

Example 2 Graph the equation $3x - 4y = 0$

x-intercept: Set $y = 0$ Then

$$\begin{aligned} 3x - 4y &= 0 \\ \text{becomes } 3x - 4(0) &= 0 \\ 3x &= 0 \\ x &= 0 \end{aligned}$$

Therefore the x-intercept is (0,0) (the *origin*). Since the line goes through the origin, the y-intercept is also (0,0).

We have found only one point on the line: (0,0). Therefore we must find another point on the line. To find another point, we can substitute any number for either letter and then solve the equation for the other letter. For example, if we choose $y = 3$

$$\begin{aligned} \text{then} \quad 3x - 4y &= 0 \\ \text{becomes } 3x - 4(3) &= 0 \quad \text{Replaced } y \text{ by } 3 \\ 3x &= 12 \\ x &= 4 \end{aligned}$$

This gives a second point (4,3) on the line. Plot the points (0,0) and (4,3). Then draw the straight line through them (Figure 7–2C).

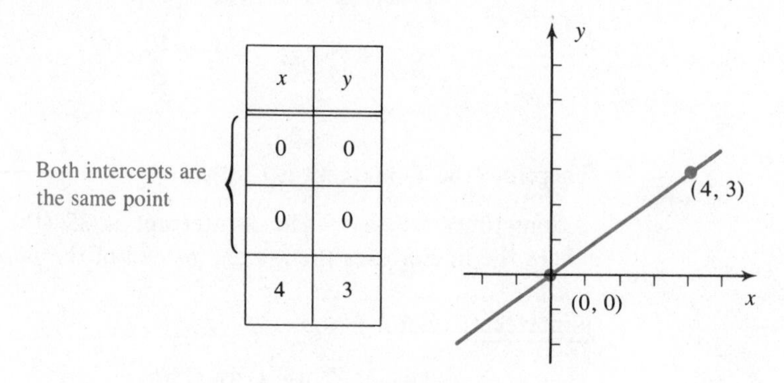

x	y
0	0
0	0
4	3

Figure 7–2C

The word "line" is ordinarily understood to mean "straight line."

Note: Sometimes it is not convenient to use the intercepts to graph an equation. The intercepts may be too close together or either one may be so large as to be off the graph. In either case, find other convenient points by setting either variable equal to *any* number, and then solving the equation for the other variable. ■

Some equations of a line have *only one variable.* Such equations have graphs that are either *vertical* or *horizontal* lines.

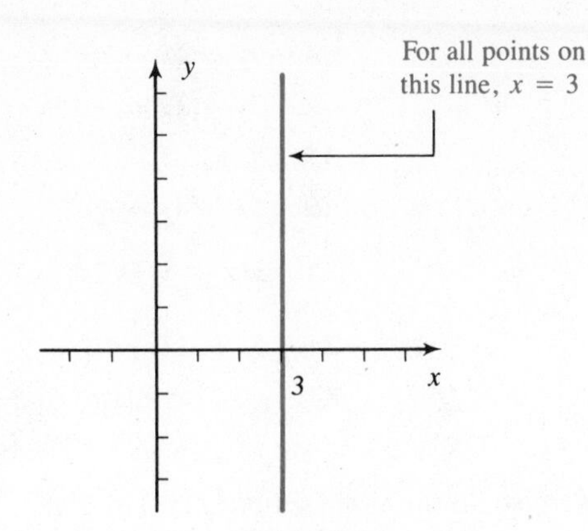

Figure 7–2D

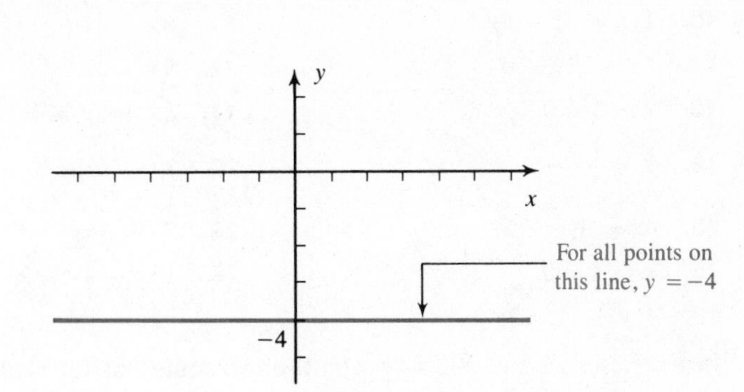

Figure 7–2E

Example 3 Graph the equation $x = 3$

The equation $x = 3$ is equivalent to $x + 0(y) = 3$. Therefore no matter what value of y is used in this equation, since y is multiplied by 0, x is always equal to 3 (Figure 7–2D).

Example 4 Graph the equation $y + 4 = 0$

$$y = -4$$

In the equation $y + 4 = 0$, no matter what value x has, y is always -4 (Figure 7–2E).

TO GRAPH A STRAIGHT LINE (INTERCEPT METHOD)

1. Find the x-intercept: Set $y = 0$; then solve for x.
2. Find the y-intercept: Set $x = 0$; then solve for y.
3. Draw a straight line through the x- and y-intercepts.
4. If both intercepts are (0,0), an additional point must be found before the line can be drawn (Example 2).
5. The graph of $x = a$ is a vertical line (at $x = a$).
6. The graph of $y = b$ is a horizontal line (at $y = b$).

EXERCISES 7-2

In Exercises 1–28 graph each equation.

1. $3x + 2y = 6$
2. $4x - 3y = 12$
3. $5x - 3y = 15$
4. $2x + 5y = 10$
5. $9x + 5y = 18$
6. $6x - 11y = 22$

7. $10x = 21 + 7y$ **8.** $13y = 40 - 8x$ **9.** $9y = 25 - 7x$

10. $17x = 31 + 6y$ **11.** $8x - 41 = 14y$ **12.** $5y - 33 = -15x$

13. $6x + 11y = 0$ **14.** $9x = 3y$ **15.** $x = -5$

16. $y - 2 = 0$ **17.** $7x + 5y = 2$ **18.** $3x + 8y = 4$

19. $y = \frac{1}{2}x - 1$ **20.** $y = \frac{1}{3}x + 2$ **21.** $3(x - 5) = 7y$

22. $4x = 3(y - 6)$ **23.** $7y = 16x - 47$ **24.** $5x - 13y = 0$

25. $x + 4 = 0$ **26.** $11x - 4y = 2$ **27.** $y = \frac{2}{5}x - 3$

28. $4(y + 7) = 9x$

In Exercises 29 and 30: (a) graph the two equations for each exercise on the same set of axes; (b) give the coordinates of the point where the two lines cross.

29. $5x - 7y = 18$
$2x + 3y = -16$

30. $4x + 9y = 3$
$2x - 5y = 11$

7–3 Equations of Straight Lines

In Section 7–2 we discussed the *graph* of a straight line. In this section we show how to write the *equation* of a line when certain facts about the line are known.

SLOPE

If we imagine the line as representing a hill, then the **slope** of the line is a measure of the steepness of the hill. To measure the slope of a line, we choose any two points on the line, $P_1\ (x_1, y_1)$ and $P_2(x_2, y_2)$ (Figure 7–3A).

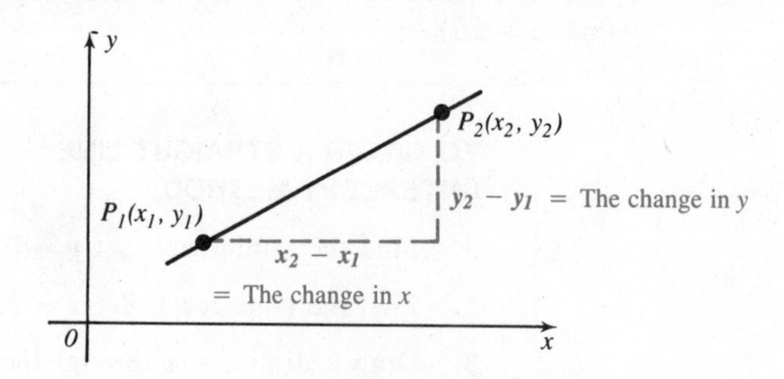

Figure 7–3A

The letter m is used to represent the slope of a line. The slope is defined as follows:

SLOPE OF A LINE

$$\text{Slope} = \frac{\text{The change in } y}{\text{The change in } x}$$

$$m = \frac{y_2 - y_1}{x_2 - x_1}$$

Example 1 Find the slope of the line through the points $(-3,5)$ and $(6,-1)$ (Figure 7–3B)

Let $P_1 = (-3,5)$

and $P_2 = (6,-1)$

$$m = \frac{y_2 - y_1}{x_2 - x_1}$$

$$= \frac{(-1) - (5)}{(6) - (-3)} = \frac{-6}{9} = -\frac{2}{3}$$

The slope is not changed if the points P_1 and P_2 are interchanged.

Let $P_1 = (6,-1)$

and $P_2 = (-3,5)$

then $m = \dfrac{y_2 - y_1}{x_2 - x_1}$

$$= \frac{(5) - (-1)}{(-3) - (6)} = \frac{6}{-9} = -\frac{2}{3}$$

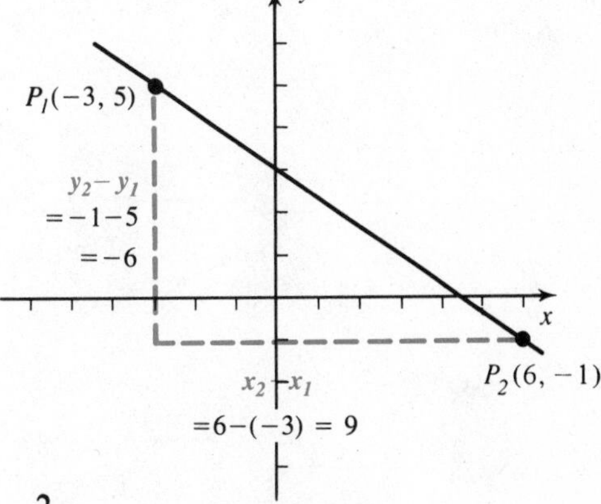

Figure 7–3B

Example 2 Find the slope of the line through the points $A(-2,-4)$ and $B(5,1)$ (Figure 7–3C)

$$m = \frac{y_2 - y_1}{x_2 - x_1}$$

$$= \frac{(1) - (-4)}{(5) - (-2)} = \frac{5}{7}$$

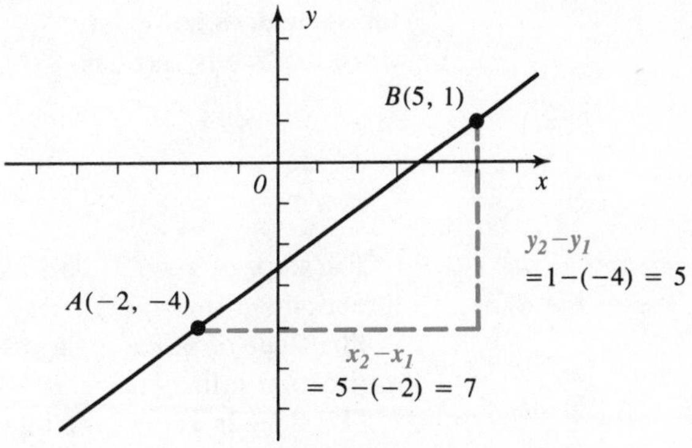

Figure 7–3C

Example 3 Find the slope of the line through the points $E(-4,-3)$ and $F(2,-3)$ (Figure 7–3D)

$$m = \frac{y_2 - y_1}{x_2 - x_1}$$

$$= \frac{(-3) - (-3)}{(2) - (-4)} = \frac{0}{6} = 0$$

Whenever the slope is zero, the line is horizontal.

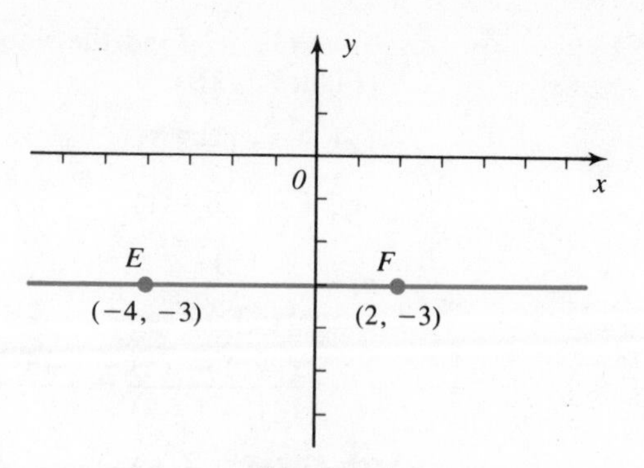

Figure 7–3D

Example 4 Find the slope of the line through the points $R(4,5)$ and $S(4,-2)$ (Figure 7–3E)

$$m = \frac{y_2 - y_1}{x_2 - x_1}$$

$$= \frac{(-2) - (5)}{(4) - (4)} = \frac{-7}{0}$$

Note that $\frac{-7}{0}$ *is not a real number. Therefore* **the slope does not exist when the line is vertical.**

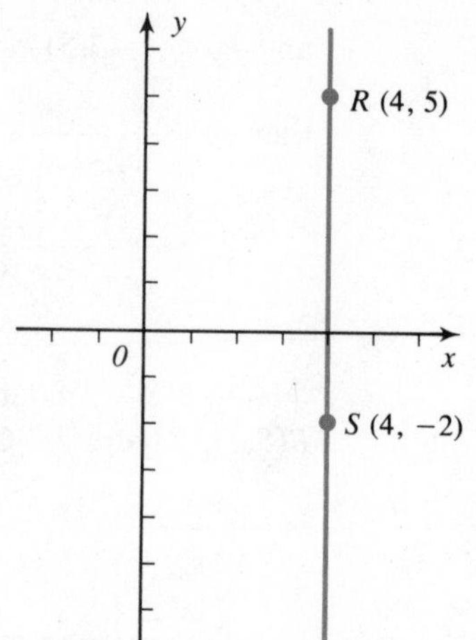

Figure 7–3E

Meaning of the Signs of the Slope

The slope of a line is positive if a point moving along the line in the positive x-direction (to the right) rises (Figure 7–3C).

The slope of a line is negative if a point moving along the line in the positive x-direction falls (Figure 7–3B).

The slope is zero if the line is horizontal (Figure 7–3D).

The slope does not exist if the line is vertical (Figure 7–3E).

EQUATIONS OF LINES

General Form

GENERAL FORM OF THE EQUATION OF A LINE

$$Ax + By + C = 0$$

where A, B, and C are real numbers, and A and B are not both 0.

Whenever possible, write the general form having *A positive* and *A*, *B*, and *C* *integers*.

Example 5 Write $-\frac{2}{3}x + \frac{1}{2}y = 1$ in the general form

$$\text{LCD} = 6 \quad \frac{6}{1}\left(-\frac{2}{3}x\right) + \frac{6}{1}\left(\frac{1}{2}y\right) = \frac{6}{1}\left(\frac{1}{1}\right)$$

$$-4x + 3y = 6$$

$$4x - 3y + 6 = 0 \quad \text{General form}$$

Point-Slope Form

Let $P_1(x_1,y_1)$ be a known point on the line with slope m. Let $P(x,y)$ represent any other point on the line (Figure 7–3F). Then

$$m = \frac{y - y_1}{x - x_1}$$

Therefore

$$\frac{(x - x_1)}{1} \cdot \frac{m}{1} = \frac{(x - x_1)}{1} \cdot \frac{y - y_1}{x - x_1}$$

$$(x - x_1)m = y - y_1$$

Point-slope form

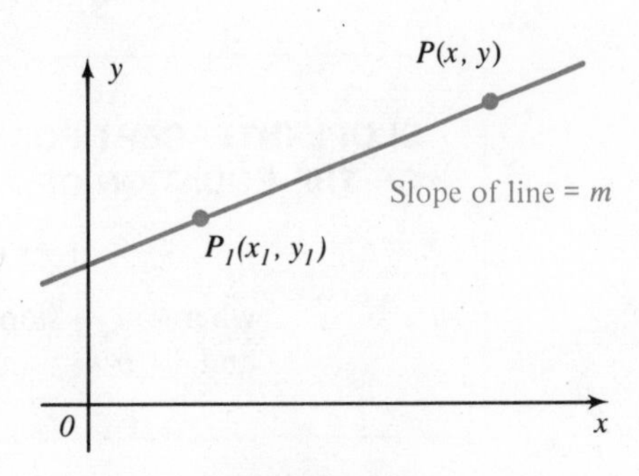

Figure 7–3F

This can be written as follows:

> **POINT-SLOPE FORM OF THE EQUATION OF A LINE**
>
> $$y - y_1 = m(x - x_1)$$
>
> where m = slope of the line,
> and $P_1(x_1,y_1)$ is a known point on the line.

Example 6 Write the equation of the line (in general form) that passes through $(-1,4)$ and has a slope of $-\frac{2}{3}$

$$y - y_1 = m(x - x_1)$$

$$y - (4) = -\frac{2}{3}[x - (-1)] \quad \text{Point-slope form}$$

$$3y - 12 = -2(x + 1)$$

$$3y - 12 = -2x - 2$$

$$2x + 3y - 10 = 0 \quad \text{General form}$$

Slope-Intercept Form

Let $(0,b)$ be the *y-intercept* of a line with slope m (Figure 7–3G). Then

$$y - y_1 = m(x - x_1)$$

$$y - b = m(x - 0)$$

$$y - b = mx$$

$$y = mx + b$$

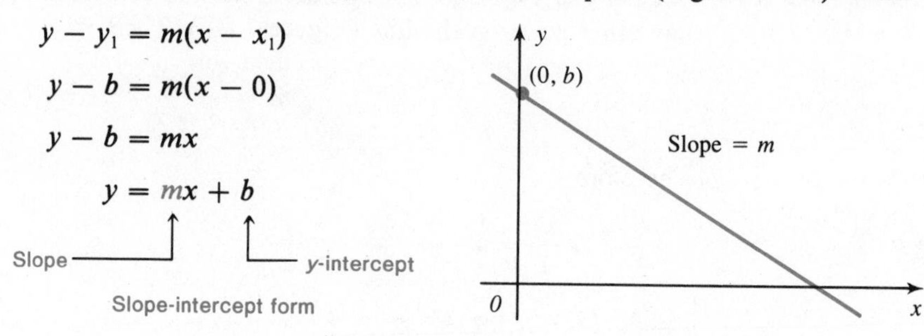

Figure 7–3G

> **SLOPE-INTERCEPT FORM**
> **OF THE EQUATION OF A LINE**
>
> $$y = mx + b$$
>
> where m = slope of the line,
> and b = y-intercept of the line.

Example 7 Write the equation of the line (in general form) that has a slope of $-\frac{3}{4}$ and a y-intercept of -2

$$y = mx + b$$

$$y = \left(-\frac{3}{4}\right)x + (-2) \quad \text{Slope-intercept form}$$

$$4y = -3x - 8 \quad \text{Multiplied by LCD, 4}$$

$$3x + 4y + 8 = 0 \quad \text{General form}$$

In the following examples we choose the particular form of the equation of the line that makes the best use of the given information.

Example 8 Find the equation of the line that passes through $(2,-4)$ and has a slope of 0

$$y - y_1 = m(x - x_1)$$

$$y - (-4) = 0(x - 2) \quad \text{Point-slope form}$$

$$y + 4 = 0 \quad \text{General form}$$

$$y = -4$$

(See Section 7-2, Example 4, and Figure 7-2E)

Example 9 Find the equation of the vertical line through $(3,-5)$

Since the line is vertical, the slope does not exist. Therefore we cannot use the point-*slope* form or the *slope*-intercept form. The x-coordinate of $(3,-5)$ is 3. Therefore every point on this vertical line has $x = 3$. This means that the equation of the line is $x = 3$. (See Section 7–2, Example 3, and Figure 7–2D.)

Example 10 Find the equation of the line that passes through the points $(-15,-9)$ and $(-5,3)$

1. Find the slope from the two given points.

$$m = \frac{(-9) - (3)}{(-15) - (-5)} = \frac{-12}{-10} = \frac{6}{5}$$

2. Use this slope with *either* given point to find the equation of the line.

Using the point $(-5,3)$	Using the point $(-15,-9)$
$y - y_1 = m(x - x_1)$	$y - y_1 = m(x - x_1)$
$y - (3) = \frac{6}{5}[x - (-5)]$	$y - (-9) = \frac{6}{5}[x - (-15)]$
$5y - 15 = 6(x + 5)$	$5(y + 9) = 6(x + 15)$
$5y - 15 = 6x + 30$	$5y + 45 = 6x + 90$
$0 = 6x - 5y + 45$	$0 = 6x - 5y + 45$

This shows that the same equation is obtained no matter which of the two given points is used.

It is sometimes useful to change the general form into the slope-intercept form. (See Example 11.)

Example 11 Write the equation of the line, in general form, through $(-4,7)$ and parallel to the line $2x + 3y + 6 = 0$.

Parallel lines have the same slope. Therefore the line we are trying to find has the same slope as $2x + 3y + 6 = 0$. To find the slope of $2x + 3y + 6 = 0$, write $2x + 3y + 6 = 0$ in the slope-intercept form by solving it for y.

$$2x + 3y + 6 = 0 \qquad \text{General form}$$

$$3y = -2x - 6 \qquad \text{Divide both sides by 3}$$

$$y = -\frac{2}{3}x - 2 \qquad \text{Slope-intercept form}$$

Slope (points to $-\frac{2}{3}$)

Therefore, the line we are trying to find has a slope of $-\frac{2}{3}$ and passes through $(-4,7)$.

$$y - y_1 = m(x - x_1) \qquad \text{Point-slope form}$$

$$y - 7 = \frac{-2}{3}[x - (-4)]$$

$$3y - 21 = -2x - 8$$

$$2x + 3y - 13 = 0 \qquad \text{General form}$$

EXERCISES 7-3

In Exercises 1–8 find the slope of the line through the given pair of points.

1. $(1,4)$ and $(10,6)$
2. $(-3,-5)$ and $(3,0)$
3. $(-5,-5)$ and $(1,-7)$
4. $(-1,0)$ and $(7,-4)$
5. $(-7,-5)$ and $(2,-5)$
6. $(0,-2)$ and $(5,-2)$
7. $(-4,3)$ and $(-4,-2)$
8. $(-5,2)$ and $(-5,7)$

In Exercises 9–16 write each equation in general form.

9. $5x = 3y - 7$
10. $4x = -9y + 3$
11. $\frac{x}{2} - \frac{y}{5} = 1$
12. $\frac{y}{6} - \frac{x}{7} = 1$
13. $y = -\frac{5}{3}x + 4$
14. $y = -\frac{3}{8}x - 5$
15. $4(x - y) = 11 - 2(x + 3y)$
16. $15 - 6(3x + y) = 7(x - 2y)$
17. Write the equation of the horizontal line that passes through the point $(-4,3)$.
18. Write the equation of the horizontal line that passes through the point $(2,-5)$.
19. Write the equation of the vertical line that passes through the point $(7,-2)$.
20. Write the equation of the vertical line that passes through the point $(-6,4)$.

In Exercises 21–26 write the equation of the line that has the indicated slope and y-intercept. Write the equation in general form.

21. $m = \frac{5}{7}$, y-intercept $= -3$
22. $m = \frac{1}{4}$, y-intercept $= -2$
23. $m = -\frac{4}{3}$, y-intercept $= \frac{1}{2}$
24. $m = -\frac{3}{5}$, y-intercept $= \frac{3}{4}$
25. $m = 0$, y-intercept $= 5$
26. $m = 0$, y-intercept $= 7$

In Exercises 27–30 write the equation of the line that passes through the given point and has the indicated slope. Write the equation in general form.

27. $(4,-3)$, $m = \frac{1}{5}$

28. $(-2,-1)$, $m = -\frac{5}{6}$

29. $(-6,5)$, $m = \frac{1}{4}$

30. $(-3,-2)$, $m = -\frac{4}{5}$

In Exercises 31–34: (a) write the given equation in the slope-intercept form; (b) give the slope of the line; (c) give the y-intercept of the line.

31. $4x - 5y + 20 = 0$

32. $8x + 3y - 24 = 0$

33. $\frac{2}{3}x + 3y + 5 = 0$

34. $3x - \frac{5}{6}y - 2 = 0$

In Exercises 35–40 find the equation of the line that passes through the given points. Write the equation in general form.

35. $(8,-1)$ and $(6,4)$

36. $(7,-2)$ and $(5,1)$

37. $(10,0)$ and $(7,4)$

38. $(-4,0)$ and $(-6,-5)$

39. $(-9,3)$ and $(-3,-1)$

40. $(-11,4)$ and $(-3,-2)$

In Exercises 41–46, write each answer in general form.

41. Write the equation of the line through $(-4,7)$ and parallel to $3x - 5y = 6$.

42. Write the equation of the line through $(8,-5)$ and parallel to $7x + 4y + 3 = 0$.

43. Write the equation of the line that has an x-intercept of 4 and is parallel to the line $3x + 5y - 12 = 0$.

44. Write the equation of the line that has an x-intercept of -3 and is parallel to the line $9x - 14y + 6 = 0$.

45. Write the equation of the line that has an x-intercept of -6 and a y-intercept of 4.

46. Write the equation of the line that has an x-intercept of 15 and a y-intercept of -12.

7–4 Graphing First-Degree Inequalities in the Plane

Half-Planes

Any line in a plane divides that plane into two **half-planes.** For example, in Figure 7–4A the line AB divides the plane into the two half-planes shown.

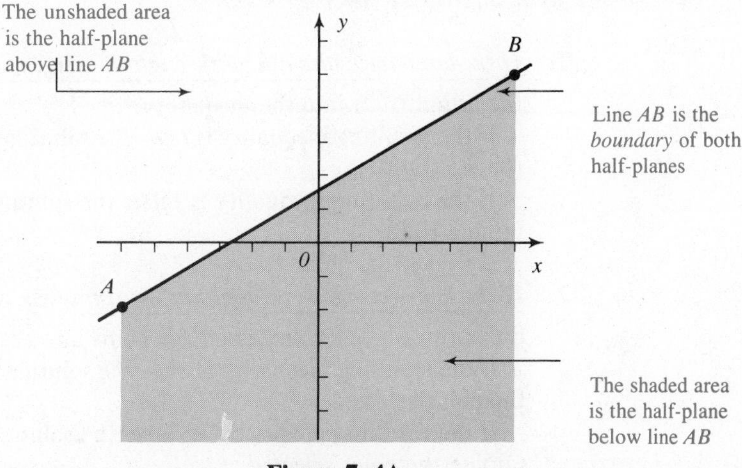

Figure 7–4A

Any first-degree *inequality* (in no more than two variables) has a graph that is a half-plane.

Boundary Line

The equation of the **boundary line** of the half-plane is obtained by replacing the inequality sign by an equal sign.

HOW TO DETERMINE WHEN THE BOUNDARY IS A DASHED OR A SOLID LINE

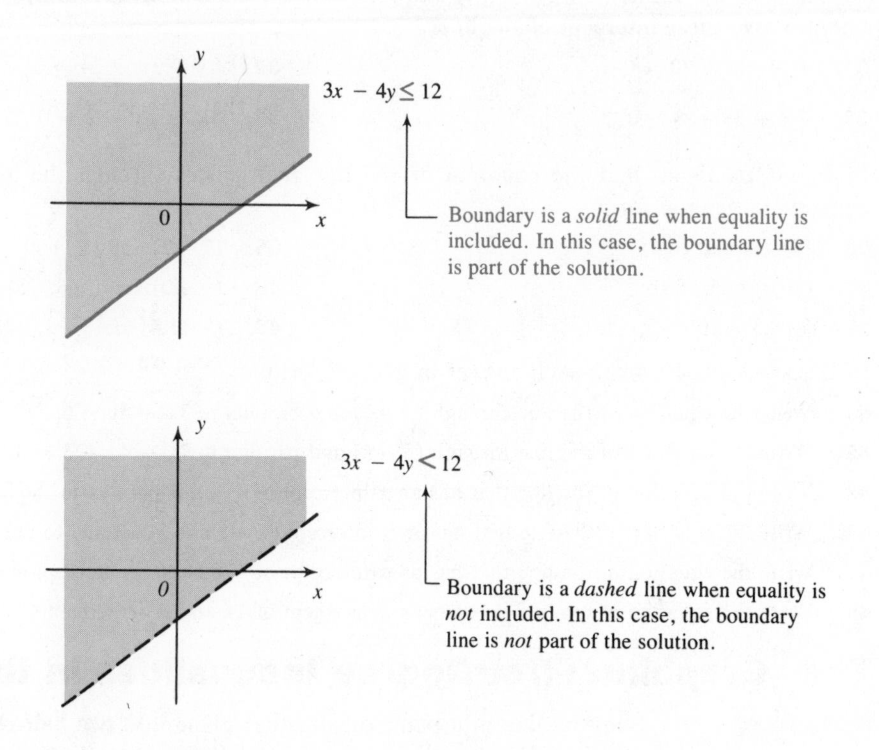

HOW TO DETERMINE THE CORRECT HALF-PLANE

1. ***If the boundary does not go through the origin,*** substitute the coordinates of the origin (0,0) into the inequality.
   If the resulting inequality is *true,* the solution is the half-plane containing (0,0).
   If the resulting inequality is *false,* the solution is the half-plane *not* containing (0,0).

2. ***If the boundary goes through the origin,*** select a point *not* on the boundary. Substitute the coordinates of this point into the inequality.
   If the resulting inequality is *true,* the solution is the half-plane containing the point selected.
   If the resulting inequality is *false,* the solution is the half-plane *not* containing the point selected.

TO GRAPH A FIRST-DEGREE INEQUALITY (IN THE PLANE)

1. The boundary line is *solid* if equality is included ($\leq, \geq$).
   The boundary line is *dashed* if equality is *not* included ($<, >$).
2. Graph the boundary line.
3. Select and shade the correct half-plane.

Example 1 Graph the inequality $2x - 3y < 6$

Change $<$ to $=$:

Boundary line: $2x - 3y = 6$

1. The boundary is a *dashed* line because the equality is *not* included.

$$2x - 3y < 6$$

2. Plot the graph of the boundary line by the intercept method.

<u>x-intercept:</u> Set $y = 0$ in $2x - 3y = 6$

$$\text{Then } 2x - 3(0) = 6$$
$$2x = 6$$
$$x = 3$$

<u>y-intercept:</u> Set $x = 0$ in $2x - 3y = 6$

$$\text{Then } 2(0) - 3y = 6$$
$$-3y = 6$$
$$y = -2$$

x	y
3	0
0	−2

Therefore the boundary line goes through (3,0) and (0,−2).

3. *Select the correct half-plane.* The solution of the inequality is only one of the two half-planes determined by the boundary line. Substitute the coordinates of the origin (0,0) into the inequality:

$$2x - 3y < 6$$
$$2(0) - 3(0) < 6$$
$$0 < 6 \quad \textit{True}$$

Therefore the half-plane containing the origin is the solution. The solution is the shaded area in Figure 7–4B.

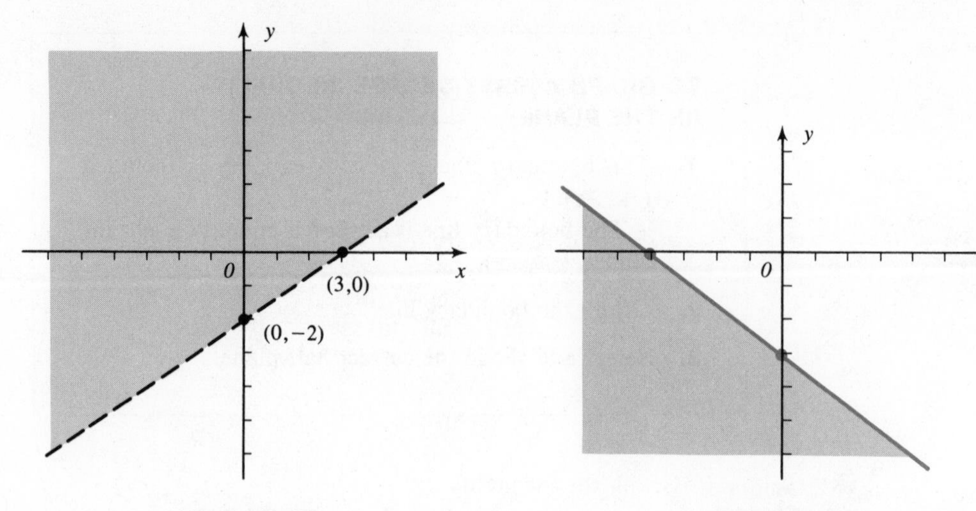

Figure 7–4B

Figure 7–4C

Example 2 Graph the inequality $3x + 4y \leq -12$

1. The boundary is a *solid* line because the equality is included.

$$3x + 4y \leq -12$$

2. Plot the graph of the boundary line: $3x + 4y = -12$ (Figure 7–4C).

x-intercept: Set $y = 0$ in $3x + 4y = -12$

$$\text{Then } 3x + 4(0) = -12$$
$$3x = -12$$
$$x = -4$$

y-intercept: Set $x = 0$ Then $3(0) + 4y = -12$

$$4y = -12$$
$$y = -3$$

x	y
−4	0
0	−3

3. Select the correct half-plane. Substitute the coordinates of the origin (0,0):

$$3x + 4y \leq -12$$
$$3(0) + 4(0) \leq -12$$
$$0 \leq -12 \quad \textit{False}$$

Therefore the solution is the half-plane *not* containing (0,0).

Some inequalities have equations with *only one variable*. Such inequalities have graphs whose *boundaries* are either *vertical* or *horizontal* lines.

Example 3 Graph the inequality $x + 4 < 0$

1. The boundary is a *dashed* line because the equality is *not* included.

$$x + 4 < 0$$

2. Plot the graph of the boundary line: $x + 4 = 0$
(See Figure 7–4D.) $x = -4$

3. Select the correct half-plane. Substitute the coordinates of the origin (0,0):

$$x + 4 < 0$$

$$(0) + 4 < 0$$

$$4 < 0 \quad \textit{False}$$

Therefore the solution is the half-plane *not* containing (0,0).

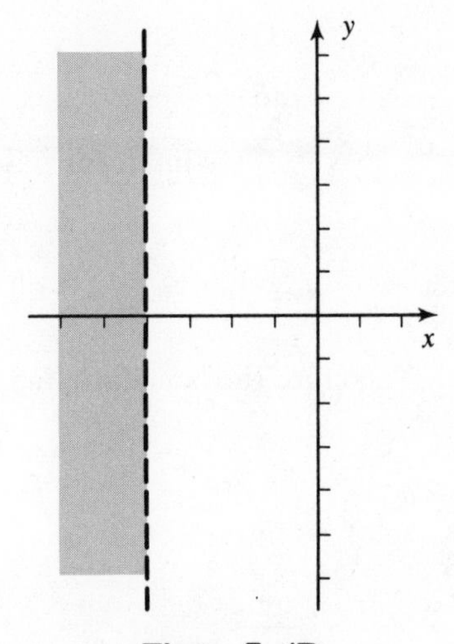

Figure 7–4D

In Section 2–3 we discussed how to graph an inequality such as $x + 4 < 0$ on a *single number line.*

Since $x + 4 < 0$

then $x < -4$

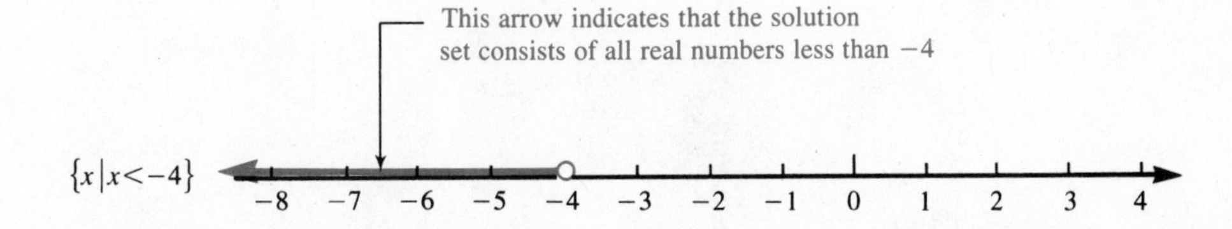

Example 3 of *this* section shows that the solution set of this same inequality, $x + 4 < 0$, represents an entire half-plane when it is plotted in the rectangular coordinate system (Figure 7–4D).

In Section 8–5 we will discuss how to graph higher-degree inequalities in a single letter on the number line.

Example 4 Graph the inequality $2y - 5x \geq 0$

1. The boundary is a *solid* line because the equality is included.

$$2y - 5x \geq 0$$

2. Plot the graph of the boundary line: $2y - 5x = 0$ (Figure 7–4E).

<u>x-intercept:</u> Set $y = 0$. Then $2(0) - 5x = 0$

$$-5x = 0$$
$$x = 0$$

<u>y-intercept:</u> Set $x = 0$. Then $2y - 5(0) = 0$

$$2y = 0$$
$$y = 0$$

Therefore the boundary line passes through the origin (0,0).

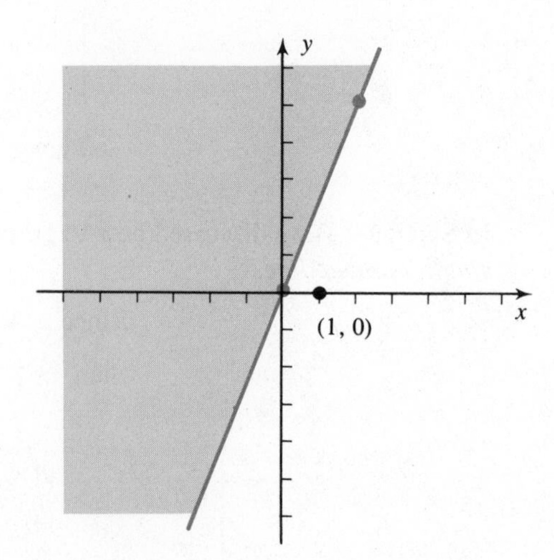

To find another point on the line: $2y - 5x = 0$

$$\text{Set } x = 2. \quad \text{Then } 2y - 5(2) = 0$$
$$2y - 10 = 0$$
$$2y = 10$$
$$y = 5$$

This gives the point (2,5) on the line.

x	y
0	0
0	0
2	5

3. Select the correct half-plane. Since the boundary goes through the origin (0,0), select a point *not* on the boundary, say (1,0). Substitute the coordinates of (1,0) into $2y - 5x \geq 0$.

$$2(0) - 5(1) \geq 0$$
$$-5 \geq 0 \quad \textit{False}$$

Therefore the solution is the half-plane *not* containing (1,0) (Figure 7–4E).

EXERCISES 7-4

Graph each inequality in the plane.

1. $4x + 5y < 20$
2. $5x - 3y > 15$
3. $3x - 8y > -16$
4. $6x + 5y < -18$
5. $9x + 7y \leq -27$
6. $5x - 14y \geq -28$
7. $x \geq -1$
8. $y \leq -4$
9. $6x - 13y > 0$
10. $4x + 9y < 0$
11. $14x + 3y \leq 17$
12. $10x - 4y \geq 23$
13. $\frac{x}{4} - \frac{y}{2} > 1$
14. $\frac{x}{3} + \frac{y}{5} < 1$
15. $4(x + 2) + 7 \leq 3(5 - 2x)$
16. $3(y + 4) - 8 > 4(1 - 2y)$
17. $\frac{2x + y}{3} - \frac{x - y}{2} \geq \frac{5}{6}$
18. $\frac{4x - 3y}{5} - \frac{2x - y}{2} \geq \frac{2}{5}$

7–5 Functions and Relations

RELATIONS

A **mathematical relation** is a set of ordered pairs (x,y). In this book we consider ordered pairs of real numbers only.

Example 1 Relations

$\mathcal{R}$ represents the relation

(a) $\mathcal{R} = \{(-1,2), (3,-4), (0,5), (4,3)\}$ Finite relation

(b) $\mathcal{R} = \{(x,y) | 2x + 3y = 6; x,y \in R\}$ Infinite relation

Domain, Range, and Graph of a Relation

The domain of a relation is the set of all the first coordinates of the ordered pairs of that relation. We represent the domain of the relation $\mathcal{R}$ by the symbol $D_{\mathcal{R}}$.

The range of a relation is the set of all the second coordinates of the ordered pairs of that relation. We represent the range of the relation $\mathcal{R}$ by the symbol $R_{\mathcal{R}}$.

The graph of a relation is the graph of all the ordered pairs of that relation.

Example 2 Find the domain, range, and graph of the relation

$$\mathcal{R} = \{(-1,2), (3,-4), (0,5), (4,3)\}$$

The domain of $\{(-1,2), (3,-4), (0,5), (4,3)\} = \{-1, 3, 0, 4\} = D_{\mathcal{R}}$
The range of $\{(-1,2), (3,-4), (0,5), (4,3)\} = \{2, -4, 5, 3\} = R_{\mathcal{R}}$
The graph of $\{(-1,2), (3,-4), (0,5), (4,3)\}$ is shown in Figure 7–5A.

Example 3 Find the range and graph of the relation
$\mathcal{R} = \{(x,y) \mid 2x + 3y = 6\}$ with domain $D_{\mathcal{R}} = \{-3, 0, 3, 6\}$

If $x = -3$, $\quad 2(-3) + 3y = 6$

$$-6 + 3y = 6$$

$$y = 4$$

If $x = 0$, $\quad 2(0) + 3y = 6$

$$0 + 3y = 6$$

$$y = 2$$

If $x = 3$, $\quad 2(3) + 3y = 6$

$$6 + 3y = 6$$

$$y = 0$$

If $x = 6$, $\quad 2(6) + 3y = 6$

$$12 + 3y = 6$$

$$y = -2$$

x	y
−3	4
0	2
3	0
6	−2
Domain	Range

Therefore the range $R_{\mathcal{R}} = \{4, 2, 0, -2\}$. The graph of the relation is shown in Figure 7–5B.

This relation, $\{(x,y) \mid 2x + 3y = 6\}$ with domain $\{-3, 0, 3, 6\}$, can also be written $\{(-3,4), (0,2), (3,0), (6,-2)\}$.

If the domain of this relation were changed from $\{-3, 0, 3, 6\}$ to the set of real numbers, then the graph would become the straight line through the four points shown in Figure 7–5B.

FUNCTIONS

A **function** is a relation such that no two of its ordered pairs have the *same first coordinate* and *different second coordinates*. This means that *no vertical line can meet the graph of a function in more than one point.*

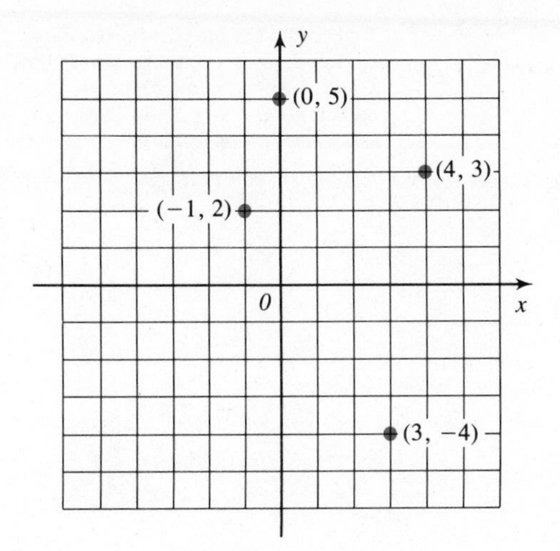

Figure 7–5A

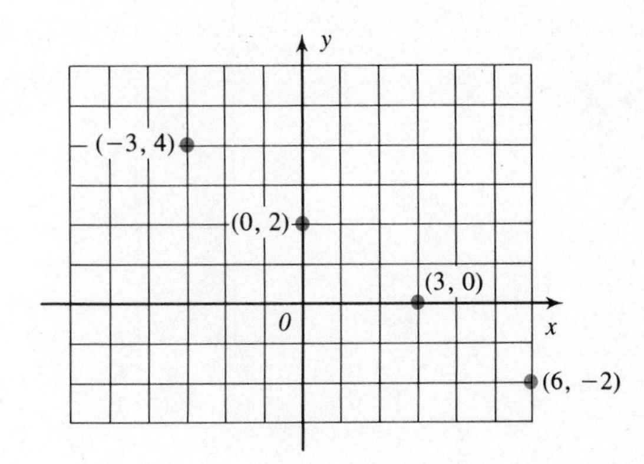

Figure 7–5B

Example 4 Determine whether each of the given relations is a function by seeing if any vertical line intersects its graph in more than one point.

(a)

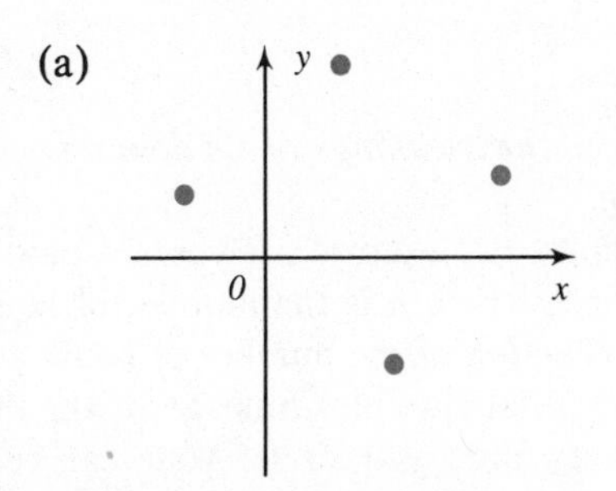

This *is* the graph of a function because any vertical line meets the graph in no more than one point.

(b)

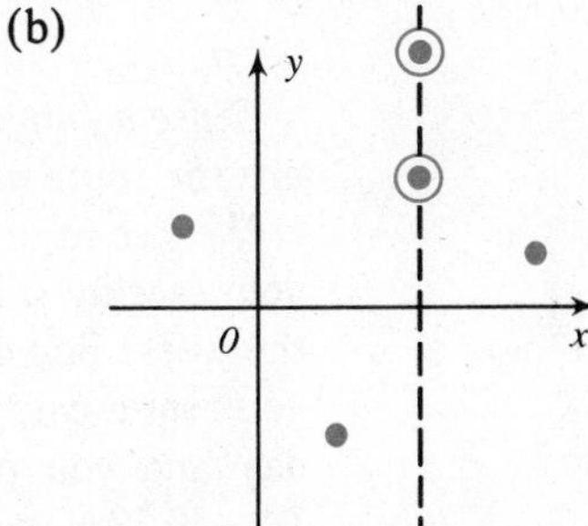

This *is not* the graph of a function because one vertical line meets the graph in two points.

(c)

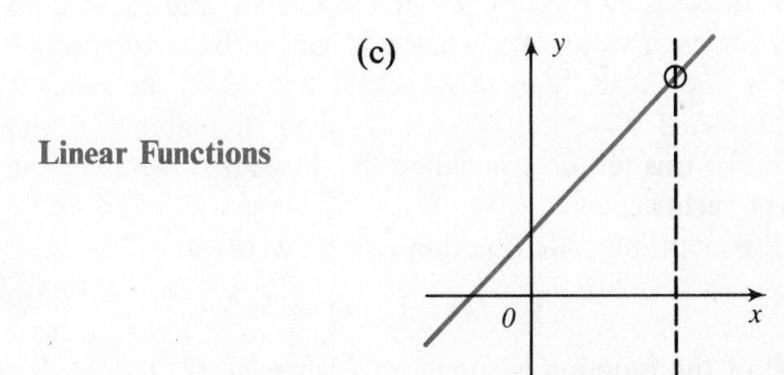

This *is* the graph of a function because any vertical line meets the graph in no more than one point.

Linear Functions

Straight lines (other than vertical ones) are graphs of **linear functions.** A linear function is one with a formula that is of the form $y = mx + b$, where m and b are constants (see Section 7-3).

(d)

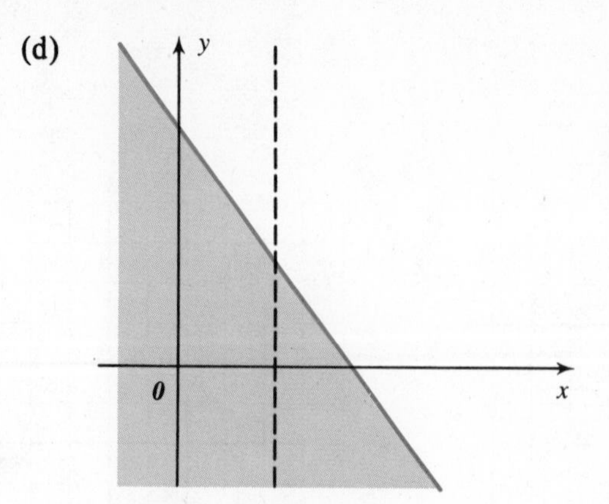

This *is not* the graph of a function because any vertical line meets the graph in an infinite number of points. Linear inequalities are **linear relations.**

(e)

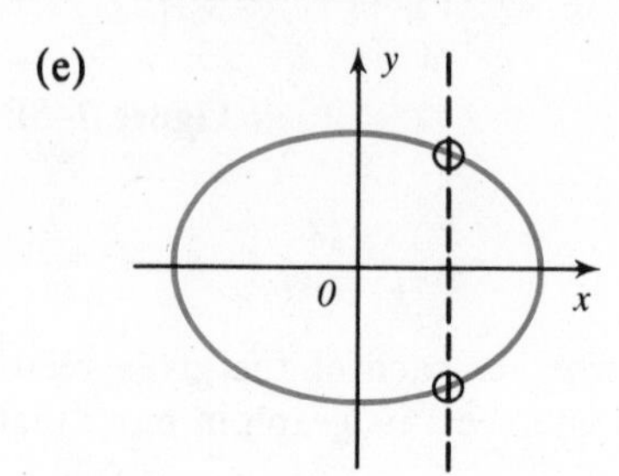

This *is not* the graph of a function because the vertical line shown meets the graph in two points.

Since a function is a relation, the meanings of its domain, range, and graph are the same as for a relation.

Typical Linear Functions

We live in a world of functions. For example, if you are paid \$5.65 an hour, your weekly salary is $S = 5.65h$, where h is the number of hours you work in the week. So your salary is a *function* of the number of hours you work. When you rent a car, you usually pay a flat rate plus mileage. If the flat rate is \$12 a day and you pay 18¢ for every mile you drive, your car rental cost $C = 12 + 0.18m$, where m is the number of miles you drive. Therefore the cost is a *function* of the miles driven. Variations are also examples of functions (or relations). See Section 3–4.

FUNCTIONAL NOTATION

If a rule or formula relating y and x is known, and there is no more than one value of y for each value of x, then y is said to be a **function of x.** This can be written $y = f(x)$, read "y equals f of x." y is called **the value of the function.**

Dependent and Independent Variables

For example, if $y = f(x) = 5x - 7$, then the value of y depends upon the value of x. For this reason y is called the **dependent variable,** and x is called the **independent variable.**

Using this notation, this function can be written

$$\{(x,f(x)) \mid f(x) = 5x - 7\}$$

The graph of the function is shown in Figure 7–5C.

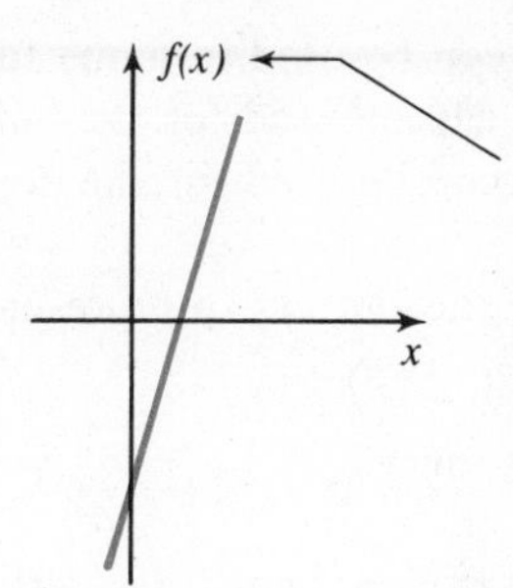

y and $f(x)$ are two ways of writing the value of a function; the vertical coordinates of points (ordinates) can be written as $f(x)$ as well as y, so the vertical axis can be considered the $f(x)$-axis.

Figure 7–5C

EVALUATING A FUNCTION

It is sometimes helpful to enclose x in parentheses wherever it appears in the function formula. If the function is to be evaluated for a particular value of x, then x is replaced by that value within each set of parentheses.

Suppose the function $f(x) = 5x - 7$ is to be evaluated when $x = 2$.

Then $f(x) = 5(x) - 7$

and $f(2) = 5(2) - 7 = 10 - 7 = 3$

The expression $f(2)$, read "f of 2," means that x has been replaced by 2. (It does *not* mean "f multiplied by 2.")

Example 5 Evaluate the function $f(x) = \dfrac{2x^2 - 7x}{5x - 3}$ when $x = -2, 0, 1, h,$ and $a + b$

$$f(x) = \frac{2(x)^2 - 7(x)}{5(x) - 3}$$

$$f(\ ) = \frac{2(\ )^2 - 7(\ )}{5(\ ) - 3}$$

(a) $$f(-2) = \frac{2(-2)^2 - 7(-2)}{5(-2) - 3} = \frac{8 + 14}{-10 - 3} = -\frac{22}{13}$$

(b) $$f(0) = \frac{2(0)^2 - 7(0)}{5(0) - 3} = \frac{0 - 0}{0 - 3} = 0$$

(c) $$f(h) = \frac{2(h)^2 - 7(h)}{5(h) - 3} = \frac{2h^2 - 7h}{5h - 3}$$

(d) $$f(a + b) = \frac{2(a + b)^2 - 7(a + b)}{5(a + b) - 3} = \frac{2a^2 + 4ab + 2b^2 - 7a - 7b}{5a + 5b - 3}$$

Example 6 Given $f(x) = 2x + 1$, evaluate $\dfrac{f(-5) - f(2)}{4}$

$$f(x) = 2(x) + 1$$

$$f(-5) = 2(-5) + 1 = -10 + 1 = -9$$

$$f(2) = 2(2) + 1 = 4 + 1 = 5$$

Therefore $\dfrac{f(-5) - f(2)}{4} = \dfrac{-9 - 5}{4} = \dfrac{-14}{4} = -\dfrac{7}{2}$

Letters other than f can be used to name functions.
Letters other than x can be used for the independent variable.
Letters other than y can be used for the dependent variable.

Example 7 Using other letters to represent functions and variables

(a) $g(x)$, $h(x)$, $F(x)$, $G(x)$

(b) $A(r)$, $P(i)$, $g(t)$, $h(s)$

Example 8 Graphs of functions using letters other than x, y, and f

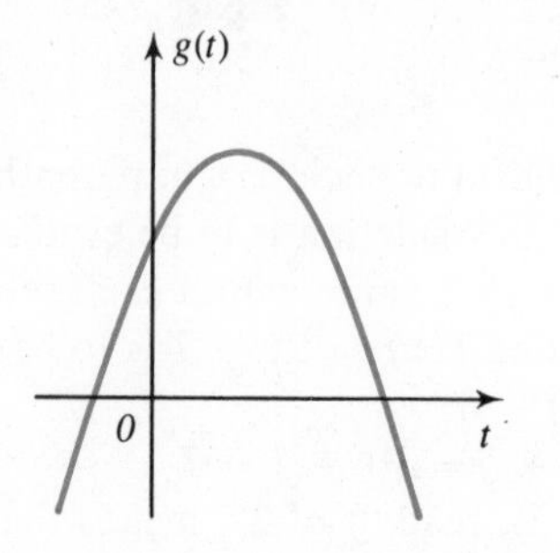

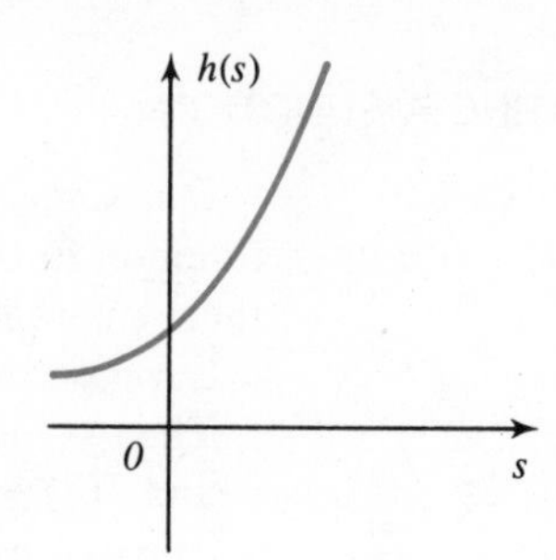

Example 9 Given $h(r) = 5r^2$ and $g(s) = 4s - 1$, find $h(a + 1) - g(3a)$

$$h(r) = 5(r)^2 = 5r^2$$

$$h(a + 1) = 5(a + 1)^2 = 5(a^2 + 2a + 1) = 5a^2 + 10a + 5$$

$$g(s) = 4(s) - 1 = 4s - 1$$

$$g(3a) = 4(3a) - 1 = 12a - 1$$

Therefore

$$h(a + 1) - g(3a) = (5a^2 + 10a + 5) - (12a - 1)$$
$$= 5a^2 - 2a + 6$$

FUNCTIONS AND RELATIONS OF MORE THAN ONE VARIABLE

All the functions and relations discussed so far have been functions of *one* variable. Many functions and relations have more than one variable. For example, your weight is a function of three variables: your mass, the mass of the earth, and your distance from the center of the earth. This is an example of a variation that is a function.

Example 10 Functions and relations of more than one variable

(a) The force (F) of the wind on a surface is a joint variation of the wind velocity (v) and the surface area (A) according to the formula

$F = kAv^2$ **This is a function of two variables:**
$F = f(A,v)$, read "F is a function of A and v."
In this case the *dependent variable* F is a function of *two independent variables*, A and v (k is constant).

(b) If you put \$100 (P) in a bank that pays $\frac{1}{2}$% a month interest (i) and leave it there for 24 months (n), the amount of money (A) in your account after that time is found by using the fcrmula

$A = P(1 + i)^n$ **This is a function of three variables:** $A = f(P,i,n)$, read "A is a function of P, i, and n." In this case the *dependent variable* A is a function of *three independent variables, P, i,* and *n.*

EXERCISES 7-5

1. Find the domain, range, and graph of the relation $\{(2,-1), (3,4), (0,2), (-3,-2)\}$.
2. Find the domain, range, and graph of the relation $\{(-4,0), (0,0), (3,-2), (1,5), (-3,-3)\}$.
3. Find the range and graph of the function $\{(x,y) \mid 2x + 5y = 10\}$ with domain $\{-5, -1, 0, 2\}$.
4. Find the range and graph of the function $\{(x,y) \mid 3x - 6 = 2y\}$ with domain $\{-2, 0, 2, 4\}$.
5. Graph the function $\{(x,y) \mid y = 2x - 3\}$ with domain the set of real numbers.
6. Graph the function $\left\{(x,y) \middle| \frac{x}{2} - \frac{y}{3} = 1\right\}$ with domain the set of real numbers.
7. Which of the following are graphs of functions?

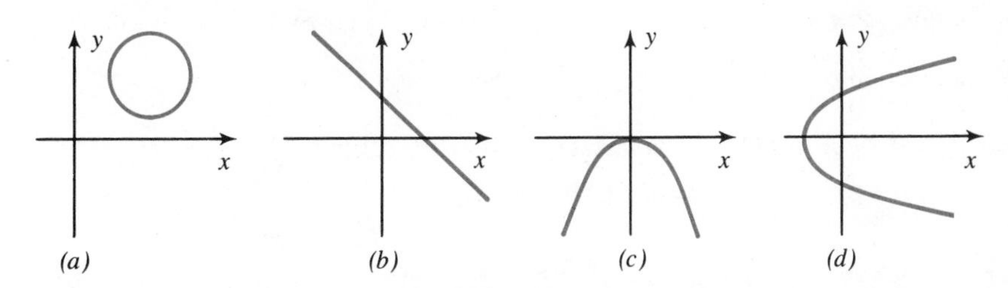

8. Which of the following are graphs of functions?

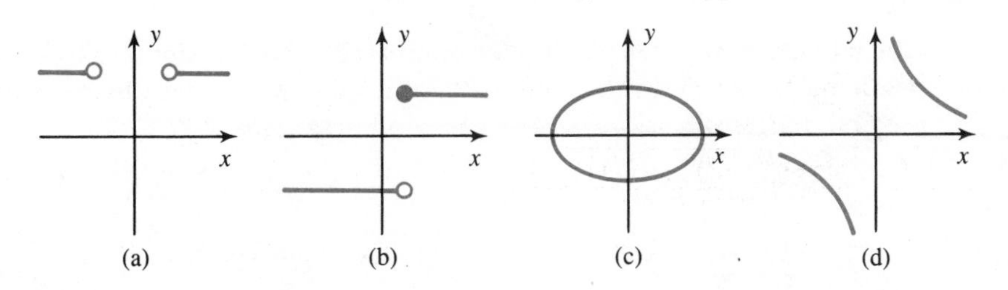

9. Given $f(x) = 3x - 1$, find:
   (a) $f(2)$ (b) $f(0)$ (c) $f(a - 2)$ (d) $f(x + 2)$
10. Given $f(x) = 4x + 1$, find:
   (a) $f(3)$ (b) $f(-5)$ (c) $f(0)$ (d) $f(x - 2)$
11. Given $f(x) = 2x^2 - 3$, evaluate $\frac{f(5) - f(2)}{6}$.

12. Given $f(x) = 8 - 3x^2$, evaluate $\dfrac{f(4) - f(-6)}{12}$.

13. Given $f(x) = 3x^2 - 2x + 4$, find $2f(3) + 4f(1) - 3f(0)$.

14. Given $f(x) = (x + 1)(x^2 - x + 1)$, find $3f(0) + 5f(2) - f(1)$.

15. If $f(x) = x^3$ and $g(x) = \dfrac{1}{x}$, find $f(-3) - 6g(2)$.

16. If $f(x) = x^2$ and $g(x) = \dfrac{1}{x}$, find $2f(-4) - 9g(3)$.

17. If $H(x) = 3x^2 - 2x + 4$ and $K(x) = x - x^2$, find $2H(2) - 3K(3)$.

18. If $F(e) = \dfrac{3e - 1}{e^2}$ and $G(w) = \dfrac{1}{5 - w}$, find $\dfrac{4F(-3)}{5G(4)}$.

19. Find the domain and range of the function $f(x) = \sqrt{x - 4}$.

20. Find the domain and range of the function $f(x) = -\sqrt{9x - 16}$.

21. Given $f(x) = x^2 - x$, find $\dfrac{f(x + h) - f(x)}{h}$.

22. Given $f(x) = 5x^2 - 3$, find $\dfrac{f(x + h) - f(x)}{h}$.

23. If $A(r) = \pi r^2$ and $C(r) = 2\pi r$, find $\dfrac{3A(r) - 2C(r)}{\pi r}$.

24. If $A(r) = \pi r^2$ and $C(r) = 2\pi r$, find $\dfrac{5A(r) + 3C(r)}{\pi r}$.

25. Evaluate the function $z = g(x,y) = 5x^2 - 2y^2 + 7x - 4y$ when $x = 3$ and $y = -4$.

26. Evaluate the function $w = f(x,y,z) = \dfrac{5xyz - x^2}{2z}$ when $x = -3$, $y = 4$, and $z = -15$.

27. Given that $A = f(P,i,n) = P(1 + i)^n$, evaluate $f(100, 0.08, 12)$.

28. Given that $F = f(A,v) = 58.6Av^2$, evaluate $f(126.5, 634)$.

7–6 Inverse Functions and Relations

INVERSE RELATION

A relation $\mathcal{R}$ was defined as a set of ordered pairs (Section 7–5). The **inverse relation** of $\mathcal{R}$ is defined to be the set of ordered pairs obtained by *interchanging the first and second coordinates in each ordered pair* of $\mathcal{R}$.

<u>*The relation:*</u> $\mathcal{R} = \{(3,1), (3,-2), (2,-4)\}$

<u>*The inverse relation of $\mathcal{R}$:*</u> $\mathcal{R}^{-1} = \{(1,3), (-2,3), (-4,2)\}$

$\mathcal{R}^{-1}$ represents the inverse relation of $\mathcal{R}$

It follows from this definition of the inverse relation that:

The **domain** of $\mathcal{R}^{-1}$ is $D_{\mathcal{R}^{-1}} = R_{\mathcal{R}}$ (the *range* of $\mathcal{R}$)

The **range** of $\mathcal{R}^{-1}$ is $R_{\mathcal{R}^{-1}} = D_{\mathcal{R}}$ (the *domain* of $\mathcal{R}$)

GRAPH OF AN INVERSE RELATION

In Figure 7–6A the points of $\mathcal{R}$ and the points of $\mathcal{R}^{-1}$ are plotted.

The points of $\mathcal{R}$		The points of $\mathcal{R}^{-1}$	
$A(3,1)$	$\longrightarrow$	$A'(1,3)$	A' is the image of A
$B(3,-2)$	$\longrightarrow$	$B'(-2,3)$	B' is the image of B
$C(2,-4)$	$\longrightarrow$	$C'(-4,2)$	C' is the image of C

Mirror Image

Each point of the inverse relation $\mathcal{R}^{-1}$ is the **mirror image** of the corresponding point of the relation $\mathcal{R}$. The mirror image of a point appears to be the same distance behind the mirror ($y = x$) as the actual point is in front of the mirror. Therefore in Figure 7–6A, $AM_1 = M_1A'$, $BM_2 = M_2B'$, and $CM_3 = M_3C'$.

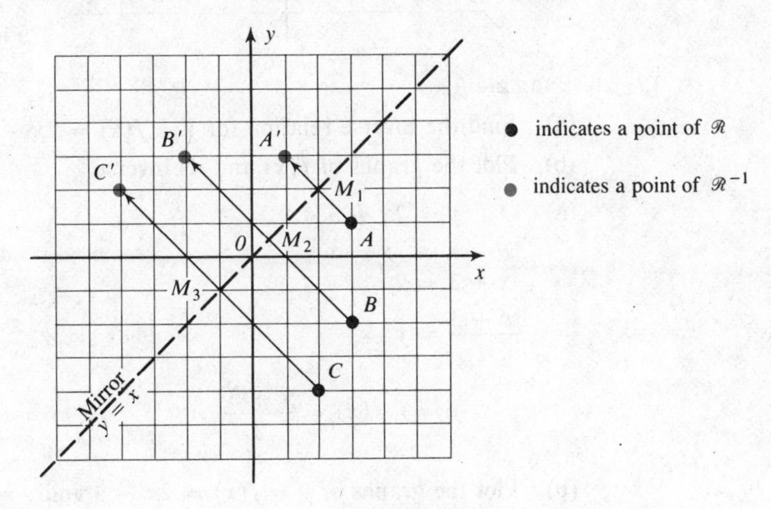

Figure 7–6A

Example 1 Given $\mathcal{R} = \{(2,-3), (-3,-1), (-4,5), (4,0)\}$, find $\mathcal{R}^{-1}$, $D_{\mathcal{R}^{-1}}$, $R_{\mathcal{R}^{-1}}$, and graph $\mathcal{R}$ and $\mathcal{R}^{-1}$.

$$\mathcal{R} = \{(2,-3), (-3,-1), (-4,5), (4,0)\} \quad \mathcal{R} \text{ is a function}$$

$$\mathcal{R}^{-1} = \{(-3,2), (-1,-3), (5,-4), (0,4)\} \quad \mathcal{R}^{-1} \text{ is a function}$$

$$D_{\mathcal{R}^{-1}} = \{-3, -1, 5, 0\} = R_{\mathcal{R}}$$

$$R_{\mathcal{R}^{-1}} = \{2, -3, -4, 4\} = D_{\mathcal{R}}$$

See Figure 7–6B.

Inverse Functions

If $\mathcal{R}^{-1}$ is itself a function, it is called an inverse function (Example 1). When the relation $\mathcal{R}$ is given by a formula $y = f(x)$, the inverse relation is found by interchanging x and y in the formula (Examples 2 and 3).

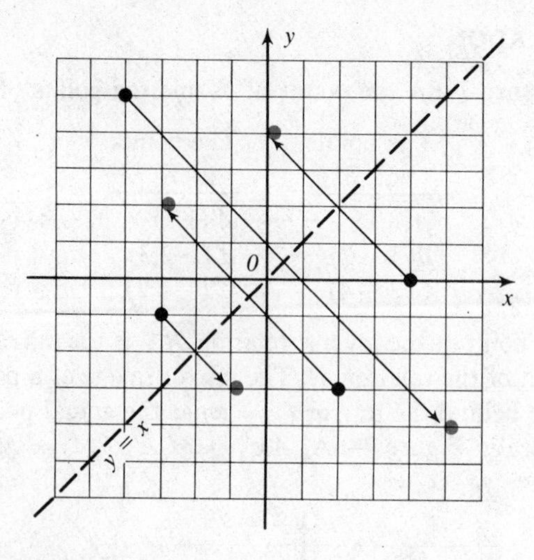

Figure 7–6B

Example 2

(a) Find the inverse relation for $y = f(x) = 2x - 3$

(b) Plot the graphs of $f(x)$ and its inverse

(a)
$$y = 2x - 3$$
$$x = 2y - 3 \qquad \text{Interchanged } x \text{ and } y$$
$$x + 3 = 2y$$
$$\frac{x+3}{2} = y \qquad \text{Solved for } y$$
$$y = f^{-1}(x) = \frac{x+3}{2} \qquad f^{-1}(x) \text{ is the inverse relation for } f(x) = 2x + 3$$

(b) Plot the graphs of $y = f(x) = 2x - 3$ and $y = f^{-1}(x) = \dfrac{x+3}{2}$.

Graph the relation

$y = f(x) = 2x - 3$

x-intercept:

$y = 2x - 3$

$0 = 2x - 3$

$\frac{3}{2} = x$

x	y
$\frac{3}{2}$	0
0	-3

y-intercept:

$y = 2x - 3$

$y = 2(0) - 3 = -3$

Graph the inverse relation

$y = f^{-1}(x) = \dfrac{x+3}{2}$

x-intercept:

$y = \dfrac{x+3}{2}$

$0 = \dfrac{x+3}{2}$

$-3 = x$

x	y
-3	0
0	$\frac{3}{2}$

y-intercept:

$y = \dfrac{x+3}{2} = \dfrac{0+3}{2} = \dfrac{3}{2}$

In Example 2 it is apparent from the graphs in Figure 7–6C that $f(x)$ and $f^{-1}(x)$ are both functions (Section 7–5).

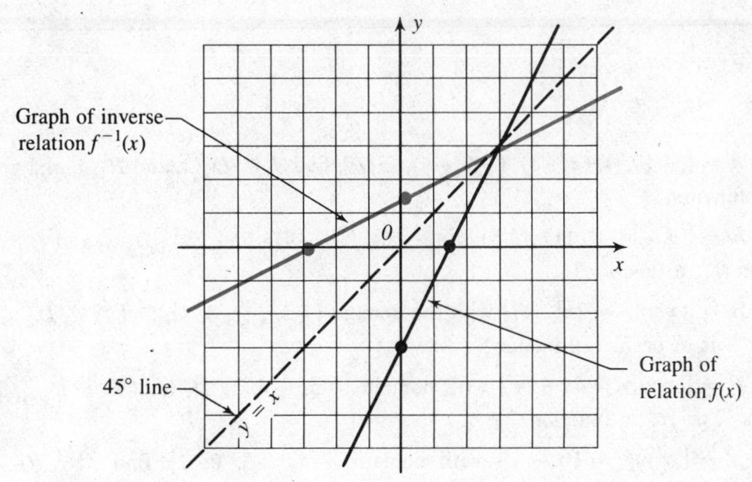

Figure 7–6C

Example 3 Find the inverse relation for $y = f(x) = \dfrac{5x - 1}{x}$

$$y = \frac{5x - 1}{x}$$

$$x = \frac{5y - 1}{y} \quad \text{Interchanged } x \text{ and } y$$

$$xy = 5y - 1$$

$$xy - 5y = -1$$

$$y(x - 5) = -1 \quad \text{Solve for } y$$

$$y = \frac{1}{5 - x} = f^{-1}(x)$$

$f^{-1}(x)$ is the inverse relation for $f(x) = \dfrac{5x - 1}{x}$

TO FIND THE INVERSE RELATION

1. *If the relation $\mathcal{R}$ is given as a set of ordered pairs:*

(a) Interchange the first and second coordinates of each ordered pair in $\mathcal{R}$.

(b) The inverse relation $\mathcal{R}^{-1}$ is the set of ordered pairs obtained in Step a.

2. *If the relation $\mathcal{R}$ is given by the formula $y = f(x)$:*

(a) Interchange x and y in the relation formula.

(b) Solve the resulting equation for y.

(c) The inverse relation $\mathcal{R}^{-1}$ has the formula obtained in Step b.

If $\mathcal{R}^{-1}$ is a function, it is called an *inverse function.*

EXERCISES 7-6

1. Given $\mathcal{R} = \{(-10,7), (3,-8), (-5,-4), (3,9)\}$, find $\mathcal{R}^{-1}$, $D_{\mathcal{R}^{-1}}$, and $R_{\mathcal{R}^{-1}}$, and graph $\mathcal{R}$ and $\mathcal{R}^{-1}$. Is $\mathcal{R}$ or $\mathcal{R}^{-1}$ a function?
2. Given $\mathcal{R} = \{(9,-6), (0,11), (3,8), (-2,-6), (10,-4)\}$, find $\mathcal{R}^{-1}$, $D_{\mathcal{R}^{-1}}$, and $R_{\mathcal{R}^{-1}}$, and graph $\mathcal{R}$ and $\mathcal{R}^{-1}$. Is $\mathcal{R}$ or $\mathcal{R}^{-1}$ a function?
3. Given $\mathcal{R} = \{(x,y)|y = 2x^2 - 7\}$ with the domain $\{-3, -2, 0, 1\}$, find $\mathcal{R}^{-1}$, $D_{\mathcal{R}^{-1}}$, and $R_{\mathcal{R}^{-1}}$, and graph $\mathcal{R}$ and $\mathcal{R}^{-1}$. Is $\mathcal{R}$ or $\mathcal{R}^{-1}$a function?
4. Given $\mathcal{R} = \{(x,y)|y = 4x + x^2\}$ with domain $\{-5, -4, 1, 2\}$, find $\mathcal{R}^{-1}$, $D_{\mathcal{R}^{-1}}$, $R_{\mathcal{R}^{-1}}$, and graph $\mathcal{R}$ and $\mathcal{R}^{-1}$. Is $\mathcal{R}$ or $\mathcal{R}^{-1}$ a function?
5. Given $\mathcal{R} = \{(x,y)|y = 10 - x^2\}$ with domain $\{-4, -3, 0, 2\}$, find $\mathcal{R}^{-1}$, $D_{\mathcal{R}^{-1}}$, $R_{\mathcal{R}^{-1}}$, and graph $\mathcal{R}$ and $\mathcal{R}^{-1}$. Is $\mathcal{R}$ or $\mathcal{R}^{-1}$ a function?
6. Find the inverse relation for $y = f(x) = \frac{1}{5}x + 2$. Plot the graph of $f(x)$ and its inverse.
7. Find the inverse relation for $y = f(x) = 5 - 2x$. Plot the graph of $f(x)$ and its inverse.
8. Find the inverse relation for $y = f(x) = 3x - 10$. Plot the graph of $f(x)$ and its inverse.
9. Find the inverse relation for $y = f(x) = \dfrac{4x - 3}{5}$.
10. Find the inverse relation for $y = f(x) = \dfrac{2x - 7}{3}$.
11. Find the inverse relation for $y = f(x) = \dfrac{3x - 8}{4}$.
12. Find the inverse relation for $y = f(x) = \dfrac{6}{4 - 3x}$.
13. Find the inverse relation for $y = f(x) = \dfrac{5}{x + 2}$.
14. Find the inverse relation for $y = f(x) = \dfrac{10}{2x - 1}$.

7-7 Graphing Polynomial Functions

When the function formula is a polynomial, we call the function a **polynomial function.**

Example 1 Polynomial functions

(a) $f(x) = 3x - 5$ — *Linear* polynomial function

(b) $f(x) = x^2 - x - 2$ — *Quadratic* polynomial function

(c) $f(x) = x^3 - 4x$ — *Cubic* polynomial function

(d) $f(x) = 7x^5 - 2x^3 + 6$ — *Fifth-degree* polynomial function

In Section 7–2 we showed how to graph straight lines (linear polynomial functions). In this section we show how to graph *curved* lines (polynomial functions other than linear).

Two points are all that we need to draw a straight line. To draw a curved line, we must find more than two points (Examples 2 and 3).

Example 2 Graph the equation $y = x^2 - x - 2$ (a quadratic function)

Make a table of values by substituting values of x in the equation and finding the corresponding values for y.

If $x = -2$, then $y = (-2)^2 - (-2) - 2 = 4 + 2 - 2 = 4$

If $x = -1$, then $y = (-1)^2 - (-1) - 2 = 1 + 1 - 2 = 0$

If $x = 0$, then $y = (0)^2 - (0) - 2 = -2$

If $x = 1$, then $y = (1)^2 - (1) - 2 = 1 - 1 - 2 = -2$

If $x = 2$, then $y = (2)^2 - (2) - 2 = 4 - 2 - 2 = 0$

If $x = 3$, then $y = (3)^2 - (3) - 2 = 9 - 3 - 2 = 4$

x	y
-2	4
-1	0
0	-2
1	-2
2	0
3	4

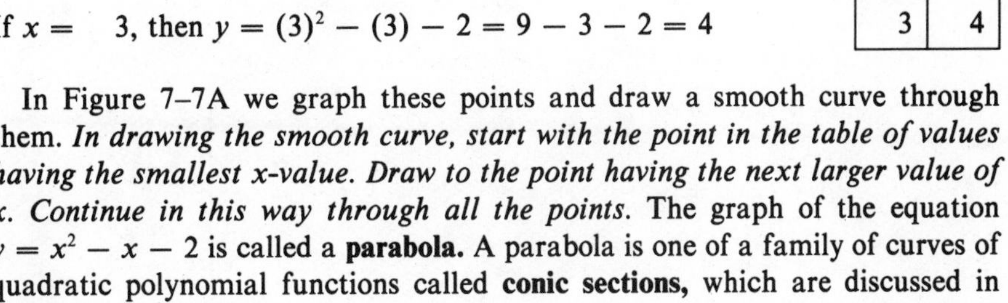

In Figure 7–7A we graph these points and draw a smooth curve through them. *In drawing the smooth curve, start with the point in the table of values having the smallest x-value. Draw to the point having the next larger value of x. Continue in this way through all the points.* The graph of the equation $y = x^2 - x - 2$ is called a **parabola.** A parabola is one of a family of curves of quadratic polynomial functions called **conic sections,** which are discussed in Section 8–6.

Parabola

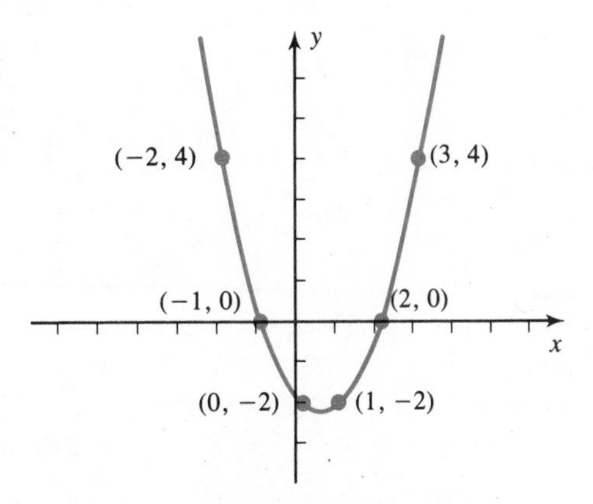

Figure 7–7A

Since no vertical line meets the graph in more than one point, $y = f(x) = x^2 - x - 2$ is a function.

Example 3 Graph the equation $y = x^3 - 4x$ (a cubic function)

First, make a table of values.

If $x = -3$, then $y = (-3)^3 - 4(-3) = -27 + 12 = -15$

If $x = -2$, then $y = (-2)^3 - 4(-2) = -8 + 8 = 0$

If $x = -1$, then $y = (-1)^3 - 4(-1) = -1 + 4 = 3$

If $x = 0$, then $y = (0)^3 - 4(0) = 0$

If $x = 1$, then $y = (1)^3 - 4(1) = 1 - 4 = -3$

If $x = 2$, then $y = (2)^3 - 4(2) = 8 - 8 = 0$

If $x = 3$, then $y = (3)^3 - 4(3) = 27 - 12 = 15$

x	y
-3	-15
-2	0
-1	3
0	0
1	-3
2	0
3	15

In Figure 7–7B we graph these points and draw a smooth curve through them, joining the points in order from left to right.

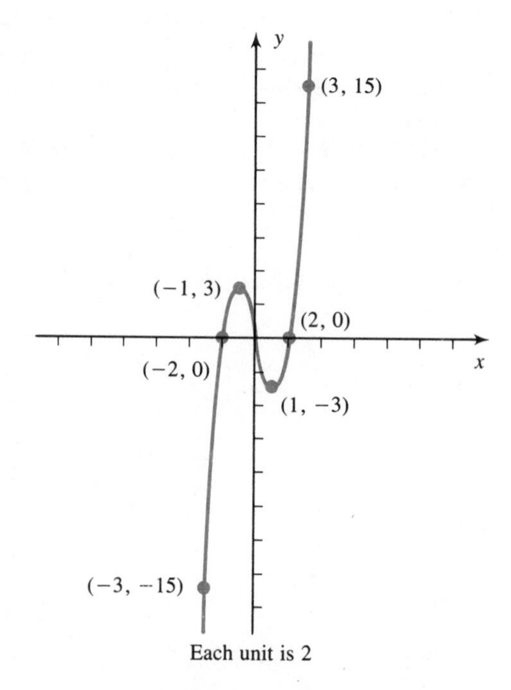

Figure 7–7B

Since no vertical line meets the graph in more than one point, $y = f(x) = x^3 - 4x$ is a function.

FINDING THE TABLE OF VALUES BY USING SYNTHETIC DIVISION

Graphing polynomial functions using synthetic division to find the table of values (Example 4) is often simpler than finding the table by direct substitution. A typical synthetic division problem is of this form:

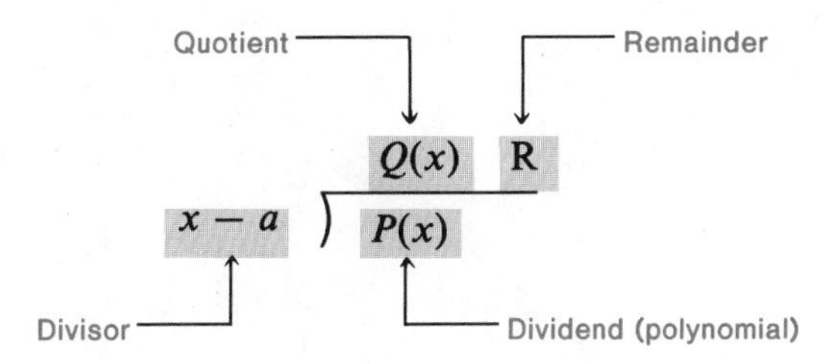

This division is equivalent to

$$P(x) = (x - a) \cdot Q(x) + \text{R}$$

Dividend = (Divisor) (Quotient) + Remainder

If $x = a$, then

$$P(x) = (x - a) \cdot Q(x) + \text{R}$$

$$P(a) = (a - a) \cdot Q(a) + \text{R}$$

Becomes $$P(a) = 0 \cdot Q(a) + \text{R}$$

$$P(a) = \text{R}$$ The **remainder theorem**

The Remainder Theorem If we divide a polynomial in x by $x - a$, the remainder is equal to the value of the polynomial when $x = a$.

Example 4 Given $f(x) = 2x^4 - 11x^3 + 7x^2 - 12x + 14$, find $f(5)$

First Method. Using substitution,

$$f(x) = 2x^4 - 11x^3 + 7x^2 - 12x + 14$$

$$f(5) = 2(5)^4 - 11(5)^3 + 7(5)^2 - 12(5) + 14$$

$$= 1250 - 1375 + 175 - 60 + 14 = 4$$

↑ $f(5)$

Second Method. Using synthetic division,

	2	−11	7	−12	14
		10	−5	10	−10 ← These products can be kept in mind and need not be written
5	2	−1	2	−2	4

↑ $f(5)$

Example 5 Use synthetic division to find the table of values in Example 3

$$y = f(x) = x^3 - 4x$$

$$= 1x^3 + 0x^2 - 4x + 0$$

x	1	0	-4	0 (y)
-3	1	-3	5	-15
-2	1	-2	0	0
-1	1	-1	-3	3
0	1	0	-4	0
1	1	1	-3	-3
2	1	2	0	0
3	1	3	5	15

To simplify the writing of this table, the products are not written

TO GRAPH A CURVE

1. Use the equation to make a table of values.
2. Plot the points from the table of values.
3. Draw a smooth curve through the points, joining them in order from left to right.

EXERCISES 7-7

In Exercises 1–4 complete the table of values for each equation and then draw its graph.

1. $y = x^2$

x	y
-2	
-1	
0	
1	
2	

2. $y = \dfrac{x^2}{4}$

x	y
-3	
-2	
-1	
0	
1	
2	
3	

3. $y = x^2 - 2x$

x	y
-2	
-1	
0	
1	
2	
3	
4	

4. $y = x - \frac{1}{2}x^2$

x	y
-2	
-1	
0	
1	
2	
3	
4	

5. Use integer values of x from -2 to $+4$ to make a table of values for the equation $y = 2x - x^2$. Graph the points and draw a smooth curve through them.

6. Use integer values of x from -2 to $+4$ to make a table of values for the equation $y = x^2 - 2x - 3$. Graph the points and draw a smooth curve through them.

7. Use integer values of x from -2 to $+2$ to make a table of values for the equation $y = x^3$. Graph the points and draw a smooth curve through them.

8. Use integer values of x from -4 to $+3$ to make a table of values for the equation $y = \frac{1}{2}(x^3 + x^2 - 6x)$. Graph the points and draw a smooth curve through them.

9. Use synthetic division to find the range of the function $f(x) = x^3 - 2x^2 - 13x + 20$ whose domain is $\{-4, -3, -2, 0, 2, 3, 4, 5\}$.

10. Use synthetic division to find the range of the function $f(x) = x^3 - 3x^2 - 20x + 12$ whose domain is $\{-3, -2, 0, 2, 3, 4, 5, 6\}$.

11. If $x^9 - 15$ were divided by $x - 1$, what would be the remainder? (Find the answer by using the remainder theorem, not by actual division.)

12. If $x^7 + 83$ were divided by $x + 2$, what would be the remainder? (Find the answer by using the remainder theorem, not by actual division.)

Chapter Seven Summary

An Ordered Pair of numbers is used to represent a point in the plane (Section 7–1).

To Graph a Straight Line. (Intercept method; Section 7–2).

1. Find the x-intercept: Set $y = 0$; then solve for x.
2. Find the y-intercept: Set $x = 0$; then solve for y.
3. Draw a straight line through the x- and y-intercepts.
4. If both intercepts are (0,0), an additional point must be found before the line can be drawn (Example 2, Section 7–2).

To Graph a Curve. (Polynomial functions other than linear; Section 7–7).

1. Use the equation to make a table of values.

2. Plot the points from the table of values.
3. Draw a smooth curve through the points, joining them in order from left to right.

To Graph a First-Degree Inequality in the Plane. (Section 7–4)

1. The boundary line is *solid* if equality *is* included ($\leq, \geq$). The boundary line is *dashed* if equality *is not* included ($<, >$).
2. Graph the boundary line.
3. Select and shade the correct half-plane.

Distance between Two Points $P_1(x_1, y_1)$ and $P_2(x_2, y_2)$. (Section 7–1)

$$d = \sqrt{(x_2 - x_1)^2 + (y_2 - y_1)^2}$$

Slope of the Line through Points $P_1(x_1, y_1)$ and $P_2(x_2, y_2)$. (Section 7–3)

$$m = \frac{y_2 - y_1}{x_2 - x_1}$$

Equations of a Line. (Section 7–3)

1. *General Form:* $Ax + By + C = 0$, where A and B are not both 0.
2. *Point-Slope Form:* $y - y_1 = m(x - x_1)$, where (x_1, y_1) is a known point on the line and m = slope of the line.
3. *Slope-Intercept Form:* $y = mx + b$, where m = slope and b is the y-intercept of the line.

Relations. (Section 7–5). A *mathematical relation* is a set of ordered pairs(x,y).

The domain of a relation is the set of all the first coordinates of the ordered pairs of that relation.

The range of a relation is the set of all the second coordinates of the ordered pairs of that relation.

The graph of a relation is the graph of all the ordered pairs of that relation.

The inverse relation $\mathcal{R}^{-1}$ of a relation $\mathcal{R}$ is:

1. The set of ordered pairs obtained by interchanging the first and second coordinates in each ordered pair of $\mathcal{R}$. Or:
2. It is found by interchanging y and x in the relation formula $y = f(x)$, then solving the resulting formula for y.

The graph of the inverse relation $\mathcal{R}^{-1}$ is the mirror image of the relation $\mathcal{R}$, when considering the mirror to be the 45° line, $y = x$.

Functions. (Section 7–5). A *function* is a relation such that no two of its ordered pairs can have the *same first coordinate* and *different second coordinates.* This means that no vertical line can meet the graph of a function in more than one point.

Linear function: $f(x) = mx + b$ Graph is a *straight line*

Quadratic function: $f(x) = ax^2 + bx + c$ Graph is a *parabola*

Cubic function: $f(x) = ax^3 + bx^2 + cx + d$

The value of a function: $f(a)$, read "f of a," is the value of $f(x)$ when x is replaced by a.

The inverse function: If the inverse relation of a function is itself a function, then that inverse relation is called an *inverse function.*

Chapter Seven Diagnostic Test or Review

Allow yourself about one hour to do these problems. Complete solutions for every problem, together with section references, are given at the end of the book.

---

1. (a) Draw the triangle with vertices at $A(-4,2)$, $B(1,-3)$, and $C(5,3)$.
(b) Find the length of the side AB.
(c) Find the slope of the line through B and C.
(d) Write the equation (in general form) of the line through A and B.
(e) Find the x-intercept of the line through A and B.
(f) Find the y-intercept of the line through A and B.

2. Find the equation of the line that has a slope of $\frac{6}{5}$ and a y-intercept of -4. Write the answer in general form.

3. Draw the graph of the equation $x - 2y = 6$.

4. Complete the table of values and draw the graph of $y = 1 + x - x^2$.

x	y
-2	
-1	
0	
1	
2	
3	

5. Graph the inequality $4x - 3y \leq -12$.

6. Given the relation $\{(-4,-5), (2,4), (4,-2), (-2,3), (2,-1)\}$:
(a) Find its domain.
(b) Find its range.
(c) Draw the graph of the relation.
(d) Is this relation a function?

7. Given $f(x) = 3x^2 - 5$. Find:
(a) $f(-2)$ (b) $f(4)$ (c) $\dfrac{f(4) - f(-2)}{6}$

8. Given $\mathcal{R} = \{(1,4), (-5,-3), (4,-2), (-5,2)\}$, find $\mathcal{R}^{-1}$, $D_{\mathcal{R}^{-1}}$, and $R_{\mathcal{R}^{-1}}$, and graph $\mathcal{R}$ and $\mathcal{R}^{-1}$. Is $\mathcal{R}$ or $\mathcal{R}^{-1}$ a function?

9. Find the inverse relation for $y = f(x) = -\frac{3}{2}x + 1$. Plot the graph of $f(x)$ and its inverse.

10. Find the inverse relation for $y = f(x) = \dfrac{15}{4(2 - 5x)}$.

8 QUADRATIC EQUATIONS, INEQUALITIES, AND FUNCTIONS

In this chapter we discuss several methods for solving quadratic equations and inequalities, as well as the graphing of quadratic functions and conic sections.

8–1 Solving Quadratic Equations by Factoring

In this section we review the solution of quadratic equations by factoring (Section 4–10) and extend it to more complicated equations.

Quadratic Equations

A **quadratic equation** (*second-degree equation*) is a polynomial equation with a *second-degree* term as its highest-degree term. The number of roots of a polynomial equation is at most equal to the degree of the polynomial. For this reason we expect to get two roots when solving quadratic equations. (Refer to the note in the box below.)

Example 1 Quadratic equations

(a) $3x^2 + 7x + 2 = 0$ (b) $\frac{1}{2}x^2 = \frac{2}{3}x - 4$

(c) $x^2 - 4 = 0$ (d) $5x^2 - 15x = 0$

General Form (Standard Form)

THE GENERAL FORM OF A QUADRATIC EQUATION

$$ax^2 + bx + c = 0$$

where a, b, and c are real numbers ($a \neq 0$).

In this text *when we write the general form of a quadratic equation, all coefficients will be integers. It is also helpful to write the general form in such a way that a is positive.*

TO CHANGE A QUADRATIC EQUATION INTO GENERAL FORM

1. Remove fractions by multiplying each term by the LCD.
2. Remove grouping symbols.
3. Combine like terms.
4. Arrange all nonzero terms in desending powers on one side, leaving only zero on the other side.

Example 2 Changing quadratic equations into general form

(a)
$$7x = 5 - 2x^2$$
$$2x^2 + 7x - 5 = 0 \quad \text{General form} \quad \begin{cases} a = 2 \\ b = 7 \\ c = -5 \end{cases}$$

(b)
$$5x^2 = 3$$
$$5x^2 + 0x - 3 = 0 \quad \text{General form} \quad \begin{cases} a = 5 \\ b = 0 \\ c = -3 \end{cases}$$

(c)
$$6x = 11x^2$$
$$0 = 11x^2 - 6x + 0$$
$$11x^2 - 6x + 0 = 0 \quad \text{General form} \quad \begin{cases} a = 11 \\ b = -6 \\ c = 0 \end{cases}$$

(d)
$$\frac{2}{3}x^2 - 5x = \frac{1}{2}$$
$$\frac{6}{1} \cdot \frac{2}{3}x^2 + \frac{6}{1} \cdot (-5x) = \frac{6}{1} \cdot \frac{1}{2} \quad \text{Multiplied each term by the LCD, 6}$$
$$4x^2 - 30x = 3$$
$$4x^2 - 30x - 3 = 0 \quad \text{General form} \quad \begin{cases} a = 4 \\ b = -30 \\ c = -3 \end{cases}$$

(e)
$$(x + 2)(2x - 3) = 3x - 7$$
$$2x^2 + x - 6 = 3x - 7$$
$$2x^2 - 2x + 1 = 0 \quad \text{General form} \quad \begin{cases} a = 2 \\ b = -2 \\ c = 1 \end{cases}$$

Solving Quadratics By Factoring

> **TO SOLVE A QUADRATIC EQUATION BY FACTORING**
>
> **1.** Arrange the equation in general form:
>
> $$ax^2 + bx + c = 0$$
>
> **2.** Factor the polynomial.
>
> **3.** Set each factor equal to zero and solve for the unknown letter.

Example 3 Solve: $(6x + 2)(x - 4) = 2 - 11x$

$$6x^2 - 22x - 8 = 2 - 11x$$
$$6x^2 - 11x - 10 = 0 \quad \text{General form}$$
$$(3x + 2)(2x - 5) = 0$$

$$\begin{array}{r|r} 3x + 2 = 0 & 2x - 5 = 0 \\ 3x = -2 & 2x = 5 \\ x = -\dfrac{2}{3} & x = \dfrac{5}{2} \end{array}$$

Example 4 Solve: $\dfrac{x - 1}{x - 3} = \dfrac{12}{x + 1}$

This is a proportion.

$$(x - 1)(x + 1) = 12(x - 3) \quad \text{Product of means} = \text{Product of extremes}$$
$$x^2 - 1 = 12x - 36$$
$$x^2 - 12x + 35 = 0$$
$$(x - 7)(x - 5) = 0$$

$$\begin{array}{r|r} x - 7 = 0 & x - 5 = 0 \\ x = 7 & x = 5 \end{array}$$

Example 5 The width of a rectangle is 5 centimeters less than its length. Its area is 10 more (numerically) than its perimeter. What are the dimensions of the rectangle?

Let L = length;
then $L - 5$ = width.

Its area	is	10 more than	its perimeter.
$L(L - 5)$	$=$	$10 +$	$2L + 2(L - 5)$

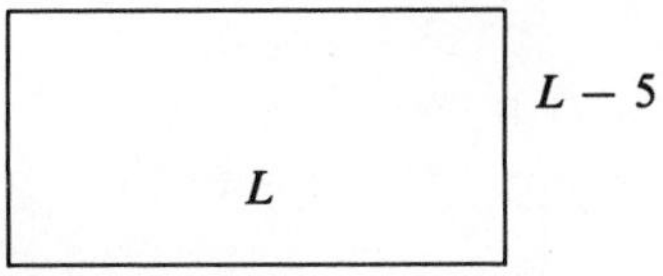

$$L(L - 5) = 10 + 2L + 2(L - 5)$$
$$L^2 - 5L = 10 + 2L + 2L - 10$$
$$L^2 - 9L = 0$$
$$L(L - 9) = 0$$

Area $= LW = L(L - 5)$
Perimeter $= 2L + 2W$
$= 2L + 2(L - 5)$

$$\begin{array}{r|rl} L = 0 & L - 9 = 0 & \\ & L = 9 & \text{Length} \\ \text{not a solution} & L - 5 = 4 & \text{Width} \end{array}$$

Therefore the rectangle has a length of 9 centimeters and a width of 4 centimeters.

Equations of Quadratic Form

Sometimes equations that are not quadratics can be solved like quadratics after making an appropriate substitution. (See Examples 6, 8, and 9.)

Example 6 Solve: $x^4 - 11x^2 + 28 = 0$

Let $z = x^2$

$z^2 = x^4$

Therefore $x^4 - 11x^2 + 28 = 0$

becomes $z^2 - 11z + 28 = 0$ This is a quadratic in z

$(z - 4)(z - 7) = 0$

$z - 4 = 0$	$z - 7 = 0$
$z = 4$	$z = 7$
$x^2 = 4$	$x^2 = 7$
$x = \pm 2$	$x = \pm\sqrt{7}$

Replaced z by x^2

Check for $x = 2$:

$x^4 - 11x^2 + 28 = 0$

$(2)^4 - 11(2)^2 + 28 \stackrel{?}{=} 0$

$16 - 44 + 28 \stackrel{?}{=} 0$

$0 = 0$

Check for $x = \sqrt{7}$:

$x^4 - 11x^2 + 28 = 0$

$(\sqrt{7})^4 - 11(\sqrt{7})^2 + 28 \stackrel{?}{=} 0$

$49 - 77 + 28 \stackrel{?}{=} 0$

$0 = 0$

The values -2 and $-\sqrt{7}$ also check. Since all powers in the equation are even, these checks will give the same results as the two checks shown.

Equations of quadratic form can be recognized by noticing that the power of an expression in one term is twice the power of the same expression in another term. (See Example 7.)

Example 7 Recognizing equations of quadratic form

(a) $y^{-8} - 10y^{-4} + 9 = 0$ $\quad -8$ is twice -4

(b) $h^{-2/3} - h^{-1/3} = 0$ $\quad -\frac{2}{3}$ is twice $-\frac{1}{3}$

(c) $(x^2 - 2x)^2 - 7(x^2 - 2x)^1 + 12 = 0$ $\quad$ 2 is twice 1

Example 8 Solve: $h^{-2/3} - h^{-1/3} = 0$

You may need to refer to Section 6–1 in order to do this problem.

Let $z = h^{-1/3}$

$z^2 = h^{-2/3}$

Therefore $h^{-2/3} - h^{-1/3} = 0$

becomes $z^2 - z = 0$

$z(z - 1) = 0$

$z = 0$	$z - 1 = 0$
	$z = 1$
$h^{-1/3} = 0$	$h^{-1/3} = 1$
$(h^{-1/3})^{-3} = (0)^{-3}$	$(h^{-1/3})^{-3} = (1)^{-3}$
Does not exist ↑	$h = 1$

Replaced z by $h^{-1/3}$

The solution $h = 1$ checks. (We leave the check to the student.)
The solution set is $\{1\}$.

Example 9 Solve: $(x^2 - 2x)^2 - 11(x^2 - 2x) + 24 = 0$

Let $z = x^2 - 2x$

$z^2 = (x^2 - 2x)^2$

Therefore $(x^2 - 2x)^2 - 11(x^2 - 2x) + 24 = 0$

becomes $z^2 - 11z + 24 = 0$

$(z - 3)(z - 8) = 0$

$z - 3 = 0$	$z - 8 = 0$
$z = 3$	$z = 8$
$x^2 - 2x = 3$	$x^2 - 2x = 8$
$x^2 - 2x - 3 = 0$	$x^2 - 2x - 8 = 0$
$(x + 1)(x - 3) = 0$	$(x + 2)(x - 4) = 0$

Replaced z by $x^2 - 2x$

$x + 1 = 0$	$x - 3 = 0$	$x + 2 = 0$	$x - 4 = 0$
$x = -1$	$x = 3$	$x = -2$	$x = 4$

All these answers check. (We leave the checks to the student.)
The solution set is $\{-1, 3, -2, 4\}$.

EXERCISES 8-1

In Exercises 1–8 write each of the quadratic equations in general form; then identify a, b, and c.

1. $3x^2 + 5x = 2$

2. $3x + 5 = 2x^2$

3. $3x^2 = 4$

4. $16x = x^2$

5. $\frac{4x}{3} = 4 + x^2$

6. $\frac{3x}{2} - 5 = x^2$

7. $3x(x - 2) = (x + 1)(x - 5)$

8. $7x(2x + 3) = (x - 3)(x + 4)$

In Exercises 9–22 solve by factoring.

9. $\frac{x}{2} + \frac{2}{x} = \frac{5}{2}$

10. $\frac{x}{3} + \frac{2}{x} = \frac{7}{3}$

11. $\frac{x - 1}{4} + \frac{6}{x + 1} = 2$

12. $\frac{3x + 1}{5} + \frac{8}{3x - 1} = 3$

13. $2x^2 = \frac{2 - x}{3}$

14. $4x^2 = \frac{14x - 3}{2}$

15. $\frac{3x - 2}{6} = \frac{2}{x + 1}$

16. $\frac{1}{2x - 3} = \frac{x - 1}{15}$

17. $z^{-4} - 4z^{-2} = 0$

18. $x^4 + 5x^2 - 36 = 0$

19. $K^{-2/3} + 2K^{-1/3} + 1 = 0$

20. $M^{-1} - 2M^{-1/2} + 1 = 0$

21. $(x^2 - 4x)^2 - (x^2 - 4x) - 20 = 0$

22. $(x^2 - 2x)^2 - 2(x^2 - 2x) - 3 = 0$

23. The length of a rectangle is twice its width. If the numerical sum of its area and perimeter is 80, find the length and width.

24. The length of a rectangle is three times its width. If the numerical sum of its area and perimeter is 80, find its dimensions.

25. Bruce drives from Los Angeles to the Mexican border and back to Los Angeles, a total distance of 240 miles. His average speed returning to Los Angeles was 20 mph faster than his average speed going to Mexico. If his total driving time was 5 hr, what was his average speed driving from Los Angeles to Mexico?

26. Ruth drives from Creston to Des Moines, a distance of 90 miles. Then she continues on from Des Moines to Omaha, a distance of 120 miles. Her average speed was 10 mph faster on the second part of the journey than on the first part. If the total driving time was 6 hr, what was her average speed on the first leg of the journey?

27. If the product of two consecutive even integers is increased by 4, the result is 84. Find the integers.

28. The tens digit of a two-digit number is 4 more than the units digit. If the product of the units digit and tens digit is 21, find the number.

29. The length of a rectangle is 2 more than twice its width. If its diagonal is 3 more than twice its width, what are its dimensions?

30. *Brainteaser.* How can two mothers and two daughters divide 21 one-dollar bills exactly among them? Each must receive an equal number of dollar bills.

8–2 Incomplete Quadratic Equations

An **incomplete quadratic equation** is one in which b or c (or both) is zero. The only letter that *cannot* be zero is a. If a were zero, the equation would not be a quadratic.

Example 1 Incomplete quadratic equations

(a) $12x^2 + 5 = 0 \quad (b = 0)$ (b) $7x^2 - 2x = 0 \quad (c = 0)$

(c) $3x^2 = 0 \quad (b \text{ and } c = 0)$

TO SOLVE A QUADRATIC EQUATION WHEN $c = 0$

1. Find the greatest common factor (GCF) (Section 4–2).
2. Then solve by factoring.

$$ax^2 + bx = 0$$
$$x(ax + b) = 0$$

$$x = 0 \quad \Big| \quad ax + b = 0$$
$$ax = -b$$
$$x = -\frac{b}{a}$$

Example 2 Solve: $12x^2 = 3x$

$$12x^2 - 3x = 0 \quad \text{General form}$$
$$3x(4x - 1) = 0 \quad \text{GCF} = 3x$$

$$3x = 0 \quad \Big| \quad 4x - 1 = 0$$
$$x = 0 \quad \Big| \quad 4x = 1$$
$$x = \frac{1}{4}$$

A Word of Caution. In doing Example 2, a common mistake students make is to divide both sides of the equation by x.

$$12x^2 = 3x$$
$$12x = 3 \quad \text{Divided both sides by } x$$
$$x = \frac{1}{4}$$

Using this method, we found only the solution $x = \frac{1}{4}$, *not* $x = 0$.

By dividing both sides of the equation by x, we lost the solution $x = 0$. *Do not divide both sides of an equation by an expression containing the unknown letter because you may lose solutions.* ■

The ± Symbol

A positive real number N has two square roots: a positive root called the **principal square root,** written $\sqrt{N}$, and a negative square root, written $-\sqrt{N}$. We can represent these two square roots by the symbol $\pm\sqrt{N}$. For example:

"$x = \pm 2$" is read "x equals plus or minus 2"

This means $x = +2$ or $x = -2$.

Example 3 Solve: $x^2 - 4 = 0$
(Here, $b = 0$.) Refer to the box below for the method of solution.

$$x^2 - 4 = 0$$
$$x^2 = 4$$
$$x = \pm 2 \quad \text{Took square root of both sides}$$

Justification for the ± Sign: This equation can be solved by factoring.

$$x^2 - 4 = 0$$
$$(x + 2)(x - 2) = 0$$

$$x + 2 = 0 \quad \Big| \quad x - 2 = 0$$
$$x = -2 \quad \Big| \quad x = 2$$

This shows why we must use ±.

TO SOLVE A QUADRATIC EQUATION WHEN $b = 0$

1. Arrange the equation so that the second-degree term is on one side and the constant term c is on the other side.
2. Divide both sides by the coefficient a of x^2.
3. Take the square root of both sides and simplify the square root.
4. There is both a positive and a negative answer (written $\pm$).

$$\begin{aligned} ax^2 + c &= 0 \\ ax^2 &= -c \\ x^2 &= -\frac{c}{a} \\ x &= \pm\sqrt{-\frac{c}{a}} \end{aligned}$$

Take the square root of both sides. There is a positive and negative answer.

5. When the radicand $-\frac{c}{a}$ is:

Positive: The square roots are real numbers.
Negative: The square roots are complex numbers (see Section 6–8).

Example 4 Solve: $3x^2 - 5 = 0$

$$\begin{aligned} 3x^2 &= 5 \\ x^2 &= \frac{5}{3} \\ x &= \pm\sqrt{\frac{5}{3}} = \pm\frac{\sqrt{5}}{\sqrt{3}} \cdot \frac{\sqrt{3}}{\sqrt{3}} = \pm\frac{\sqrt{15}}{3} \\ x &= \pm\frac{\sqrt{15}}{3} \end{aligned}$$

We can use Table I at the back of the book, or a calculator, to evaluate $\sqrt{15}$ and express the answers as approximate decimals.

$$x = \pm\frac{\sqrt{15}}{3} \doteq \pm\frac{3.873}{3} = \pm 1.291 \doteq \pm 1.29$$

Rounded off to two decimal places

Example 5 Solve:

$$x^2 + 25 = 0$$
$$x^2 = -25$$
$$x = \pm\sqrt{-25}$$
$$x = \pm 5i$$

When the radicand is negative, the roots are complex numbers (Section 6-8)

EXERCISES 8-2

In Exercises 1–16 solve each equation.

1. $x^2 - 16 = 0$ **2.** $x^2 = 81$ **3.** $5x^2 + 4 = 0$

4. $3x^2 + 25 = 0$ **5.** $\frac{2x^2}{3} = 4x$ **6.** $\frac{3x}{5} = 6x^2$

7. $\frac{5x}{2} = 15x^2$ **8.** $\frac{2x-1}{3x} = \frac{x-3}{x}$ **9.** $\frac{x+2}{3x} = \frac{x+1}{x}$

10. $\frac{3x-2}{4x} = \frac{x+1}{3x}$ **11.** $x(x-2) = (2x+3)x$ **12.** $2x(x-1) = 3x(2x+1)$

13. $12x = 8x^3$ **14.** $9x = 12x^3$ **15.** $x^3 = 7x$

16. $5(x-3)^2 - 40 = 0$

17. The length of the diagonal of a square is $\sqrt{32}$. What is the length of its sides?

18. The length of the diagonal of a square is 18. What is the length of its sides?

19. A rectangle is 12 cm long and 8 cm wide. Find the length of its diagonal.

20. A rectangle is 8.6 cm long and 5.7 cm wide. Find the length of its diagonal.

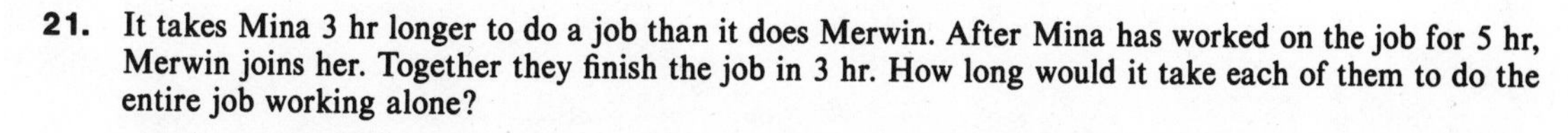
21. It takes Mina 3 hr longer to do a job than it does Merwin. After Mina has worked on the job for 5 hr, Merwin joins her. Together they finish the job in 3 hr. How long would it take each of them to do the entire job working alone?

22. It takes Darryl 2 hr longer to do a job than it does Jeannie. After Darryl has worked on the job for 1 hr, Jeannie joins him. Together they finish the job in 3 hr. How long would it take each of them to do the entire job working alone?

8–3 The Quadratic Formula and Completing the Square

In Section 4–7 we used **completing the square** as a method of factoring. In this section we use completing the square to solve quadratic equations and to derive the **quadratic formula.** The methods we have used in this chapter so far can be used to solve *some* quadratic equations. The quadratic formula (or completing the square) can be used to solve *any* quadratic equation.

Example 1 Solve: $x^2 - 4x + 1 = 0$

$$x^2 - 4x = -1$$

Take $\frac{1}{2}(-4) = -2$

Then $(-2)^2 = 4$

$$x^2 - 4x + 4 = -1 + 4$$ Added 4 to both sides; this made the left side a trinomial square

$$(x-2)^2 = 3$$ Factored the left side

$$\sqrt{(x-2)^2} = \pm\sqrt{3}$$

$$x - 2 = \pm\sqrt{3}$$ Took the square root of both sides

$$x = 2 \pm \sqrt{3}$$ Added 2 to both sides

Check for $x = 2 + \sqrt{3}$:

$$x^2 - 4x + 1 = 0$$
$$(2+\sqrt{3})^2 - 4(2+\sqrt{3}) + 1 \stackrel{?}{=} 0$$
$$4 + 4\sqrt{3} + 3 - 8 - 4\sqrt{3} + 1 \stackrel{?}{=} 0$$
$$0 = 0$$

We leave the check for $x = 2 - \sqrt{3}$ for the student.

Example 2 Solve: $25x^2 - 30x + 11 = 0$

$$25x^2 - 30x = -11$$

$$x^2 - \frac{30}{25}x = -\frac{11}{25}$$ Divided both sides by 25

Take $\frac{1}{2}\left(-\frac{30}{25}\right) = -\frac{15}{25} = -\frac{3}{5}$

Then $\left(-\frac{3}{5}\right)^2 = \frac{9}{25}$

$$x^2 - \frac{6}{5}x + \frac{9}{25} = -\frac{11}{25} + \frac{9}{25}$$ Added $\frac{9}{25}$ to both sides. This made the left side a trinomial square

$$\left(x - \frac{3}{5}\right)^2 = -\frac{2}{25}$$ Factored the left side

$$x - \frac{3}{5} = \pm\sqrt{-\frac{2}{25}}$$

$$x - \frac{3}{5} = \pm\frac{\sqrt{2}}{5}i$$ Simplified the radicals

$$x = \frac{3}{5} \pm \frac{\sqrt{2}}{5}i$$ Added $\frac{3}{5}$ to both sides

Deriving the Quadratic Formula

The method of completing the square can be used to solve *any* quadratic equation. We now use it to solve the general form of the quadratic equation and in this way derive the *quadratic formula*.

$$ax^2 + bx + c = 0$$ General form

$$ax^2 + bx = 0 - c$$ Subtracted c from both sides

$$x^2 + \frac{b}{a}x = -\frac{c}{a}$$ Divided both sides by a

Take $\frac{1}{2}\left(\frac{b}{a}\right) = \frac{b}{2a}$

Then $\left(\frac{b}{2a}\right)^2 = \frac{b^2}{4a^2}$

$$x^2 + \frac{b}{a}x + \frac{b^2}{4a^2} = \frac{b^2}{4a^2} - \frac{c}{a} = \frac{b^2}{4a^2} - \frac{4ac}{4a^2}$$ Added $\frac{b^2}{4a^2}$ to both sides. This made the left side a trinomial square

$$\left(x + \frac{b}{2a}\right)^2 = \frac{b^2 - 4ac}{4a^2}$$ Factored the left side and added the fractions on the right side

$$x + \frac{b}{2a} = \pm\sqrt{\frac{b^2 - 4ac}{4a^2}}$$ Took the square root of both sides

$$x + \frac{b}{2a} = \pm\frac{\sqrt{b^2 - 4ac}}{\sqrt{4a^2}} = \pm\frac{\sqrt{b^2 - 4ac}}{2a}$$ Simplified radicals

$$x = -\frac{b}{2a} \pm \frac{\sqrt{b^2 - 4ac}}{2a}$$ Added $-\frac{b}{2a}$ to both sides

Therefore $$x = \frac{-b \pm \sqrt{b^2 - 4ac}}{2a}$$ Quadratic formula

TO SOLVE A QUADRATIC EQUATION BY FORMULA

1. Arrange the equation in general form.

$$ax^2 + bx + c = 0$$

2. Substitute the values of a, b, and c into the **quadratic formula.**

$$x = \frac{-b \pm \sqrt{b^2 - 4ac}}{2a} \qquad (a \neq 0)$$

3. Simplify your answers.

Check your answers by substituting them in the *original* equation.

Example 3 Solve $x^2 - 5x + 6 = 0$ by formula

Substitute $\begin{Bmatrix} a = 1 \\ b = -5 \\ c = 6 \end{Bmatrix}$ in the formula $x = \dfrac{-b \pm \sqrt{b^2 - 4ac}}{2a}$.

$$x = \frac{-(-5) \pm \sqrt{(-5)^2 - 4(1)(6)}}{2(1)}$$

$$x = \frac{5 \pm \sqrt{25 - 24}}{2} = \frac{5 \pm \sqrt{1}}{2}$$

$$x = \frac{5 \pm 1}{2} = \begin{cases} \dfrac{5+1}{2} = \dfrac{6}{2} = 3 \\ \dfrac{5-1}{2} = \dfrac{4}{2} = 2 \end{cases}$$

This equation can also be solved by factoring.

$$x^2 - 5x + 6 = 0$$
$$(x - 2)(x - 3) = 0$$

$$x - 2 = 0 \quad \Big| \quad x - 3 = 0$$
$$x = 2 \quad \Big| \quad x = 3$$

Solving a quadratic equation by factoring is ordinarily shorter than using the formula. Therefore first check to see if the equation can be factored by any of the methods discussed in Chapter Four. If it cannot, use the formula.

Example 4 Solve: $\frac{1}{4}x^2 = 1 - x$

$$\frac{4}{1} \cdot \frac{1}{4}x^2 = 4 \cdot (1 - x)$$ Multiplied each term by the LCD, 4

$$x^2 = 4 - 4x$$

$$x^2 + 4x - 4 = 0$$ Changed the equation to general form

Substitute $\begin{Bmatrix} a = 1 \\ b = 4 \\ c = -4 \end{Bmatrix}$ into $x = \dfrac{-b \pm \sqrt{b^2 - 4ac}}{2a}$.

$$x = \frac{-(4) \pm \sqrt{(4)^2 - 4(1)(-4)}}{2(1)}$$

$$= \frac{-4 \pm \sqrt{16 + 16}}{2} = \frac{-4 \pm \sqrt{32}}{2}$$

$$= \frac{-4 \pm 4\sqrt{2}}{2} = -2 \pm 2\sqrt{2}$$

Example 5 Solve: $x^2 - 6x + 13 = 0$

Substitute $\begin{Bmatrix} a = 1 \\ b = -6 \\ c = 13 \end{Bmatrix}$ into $x = \dfrac{-b \pm \sqrt{b^2 - 4ac}}{2a}$.

$$x = \frac{-(-6) \pm \sqrt{(-6)^2 - 4(1)(13)}}{2(1)}$$

$$= \frac{6 \pm \sqrt{36 - 52}}{2} = \frac{6 \pm \sqrt{-16}}{2}$$

$$= \frac{6 \pm 4i}{2} = 3 \pm 2i$$

Check for $x = 3 + 2i$:

$$\begin{aligned} x^2 \quad - \quad 6x \quad + 13 &= 0 \\ (3 + 2i)^2 - 6(3 + 2i) + 13 &\stackrel{?}{=} 0 \\ 9 + 12i + 4i^2 - 18 - 12i + 13 &\stackrel{?}{=} 0 \\ 9 + 12i - 4 - 18 - 12i + 13 &\stackrel{?}{=} 0 \\ 0 &= 0 \end{aligned}$$

We leave the check for $x = 3 - 2i$ for the student.

Example 6 Solve: $x^2 - 5x + 3 = 0$

Express the answers as decimals correct to two decimal places.

Substitute $\begin{Bmatrix} a = 1 \\ b = -5 \\ c = 3 \end{Bmatrix}$ into $x = \dfrac{-b \pm \sqrt{b^2 - 4ac}}{2a}$.

$$x = \frac{-(-5) \pm \sqrt{(-5)^2 - 4(1)(3)}}{2(1)}$$

$$= \frac{5 \pm \sqrt{25 - 12}}{2} = \frac{5 \pm \sqrt{13}}{2}$$

$\sqrt{13} \doteq 3.606$ from Table I

$$\doteq \frac{5 \pm 3.606}{2} = \begin{cases} \dfrac{5 + 3.606}{2} = \dfrac{8.606}{2} = 4.303 \doteq 4.30 \\[2ex] \dfrac{5 - 3.606}{2} = \dfrac{1.394}{2} = 0.697 \doteq 0.70 \end{cases}$$

EXERCISES 8-3

In Exercises 1–8 use the method of completing the square to solve each equation (see Examples 1 and 2).

1. $x^2 = 6x + 11$
2. $x^2 = 10x - 13$
3. $x^2 - 13 = 4x$
4. $x^2 + 20 = 8x$
5. $x^2 - 2x - 2 = 0$
6. $4x^2 - 8x + 1 = 0$
7. $x^2 - 6x - 5 = 0$
8. $9x^2 - 12x + 5 = 0$

In Exercises 9–22 use the quadratic formula to solve each equation.

9. $3x^2 - x - 2 = 0$ **10.** $2x^2 + 3x - 2 = 0$ **11.** $x^2 - 4x + 1 = 0$

12. $x^2 - 4x - 1 = 0$ **13.** $x^2 - 4x + 2 = 0$ **14.** $x^2 - 2x - 2 = 0$

15. $3x^2 + 2x + 1 = 0$ **16.** $4x^2 + 3x + 2 = 0$ **17.** $2x^2 = 8x - 9$

18. $3x^2 = 6x - 4$ **19.** $x^2 = \frac{3 - 5x}{2}$ **20.** $2x^2 = \frac{2 - x}{3}$

21. $\frac{x}{2} + \frac{6}{x} = \frac{5}{2}$ **22.** $\frac{x}{3} + \frac{2}{x} = \frac{2}{3}$

Solve Exercises 23–26 by the quadratic formula and show the check (see Example 5).

23. $x^2 - 4x + 5 = 0$ **24.** $x^2 - 6x + 10 = 0$

25. $12x - 7 = 4x^2$ **26.** $4x + 1 = 3x^2$

In Exercises 27–32 use Table I or a calculator if necessary to complete the solutions. Give answers correct to two decimal places.

27. The length of a rectangle is 2 more than its width. If its area is 2, find its dimensions.

28. The length of a rectangle is 4 more than its width. If its area is 6, find its dimensions.

29. The perimeter of a square is numerically 4 more than its area. Find the length of its side.

30. The area of a square is numerically 2 more than its perimeter. Find the length of its side.

31. $x^3 - 8x^2 + 16x - 8 = 0$ **32.** $x^3 - 3x^2 - 3x + 14 = 0$

8–4 The Nature of Quadratic Roots

Of the quadratic equations solved so far, some have had real roots, some complex roots, some rational roots, and some irrational roots. Some have had equal roots, and some have had unequal roots. In this section we will show how to determine what kinds of roots a quadratic equation has *without actually solving* the equation.

THE QUADRATIC DISCRIMINANT $b^2 - 4ac$

The roots of the quadratic equation $ax^2 + bx + c = 0$ are

$$x = \frac{-b \pm \sqrt{b^2 - 4ac}}{2a}$$

For the equation $5x^2 - 4x + 2 = 0$ we have $\begin{cases} a = 5 \\ b = -4 \\ c = 2 \end{cases}$

Therefore $b^2 - 4ac = (-4)^2 - 4(5)(2) = 16 - 40 = -24$

This means $\sqrt{b^2 - 4ac} = \sqrt{-24}$, so the roots are complex conjugates.

If, for another equation, $\sqrt{b^2 - 4ac} = \sqrt{13}$, then its roots must be real and irrational conjugates. If, for still another equation, $\sqrt{b^2 - 4ac} = \sqrt{25} = 5$, then its roots must be real and rational. The value of $b^2 - 4ac$ determines the nature of the roots. For this reason $b^2 - 4ac$ is called the **quadratic discriminant.**

THE QUADRATIC DISCRIMINANT
$b^2 - 4ac$ AND THE ROOTS

For the quadratic equation $ax^2 + bx + c = 0$

where $a, b, c \in R$, and $a \neq 0$

1. **If $b^2 - 4ac < 0$,** there are two complex conjugate roots.
2. **If $b^2 - 4ac = 0$,** there is one real, rational root. *
3. **If $b^2 - 4ac > 0$, and is not a perfect square,** there are two real, irrational conjugate roots.
   If $b^2 - 4ac > 0$, and is a perfect square, there are two real, rational, unequal roots.

Example 1 Determine the nature of the roots of $2x^2 + 5x - 12 = 0$ without solving the equation

$$\begin{Bmatrix} a = 2 \\ b = 5 \\ c = -12 \end{Bmatrix} \qquad \begin{aligned} b^2 - 4ac &= (5)^2 - 4(2)(-12) \\ &= 25 + 96 = 121 \end{aligned}$$

$b^2 - 4ac = 121 > 0$ and perfect square

Therefore the roots are real, rational, and unequal.

Example 2 Determine the nature of the roots of $x^2 - 2x - 2 = 0$ without solving the equation

$$\begin{Bmatrix} a = 1 \\ b = -2 \\ c = -2 \end{Bmatrix} \qquad \begin{aligned} b^2 - 4ac &= (-2)^2 - 4(1)(-2) \\ &= 4 + 8 = 12 \end{aligned}$$

$b^2 - 4ac = 12 > 0$ and *not* perfect square

Therefore the roots are real, irrational conjugates, and unequal.

Example 3 Determine the nature of the roots of $9x^2 - 6x + 1 = 0$ without solving the equation

$$\begin{Bmatrix} a = 9 \\ b = -6 \\ c = 1 \end{Bmatrix} \qquad \begin{aligned} b^2 - 4ac &= (-6)^2 - 4(9)(1) \\ &= 36 - 36 = 0 \end{aligned}$$

$b^2 - 4ac = 0$

Therefore there is one real, rational root.

Example 4 Determine the nature of the roots of $x^2 - 6x + 11 = 0$ without solving the equation

$$\begin{Bmatrix} a = 1 \\ b = -6 \\ c = 11 \end{Bmatrix} \qquad \begin{aligned} b^2 - 4ac &= (-6)^2 - 4(1)(11) \\ &= 36 - 44 = -8 \end{aligned}$$

$b^2 - 4ac = -8 < 0$

Therefore the roots are complex conjugates.

---

*When $b^2 - 4ac = 0$, the one real root is sometimes considered as two equal roots, or a *root of multiplicity two.* This is done so that *all* quadratic equations can be considered to have two roots.

USING ROOTS TO FIND THE EQUATION

Sometimes we want to find an equation that has a given set of numbers as its roots. This can be done by reversing the procedure used to solve an equation by factoring.

Example 5 Find a quadratic equation that has roots -3 and 5

$$x = -3 \quad \text{and } x = 5$$
$$x + 3 = 0 \quad x - 5 = 0$$
Rational roots
$$(x + 3) \cdot (x - 5) = 0$$
$$x^2 - 2x - 15 = 0$$
Solution set $= \{-3, 5\}$

Example 6 Find a quadratic equation that has roots $1 + \sqrt{2}$ and $1 - \sqrt{2}$

$$x = 1 + \sqrt{2} \quad x = 1 - \sqrt{2}$$
$$x - 1 - \sqrt{2} = 0 \quad x - 1 + \sqrt{2} = 0$$
Irrational conjugate roots
$$(x - 1 - \sqrt{2}) \cdot (x - 1 + \sqrt{2}) = 0$$
$$[(x - 1) - \sqrt{2}]\,[(x - 1) + \sqrt{2}] = 0$$
$$(x - 1)^2 - (\sqrt{2})^2 = 0$$
$$x^2 - 2x + 1 - 2 = 0$$
$$x^2 - 2x - 1 = 0$$
Solution set $= \{1 \pm \sqrt{2}\}$

Example 7 Find a quadratic equation that has roots $2 + 3i$ and $2 - 3i$

$$x = 2 + 3i \quad x = 2 - 3i$$
$$x - 2 - 3i = 0 \quad x - 2 + 3i = 0$$
Complex conjugate roots
$$(x - 2 - 3i) \cdot (x - 2 + 3i) = 0$$
$$[(x - 2) - 3i]\,[(x - 2) + 3i] = 0$$
$$(x - 2)^2 - (3i)^2 = 0$$
$$x^2 - 4x + 4 - 9i^2 = 0$$
$$x^2 - 4x + 13 = 0$$
Solution set $= \{2 \pm 3i\}$

Example 8 Find a *cubic* equation that has roots $\frac{1}{2}$, -3, and $\frac{2}{3}$

$$x = \frac{1}{2} \quad x = -3 \quad x = \frac{2}{3}$$
$$2x - 1 = 0 \quad x + 3 = 0 \quad 3x - 2 = 0$$
$$(2x - 1) \cdot (x + 3) \cdot (3x - 2) = 0$$
$$(2x^2 + 5x - 3)\,(3x - 2) = 0$$
$$6x^3 + 11x^2 - 19x + 6 = 0$$
Solution set $= \left\{\frac{1}{2}, -3, \frac{2}{3}\right\}$

EXERCISES 8-4

In Exercises 1–10 use the quadratic discriminant to determine the nature of the roots without solving the equation.

1. $x^2 - x - 12 = 0$
2. $x^2 + 3x - 10 = 0$
3. $6x^2 - 7x = 2$
4. $10x^2 - 11x = 5$
5. $x^2 - 4x = 0$
6. $x^2 + 9x = 0$
7. $9x^2 + 2 = 6x$
8. $2x^2 + 6x + 5 = 0$
9. $x^2 + 25 = 10x$
10. $x^2 - 2x = 2$

In Exercises 11–22 find a quadratic equation that has the given roots.

11. 4 and -2
12. -3 and 2
13. 0 and 5
14. 6 and 0
15. $\frac{1}{2}$ and $\frac{2}{3}$
16. $\frac{3}{5}$ and $\frac{2}{3}$
17. $2 + \sqrt{3}$ and $2 - \sqrt{3}$
18. $3 + \sqrt{5}$ and $3 - \sqrt{5}$
19. $3\sqrt{2}$ and $-3\sqrt{2}$
20. $1 + 3i$ and $1 - 3i$
21. $\frac{1 + \sqrt{3}i}{2}$ and $\frac{1 - \sqrt{3}i}{2}$
22. $\frac{3 + \sqrt{5}i}{2}$ and $\frac{3 - \sqrt{5}i}{2}$

In Exercises 23–28 find a cubic equation that has the given roots.

23. 1, 3, and 4
24. 2, 1, and 5
25. $\frac{1}{3}$, 1, and $\frac{2}{5}$
26. 1, $\sqrt{2}$, and $-\sqrt{2}$
27. 3, $-2i$, and $+2i$
28. -1, $\sqrt{2}i$, and $-\sqrt{2}i$

8–5 Graphing Quadratic Functions of One Variable

We have already discussed graphing quadratic functions of one variable in Section 7–7. In this section we introduce additional information about quadratic functions of one variable which will simplify drawing their graphs.

Quadratic Function A **quadratic function** is a function whose formula is a quadratic polynomial.

GENERAL FORM OF A QUADRATIC FUNCTION OF ONE VARIABLE

$$y = f(x) = ax^2 + bx + c$$

Its graph is a curve called a *parabola* * (Figure 8–5A).

y-intercept The **y-intercept** is found by setting $x = 0$.

$$y = f(x) = ax^2 + bx + c$$
$$y = f(0) = a(0)^2 + b(0) + c$$
$$y = c$$

The y-intercept of a quadratic function of one variable is *always equal to the constant term.*

x-intercepts The **x-intercepts** are found by setting $y = 0$ and then solving the resulting quadratic equation for x.

$$y = f(x) = ax^2 + bx + c = 0$$

---

*A parabola is a member of a set of curves called *conic sections* (Section 8–6).

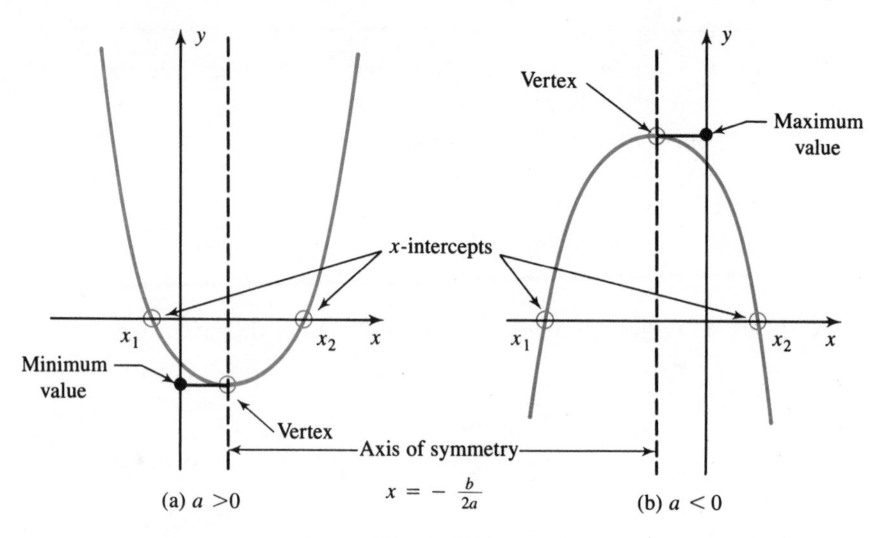

Figure 8–5A

Example 1 Find the intercepts of $y = f(x) = x^2 + x - 6$

y-intercept: $c = -6$

x-intercepts: $y = f(x) = x^2 + x - 6 = 0$ Set $y = 0$

$$(x - 2)(x + 3) = 0$$

$$x - 2 = 0 \quad | \quad x + 3 = 0$$
$$x = 2 \quad | \quad x = -3$$

The x-intercepts are 2 and -3.

Example 2 Find the intercepts of $y = f(x) = x^2 - 2x - 1$

y-intercept: $c = -1$

x-intercepts: $y = f(x) = x^2 - 2x - 1 = 0$ Set $y = 0$

Substitute $\begin{Bmatrix} a = 1 \\ b = -2 \\ c = -1 \end{Bmatrix}$ into $x = \dfrac{-b \pm \sqrt{b^2 - 4ac}}{2a}$.

$$x = \frac{-(-2) \pm \sqrt{(-2)^2 - 4(1)(-1)}}{2(1)}$$

$$= \frac{2 \pm \sqrt{4 + 4}}{2} = \frac{2 \pm \sqrt{8}}{2}$$

$$= \frac{2 \pm 2\sqrt{2}}{2} = 1 \pm \sqrt{2}$$

The x-intercepts are $1 + \sqrt{2}$ and $1 - \sqrt{2}$.

These can be approximated by using Table I or a calculator to find $\sqrt{2} \doteq 1.4$.

$$\text{Therefore}\begin{cases}1+\sqrt{2}\doteq 1+1.4=2.4\\ 1-\sqrt{2}\doteq 1-1.4=-0.4\end{cases}$$

Axis of Symmetry

The x-intercepts of a parabola are symmetrical points, (if they exist). The **axis of symmetry** of a parabola is a vertical line midway between *any* pair of symmetrical points on that parabola. To find the axis of symmetry we take the average of the two x-intercepts.

$$\text{Axis of symmetry:} \quad x=\frac{x_1+x_2}{2}$$ where x_1 and x_2 are the x-intercepts

In Example 1, the x-intercepts are 2 and -3.

Therefore, the axis of symmetry is $x=\frac{x_1+x_2}{2}=\frac{2+(-3)}{2}=-\frac{1}{2}$.

We can also find the axis of symmetry as follows:

$$x=\frac{x_1+x_2}{2}=\frac{\dfrac{-b+\sqrt{b^2-4ac}}{2a}+\dfrac{-b-\sqrt{b^2-4ac}}{2a}}{2}=-\frac{b}{2a}.$$

Symmetrical Points

Symmetrical Points of a parabola are shown in Figure 8–5B.

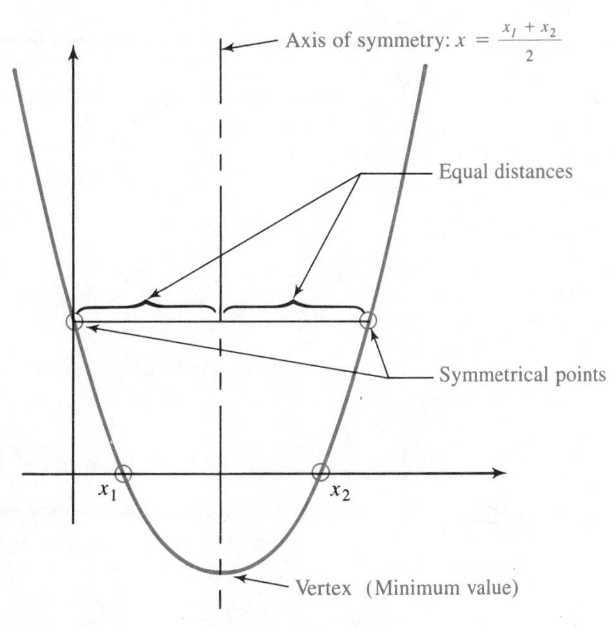

Figure 8–5B

Vertex

The **vertex** of a parabola is the point where the axis of symmetry intersects the parabola. Its coordinates are

$$\left(-\frac{b}{2a}, f\left(-\frac{b}{2a}\right)\right)$$

Maximum and Minimum Values

If the parabola opens upward, the quadratic function has its **minimum value** at the vertex. *If the parabola opens downward,* the quadratic function has its **maximum value** at the vertex.

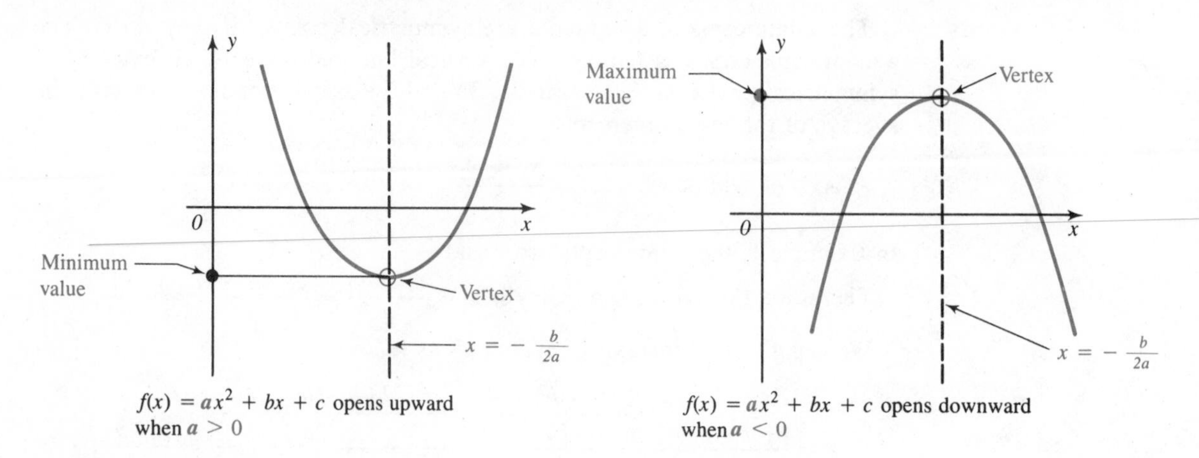

GRAPHING QUADRATIC FUNCTIONS OF ONE VARIABLE

TO GRAPH A QUADRATIC FUNCTION OF ONE VARIABLE

$$y = f(x) = ax^2 + bx + c$$

1. *y-intercept* $= c$.
2. *x-intercepts*: Set $y = 0$ and solve the resulting equation for x. The x-intercepts are x_1 and x_2.
3. *Axis of symmetry*: $x = \dfrac{-b}{2a}$ or $\dfrac{x_1 + x_2}{2}$.
4. *Vertex*: $x = \dfrac{-b}{2a}$; $y = f\left(\dfrac{-b}{2a}\right)$.
5. Plot points found in Steps 1–4. Then plot points symmetrical to those points (with respect to the axis of symmetry).
6. Draw a smooth curve through the points found in Steps 1–5, joining them in order from left to right. If necessary, calculate some additional points.

Example 3 Graph the function $f(x) = x^2 - x - 2$

1. *y-intercept:* $c = -2$.

2. *x-intercepts:* $y = f(x) = x^2 - x - 2 = 0$

$$(x + 1)(x - 2) = 0$$

$$x + 1 = 0 \quad\Big|\quad x - 2 = 0$$
$$x = -1 \quad\Big|\quad x = 2$$

3. *Axis of symmetry:* $x = \dfrac{x_1 + x_2}{2} = \dfrac{-1 + 2}{2} = \dfrac{1}{2}$

4. *Vertex:* The vertex is found by setting $x = \dfrac{1}{2}$ in

$$f(x) = x^2 - x - 2.$$

$$f\left(\frac{1}{2}\right) = \left(\frac{1}{2}\right)^2 - \left(\frac{1}{2}\right) - 2 = -2\frac{1}{4}$$

$$\text{Vertex} = \left(\frac{1}{2}, -2\frac{1}{4}\right)$$

5. Plot the points found so far:

y-intercept: $(0,-2)$ — Symmetrical point: $(1,-2)$ (Figure 8-5C)

x-intercepts: $(-1,0)$, $(2,0)$

Vertex: $\left(\dfrac{1}{2}, -2\dfrac{1}{4}\right)$

6. Calculate an additional point to help draw the graph:

$$y = f(x) = x^2 - x - 2$$
$$f(3) = (3)^2 - (3) - 2 = 9 - 3 - 2 = 4 \qquad \text{Chose } x = 3$$

Therefore an additional point is $(3,4)$

Symmetrical point: $(-2,4)$ (Figure 8–5C)

Draw a smooth curve through all points mentioned in Steps 5 and 6, taking them in order from left to right (Figure 8–5C).

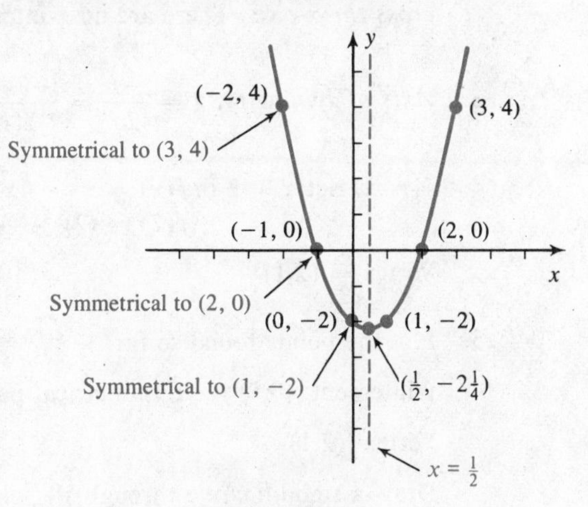

Figure 8–5C

The table of values can also be found by using synethetic division (Section 7–7). To illustrate this, we make up the table of values for the quadratic function in Example 3.

$$f(x) = x^2 \quad - \quad x \quad - \quad 2$$

x	1	−1	−2 = $f(x) = y$
−2	1	−3	4
−1	1	−2	0
0	1	−1	−2
1	1	0	−2
2	1	1	0
3	1	2	4

Example 4 Graph the function $y = f(x) = x^2 - 4x + 5$

1. *y-intercept:* $c = 5$

2. *x-intercepts:* $y = f(x) = x^2 - 4x + 5 = 0$

Substitute $\begin{Bmatrix} a = 1 \\ b = -4 \\ c = 5 \end{Bmatrix}$ into $x = \dfrac{-b \pm \sqrt{b^2 - 4ac}}{2a}$.

$$= \frac{-(-4) \pm \sqrt{(-4)^2 - 4(1)(5)}}{2(1)}$$

$$= \frac{4 \pm \sqrt{16 - 20}}{2} = \frac{4 \pm \sqrt{-4}}{2}$$

$$= \frac{4 \pm 2i}{2} = 2 \pm i$$

Since $2 \pm i$ are not real numbers, this means that *the parabola does not cross the x-axis.* There are no x-intercepts in this case (see Figure 8–5D).

3. *Axis of symmetry:* $x = \dfrac{-b}{2a} = \dfrac{-(-4)}{2(1)} = \dfrac{4}{2} = 2$

4. *Vertex:* Set $x = 2$ in $f(x) = x^2 - 4x + 5$

$$f(2) = (2)^2 - 4(2) + 5 = 1$$

Vertex = (2,1)

5. Plot the points found so far:

y-intercept: (0,5) Symmetrical point: (4,5)

Vertex: (2,1)

Draw a smooth curve through all points mentioned in Step 5.

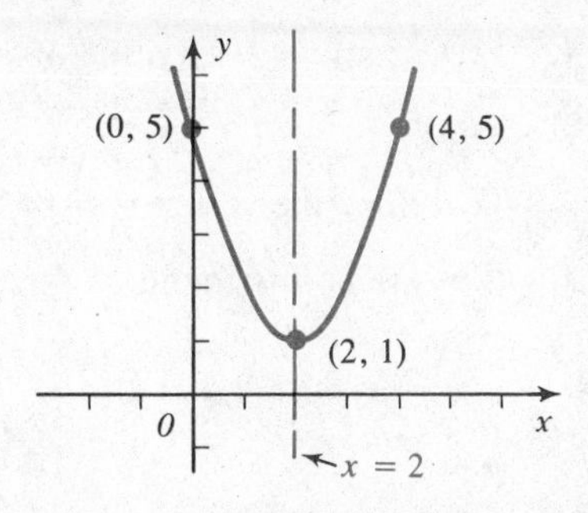

Figure 8–5D No Real-Number x-Intercepts

INVERSE QUADRATIC RELATIONS

We have discussed quadratic functions of the form:

$$y = ax^2 + bx + c$$

The inverse relation for this quadratic function is obtained by interchanging x and y. (Section 7–6)

$$x = ay^2 + by + c$$

The information already obtained in this section for $y = ax^2 + bx + c$ carries over to $x = ay^2 + by + c$ in the following way:

Quadratic Function		*Inverse Quadratic Relation*
$y = ax^2 + bx + c$	becomes	$x = ay^2 + by + c$
axis of symmetry: $x = \frac{-b}{2a}$	becomes	$y = \frac{-b}{2a}$
Two x-intercepts	become	two y-intercepts
One y-intercept	becomes	one x-intercept
Vertex: $\left(\frac{-b}{2a}, f\left(\frac{-b}{2a}\right)\right)$	becomes	$\left(\frac{-b}{2a}, f\left(\frac{-b}{2a}\right)\right)$

Example 5 Graph the quadratic relation $f(y) = y^2 + 2y - 3$

1. *x-intercept:* Set $y = 0$ in $x = f(y) = y^2 + 2y - 3$
$$x = f(0) = (0)^2 + 2(0) - 3 = -3$$

2. *y-intercepts:* Set $x = 0$ in $x = f(y) = y^2 + 2y - 3$
$$0 = f(y) = y^2 + 2y - 3$$
$$0 = (y - 1)\ (y + 3)$$
$$y = 1 \mid y = -3$$

3. *Axis of symmetry:* $y = \dfrac{-b}{2a} = \dfrac{-2}{2(1)} = -1$

4. *Vertex:* $y = -1$ (value of y at axis of symmetry)
$x = f(y) = y^2 + 2y - 3$
$x = f(-1) = (-1)^2 + 2(-1) - 3 = -4$
Therefore the vertex is $(-4, -1)$

5. *Plot the points* found so far (Figure 8–5E).

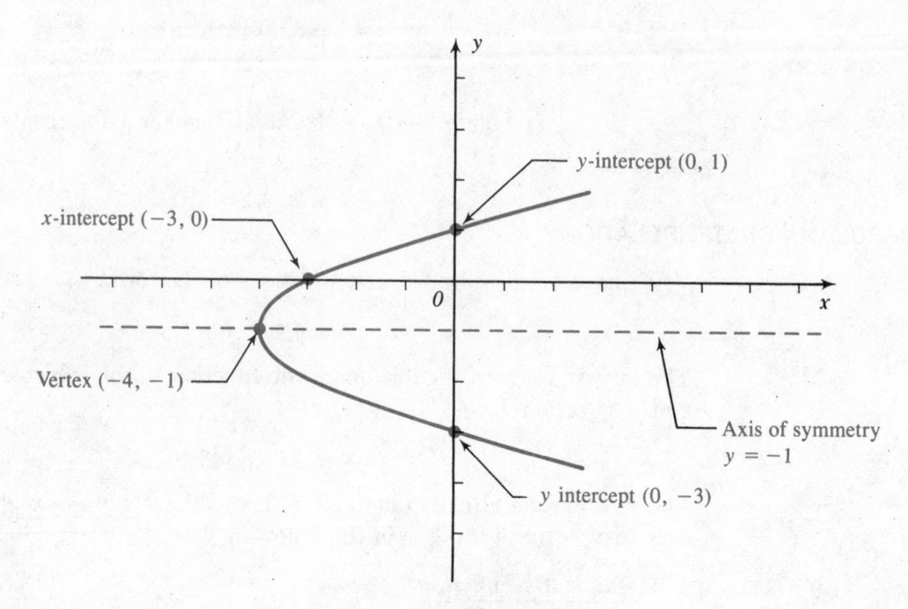

Figure 8–5E

EXERCISES 8-5

In Exercises 1–8 find the x-intercepts, if they exist.

1. $y = f(x) = x^2 - 2x - 3$
2. $y = f(x) = x^2 - 2x - 15$
3. $y = f(x) = x^2 + x - 2$
4. $y = f(x) = 2x^2 - x - 6$
5. $y = f(x) = x^2 - 2x - 13$
6. $y = f(x) = 3x^2 - 10x + 6$
7. $y = f(x) = x^2 - 6x + 10$
8. $y = f(x) = x^2 - 6x + 11$

In Exercises 9–14 find the equation of the axis of symmetry for the graph of each quadratic function.

9. $y = f(x) = 3 + x^2 - 4x$
10. $y = f(x) = 2x - 8 + x^2$
11. $y = f(x) = x^2 + 3x - 10$
12. $y = f(x) = 2x^2 - 7x - 4$
13. $y = f(x) = 4x^2 - 9$
14. $y = f(x) = 9x^2 - 4$

In Exercises 15–20 find the maximum or minimum value and the vertex of each quadratic function.

15. $f(x) = x^2 - 6x + 7$
16. $f(x) = x^2 - 4x + 5$
17. $f(x) = 8x - 5x^2 - 3$
18. $f(x) = 4 + 2x - 3x^2$
19. $f(x) = -\frac{1}{2}x^2 + x + \frac{3}{2}$
20. $f(x) = -\frac{2}{3}x^2 - \frac{8}{3}x + \frac{1}{3}$

In Exercises 21–32 find: (a) x-intercept(s); (b) y-intercept(s); (c) axis of symmetry; (d) vertex; and (e) plot the graph.

21. $f(x) = x^2 - 8x + 12$

22. $f(x) = x^2 - 4x - 6$

23. $f(x) = 2x - x^2 + 3$

24. $f(x) = 5 - 4x - x^2$

25. $f(x) = 6 + x^2 - 4x$

26. $f(x) = 2x - x^2 - 2$

27. $f(x) = 30x - 9x^2 - 25$

28. $f(x) = 2x^2 - 4x + 1$

29. $f(y) = y^2 + y - 2$

30. $f(y) = -y^2 + 3y + 4$

31. $f(y) = -\frac{3}{4}y^2 + 3y - 3$

32. $f(y) = \frac{1}{3}y^2 + \frac{4}{3}y + 3$

8–6 Conic Sections

Conic sections are curves formed when a plane cuts through a cone. Some examples of conic sections are shown in Figures 8–6A–D.

Parabola

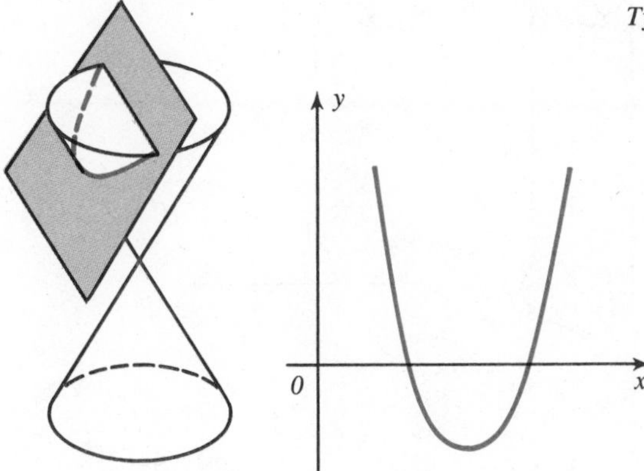

Typical Equations of Parabolas

(vertical axis of symmetry)

$y = 2x^2 - 5x + 1$

$y = 5 - 2x - 3x^2$ (Example 3)

Circle

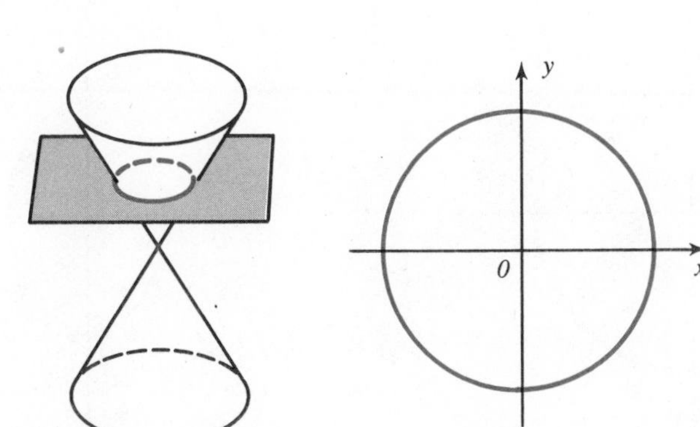

Typical Equations of Circles

(Center at the origin)

$x^2 + y^2 = 4$

$x^2 + y^2 = 16$ (Example 1)

$2x^2 + 2y^2 = 7$

Figure 8–6B

Ellipse

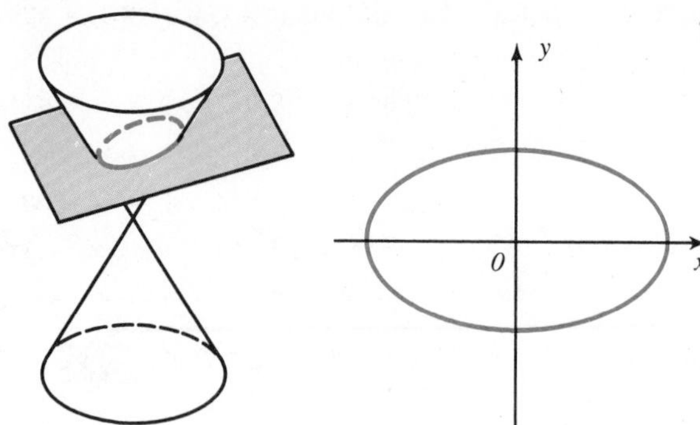

Typical Equations of Ellipses

(Center at the origin)

$x^2 + 2y^2 = 4$

$9x^2 + 4y^2 = 36$ (Example 2)

Figure 8–6C

Hyperbola

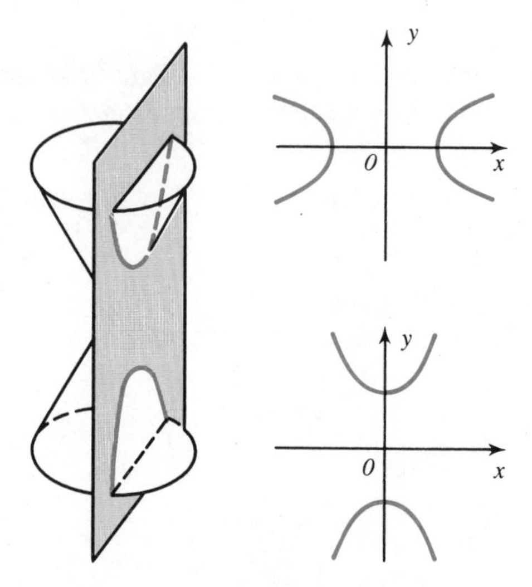

Typical Equations of Hyperbolas

(center at the origin)

$2x^2 - y^2 = 4$

$3y^2 - 4x^2 = 12$ (Example 4)

Figure 8–6D

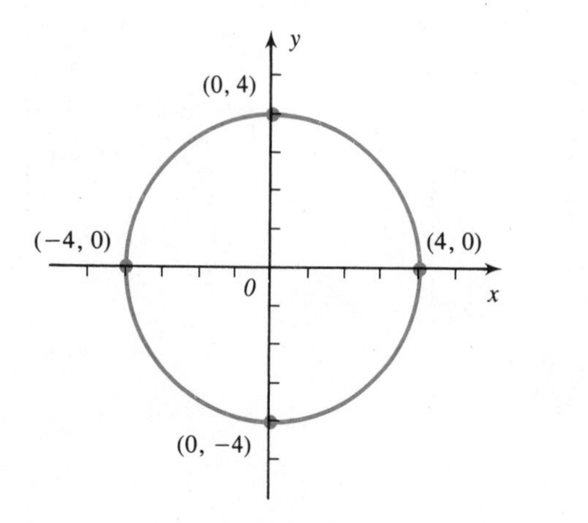

Figure 8–6E

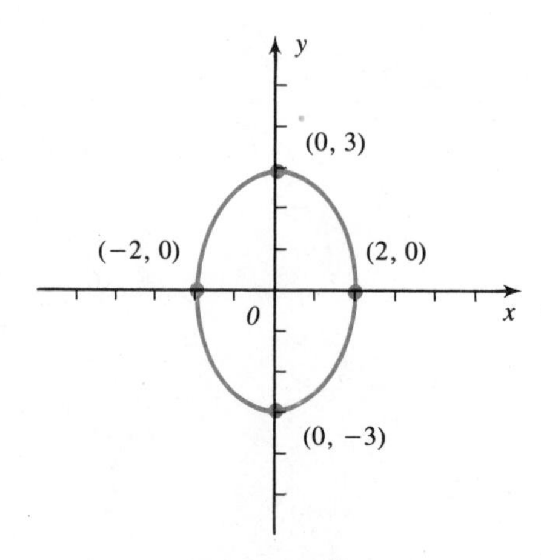

Figure 8–6F

The equations given in this section for the different kinds of conic sections are developed in *Analytic Geometry.*

Graphing Conic Sections

In this section we sketch conics using only the **intercept method** (Section 7–2).

Circle

Example 1 Sketch $x^2 + y^2 = 16$. See Figure 8–6B for this typical equation of a conic

x-intercepts: Set $y = 0$ in $x^2 + y^2 = 16$

$$x^2 + 0^2 = 16$$
$$x^2 = 16$$
$$x = \pm 4$$

y-intercepts: Set $x = 0$ in $x^2 + y^2 = 16$

$$0^2 + y^2 = 16$$
$$y = \pm 4$$

x	y
4	0
−4	0
0	4
0	−4

Because of the form of the equation, we know the graph must be a *circle* going through the intercepts just found (Figure 8–6E).

Ellipse

Example 2 Sketch $9x^2 + 4y^2 = 36$. See Figure 8–6C for this typical equation of a conic

x-intercepts: Set $y = 0$ in $9x^2 + 4y^2 = 36$

$$9x^2 + 4(0)^2 = 36$$
$$9x^2 = 36$$
$$x^2 = 4$$
$$x = \pm 2$$

y-intercepts: Set $x = 0$ in $9x^2 + 4y^2 = 36$

$$9(0)^2 + 4y^2 = 36$$
$$4y^2 = 36$$
$$y^2 = 9$$
$$y = \pm 3$$

x	y
2	0
−2	0
0	3
0	−3

Because of the form of the equation, we know the graph must be an *ellipse* going through the intercepts just found (Figure 8–6F).

Parabola

Example 3 Sketch $y = 5 - 2x - 3x^2$. See Figure 8–6A for this typical equation of a conic

x-intercepts: Set $y = 0$ in $y = 5 - 2x - 3x^2$

$$0 = 5 - 3x - 3x^2$$
$$0 = (5 + 3x)(1 - x)$$
$$5 + 3x = 0 \quad\Big|\quad 1 - x = 0$$
$$x = -\frac{5}{3} \quad\Big|\quad x = 1$$

y-intercepts: Set $x = 0$ in $y = 5 - 2x - 3x^2$

$$y = 5 - 2(0) - 3(0)^2$$
$$y = 5$$

x	y
1	0
$-\frac{5}{3}$	0
0	5

Because of the form of the equation, we know the graph must be a *parabola* going through the intercepts just found (Figure 8–6G).

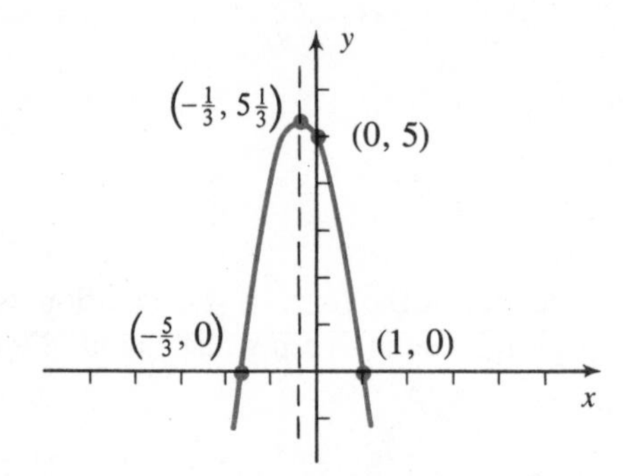

Figure 8–6G

The axis of symmetry of a parabola (Section 8–5) is the vertical line halfway between its x-intercepts.

$$x = \frac{1 + \left(-\frac{5}{3}\right)}{2} = \frac{-\frac{2}{3}}{2} = -\frac{1}{3}$$

Or, by formula (Section 8–5): $x = \dfrac{-b}{2a} = \dfrac{-(-2)}{2(-3)} = -\dfrac{1}{3}$

The vertex is the point of the parabola that lies on the axis of symmetry. Therefore to find the vertex:

Let $x = -\dfrac{1}{3}$. $\quad y = 5 - 2x - 3x^2$

$$y = 5 - 2\left(-\frac{1}{3}\right) - 3\left(-\frac{1}{3}\right)^2 = 5\frac{1}{3}$$

In case more points are needed to sketch the graph, for each point found on one side of the axis of symmetry there is a symmetrical point on the other side of the axis of symmetry.

Hyperbola

Example 4 Sketch $3y^2 - 4x^2 = 12$. See Figure 8–6D for this typical equation of a conic

<u>*x-intercepts:*</u> Set $y = 0$ in $3y^2 - 4x^2 = 12$

$$3(0)^2 - 4x^2 = 12$$
$$x^2 = -3$$
$$x = \pm\sqrt{3}i$$

Since $\pm\sqrt{3}i$ are complex numbers, they cannot be plotted in a coordinate system in which points on both axes represent only real numbers. This means this graph does not intersect the x-axis.

<u>*y-intercepts:*</u> Set $x = 0$ in $3y^2 - 4x^2 = 12$

$$3y^2 - 4(0)^2 = 12$$
$$y^2 = 4$$
$$y = \pm 2$$

x	y
0	2
0	−2

Rectangle of Reference

Because of the form of the equation, we know the graph must be a *hyperbola* going through the intercepts just found (Figure 8–6H). We don't have enough points to draw the graph, so we need more information. One way to complete the graph is to draw a **rectangle of reference,** *ABCD* (Figure 8–6H). Lines *CB* and *DA* are drawn through the y-intercepts $y = \pm 2$ (and parallel to the x-axis). Lines *DC* and *AB* are drawn through $x = \pm\sqrt{3}$ (and parallel to the y-axis). Draw and extend the diagonals *DB* and *AC* of the reference rectangle. The graph of the hyperbola $3y^2 - 4x^2 = 12$ gets closer and closer to these extended diagonals as the points of the graph get further and further from the origin (Figure 8–6I). The extended diagonals *DB* and *AC* are called the **asymptotes** of the hyperbola.

Asymptotes

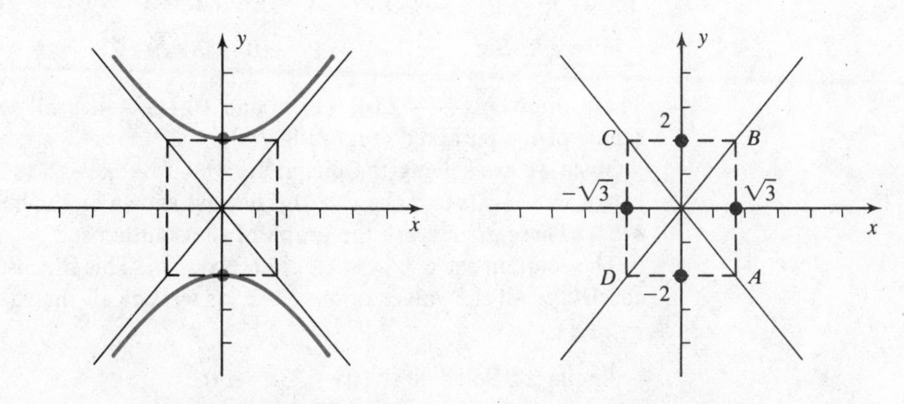

Figure 8–6H **Figure 8–6I**

EXERCISES 8-6

Sketch and identify each conic section.

1. $x^2 + y^2 = 9$
2. $x^2 + y^2 = 4$
3. $3x^2 + 3y^2 = 21$
4. $x^2 + y^2 = 20$
5. $4x^2 + 9y^2 = 36$
6. $16x^2 + y^2 = 16$
7. $9x^2 + 4y^2 = 36$
8. $3x^2 + 4y^2 = 12$
9. $y = x^2 + 2x - 3$
10. $y = 6 - x - x^2$
11. $y^2 = 8x$
12. $y^2 = -4x$
13. $x^2 = 6y$
14. $x = -\frac{1}{4}y^2 - y$
15. $9x^2 - 4y^2 = 36$
16. $x^2 - 4y^2 = 4$
17. $4x^2 - 9y^2 = 36$
18. $4y^2 - 16x^2 = 64$

8-7 Solving Quadratic Inequalities in One Unknown

In Sections 2–2 and 2–3 we discussed solving first-degree inequalities that have one unknown and how to plot their solution sets on a single number line. In this section we extend this to second-degree inequalities that have one unknown.

Quadratic inequalities in one unknown have the form:

$$ax^2 + bx + c > 0$$

This can be any inequality symbol: $>$, $<$, $\geq$, $\leq$

Example 1 Solve $x^2 + x - 2 > 0$ and plot the graph of its solution set on the number line

We know from Section 8–5 that this is the equation of a parabola that opens upward ($a > 0$).

One method of solving the quadratic inequality $x^2 + x - 2 > 0$ is to sketch the graph of the quadratic function $y = f(x) = x^2 + x - 2$.

<u>*x-intercept:*</u> Set $y = 0$: $0 = x^2 + x - 2$

$$0 = (x + 2)\ (x - 1)$$
$$x = -2 \mid x = 1$$

<u>*y-intercept:*</u> Set $x = 0$: $y = 0^2 + 0 - 2 = -2$

These intercepts—$(-2,0)$, $(1,0)$, and $(0,-2)$—are all we need for a rough sketch of the parabola (Figure 8–7A).

Since we are solving the inequality $y = x^2 + x - 2 > 0$, we need only indicate which x-values make $y > 0$. They are shown by the heavy arrows in Figure 8–7A. These arrows are the graph of the solution set.

The solution set is $\{x|x < -2\} \cup \{x|x > 1\}$. This means that the solution set consists of all the values of $x < -2$, *as well as* all the values of $x > 1$.

Example 2 Solve: $8 - 10x - 3x^2 \geq 0$

$$y = f(x) = 8 - 10x - 3x^2 \geq 0$$

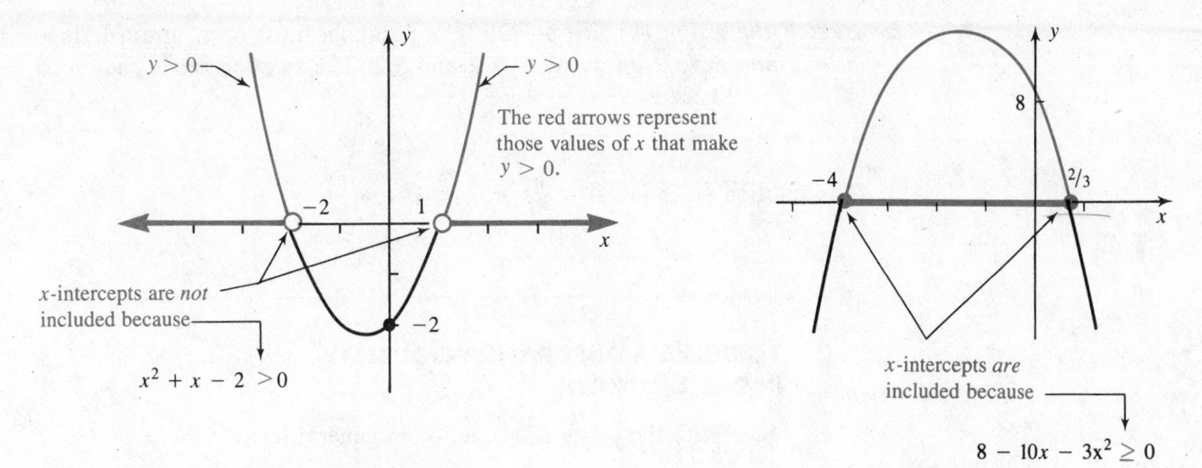

Figure 8–7A **Figure 8–7B**

To find *x-intercepts:*

$$(2 - 3x)(4 + x) = 0$$

$$2 - 3x = 0 \quad \Big| \quad 4 + x = 0$$

$$x = \frac{2}{3} \quad \Big| \quad x = -4$$

The parabola must open downward since $a < 0$, and have x-intercepts $\frac{2}{3}$ and -4.

To find its y-intercept, set $x = 0$:

$$f(0) = 8 - 10(0) - 3(0)^2 = 8$$

This means that the parabola must go through (0,8).

The heavy line in Figure 8–7B is the graph of the solution set, $\left\{x \,\middle|\, -4 \leq x \leq \frac{2}{3}\right\}$. These are the x-values that make $y \geq 0$.

Sometimes the quadratic function cannot be factored. In this case the quadratic formula can be used to find the x-intercepts.

Example 3 Solve: $x^2 - 2x - 4 < 0$

1. *Find x-intercepts:* $x^2 - 2x - 4 = 0 \begin{cases} a = 1 \\ b = -2 \\ c = -4 \end{cases}$

$$x = \frac{-b \pm \sqrt{b^2 - 4ac}}{2a} = \frac{-(-2) \pm \sqrt{(-2)^2 - 4(1)(-4)}}{2(1)}$$

$$= \frac{2 \pm \sqrt{4 + 16}}{2} = \frac{2 \pm \sqrt{20}}{2} = \frac{2 \pm 2\sqrt{5}}{2} = 1 \pm \sqrt{5}$$

$$x_1 = 1 - \sqrt{5} \doteq 1 - 2.2 = -1.2$$ ($\sqrt{5} \doteq 2.236$ from Table I or a calculator)

$$x_2 = 1 + \sqrt{5} \doteq 1 + 2.2 = 3.2$$

2. *Sketch the graph* (Figure 8–7C): The parabola must open upward since $a > 0$, and have x-intercepts -1.2 and 3.2. The solution set is chosen to make $y < 0$ because

$$x^2 - 2x - 4 < 0$$

The solution set is $\{x \mid 1 - \sqrt{5} < x < 1 + \sqrt{5}\}$.

> **TO SOLVE A QUADRATIC INEQUALITY IN ONE UNKNOWN**
>
> 1. Write the quadratic function in general form.
> 2. Find the x-intercepts by factoring or by the quadratic formula.
> 3. Sketch the parabola: $\begin{cases} a > 0 & \text{Opens upward} \\ a < 0 & \text{Opens downward} \end{cases}$
> 4. Read the solution set from the graph.

Other Kinds of Inequalities

The method used to solve quadratic inequalities can also be used to solve other kinds of inequalities.

Example 4 Solve: $x < \dfrac{3}{x+2}$

$$\frac{(x+2)^2}{1} \cdot \frac{x}{1} < \frac{(x+2)^2}{1} \cdot \frac{3}{(x+2)}$$

$$x(x+2)^2 < 3(x+2)$$

$$x(x+2)^2 - 3(x+2) < 0$$

$$(x+2)[x(x+2) - 3] < 0$$

$$(x+2)[x^2 + 2x - 3] < 0$$

$$(x+2)\ (x+3)\ (x-1) < 0$$

$$x = -2 \mid x = -3 \mid x = 1$$

To remove fractions, we usually multiply both sides by the LCD, $x + 2$. Since this is an inequality, we must multiply by $(x + 2)^2$ to make sure both sides are being multiplied by a positive number so that the sense of the inequality does not change.

$(x + 2)$ is a common factor

Therefore -2, -3, and 1 are the x-intercepts.

Also, since $f(x) = (x+2)\ (x+3)\ (x-1)$
then $f(0) = (2) \cdot (3) \cdot (-1) = -6$

Therefore the curve must go through $(0, -6)$.

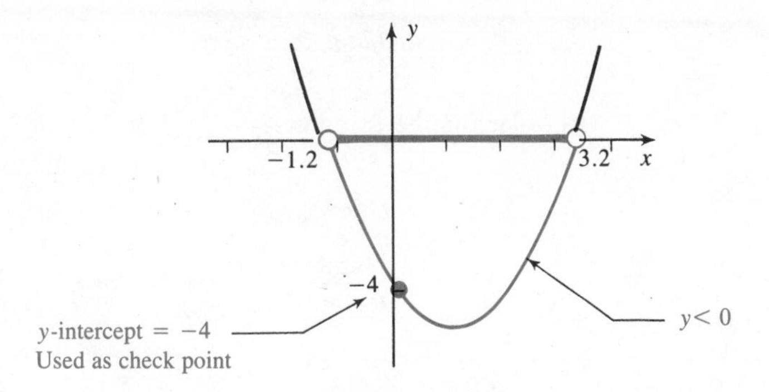

Figure 8–7C

Sketch the graph: Polynomials in one letter are functions. Therefore $y = (x + 2)(x + 3)(x - 1)$ is a function. The curve must be drawn through the x-intercepts -3, -2, and 1. How high or low the graph goes is unimportant in this application. There are two possible curves, Figure 8–7D or Figure 8–7E.

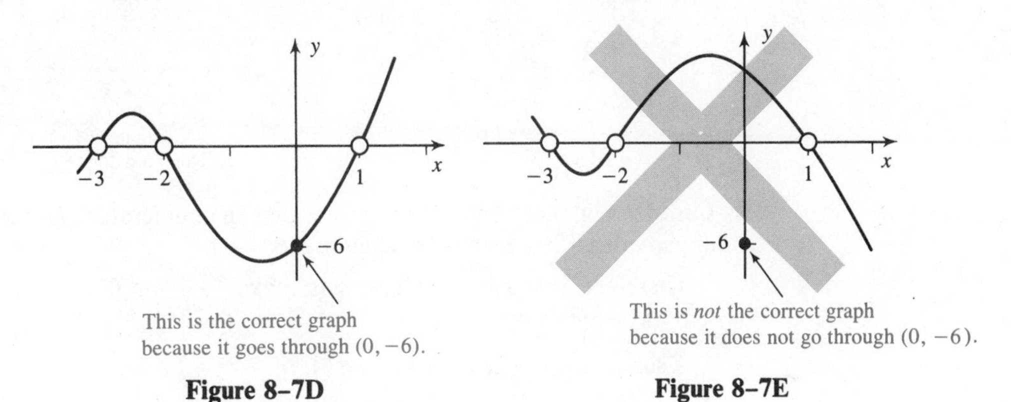

Figure 8–7D **Figure 8–7E**

Now we can draw the graph of the solution set (Figure 8–7F).

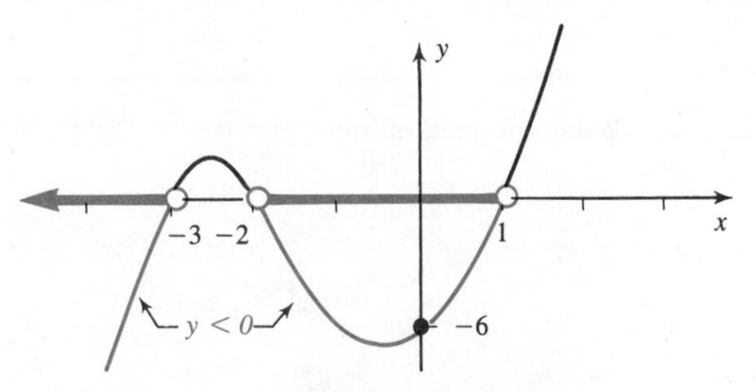

Figure 8–7F

The solution set is $\{x|x < -3\} \cup \{x|-2 < x < 1\}$. This means that the solution set consists of all the values of $x < -3$, *as well as* all values of x between -2 and 1.

Example 5 Solve: $\dfrac{x+5}{x-3} < 0$

In order for the fraction $\dfrac{x+5}{x-3}$ to be negative, either

Case I		*Case II*
$\dfrac{\text{Numerator } (+)}{\text{Denominator } (-)}$	*or*	$\dfrac{\text{Numerator } (-)}{\text{Denominator } (+)}$

Case I:

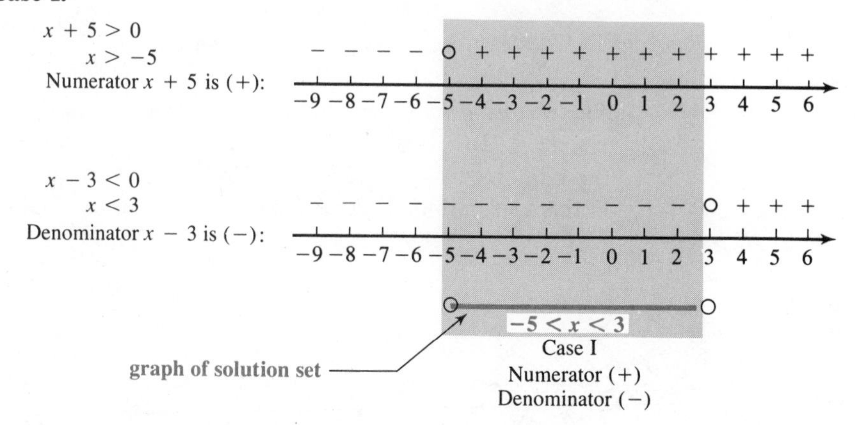

Case II: A glance at the figure shows that the numerator can never be $(-)$ for any x-values that make the denominator $(+)$.

Therefore the only x-values that make $\dfrac{x+5}{x-3} < 0$ are those that satisfy Case I: $-5 < x < 3$.

The solution set is $\{x \mid -5 < x < 3\}$.

Note. All the examples done so far could have been solved by the method used in Example 5. However, we think the method shown in Examples 1–4 is faster and easier to understand in most cases. ■

EXERCISES 8-7

Solve the inequalities.

1. $(x+1)(x-2) < 0$
2. $(x+1)(x-2) > 0$
3. $x^2 + 4x - 5 > 0$
4. $x^2 - 3x - 4 < 0$
5. $x^2 + 7 > 6x$
6. $x^2 + 13 < 8x$
7. $x^2 \leq 5x$
8. $x^2 \geq 3x$
9. $\dfrac{x^2}{2} + \dfrac{5x}{4} \leq 3$
10. $4x^2 + 4x \geq 15$
11. $x^2 + 2x + 3 < 0$
12. $x^2 + x + 2 > 0$
13. $x < \dfrac{2}{x+1}$
14. $x > \dfrac{4}{x+3}$
15. $(x+3)(x+2)(x-1)(x+1)(x-3) > 0$
16. $(x+2)(x-1)(x+1)(x-2)(x-4) < 0$

8–8 Graphing Quadratic Inequalities in the Plane

In Section 7–4 we graphed *linear* (first-degree) inequalities like $3x - 5y > 7$ in the plane. In those inequalities the boundary was a *straight line*. In this section we graph *quadratic* inequalities. They have boundaries which are *conic sections*.

A quadratic inequality is an inequality in which one or both of its letters are raised to the second power.

Example 1 Quadratic inequalities

(a) $x^2 + y^2 < 16$ (b) $9x^2 + 4y^2 \geq 36$

(c) $2x^2 - 5y^2 \leq 4$ (d) $y > x^2 + x - 6$

TO GRAPH A QUADRATIC INEQUALITY IN THE PLANE

1. Sketch the conic section boundary (Section 8–6).
2. Test a point [usually (0,0)] in the inequality to see on which side of the boundary the solution lies.

Example 2 Graph: $x^2 + y^2 < 16$

1. *Boundary:* $x^2 + y^2 = 16$

   This is a circle of radius 4 with center at the origin. (See Section 8–6, Example 1.) Boundary is dashed because $=$ sign *is not included* in $x^2 + y^2 < 16$.

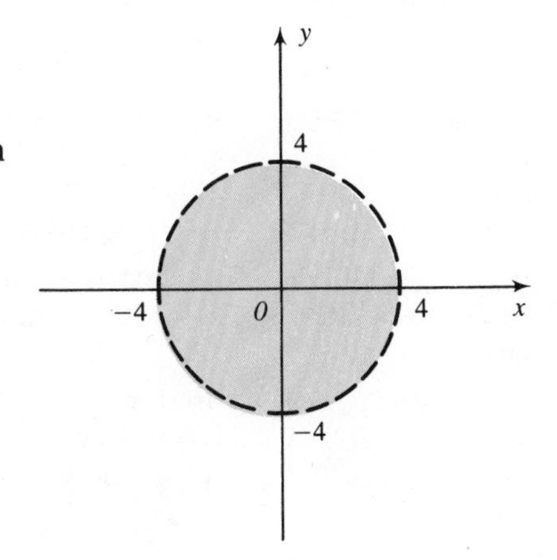

2. *Test point (0,0):*

   $x^2 + y^2 < 16$

   $(0)^2 + (0)^2 < 16$ *True*

   Therefore the solution lies inside the circle, along with (0,0).

Example 3 Graph: $9x^2 + 4y^2 \geq 36$

1. *Boundary:* $9x^2 + 4y^2 = 36$

   This is an ellipse with x-intercepts ± 2 and y-intercepts ± 3. (See Section 8–6, Example 2.) Boundary is solid because $=$ sign *is included* in $9x^2 + 4y^2 \leq 16$.

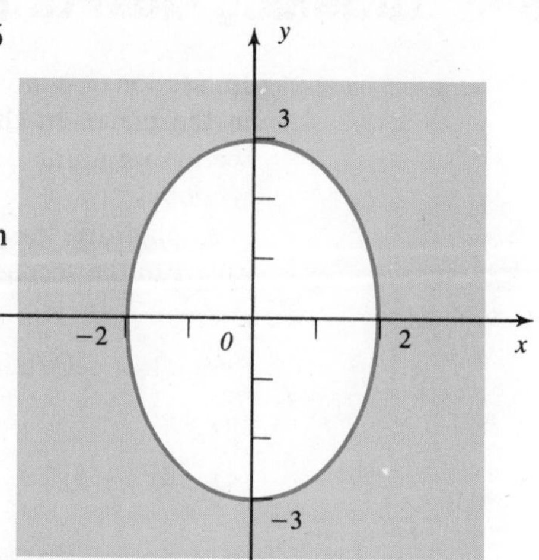

2. *Test point (0,0):*

   $9x^2 + 4y^2 \geq 16$

   $9(0)^2 + 4(0)^2 \geq 16$ *False*

   Therefore the solution lies outside of the ellipse, since (0,0) lies inside the ellipse.

EXERCISES 8–8

Graph the inequalities in the plane.

1. $x^2 + y^2 < 9$ **2.** $x^2 + y^2 > 4$ **3.** $4x^2 + 9y^2 > 36$ **4.** $16x^2 + y^2 < 16$

5. $y \geq x^2 + 2x - 3$ **6.** $y \geq 6 - x - x^2$ **7.** $y^2 \geq 8x$ **8.** $x \leq -\frac{1}{4}y^2 - y$

9. $9x^2 - 4y^2 < 36$ **10.** $x^2 - 4y^2 > 4$

Chapter Eight Summary

Quadratic Equations. (Section 8–1). A *quadratic equation* is a polynomial equation that has a second-degree term as its highest-degree term.

The General Form (Section 8–1) of a quadratic equation is:

$$ax^2 + bx + c = 0$$

where a, b, and c are real numbers ($a \neq 0$).

Methods of Solving Quadratic Equations. First try to solve by factoring. If this fails, use the quadratic formula.

1. *Factoring* (Section 8–2)
   (1) Arrange in general form.
   (2) Factor the polynomial.
   (3) Set each factor equal to zero and solve for the letter.

2. *Formula* (Section 8–3)
   (1) Arrange in general form.
   (2) Substitute values of a, b, and c into

$$x = \frac{-b \pm \sqrt{b^2 - 4ac}}{2a} \qquad (a \neq 0)$$

   (3) Simplify the answers.

3. *Incomplete quadratic* (Section 8–2)
   (a) *When $c = 0$:* Find the greatest common factor (GCF), then solve by factoring.
   (b) *When $b = 0$:* (1) Write the equation as $ax^2 = -c$.
   (2) Divide both sides by a.
   (3) Take the square root of both sides and simplify.
   (4) There are two answers, $\pm$.

The Relation between the Quadratic Discriminant and Roots. (Section 8–4)

$b^2 - 4ac > 0$ and a perfect square: The roots are real, rational, and unequal

$b^2 - 4ac > 0$ and not a perfect square: The roots are real, unequal, irrational conjugates

$b^2 - 4ac = 0$: One real, rational root

$b^2 - 4ac < 0$: The roots are complex conjugates

Using Roots to Find the Equation. (Section 8–4). Reverse the procedure used to solve an equation by factoring.

Quadratic Functions. (Section 8–5). A *quadratic function* is a function whose formula is a quadratic polynomial.

$$y = f(x) = ax^2 + bx + c \quad \text{General form of a quadratic function}$$

Its graph is called a *parabola*.

Conic Sections (Section 8–6) are curves formed when a plane cuts through a cone. They are the parabola, circle, ellipse, and hyperbola. A quick sketch of the conics covered in this section can usually be made by using the *intercept method*.

Quadratic Inequalities in one unknown can be solved by sketching the graph and reading the solution on the number line (Section 8–7).

Quadratic Inequalities in two unknowns can be solved by sketching its boundary graph in the plane and testing to see whether the solution lies inside or outside the boundary (Section 8–8).

Chapter Eight Diagnostic Test or Review

Allow yourself about one hour to do these problems. Complete solutions for every problem, together with section references, are given in the answer section at the end of the book.

In Problems 1–6 solve by any convenient method.

1. $4x^2 = 3 + 4x$

2. $2x^2 = 6x$

3. $2x^2 = 18$

4. $\dfrac{x-1}{2} + \dfrac{4}{x+1} = 2$

5. $x^2 = 6x - 7$

6. $2x^{2/3} + 3x^{1/3} = 2$

7. Use the quadratic discriminant to determine the nature of the roots of the equation $25x^2 - 20x + 7 = 0$. Do not solve for the roots of the equation.

8. Write a quadratic equation that has roots $2 + \sqrt{3}\,i$ and $2 - \sqrt{3}\,i$.

9. Graph the quadratic inequality $25x^2 + 16y^2 > 400$ in the plane.

10. Use the quadratic formula to solve the equation $x^2 + 6x + 10 = 0$.

11. Solve the inequality $x^2 + 5 < 6x$.

12. Solve the inequality $8 \leq 2x + x^2$.

13. Given the quadratic function $f(x) = x^2 - 4x$, find:
(a) The equation of the axis of symmetry.
(b) The coordinates of the vertex.
(c) The x-intercepts (if they exist).
Draw the graph.

14. It takes Jan 2 hr longer to do a job than it does Oscar. After Jan works on the job for 1 hr, Oscar joins her. Together they finish the job in 3 hr. How long would it take each of them working alone to do the entire job?

9 SYSTEMS OF EQUATIONS AND INEQUALITIES

In previous chapters we showed how to solve a single equation for a single variable. In this chapter we show how to solve systems of two (or more) equations in two (or more) variables. Systems of inequalities are also discussed.

9–1 Graphical Method for Solving a Linear System of Two Equations in Two Variables

Basic Definitions

One Equation in One Variable. *A solution of one equation in one variable* is a number which, when put in place of the variable, makes the two sides of the equation equal.

Example 1 3 is a solution for $x + 2 = 5$
because $3 + 2 = 5$ is a true statement.

The expression *"3 satisfies the equation $x + 2 = 5$"* is often used.

Two Equations in Two Variables

$\begin{Bmatrix} x + y = 6 \\ x - y = 2 \end{Bmatrix}$ is called *a system of two equations in two variables.*

A solution of a system of two equations in two variables is an *ordered pair* which, when substituted into each equation, makes each equation a true statement.

Example 2 (4,2) is the solution for the system $\begin{Bmatrix} x + y = 6 \\ x - y = 2 \end{Bmatrix}$

because $\begin{Bmatrix} x + y = 6 & \\ 4 + 2 = 6 & \textit{True} \end{Bmatrix}$ and $\begin{Bmatrix} x - y = 2 & \\ 4 - 2 = 2 & \textit{True} \end{Bmatrix}$

Therefore *(4,2) satisfies the system* $\begin{Bmatrix} x + y = 6 \\ x - y = 2 \end{Bmatrix}$.

Three Equations in Three Variables

$\begin{Bmatrix} 2x - 3y + z = 1 \\ x + 2y + z = -1 \\ 3x - y + 3z = 4 \end{Bmatrix}$ is called *a system of three equations in three variables.*

Ordered Triple

A solution of a system of three equations in three variables is an **ordered triple** which, when substituted into each equation, makes each equation a true statement.

Example 3

The ordered triple $(-3,-1,4)$ is a solution for the system $\begin{Bmatrix} 2x - 3y + z = 1 \\ x + 2y + z = -1 \\ 3x - y + 3z = 4 \end{Bmatrix}$

Substitute $(-3,-1,4)$ into the first equation:

$$2x - 3y + z = 1$$
$$2(-3) - 3(-1) + (4) \stackrel{?}{=} 1$$
$$-6 + 3 + 4 = 1 \quad \textit{True}$$

Substitute $(-3,-1,4)$ into the second equation:

$$x + 2y + z = -1$$
$$(-3) + 2(-1) + (4) \stackrel{?}{=} -1$$
$$-3 - 2 + 4 = -1 \quad \textit{True}$$

Substitute $(-3,-1,4)$ into the third equation:

$$3x - y + 3z = 4$$
$$3(-3) - (-1) + 3(4) \stackrel{?}{=} 4$$
$$-9 + 1 + 12 = 4 \quad \textit{True}$$

Therefore the ordered triple $(-3,-1,4)$ satisfies this system.

Linear and Quadratic Systems

If each equation of a system is a first-degree equation, the system is called a *linear system.*

If the highest-degree equation of a system is second degree, then the system is called a *quadratic system.*

Example 4

(a) $\begin{Bmatrix} x + y = 6 \\ x - y = 2 \end{Bmatrix}$ is a linear system

(b) $\begin{Bmatrix} x^2 + y^2 = 25 \\ x - y = 4 \end{Bmatrix}$ is a quadratic system

Graphical Solution

TO SOLVE A LINEAR SYSTEM OF TWO EQUATIONS BY THE GRAPHICAL METHOD

1. Graph each equation of the system on the same set of axes.
2. There are three possibilities:
   (a) *The lines intersect at one point.* The solution is the ordered pair representing the point of intersection. (See Figure 9–1A.)
   (b) *The lines never cross* (they are parallel). There is no solution. (See Figure 9–1B.)
   (c) *Both equations have the same line for their graph.* Any ordered pair that represents a point on the line is a solution. (See Figure 9–1C.)

Example 5 Solve the system $\begin{cases} x + y = 6 \\ x - y = 2 \end{cases}$ graphically

Draw the graph of each equation on the same set of axes (Figure 9–1A).

Line (1): $x + y = 6$

	x	y
x-intercept: If $y = 0$, then $x = 6$	6	0
y-intercept: If $x = 0$, then $y = 6$	0	6

Therefore line (1) goes through (6,0) and (0,6).

Line (2): $x - y = 2$

	x	y
x-intercept: If $y = 0$, then $x = 2$	2	0
y-intercept: If $x = 0$, then $y = -2$	0	-2

Therefore line (2) goes through (2,0) and (0,−2).

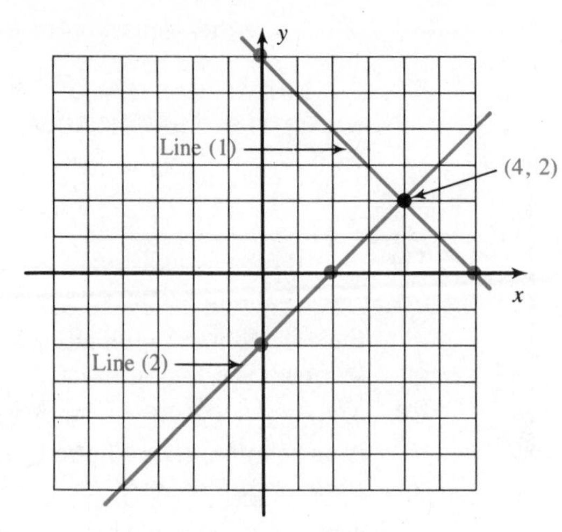

Figure 9–1A

The coordinates of any point on line (1) satisfy the equation of line (1). The coordinates of any point on line (2) satisfy the equation of line (2). The only point that lies on *both* lines is (4,2). Therefore it is the only point whose coordinates satisfy *both* equations.

Consistent System

When the system has a solution, it is called a **consistent system.**

Independent Equations

When each equation in the system has a different graph, they are called **independent equations.**

Example 6 Solve the system $\begin{Bmatrix} 2x - 3y = 6 \\ 6x - 9y = 36 \end{Bmatrix}$ graphically

Draw the graph of each equation on the same set of axes (Figure 9–1B).

Line (1): $2x - 3y = 6$ has intercepts $(3,0)$ and $(0,-2)$.

Line (2): $6x - 9y = 36$ has intercepts $(6,0)$ and $(0,-4)$.

There is no solution because these lines never meet.

Lines that never meet, such as these, are called *parallel lines*.

Inconsistent System

When the system has no solution, it is called an **inconsistent system.**

Since each equation in this system has a different graph, they are called *independent equations*.

Example 7 Solve the system $\begin{Bmatrix} 3x + 5y = 15 \\ 6x + 10y = 30 \end{Bmatrix}$ graphically

Draw the graph of each equation on the same set of axes (Figure 9–1C).

Line (1): $3x + 5y = 15$ has intercepts $(5,0)$ and $(0,3)$.

Line (2): $6x + 10y = 30$ has intercepts $(5,0)$ and $(0,3)$.

Since each line goes through the same two points, they must be the same line.

To find the solution, solve either equation for either letter in terms of the other. Solving equation (1) for y, we have:

$$3x + 5y = 15$$

$$5y = 15 - 3x$$

$$(3) \qquad y = \frac{15 - 3x}{5}$$

To find one of the infinite number of solutions, pick any value for x and substitute it into expression (3). For example, when $x = 1$:

$$y = \frac{15 - 3(1)}{5} = \frac{12}{5} = 2\frac{2}{5}$$

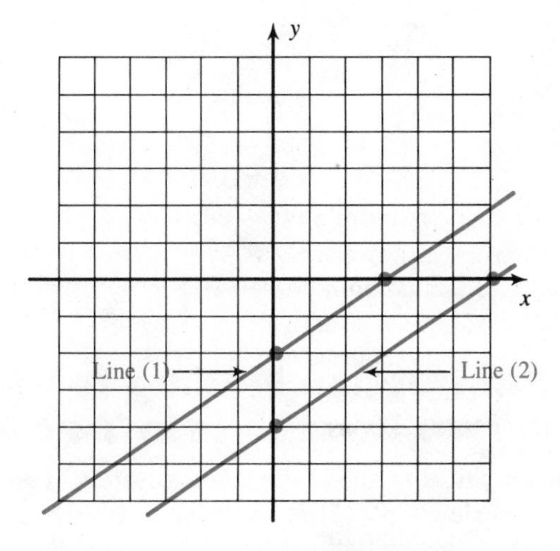

Figure 9–1B

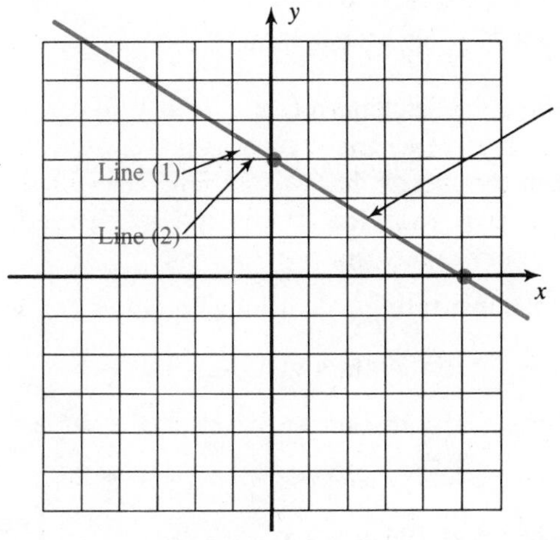

Figure 9–1C

Therefore $\left(1,2\frac{2}{5}\right)$ is a solution for this system. The intercepts (5,0) and (0,3) are also solutions.

The solutions set $= \left\{(x,y) \middle| y = \frac{15 - 3x}{5}\right\}$.

Dependent Equations

Since this system has solutions, it is a *consistent system*. When each equation in the system has the same graph, they are called **dependent equations.**

EXERCISES 9-1

Find the solution of each system graphically (if a solution exists).

1. $2x + y = 6$, $x - y = 0$

2. $2x - y = -4$, $x + y = -2$

3. $x + 2y = 3$, $3x - y = -5$

4. $5x + 4y = 12$, $x + 3y = -2$

5. $2x + 3y = -12$, $x - y = -1$

6. $3x + 8y = 9$, $9x + 5y = -30$

7. $x + 5 = 0$, $y = -2$

8. $x = -3$, $y - 4 = 0$

9. $x + 2y = 0$, $2x - y = 0$

10. $2x + y = 0$, $x - 3y = 0$

11. $8x - 5y = 15$, $10y - 16x = 16$

12. $3x + 9y = 18$, $2x + 6y = -24$

13. $8x - 10y = 16$, $15y - 12x = -24$

14. $14x + 30y = -70$, $15y + 7x = -35$

15. $5x + 8y = -22$, $4x - 3y = 20$

16. $6x + 7y = 9$, $3x - 4y = -18$

9–2 Addition-Subtraction Method for Solving a Linear System of Two Equations in Two Variables

The graphical method for solving a system of equations has two disadvantages: (1) it is slow, and (2) it is not an exact method of solution. The method we discuss in this section has neither of these disadvantages.

The addition-subtraction method for solving a system is one in which the equations are either added or subtracted to eliminate one variable.

Example 1 Solve the system: $\begin{Bmatrix} x + y = 6 \\ x - y = 2 \end{Bmatrix}$

Equation (1): $x + y = 6$

Equation (2): $\underline{x - y = 2}$

$2x \quad = 8$ Added the equations

$x = 4$

Then, substituting $x = 4$ into Equation (1), we have:

Equation (1): $x + y = 6$

$4 + y = 6$

$y = 2$

Therefore the solution for the system is $\begin{Bmatrix} x = 4 \\ y = 2 \end{Bmatrix}$, written as the ordered pair (4,2).

Check for (4,2). (4,2) must satisfy *both* equations to be a solution.

(1): $x + y = 6$ and (2): $x - y = 2$

$4 + 2 = 6$ *True* $4 - 2 = 2$ *True*

TO SOLVE A LINEAR SYSTEM OF TWO EQUATIONS BY THE ADDITION-SUBTRACTION METHOD

1. Multiply the equations by numbers that make the coefficients of one of the letters the same.
2. Add (or subtract) the equations to eliminate the letter whose coefficients were made the same in Step 1.
3. There are three possibilities:
   (a) *The resulting equation can be solved for the letter.*
   Substitute this value into either of the system's equations to find the value of the other letter.
   (b) *Both letters drop out and an inequality results.*
   There is no solution. (Inconsistent system)
   (c) *Both letters drop out and an equality results.*
   Any ordered pair that satisfies *either* of the system's equations is a solution. (Dependent equations)

SYSTEMS THAT HAVE ONLY ONE SOLUTION (CONSISTENT SYSTEMS WITH INDEPENDENT EQUATIONS)

Example 2 Solve the system: $\begin{Bmatrix} 3x + 4y = 6 \\ 2x + 3y = 5 \end{Bmatrix}$

Equation (1): $3x + 4y = 6$
Equation (2): $2x + 3y = 5$

If Equation (1) is multiplied by 2 and Equation (2) is multiplied by 3, the coefficients of x will be the same in the resulting equations

This symbol means the equation is to be multiplied by 2

$$\begin{aligned} 2\,]\quad 3x + 4y = 6 &\overset{*}{\Rightarrow} \quad 6x + 8y = 12 \\ 3\,]\quad 2x + 3y = 5 &\Rightarrow \quad 6x + 9y = 15 \\ & \qquad\quad -y = -3 \\ & \qquad\qquad y = 3 \end{aligned}$$

Subtracted the bottom equation from the top one

*The symbol ⇒, read "implies," means the second statement is true if the first statement is true. For example:

$x - 2 = 0 \Rightarrow x = 2$

is read "$x - 2 = 0$ implies $x = 2$"

and means $x = 2$ is true if $x - 2 = 0$ is true.

The value $y = 3$ may be substituted into either Equation (1) or Equation (2) to find the value of x. It is usually easier to substitute in the equation having the smaller coefficients.

We substitute $y = 3$ into Equation (2):

$$\begin{aligned}\text{Equation (2):}\quad 2x + 3y &= 5\\ 2x + 3(3) &= 5\\ 2x + 9 &= 5\\ 2x &= -4\\ x &= -2\end{aligned}$$

Therefore the solution of the system is $\begin{Bmatrix} x = -2 \\ y = 3 \end{Bmatrix}$, written as the ordered pair $(-2,3)$.

Note: In Example 2, we subtracted the bottom equation from the top equation. We could have subtracted the *top* equation from the *bottom* equation. That would have eliminated negative signs from the resulting equation.

$$\begin{aligned}6x + 8y &= 12\\ 6x + 9y &= 15\\ \hline y &= 3\end{aligned}\quad \text{No negative signs}$$

■

How to Choose the Numbers Each Equation is Multiplied By. In Example 2 we multiplied Equation (1) by 2 and Equation (2) by 3. How were these numbers found?

The coefficients of x were made the same

$$\begin{aligned}2\,]\quad 3x + 4y = 6 \quad &\Rightarrow \quad 6x + 8y = 12\\ 3\,]\quad 2x + 3y = 5 \quad &\Rightarrow \quad 6x + 9y = 15\end{aligned}$$

This pair of numbers is found by interchanging the coefficients of x

In this same system we show how to make the coefficients of y the same:

The coefficients of y were made the same

$$\begin{aligned}3\,]\quad 3x + 4y = 6 \quad &\Rightarrow \quad 9x + 12y = 18\\ 4\,]\quad 2x + 3y = 5 \quad &\Rightarrow \quad 8x + 12y = 20\end{aligned}$$

This pair of numbers is found by interchanging the coefficients of y

Example 3 Consider the system: $\begin{cases} 10x - 9y = 5 \\ 15x + 6y = 4 \end{cases}$

This pair is found by interchanging the coefficients of x

$$\begin{array}{ll|l} 15\,] & 3\,] & 10\,x - 9y = 5 \Rightarrow 30\,x - 27y = 15 \\ 10\,] & 2\,] & 15\,x + 6y = 4 \Rightarrow 30\,x + 12y = 8 \end{array}$$

This pair is found by reducing the ratio $\frac{15}{10} = \frac{3}{2}$

This pair is found by interchanging the coefficients of y

$$\begin{array}{ll|l} 6\,] & 2\,] & 10x - 9\,y = 5 \Rightarrow 20x - 18\,y = 10 \\ 9\,] & 3\,] & 15x + 6\,y = 4 \Rightarrow 45x + 18\,y = 12 \end{array}$$

This pair is found by reducing the ratio $\frac{6}{9} = \frac{2}{3}$

SYSTEMS THAT HAVE NO SOLUTION (INCONSISTENT SYSTEMS WITH INDEPENDENT EQUATIONS)

In Section 9–1 (Example 6) we found that a system whose graphs are parallel lines has no solution. Here, we show how to identify systems that have no solution by using the *addition-subtraction method.*

Example 4 Solve the system: $\begin{cases} 2x - 3y = 6 \\ 6x - 9y = 36 \end{cases}$

$$\begin{array}{lll|l} \text{Equation (1):} & 6\,] & 3\,] & 2x - 3\,y = 6 \Rightarrow 6x - 9y = 18 \\ \text{Equation (2):} & 2\,] & 1\,] & 6x - 9\,y = 36 \Rightarrow \underline{6x - 9y = 36} \\ & & & \qquad\qquad\qquad\qquad 0 \neq -18 \end{array}$$

An inequality

No values for x and y can make $0 = -18$. Therefore there is *no solution* for this system of equations.

SYSTEMS THAT HAVE MORE THAN ONE SOLUTION (CONSISTENT SYSTEMS WITH DEPENDENT EQUATIONS)

In Section 9–1 (Example 7) we found that a system whose equations have the same graph has an infinite number of solutions. Here, we show how to identify such systems using the *addition-subtraction method.*

Example 5 Solve the system: $\begin{Bmatrix} 4x + 6y = 4 \\ 6x + 9y = 6 \end{Bmatrix}$

$$\begin{array}{llllll} \text{Equation (1):} & 9\,] & 3\,] & 4x + 6y = 4 & \Rightarrow & 12x + 18y = 12 \\ \text{Equation (2):} & 6\,] & 2\,] & 6x + 9y = 6 & \Rightarrow & \underline{12x + 18y = 12} \\ & & & & & 0 = 0 \end{array}$$

An equality

This means both equations are equivalent to the same equation. Therefore, in this case, *any* ordered pair that satisfies Equation (1) *or* Equation (2) is a solution of the system. The solution set is:

$$\{(x,y) \mid 4x + 6y = 4\} \quad \textit{or} \quad \left\{(x,y) \,\middle|\, y = \frac{2 - 2x}{3}\right\}$$

EXERCISES 9-2

Find the solution of each system by the addition-subtraction method. Write "inconsistent" if no solution exists. If an infinite number of solutions exist, write "dependent" and give the solution set.

1. $3x - y = 11$; $3x + 2y = -4$

2. $6x + 5y = 2$; $2x - 5y = -26$

3. $8x + 15y = 11$; $4x - y = 31$

4. $x + 6y = 24$; $5x - 3y = 21$

5. $7x - 3y = 3$; $20x - 9y = 12$

6. $10x + 7y = -1$; $2x + y = 5$

7. $6x + 5y = 0$; $4x - 3y = 38$

8. $7x + 4y = 12$; $2x - 3y = -38$

9. $6x - 10y = 6$; $9x - 15y = -4$

10. $7x - 2y = 7$; $21x - 6y = 6$

11. $3x - 5y = -2$; $10y - 6x = 4$

12. $15x - 9y = -3$; $6y - 10x = 2$

13. $9x + 4y = -4$; $15x - 6y = 25$

14. $16x - 25y = -38$; $8x + 5y = -12$

15. $9x + 10y = -3$; $15x + 14y = 7$

16. $35x + 18y = 30$; $7x - 24y = -17$

17. $3.64x - 7.92y = 37.5$; $5.25x + 4.06y = -35.9$

18. $1.14x - 3.16y = 6.06$; $8.61x + 5.32y = -12.3$

9-3 Substitution Method for Solving a Linear System of Two Equations in Two Variables

It is important that you learn the *substitution method* of solving relatively simple systems at this time so that you can apply this method later to solve more complicated systems (Section 9-7).

TO SOLVE A LINEAR SYSTEM OF TWO EQUATIONS BY THE SUBSTITUTION METHOD

1. Solve one equation for one of the letters *in terms of the other letter.*
2. Substitute the expression obtained in Step 1 into the *other* equation (in place of the letter solved for in Step 1).
3. There are three possibilities:
   (a) *The resulting equation can be solved for the letter.*
   Substitute this value into either of the system's equations to find the value of the other letter.
   (b) *Both letters drop out and an inequality results.*
   There is no solution. (Inconsistent system)
   (c) *Both letters drop out and an equality results.*
   Any ordered pair that satisfies *either* of the system's equations is a solution. (Dependent equations)

Example 1 Solve the system: $\begin{Bmatrix} x - 2y = 11 \\ 3x + 5y = -11 \end{Bmatrix}$ Equation (1) Equation (2)

Equation (1): $x - 2y = 11$ is a literal equation.

1. Solve Equation (1) for x: $x - 2y = 11$

$$x = 11 + 2y$$

2. Substitute $11 + 2y$ in place of x in Equation (2):

Equation (2):
$$\begin{aligned} 3x + 5y &= -11 \\ 3(11 + 2y) + 5y &= -11 \\ (33 + 6y) + 5y &= -11 \\ 11y &= -44 \\ y &= -4 \end{aligned}$$

3. Substitute $y = -4$ in $x = 11 + 2y$

$$x = 11 + 2(-4)$$
$$x = 11 - 8$$
$$x = 3$$

Therefore $(3, -4)$ is the solution for this system.

Check:

$$\left\{\begin{array}{llll} (1): & x - 2y & = 11 & \\ & 3 - 2(-4) & \stackrel{?}{=} 11 & \\ & 3 + 8 & = 11 & \textit{True} \end{array}\right\}$$

and

$$\left\{\begin{array}{llll} (2): & 3x + 5y & = -11 & \\ & 3(3) + 5(-4) & \stackrel{?}{=} -11 & \\ & 9 - 20 & = -11 & \textit{True} \end{array}\right\}$$

Therefore (3,−4) is a solution.

HOW TO CHOOSE WHICH LETTER TO SOLVE FOR

(A) One of the Equations May Already Be Solved for a Letter

Example 2 $\left\{\begin{array}{l} 2x - 5y = 4 \\ y = 3x + 7 \end{array}\right\}$

Already solved for y

(B) One of the Letters May Have a Coefficient of 1

Example 3 $\left\{\begin{array}{l} 2x + 6y = 3 \\ x - 4y = 2 \Rightarrow x = 4y + 2 \end{array}\right\}$

x has a coefficient of 1

(C) Choose the Letter with the Smallest Coefficient

Example 4 $\left\{\begin{array}{l} 11x - 7y = 10 \\ 14x + 2y = 9 \Rightarrow y = \dfrac{9 - 14x}{2} \end{array}\right\}$

Smallest of the four coefficients

Smallest possible denominator in this case

SYSTEMS THAT HAVE NO SOLUTION (INCONSISTENT SYSTEMS)

We now show how to identify systems that have no solution by using the *substitution method.*

Smallest coefficient

Example 5 Solving the system: $\left\{\begin{array}{l} 2x - 3y = 6 \\ 6x - 9y = 36 \end{array}\right\}$ Equation (1) Equation (2)

1. Solve Equation (1) for x: $2x - 3y = 6 \Rightarrow 2x = 3y + 6$

$$x = \frac{3y + 6}{2}$$

2. Substitute $\frac{3y + 6}{2}$ in place of x in Equation (2):

Equation (2):
$$6x - 9y = 36$$
$$\overset{3}{\not{6}}\left(\frac{3y + 6}{\not{2}}\right) - 9y = 36$$
$$3(3y + 6) - 9y = 36$$
$$9y + 18 - 9y = 36$$
$$18 \neq 36 \quad \text{An inequality}$$

No values of x and y can make $18 = 36$. Therefore there is *no solution* for this system of equations.

SYSTEMS THAT HAVE MORE THAN ONE SOLUTION (DEPENDENT EQUATIONS)

We now show how to identify systems that have more than one solution by using the *substitution method.*

Example 6 Solve the system: $\begin{Bmatrix} 9x + 6y = 6 \\ 6x + 4y = 4 \end{Bmatrix}$ Equation (1), Equation (2)

(Smallest coefficient: the 4 in $4y$)

1. Solve Equation (2) for y: $6x + 4y = 4 \Rightarrow 4y = 4 - 6x$
$$y = \frac{4 - 6x}{4}$$
$$y = \frac{\not{2}(2 - 3x)}{\underset{2}{\not{4}}}$$
$$y = \frac{2 - 3x}{2}$$

2. Substitute $\frac{2 - 3x}{2}$ in place of y in Equation (1):

Equation (1):
$$9x + 6y = 6$$
$$9x + \overset{3}{\not{6}}\left(\frac{2 - 3x}{\not{2}}\right) = 6$$
$$9x + 3(2 - 3x) = 6$$
$$9x + 6 - 9x = 6$$
$$6 = 6 \quad \text{An equality}$$

Therefore *any* ordered pair that satisfies Equation (1) *or* Equation (2) is a solution of the system.

EXERCISES 9-3

Find the solution of each system using the substitution method. Write "inconsistent" if no solution exists. If an infinite number of solutions exist, write "dependent" and give the solution set.

1. $7x + 4y = 4$, $y = 6 - 3x$
2. $2x + 3y = -5$, $x = y - 10$
3. $5x - 4y = -1$, $3x + y = -38$
4. $3x - 5y = 5$, $x - 6y = 19$
5. $8x - 5y = 4$, $2x - 3y = -20$
6. $11x - 6y = 14$, $3x - 2y = -2$
7. $6x - 9y = 16$, $8x - 12y = 16$
8. $15x - 10y = 30$, $18x - 12y = 30$
9. $20x + 15y = 35$, $12x + 9y = 21$
10. $18x - 27y = 36$, $22x - 33y = 44$
11. $8x + 4y = 7$, $3x + 6y = 6$
12. $5x - 4y = 2$, $15x + 12y = 12$
13. $4x + 4y = 3$, $6x + 12y = -6$
14. $4x + 9y = -11$, $10x + 6y = 11$
15. $3x + 2y = -1$, $15x + 14y = -23$
16. $5.21x - 8.7y = 3.44$, $4.9x + 2.15y = 9.6$

9-4 Higher-Order Systems

So far we have considered only systems of two equations in two variables. If a system has more than two equations and two variables, it is usually called a **higher-order system.** We shall consider only systems that have the same number of equations as variables. A system having three equations and three variables is called a *third-order system;* one having four equations and four variables is a *fourth-order system;* etc.

Third-Order Systems

TO SOLVE A THIRD-ORDER LINEAR SYSTEM BY ADDITION-SUBTRACTION

1. Eliminate one letter from a pair of equations by addition-subtraction.
2. Eliminate the *same* letter from a *different* pair of equations by addition-subtraction.
3. The two equations found in Steps 1 and 2 form a second-order system which is then solved by addition-subtraction.

The solution is an *ordered triple:* (a,b,c).

Example 1 Solve the system:

$$\begin{cases} 2x - 3y + z = 1 & \text{Equation (1)} \\ x + 2y + z = -1 & \text{Equation (2)} \\ 3x - y + 3z = 4 & \text{Equation (3)} \end{cases}$$

First, eliminate z from Equations (1) and (2). Any one of the letters can be eliminated. We choose z in this case.

$$\begin{array}{lrcl} (1) & 2x - 3y + z & = & 1 \\ (2) & x + 2y + z & = & -1 \\ \hline (4) & x - 5y & = & 2 \end{array}$$

Subtract (2) from (1)

Next, eliminate z from Equations (1) and (3):

$$\begin{array}{llrcl} (1) & 3\,] & 2x - 3y + z = 1 & \Rightarrow & 6x - 9y + 3z = 3 \\ (3) & 1\,] & 3x - y + 3z = 4 & \Rightarrow & 3x - y + 3z = 4 \\ \hline & & (5) & & 3x - 8y = -1 \end{array}$$

Subtract bottom equation from top equation

Equations (4) and (5) form a second-order system which we solve by addition-subtraction.

$$\begin{array}{llrcl} (4) & 3\,] & x - 5y = 2 & \Rightarrow & 3x - 15y = 6 \\ (5) & 1\,] & 3x - 8y = -1 & \Rightarrow & 3x - 8y = -1 \\ \hline & & & & -7y = 7 \\ & & & & y = -1 \end{array}$$

Then, substituting $y = -1$ into Equation (4), we have:

$$\begin{array}{lrcl} (4) & x - 5y & = & 2 \\ & x - 5(-1) & = & 2 \\ & x + 5 & = & 2 \\ & x & = & -3 \end{array}$$

After having found the value of two variables, those values are substituted into any one of the original equations to find the value of the third variable.

Substituting $x = -3$ and $y = -1$ into Equation (2), we have:

$$\begin{array}{lrcl} (2) & x + 2y + z & = & -1 \\ & (-3) + 2(-1) + z & = & -1 \\ & -3 - 2 + z & = & -1 \\ & z & = & 4 \end{array}$$

Therefore the solution of the system is $\begin{Bmatrix} x = -3 \\ y = -1 \\ z = 4 \end{Bmatrix}$, which is written as the ordered triple $(-3,-1,4)$.

The work for Example 1 can be conveniently arranged as follows:

$$\begin{array}{rrcl} (1) & 2x - 3y + z & = & 1 \\ (2) & x + 2y + z & = & -1 \\ (3) & 3x - y + 3z & = & 4 \\ \hline (1) - (2) & x - 5y & = & 2 \\ 3(1) - (3) & 3x - 8y & = & -1 \\ \hline 3(4) - (5) & -7y & = & 7 \\ & y & = & -1 \end{array}$$

(4) = Equation (1) − Equation (2)

(5) = 3 × Equation (1) − Equation (3)

3 × Equation (4) − Equation (5)

Substitute $y = -1$ in (4): $x - 5(-1) = 2 \Rightarrow x = -3$

Substitute $x = -3$ and $y = -1$ in (2): $(-3) + 2(-1) + z = -1 \Rightarrow z = 4$

The *solution* is the *ordered triple* $(-3,-1,4)$.

Check: The check for this solution is shown in Example 3 of Section 9–1.

Fourth-Order Systems

The method just described for solving a third-order system can be extended to a fourth-order system as follows:

TO SOLVE A FOURTH-ORDER LINEAR SYSTEM BY ADDITION-SUBTRACTION

1. Eliminate one letter from a pair of equations. This gives one equation in three variables.
2. Eliminate the *same* letter from a *different* pair of equations. This gives a second equation in three variables.
3. Eliminate the *same* letter from a *third* different pair of equations. This gives a third equation in three variables. (*Each equation of the system must be used in at least one pair of equations.*)
4. The three equations obtained in Steps 1, 2, and 3 form a third-order system which is then solved by the method given in the preceding box and shown in Example 1.

The solution is an *ordered quadruple:* (a,b,c,d).

This same method can be extended to solve systems of *any* order. Because of the amount of work involved in solving higher-order systems, solutions are usually carried out by computer.

Some higher-order systems have no solution (inconsistent systems) and others have many solutions (dependent equations), as was shown for second-order systems. In this book we consider only higher-order systems that have a single solution (consistent systems with independent equations).

The graph of a linear equation in three variables is a *plane* in three-dimensional space. Graphical solutions of third-order linear systems are too complicated for practical use. Graphical solutions for *higher* than third-order linear systems are not possible.

EXERCISES 9-4

Solve each system.

1. $\begin{aligned} 2x + y + z &= 4 \\ x - y + 3z &= -2 \\ x + y + 2z &= 1 \end{aligned}$

2. $\begin{aligned} x + y + z &= 1 \\ 2x + y - 2z &= -4 \\ x + y + 2z &= 3 \end{aligned}$

3. $\begin{aligned} x + 2y + 2z &= 0 \\ 2x - y + z &= -3 \\ 4x + 2y + 3z &= 2 \end{aligned}$

4. $\begin{aligned} 2x + y - z &= 0 \\ 3x + 2y + z &= 3 \\ x - 3y - 5z &= 5 \end{aligned}$

5. $\begin{aligned} x + y - z &= 0 \\ 2x - y + 3z &= 1 \\ 3x + y + z &= 2 \end{aligned}$

6. $\begin{aligned} x + 2y - 3z &= 5 \\ x + y + z &= 0 \\ 3x + 4y + 2z &= -1 \end{aligned}$

7. $\begin{aligned} x + 2z &= 7 \\ 2x - y &= 5 \\ 2y + z &= 4 \end{aligned}$

8. $\begin{aligned} x - 2y &= 4 \\ y + 3z &= 8 \\ 2x - z &= 1 \end{aligned}$

9. $\begin{aligned} 2x + 3y + z &= 7 \\ 4x - 2z &= -6 \\ 6y - z &= 0 \end{aligned}$

10. $\begin{aligned} 4x + 5y + z &= 4 \\ 10y - 2z &= 6 \\ 8x + 3z &= 3 \end{aligned}$

11. $\begin{aligned} x + y + z + w &= 5 \\ 2x - y + 2z - w &= -2 \\ x + 2y - z - 2w &= -1 \\ -x + 3y + 3z + w &= 1 \end{aligned}$

12. $\begin{aligned} x + y + z + w &= 4 \\ x - 2y - z - 2w &= -2 \\ 3x + 2y + z + 3w &= 4 \\ 2x + y - 2z - w &= 0 \end{aligned}$

13. $\begin{aligned} 2x + 5y + 3z - 4w &= 0 \\ -x + y + 6w &= -1 \\ 3x - z - 2w &= 1 \\ 4x + 2y - z &= 3 \end{aligned}$

14. $\begin{aligned} 3x + 2y + 4z + 5w &= 5 \\ 9x - 8z + 10w &= -4 \\ 6y + 12z + 5w &= 5 \\ -6x - 4y + 15w &= -1 \end{aligned}$

15. $\begin{aligned} 6x + 4y + 9z + 5w &= -3 \\ 2x + 8y - 6z + 15w &= 8 \\ 4x - 4y + 3z - 10w &= -3 \\ 2x - 4y + 3z - 5w &= -1 \end{aligned}$

16. $\begin{aligned} 12x + 9y + 4z - 8w &= 1 \\ 6x + 15y + 2z + 4w &= 2 \\ 3x + 6y + 4z + 2w &= 5 \\ 4x + 4y + 4z + 4w &= 3 \end{aligned}$

9–5 Determinant Method for Solving a Second-Order Linear System

Determinants

A **determinant** is a square *array* of numbers enclosed between two vertical bars (Figure 9–5). A determinant represents the sum of certain products.

The **elements** of a determinant are the numbers in its array.

A **row** of a determinant is a *horizontal* line of its elements.

A **column** of a determinant is a *vertical* line of its elements.

The **principal diagonal** of a determinant is the line of its elements from the *upper left corner to the lower right corner.*

The **secondary diagonal** of a determinant is the line of elements from the *lower left corner to the upper right corner.*

Second-Order Determinants

A **second-order determinant** is a determinant that has *two rows and two columns* of elements (Figure 9–5).

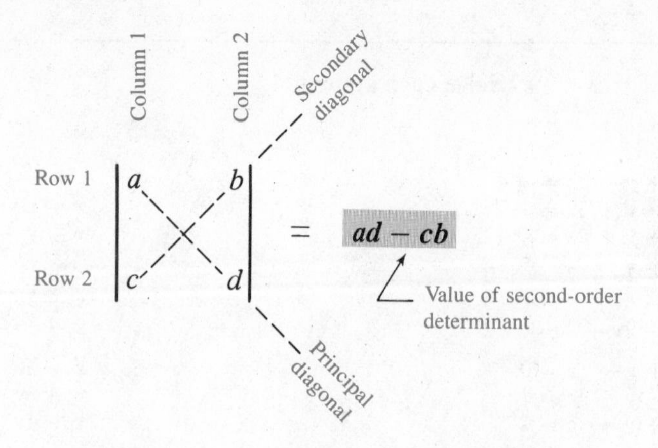

Figure 9–5 Second-Order Determinant

Value of Second-Order Determinant

The **value of a second-order determinant** is the *product of the elements in its principal diagonal, minus* the *product of the elements in its secondary diagonal.*

Example 1 Find the value of each second-order determinant

(a) $\begin{vmatrix} -5 & -6 \\ 2 & 3 \end{vmatrix} = (-5)(3) - (2)(-6) = -15 + 12 = -3$

(b) $\begin{vmatrix} 6 & -7 \\ 4 & 0 \end{vmatrix} = (6)(0) - (4)(-7) = 0 + 28 = 28$

GENERAL SOLUTION OF A SECOND-ORDER LINEAR SYSTEM

Consider the linear system

$$a_1x + b_1y = c_1$$
$$a_2x + b_2y = c_2$$

Solving for x by addition-subtraction, we have:

$$\begin{aligned} b_2\,]\quad a_1x + b_1y &= c_1 \Rightarrow & a_1b_2x + b_1b_2y &= c_1b_2 \\ b_1\,]\quad a_2x + b_2y &= c_2 \Rightarrow & a_2b_1x + b_1b_2y &= c_2b_1 \\ && (a_1b_2 - a_2b_1)x &= c_1b_2 - c_2b_1 \\ && x &= \frac{c_1b_2 - c_2b_1}{a_1b_2 - a_2b_1} \end{aligned}$$

Subtract bottom equation from top equation

This can be written:

$$x = \frac{c_1b_2 - c_2b_1}{a_1b_2 - a_2b_1} = \frac{\begin{vmatrix} c_1 & b_1 \\ c_2 & b_2 \end{vmatrix}}{\begin{vmatrix} a_1 & b_1 \\ a_2 & b_2 \end{vmatrix}}$$

Solving for y by addition-subtraction, we have:

$$\begin{aligned} a_2]\quad a_1x + b_1y = c_1 &\Rightarrow a_1a_2x + a_2b_1y = a_2c_1 \\ a_1]\quad a_2x + b_2y = c_2 &\Rightarrow \underline{a_1a_2x + a_1b_2y = a_1c_2} \\ &(a_1b_2 - a_2b_1)y = a_1c_2 - a_2c_1 \\ &y = \frac{a_1c_2 - a_2c_1}{a_1b_2 - a_2b_1} \end{aligned}$$

Subtract top equation from bottom equation

This can be written:

$$y = \frac{a_1c_2 - a_2c_1}{a_1b_2 - a_2b_1} = \frac{\begin{vmatrix} a_1 & c_1 \\ a_2 & c_2 \end{vmatrix}}{\begin{vmatrix} a_1 & b_1 \\ a_2 & b_2 \end{vmatrix}}$$

NAMES ASSOCIATED WITH A LINEAR SYSTEM

$$a_1x + b_1y = c_1$$
$$a_2x + b_2y = c_2$$

Determinant of coefficients

$$\begin{vmatrix} a_1 & b_1 \\ a_2 & b_2 \end{vmatrix} \quad \begin{matrix} c_1 \\ c_2 \end{matrix}$$

Column of constants

$$\begin{matrix} a_1 \\ a_2 \end{matrix} \quad \begin{matrix} b_1 \\ b_2 \end{matrix}$$

Column of y-coefficients

Column of x-coefficients

Cramer's Rule In the solution of a linear system of equations, each variable is equal to the ratio of two determinants.

1. The denominator of every variable is the determinant of the coefficients.

2. The numerator for each variable is the determinant of the coefficients with the column of coefficients for that letter replaced by the column of constants.

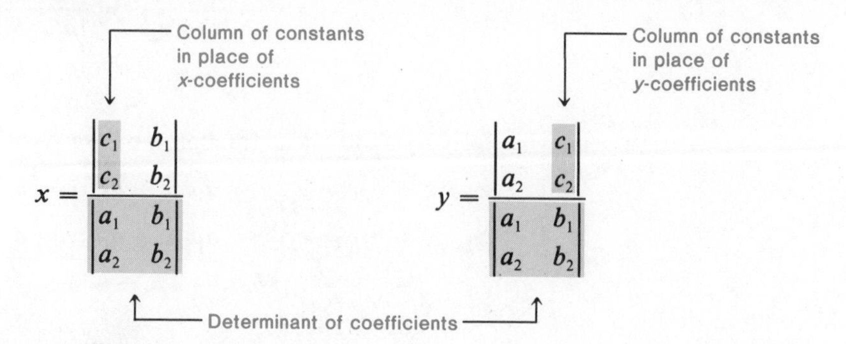

$$x = \frac{\begin{vmatrix} c_1 & b_1 \\ c_2 & b_2 \end{vmatrix}}{\begin{vmatrix} a_1 & b_1 \\ a_2 & b_2 \end{vmatrix}} \qquad y = \frac{\begin{vmatrix} a_1 & c_1 \\ a_2 & c_2 \end{vmatrix}}{\begin{vmatrix} a_1 & b_1 \\ a_2 & b_2 \end{vmatrix}}$$

The method for solving a second-order linear system by using Cramer's rule can be summarized as follows:

TO SOLVE $\begin{Bmatrix} a_1x + b_1y = c_1 \\ a_2x + b_2y = c_2 \end{Bmatrix}$

BY USING CRAMER'S RULE

The solution is $x = \dfrac{D_x}{D}$, $\quad y = \dfrac{D_y}{D}$,

where $D = \begin{vmatrix} a_1 & b_1 \\ a_2 & b_2 \end{vmatrix}$ Determinant of coefficients

$D_x = \begin{vmatrix} c_1 & b_1 \\ c_2 & b_2 \end{vmatrix}$ Column of constants in place of x-coefficients

$D_y = \begin{vmatrix} a_1 & c_1 \\ a_2 & c_2 \end{vmatrix}$ Column of constants in place of y-coefficients

When solving a system of equations by Cramer's rule, there are three possibilities:

1. *Only one solution (consistent system):* The determinant of the coefficients is not equal to zero.
2. *No solution (inconsistent system):* The determinant of the coefficients is equal to zero, *and* the determinants serving as numerators for the letters are not equal to zero.
3. *Infinite number of solutions (dependent equations):* The determinant of the coefficients is equal to zero, *and* the determinants serving as numerators for the letters are *also* equal to zero.

Example 2 Solve $\begin{Bmatrix} x + 4y = 7 \\ 3x + 5y = 0 \end{Bmatrix}$ by Cramer's rule

Column of constants in place of x-coefficients

$$x = \frac{\begin{vmatrix} 7 & 4 \\ 0 & 5 \end{vmatrix}}{\begin{vmatrix} 1 & 4 \\ 3 & 5 \end{vmatrix}} = \frac{(7)(5) - (0)(4)}{(1)(5) - (3)(4)} = \frac{35}{-7} = -5$$

Determinant of coefficients

Column of constants in place of y-coefficients

$$y = \frac{\begin{vmatrix} 1 & 7 \\ 3 & 0 \end{vmatrix}}{\begin{vmatrix} 1 & 4 \\ 3 & 5 \end{vmatrix}} = \frac{(1)(0) - (3)(7)}{(1)(5) - (3)(4)} = \frac{-21}{-7} = 3$$

Determinant of coefficients

Therefore the solution is $\begin{Bmatrix} x = -5 \\ y = 3 \end{Bmatrix}$, written $(-5,3)$. This is a consistent system made up of independent equations.

Example 3 Solve $\begin{Bmatrix} 2x - 3y = 6 \\ 6x - 9y = 36 \end{Bmatrix}$ by Cramer's rule

$$x = \frac{\begin{vmatrix} 6 & -3 \\ 36 & -9 \end{vmatrix}}{\begin{vmatrix} 2 & -3 \\ 6 & -9 \end{vmatrix}} = \frac{(6)(-9) - (36)(-3)}{(2)(-9) - (6)(-3)} = \frac{-54 + 108}{-18 + 18} = \frac{54}{0} \quad \text{Not possible}$$

$$y = \frac{\begin{vmatrix} 2 & 6 \\ 6 & 36 \end{vmatrix}}{\begin{vmatrix} 2 & -3 \\ 6 & -9 \end{vmatrix}} = \frac{(2)(36) - (6)(6)}{(2)(-9) - (6)(-3)} = \frac{72 - 36}{-18 + 18} = \frac{36}{0} \quad \text{Not possible}$$

This system has *no solution*. This is an inconsistent system made up of independent equations (see Example 6, Section 9–1).

Example 4 Solve $\begin{Bmatrix} 3x + 5y = 15 \\ 6x + 10y = 30 \end{Bmatrix}$ by Cramer's rule

$$x = \frac{\begin{vmatrix} 15 & 5 \\ 30 & 10 \end{vmatrix}}{\begin{vmatrix} 3 & 5 \\ 6 & 10 \end{vmatrix}} = \frac{(15)(10) - (30)(5)}{(3)(10) - (6)(5)} = \frac{150 - 150}{30 - 30} = \frac{0}{0} \quad \text{Cannot be determined}$$

$$y = \frac{\begin{vmatrix} 3 & 15 \\ 6 & 30 \end{vmatrix}}{\begin{vmatrix} 3 & 5 \\ 6 & 10 \end{vmatrix}} = \frac{(3)(30) - (6)(15)}{(3)(10) - (6)(5)} = \frac{90 - 90}{30 - 30} = \frac{0}{0} \quad \text{Cannot be determined}$$

Therefore the solution cannot be found by Cramer's rule. This is a consistent system made up of dependent equations. There are an infinite number of solutions (see Example 7, Section 9–1).

EXERCISES 9-5

In Exercises 1–6 find the value of each second-order determinant.

1. $\begin{vmatrix} 3 & 4 \\ 2 & 5 \end{vmatrix}$
2. $\begin{vmatrix} 4 & 3 \\ 2 & 7 \end{vmatrix}$
3. $\begin{vmatrix} 2 & -4 \\ 5 & -3 \end{vmatrix}$
4. $\begin{vmatrix} 5 & 0 \\ -9 & 8 \end{vmatrix}$
5. $\begin{vmatrix} -7 & -3 \\ 5 & 8 \end{vmatrix}$
6. $\begin{vmatrix} 2 & -4 \\ -3 & 6 \end{vmatrix}$

In Exercises 7–20 use Cramer's rule to solve each system. Write "inconsistent" if no solution exists. If an infinite number of solutions exist, write "dependent" and give the solution set.

7. $2x + y = 7$, $x + 2y = 8$
8. $x + 3y = 6$, $2x + y = 7$
9. $x - 2y = 3$, $3x + 7y = -4$
10. $5x + 7y = 1$, $3x + 4y = 1$
11. $3x + y = -2$, $5x + 4y = -1$
12. $4x + 3y = 2$, $3x + 5y = -4$
13. $3x - 4y = 5$, $9x + 8y = 0$
14. $6x - 5y = -5$, $8x + 15y = 2$
15. $2x + y = 0$, $x - 3y = 0$
16. $3x - 6y = 0$, $2x - 4y = 0$
17. $x + 3y = 1$, $2x + 6y = 3$
18. $2x - y = 2$, $6x - 3y = 4$
19. $2x - 4y = 6$, $3x - 6y = 9$
20. $2x - 6y = 2$, $3x - 9y = 3$
21. Solve for x: $\begin{vmatrix} 2 & -4 \\ 3 & x \end{vmatrix} = 20$
22. Solve for x: $\begin{vmatrix} 2x & 5 \\ 2 & 3x \end{vmatrix} = -11x$

9–6 Determinant Method for Solving a Third-Order Linear System

Third-Order Determinants

A **third-order determinant** is a determinant that has three rows and three columns (Figure 9–6A).

The minor of an element is the determinant that remains after *striking out the row and column in which that element appears.*

$$\begin{array}{r} \\ \text{Row 1} \\ \text{Row 2} \\ \text{Row 3} \end{array} \begin{array}{c} \text{Column 1} \quad \text{Column 2} \quad \text{Column 3} \\ \begin{vmatrix} a_1 & b_1 & c_1 \\ a_2 & b_2 & c_2 \\ a_3 & b_3 & c_3 \end{vmatrix} \end{array}$$

Figure 9–6A

Example 1 In the determinant $\begin{vmatrix} 2 & 0 & -1 \\ 5 & -4 & 6 \\ -3 & 1 & 7 \end{vmatrix}$:

(a) Find the minor of 5.

$$\begin{vmatrix} 2 & 0 & -1 \\ 5 & -4 & 6 \\ -3 & 1 & 7 \end{vmatrix}$$

Strike out the row and column containing 5, leaving $\begin{vmatrix} 0 & -1 \\ 1 & 7 \end{vmatrix}$

Therefore the minor of 5 is $\begin{vmatrix} 0 & -1 \\ 1 & 7 \end{vmatrix}$.

(b) The minor of 1 is $\begin{vmatrix} 2 & -1 \\ 5 & 6 \end{vmatrix}$ because $\begin{vmatrix} 2 & 0 & -1 \\ 5 & -4 & 6 \\ -3 & 1 & 7 \end{vmatrix}$

(c) The minor of 7 is $\begin{vmatrix} 2 & 0 \\ 5 & -4 \end{vmatrix}$ because $\begin{vmatrix} 2 & 0 & -1 \\ 5 & -4 & 6 \\ -3 & 1 & 7 \end{vmatrix}$

THE VALUE OF A THIRD-ORDER DETERMINANT

Figure 9–6B shows a third order determinant and its value.

Definition

$$\begin{vmatrix} a_1 & b_1 & c_1 \\ a_2 & b_2 & c_2 \\ a_3 & b_3 & c_3 \end{vmatrix} = a_1b_2c_3 - a_1b_3c_2 + a_3b_1c_2 - a_2b_1c_3 + a_2b_3c_1 - a_3b_2c_1$$

Figure 9–6B

Factoring the right side in Figure 9–6B by grouping, we have:

$$\begin{vmatrix} a_1 & b_1 & c_1 \\ a_2 & b_2 & c_2 \\ a_3 & b_3 & c_3 \end{vmatrix} = a_1(b_2c_3 - b_3c_2) - a_2(b_1c_3 - b_3c_1) + a_3(b_1c_2 - b_2c_1)$$

The expressions in parentheses can be expressed as second-order determinants.

Expansion by Minors

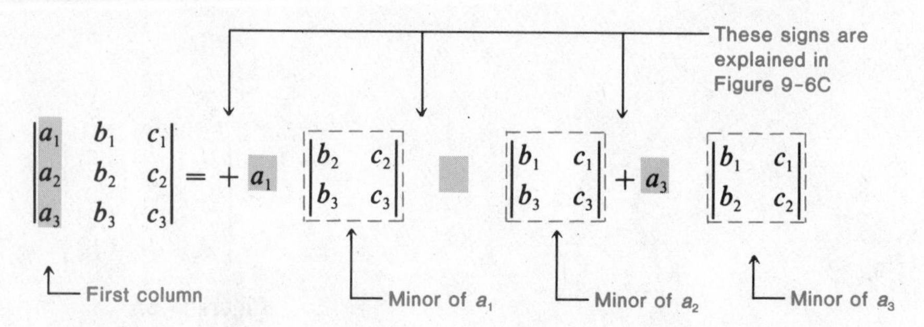

The expression on the right side is called **the expansion of the determinant by minors** of the elements of the first column. The determinant can be expanded by the minors of *any* row or column.

The signs of the terms in any expansion by minors can be conveniently determined from Figure 9–6C.

$$\begin{vmatrix} + & - & + \\ - & + & - \\ + & - & + \end{vmatrix}$$

These signs are found by choosing different groupings when factoring the right side in Figure 9-6B

Figure 9–6C

Example 2 Evaluate $\begin{vmatrix} 1 & 3 & 2 \\ 2 & -1 & 1 \\ -4 & 1 & -3 \end{vmatrix}$ by: (a) Expansion by column 1 (b) Expansion by row 2

(a) Expansion by column 1

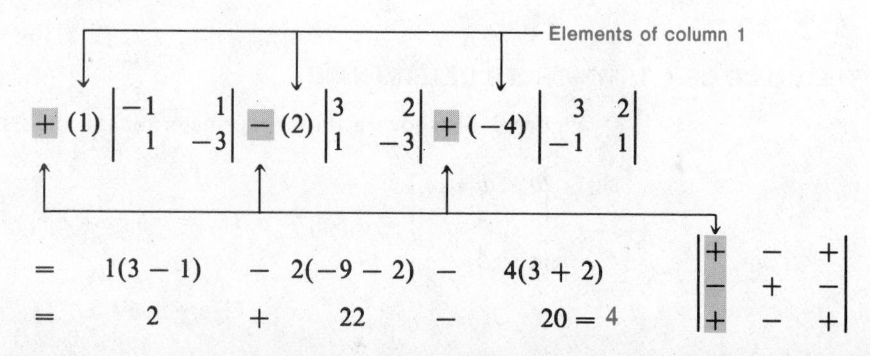

$$+ (1)\begin{vmatrix} -1 & 1 \\ 1 & -3 \end{vmatrix} - (2)\begin{vmatrix} 3 & 2 \\ 1 & -3 \end{vmatrix} + (-4)\begin{vmatrix} 3 & 2 \\ -1 & 1 \end{vmatrix}$$

$$= 1(3 - 1) - 2(-9 - 2) - 4(3 + 2)$$

$$= 2 + 22 - 20 = 4$$

(b) Expansion by row 2

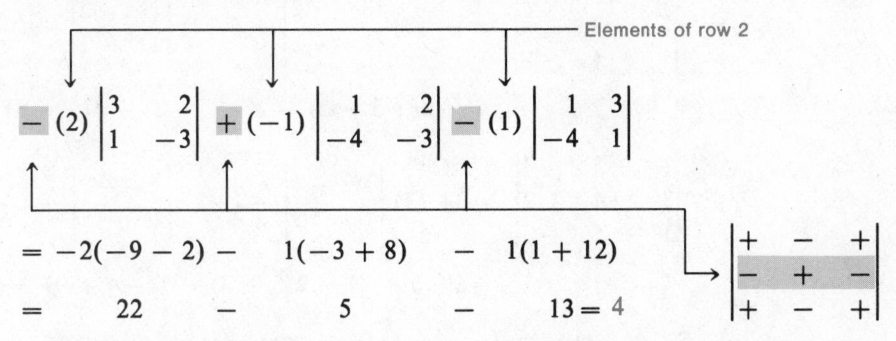

$$-(2)\begin{vmatrix} 3 & 2 \\ 1 & -3 \end{vmatrix} + (-1)\begin{vmatrix} 1 & 2 \\ -4 & -3 \end{vmatrix} - (1)\begin{vmatrix} 1 & 3 \\ -4 & 1 \end{vmatrix}$$

$$= -2(-9-2) - 1(-3+8) - 1(1+12)$$

$$= 22 - 5 - 13 = 4$$

$$\begin{vmatrix} + & - & + \\ - & + & - \\ + & - & + \end{vmatrix}$$

Note that the same value, 4, was obtained from *both* expansions. The value of the determinant is the same no matter which row or column is used.

USING CRAMER'S RULE TO SOLVE A THIRD-ORDER LINEAR SYSTEM

Cramer's rule is extended to third-order systems, as shown in Example 3.

Example 3 Solve the system $\begin{Bmatrix} x - 3y + 2z = 3 \\ 3x - 4y + 2z = -2 \\ x + 5y - z = -1 \end{Bmatrix}$ using Cramer's rule

$$\begin{Bmatrix} x - 3y + 2z = 3 \\ 3x - 4y + 2z = -2 \\ x + 5y - z = -1 \end{Bmatrix}$$

Column of constants in place of coefficients of letter being solved for

$$x = \frac{\begin{vmatrix} 3 & -3 & 2 \\ -2 & -4 & 2 \\ -1 & 5 & -1 \end{vmatrix}}{\begin{vmatrix} 1 & -3 & 2 \\ 3 & -4 & 2 \\ 1 & 5 & -1 \end{vmatrix}} = \frac{D_x}{D}, \quad y = \frac{\begin{vmatrix} 1 & 3 & 2 \\ 3 & -2 & 2 \\ 1 & -1 & -1 \end{vmatrix}}{\begin{vmatrix} 1 & -3 & 2 \\ 3 & -4 & 2 \\ 1 & 5 & -1 \end{vmatrix}} = \frac{D_y}{D}, \quad z = \frac{\begin{vmatrix} 1 & -3 & 3 \\ 3 & -4 & -2 \\ 1 & 5 & -1 \end{vmatrix}}{\begin{vmatrix} 1 & -3 & 2 \\ 3 & -4 & 2 \\ 1 & 5 & -1 \end{vmatrix}} = \frac{D_z}{D}$$

$$D = \begin{vmatrix} 1 & -3 & 2 \\ 3 & -4 & 2 \\ 1 & 5 & -1 \end{vmatrix} = +(1)\begin{vmatrix} -4 & 2 \\ 5 & -1 \end{vmatrix} - (-3)\begin{vmatrix} 3 & 2 \\ 1 & -1 \end{vmatrix} + (2)\begin{vmatrix} 3 & -4 \\ 1 & 5 \end{vmatrix}$$

$$= 1(4 - 10) + 3(-3 - 2) + 2(15 + 4) = 17$$

$$D_x = \begin{vmatrix} 3 & -3 & 2 \\ -2 & -4 & 2 \\ -1 & 5 & -1 \end{vmatrix} = +(3)\begin{vmatrix} -4 & 2 \\ 5 & -1 \end{vmatrix} - (-2)\begin{vmatrix} -3 & 2 \\ 5 & -1 \end{vmatrix} + (-1)\begin{vmatrix} -3 & 2 \\ -4 & 2 \end{vmatrix}$$

$$= 3(4 - 10) + 2(3 - 10) - 1(-6 + 8) = -34$$

$$D_y = \begin{vmatrix} 1 & 3 & 2 \\ 3 & -2 & 2 \\ 1 & -1 & -1 \end{vmatrix} = -(3)\begin{vmatrix} 3 & 2 \\ 1 & -1 \end{vmatrix} + (-2)\begin{vmatrix} 1 & 2 \\ 1 & -1 \end{vmatrix} - (-1)\begin{vmatrix} 1 & 2 \\ 3 & 2 \end{vmatrix}$$

$$= -3(-3-2) - 2(-1-2) + 1(2-6) = 17$$

$$D_z = \begin{vmatrix} 1 & -3 & 3 \\ 3 & -4 & -2 \\ 1 & 5 & -1 \end{vmatrix} = +(3)\begin{vmatrix} 3 & -4 \\ 1 & 5 \end{vmatrix} - (-2)\begin{vmatrix} 1 & -3 \\ 1 & 5 \end{vmatrix} + (-1)\begin{vmatrix} 1 & -3 \\ 3 & -4 \end{vmatrix}$$

$$= 3(15+4) + 2(5+3) - 1(-4+9) = 68$$

$$x = \frac{D_x}{D} = \frac{-34}{17} = -2, \quad y = \frac{D_y}{D} = \frac{17}{17} = 1, \quad z = \frac{D_z}{D} = \frac{68}{17} = 4$$

Therefore the solution is $(-2,1,4)$.

Zeros in a Determinant

If zeros appear anywhere in a determinant, expand the determinant by the row or column containing the most zeros. This minimizes the numerical work done in evaluating the determinant.

Example 4 **Selecting the best row or column to expand the determinant by**

(a) $\begin{vmatrix} -2 & 1 & 0 \\ 4 & -2 & 1 \\ 3 & -1 & 5 \end{vmatrix}$ Expand by row 1 (or column 3) because it contains a zero

Expanding by row 1:

$$= (-2)\begin{vmatrix} -2 & 1 \\ -1 & 5 \end{vmatrix} - (1)\begin{vmatrix} 4 & 1 \\ 3 & 5 \end{vmatrix} + (0)\begin{vmatrix} 4 & -2 \\ 3 & -1 \end{vmatrix}$$

$$= (-2)(-10+1) - 1(20-3) + 0 = 1$$

(b) $\begin{vmatrix} -4 & 0 & -1 \\ 1 & 2 & -3 \\ -2 & 0 & 5 \end{vmatrix}$ Expand by column 2 because it contains two zeros

Expanding by column 2:

It is not necessary to write the minor when it is multiplied by zero

$$= -(0)\begin{vmatrix} \blacksquare \end{vmatrix} + (2)\begin{vmatrix} -4 & -1 \\ -2 & 5 \end{vmatrix} - (0)\begin{vmatrix} \blacksquare \end{vmatrix}$$

$$= 0 + 2(-20-2) - 0 = -44$$

Cramer's rule can be used to solve linear systems of *any* order. When solving a third-order linear system of equations by Cramer's rule, there are three possibilities:

1. *Only one solution (consistent system):* $D \neq 0$.
2. *No solution (inconsistent system):* $D = 0$, and $D_x, D_y, D_z \neq 0$.
3. *Infinite number of solutions: (dependent equations)* $D = 0$, and $D_x = D_y = D_z = 0$.

These same criteria can be extended to linear systems of higher order.

EXERCISES 9-6

1. Find the minor of the element 2 in the determinant: $\begin{vmatrix} 1 & 2 & 3 \\ 4 & 5 & -1 \\ -3 & -5 & 0 \end{vmatrix}$

2. Find the minor of the element -3 in the determinant: $\begin{vmatrix} 2 & 0 & 1 \\ 4 & 1 & 5 \\ -3 & -2 & 4 \end{vmatrix}$

In Exercises 3–6 find the value of each determinant by expanding by minors of the indicated row or column.

3. $\begin{vmatrix} 1 & 2 & 1 \\ 3 & 1 & 2 \\ 4 & 2 & 0 \end{vmatrix}$ column 3

4. $\begin{vmatrix} 2 & 1 & 1 \\ 1 & 3 & 1 \\ 0 & 2 & 4 \end{vmatrix}$ column 1

5. $\begin{vmatrix} 1 & 3 & -2 \\ -1 & 2 & -3 \\ 0 & 4 & 1 \end{vmatrix}$ row 3

6. $\begin{vmatrix} 2 & -1 & 1 \\ 0 & 5 & -3 \\ 1 & 2 & -2 \end{vmatrix}$ row 2

In Exercises 7–12 find the value of each determinant by expanding by any row or column.

7. $\begin{vmatrix} 1 & -2 & 3 \\ -3 & 4 & 0 \\ 2 & 6 & 5 \end{vmatrix}$

8. $\begin{vmatrix} 2 & -4 & -1 \\ 5 & 0 & -6 \\ -3 & 4 & -2 \end{vmatrix}$

9. $\begin{vmatrix} 6 & 7 & 8 \\ -6 & 7 & -9 \\ 0 & 0 & -2 \end{vmatrix}$

10. $\begin{vmatrix} 0 & 0 & -3 \\ 5 & 6 & 9 \\ -5 & 6 & -8 \end{vmatrix}$

11. $\begin{vmatrix} -1 & 3 & -2 \\ -4 & 2 & 5 \\ 0 & 1 & -3 \end{vmatrix}$

12. $\begin{vmatrix} 4 & -5 & 2 \\ -2 & 0 & -3 \\ 6 & 0 & -1 \end{vmatrix}$

In Exercises 13–18 solve each system by using Cramer's rule.

13. $\begin{aligned} 2x + y + z &= 4 \\ x - y + 3z &= -2 \\ x + y + 2z &= 1 \end{aligned}$

14. $\begin{aligned} x + y + z &= 1 \\ 2x + y - 2z &= -4 \\ x + y + 2z &= 3 \end{aligned}$

15. $\begin{aligned} 2x + 3y + z &= 7 \\ 4x - 2z &= -6 \\ 6y - z &= 0 \end{aligned}$

16. $\begin{aligned} x - 2y &= 4 \\ y + 3z &= 8 \\ 2x - z &= 1 \end{aligned}$

17. $\begin{aligned} x + 2z &= 7 \\ 2x - y &= 5 \\ 2y + z &= 4 \end{aligned}$

18. $\begin{aligned} 4x + 5y + z &= 4 \\ 10y - 2z &= 6 \\ 8x + 3z &= 3 \end{aligned}$

19. Solve for x: $\begin{vmatrix} x & 0 & 1 \\ 0 & 2 & 3 \\ 4 & -1 & -2 \end{vmatrix} = 6$

20. Solve for x: $\begin{vmatrix} 0 & -4 & 3 \\ x & 2 & 0 \\ -1 & 5 & x \end{vmatrix} = 31$

9–7 Quadratic Systems

Quadratic System

All the systems studied in this chapter so far have been *linear* systems of equations. In this section we discuss the solution of *quadratic* systems. A **quadratic system** is one that has a *second-degree* equation as its highest-degree equation.

The quadratic *equations* that appear in the quadratic *systems* of this section are equations of conic sections. A method of graphing conic sections was discussed in Section 8–6.

SYSTEMS THAT HAVE ONE QUADRATIC EQUATION AND ONE LINEAR EQUATION

TO SOLVE A SYSTEM THAT HAS ONE QUADRATIC AND ONE LINEAR EQUATION

1. Solve the linear equation for one variable.
2. Substitute the expression obtained in Step 1 into the quadratic equation of the system; then solve the resulting equation.
3. Substitute the solutions obtained in Step 2 into the expression obtained in Step 1; then solve for the remaining variable.

Example 1 Solve the quadratic system: $\begin{Bmatrix} (1)\ x - 2y = -4 \\ (2)\quad\ \ x^2 = 4y \end{Bmatrix}$

1. Solve (1) for x: $x - 2y = -4$
$$x = 2y - 4$$
2. Substitute $2y - 4$ for x in Equation (2):
$$x^2 = 4y$$
$$(2y - 4)^2 = 4y$$
$$4y^2 - 16y + 16 = 4y$$
$$4y^2 - 20y + 16 = 0$$
$$y^2 - 5y + 4 = 0$$
$$(y - 1)(y - 4) = 0$$
$$y - 1 = 0 \quad \Big| \quad y - 4 = 0$$
$$y = 1 \quad \Big| \quad y = 4$$
3. If $y = 1$, then $x = 2y - 4 = 2(1) - 4 = -2$.
Therefore one solution is $(-2,1)$.
If $y = 4$, then $x = 2y - 4 = 2(4) - 4 = 8 - 4 = 4$.
Therefore the second solution is $(4,4)$.

When the graphs of Equations (1) and (2) are drawn on the same set of axes, we see that *the two solutions,* $(-2,1)$ and $(4,4)$, *are the points where the graphs intersect* (Figure 9–7A). [Refer to Section 7–2 for graphing (1) $x - 2y = -4$. Refer to Section 8–6 for graphing (2) $x^2 = 4y$.]

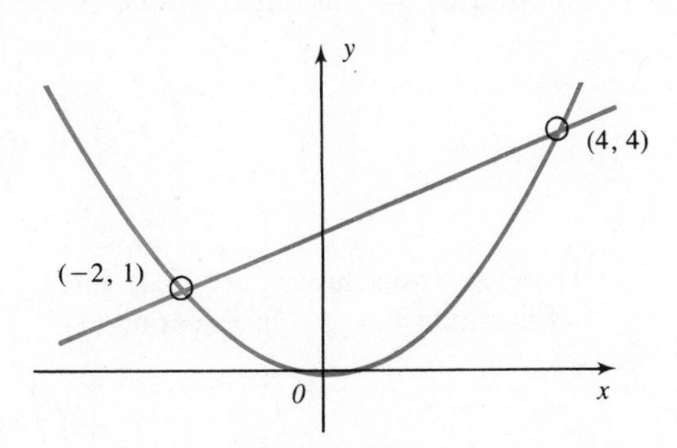

Figure 9–7A

Number of Solutions

When the system has one quadratic equation and one linear equation, there are three possible cases:

(1) *Two Points of Intersection* (two real-number solutions)

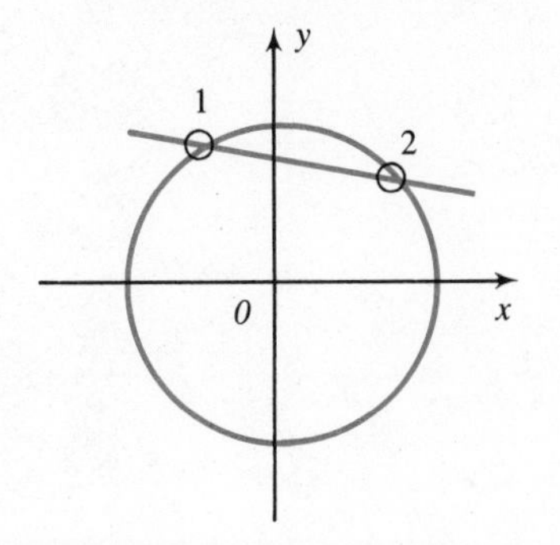

(2) *One Point of Intersection* (one real-number solution)

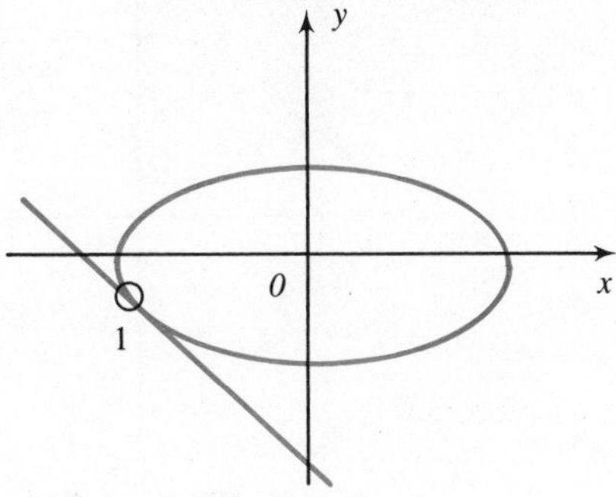

(3) *No Point of Intersection* (two complex solutions)

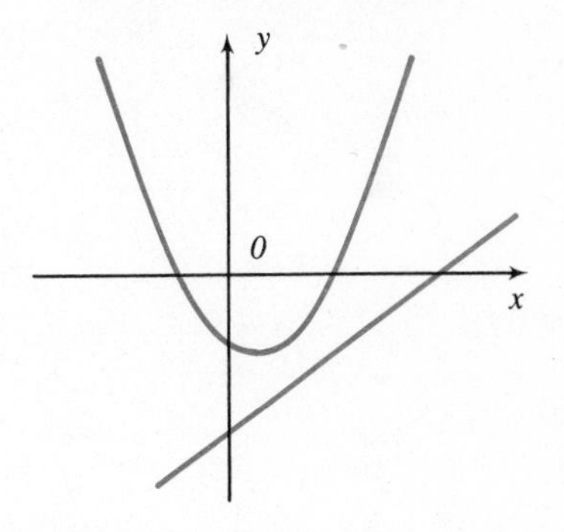

SYSTEMS THAT HAVE TWO QUADRATIC EQUATIONS

We now discuss how to solve *some* systems that have two quadratic equations.

Example 2 Solve the quadratic system: $\begin{Bmatrix} (1) & x^2 + y^2 = 13 \\ (2) & 2x^2 + 3y^2 = 30 \end{Bmatrix}$

We can use *addition-subtraction* because the two equations have like terms.

$$\begin{array}{lllll} (1) & 2\,] & x^2 + y^2 = 13 & \Rightarrow & 2x^2 + 2y^2 = 26 \\ (2) & 1\,] & 2x^2 + 3y^2 = 30 & \Rightarrow & 2x^2 + 3y^2 = 30 \\ & & & & \quad\;\; y^2 = 4 \\ & & & & \quad\;\; y = \pm 2 \end{array}$$

Subtract top equation from bottom equation

Substitute $y = 2$ in Equation (1):

$$\begin{aligned} x^2 + y^2 &= 13 \\ x^2 + (2)^2 &= 13 \\ x^2 + 4 &= 13 \\ x^2 &= 9 \\ x &= \pm 3 \end{aligned}$$

Therefore two solutions are (3,2) and (−3,2).

Substitute $y = -2$ in Equation (1):

$$\begin{aligned} x^2 + y^2 &= 13 \\ x^2 + (-2)^2 &= 13 \\ x^2 + 4 &= 13 \\ x^2 &= 9 \\ x &= \pm 3 \end{aligned}$$

Therefore two more solutions are (3,−2) and (−3,−2).

When the graphs of Equations (1) and (2) are drawn on the same set of axes, we see that *the four solutions,* (3,2), (3,−2), (−3,2) and (−3,−2), *are the points where the two graphs intersect* (Figure 9–7B).

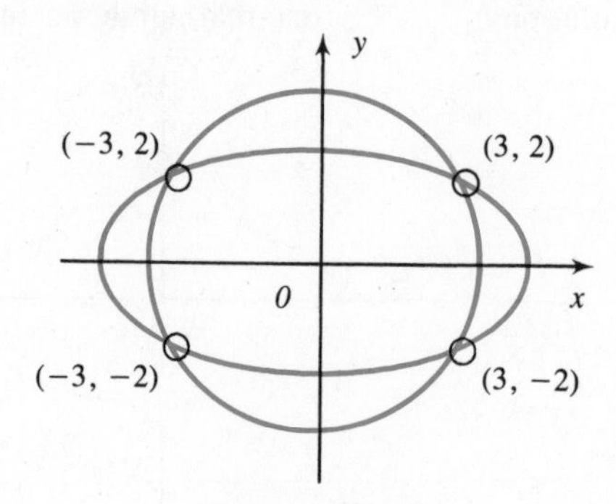

Figure 9–7B

Example 3 Solve the quadratic system: $\begin{Bmatrix} (1) & x^2 + xy + y^2 = 7 \\ (2) & x^2 + y^2 = 5 \end{Bmatrix}$

$$\begin{array}{ll} (1) & x^2 + xy + y^2 = 7 \\ (2) & x^2 \qquad\;\; + y^2 = 5 \\ & \qquad xy \qquad\;\; = 2 \end{array}$$

Equation (1) − Equation (2)

$$y = \frac{2}{x}$$

$$(3) \qquad y^2 = \frac{4}{x^2} \qquad \text{Squared both sides}$$

$$x^2 + \frac{4}{x^2} = 5 \qquad \text{Substituted } \frac{4}{x^2} \text{ in place of } y^2 \text{ in Equation (2)}$$

$$x^4 + 4 = 5x^2 \qquad \text{Multiplied each term by LCD, } x^2$$

$$x^4 - 5x^2 + 4 = 0$$

$$(x^2 - 1)(x^2 - 4) = 0$$

$$x = \pm 1 \mid x = \pm 2$$

$\underline{\text{If } x = 1}$: $y = \frac{2}{x} = \frac{2}{1} = 2$; solution (1,2)

$\underline{\text{If } x = -1}$: $y = \frac{2}{x} = \frac{2}{-1} = -2$; solution (−1,−2)

$\underline{\text{If } x = 2}$: $y = \frac{2}{x} = \frac{2}{2} = 1$; solution (2,1)

$\underline{\text{If } x = -2}$: $y = \frac{2}{x} = \frac{2}{-2} = -1$; solution (−2,−1)

Number of Solutions

When the system has two quadratic equations, **there are five possible cases:**

(1) *No Points of Intersection* (*complex* solutions)

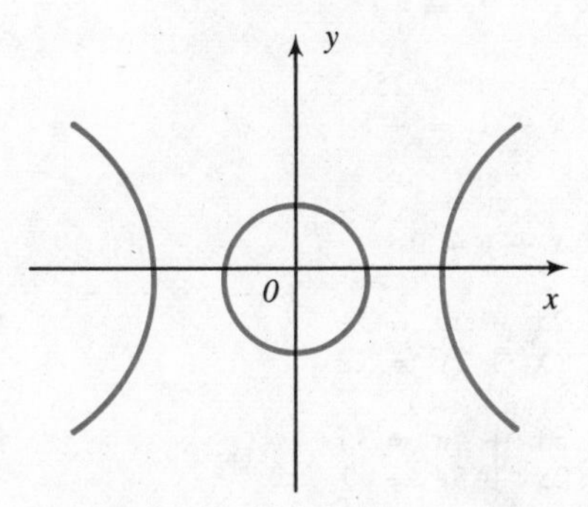

(2) *One Point of Intersection* (one real-number solution)

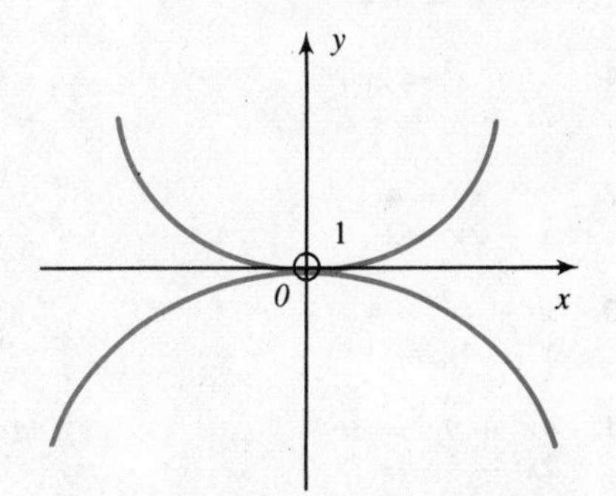

(3) *Two Points of Intersection* (two real-number solutions)

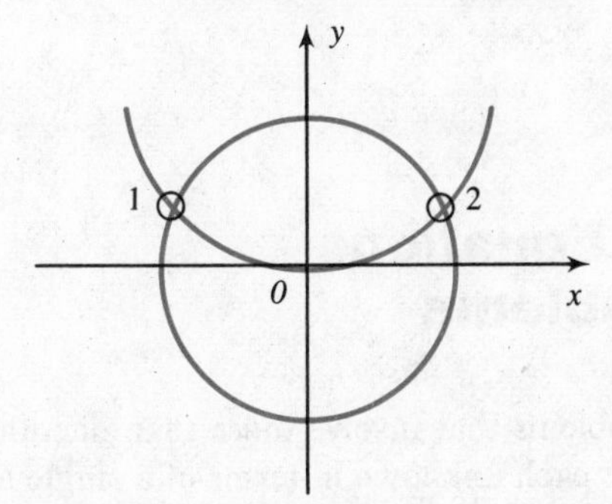

(4) *Three Points of Intersection* (three real-number solutions)

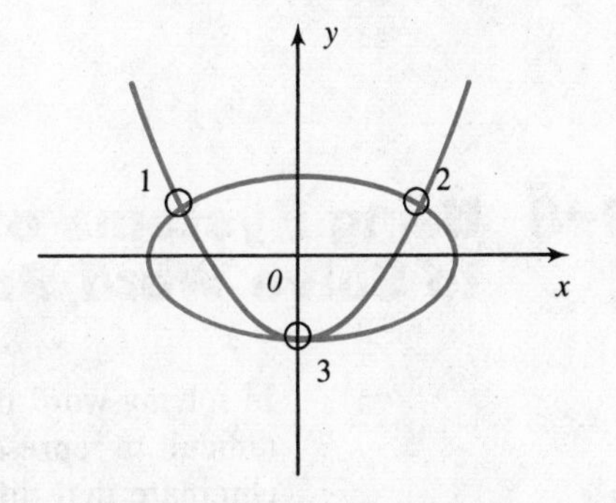

(5) *Four Points of Intersection*
(four real-number solutions)

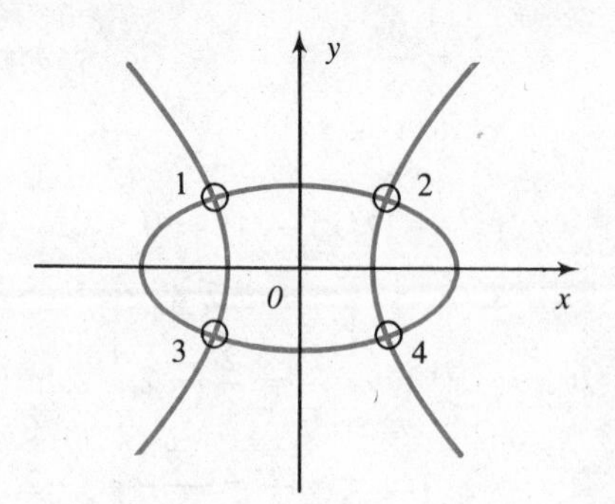

EXERCISES 9-7

Solve each system of equations.

1. $x^2 = 2y$
$x - y = -4$

2. $x^2 = 4y$
$x + 2y = 4$

3. $x^2 = 4y$
$x - y = 1$

4. $x^2 = 2y$
$2x + y = -2$

5. $x^2 + y^2 = 25$
$x - 3y = -5$

6. $xy = 6$
$x - y = 1$

7. $xy = 4$
$x - 2y = 2$

8. $xy = 3$
$x + y = 0$

9. $x^2 + y^2 = 61$
$x^2 - y^2 = 11$

10. $x^2 + 4y^2 = 4$
$x^2 - 4y = 4$

11. $2x^2 + 3y^2 = 21$
$x^2 + 2y^2 = 12$

12. $4x^2 - 5y^2 = 62$
$5x^2 + 8y^2 = 106$

13. $4x^2 + 9y^2 = 36$
$2x^2 - 9y = 18$

14. $3x^2 + 4y^2 = 35$
$2x^2 + 5y^2 = 42$

15. $x^2 - 3xy + y^2 = 1$
$x^2 + y^2 = 10$

16. $x^2 + xy - y^2 = 1$
$x^2 - y^2 = 3$

17. $xy = -4$
$y = 6x - 9 - x^2$

18. $xy = 8$
$y = 8 + x^2 - 4x$

9–8 Using Systems of Equations to Solve Word Problems

In solving word problems that involve more than one unknown, it is sometimes difficult to represent each unknown in terms of a single letter. In this section we eliminate that difficulty by using a different letter for each unknown.

TO SOLVE A WORD PROBLEM USING A SYSTEM OF EQUATIONS

1. Read the problem completely and determine *how many* unknown numbers there are.
2. Draw a diagram showing the relationships in the problem whenever possible.
3. Represent *each* unknown number by a *different* letter.
4. Use word statements to write a system of equations. *There must be as many equations as unknown letters.*
5. Solve the system of equations using one of the following:
   (a) Addition-subtraction method (Section 9–2)
   (b) Substitution method (Section 9–3)
   (c) Determinant method (Sections 9–5 and 9–6)
   (d) Graphical method (Section 9–1)

Example 1 The sum of two numbers is 20. Their difference is 6. What are the numbers?

Using One Variable

Let x = Larger number
$20 - x$ = Smaller number

Their difference is 6

$x - (20 - x) = 6$
$x - 20 + x = 6$
$2x = 26$
Larger number $x = 13$
Smaller number $20 - x = 7$

The difficulty in using the one-variable method to solve this problem is that some students cannot decide whether to represent the second unknown number by $x - 20$ or by $20 - x$.

Using Two Variables

Let x = Larger number
y = Smaller number

The sum of two numbers is 20

(1) $x + y = 20$

Their difference is 6

(2) $x - y = 6$

Using addition-subtraction method:

(1) $x + y = 20$
(2) $x - y = 6$
$2x = 26$
$x = 13$ Larger number

Substitute $x = 13$ into (1):

(1) $x + y = 20$
$13 + y = 20$
$y = 7$ Smaller number

Example 2 Doris has 17 coins in her purse that have a total value of \$1.15. If she has only nickels and dimes, how many of each are there?

Let D = Number of dimes
N = Number of nickels

Doris has 17 coins.		She has only nickels and dimes.
(1) 17	$=$	$N + D$

The coins in her purse have a total value of \$1.15 (115¢).

Amount of money in dimes	+	Amount of money in nickels	=	Total amount of money
(2) $10D$	+	$5N$	=	115

$$\begin{aligned} (1)\quad N + D &= 17 \\ (2)\quad 5N + 10D &= 115 \end{aligned}$$ Arrange the equations so that like terms are in the same column

(1) $N + D = 17 \Rightarrow N = 17 - D$

$$\begin{aligned} 5N + 10D &= 115 \\ 5(17 - D) + 10D &= 115 \\ 85 - 5D + 10D &= 115 \\ 5D &= 30 \\ D &= 6 \end{aligned}$$ Number of dimes

$N = 17 - D = 17 - 6 = 11$ Number of nickels

Example 3 The difference of two numbers is 9. If their product is -20, what are the numbers?

(1) $x - y = 9$ The difference of two numbers is 9
(2) $xy = -20$ Their product is -20

Solving (1) for x, we have $x = y + 9$.
Substituting $x = y + 9$ in (2), we have:

$$\begin{aligned} (y + 9)y &= -20 \\ y^2 + 9y &= -20 \\ y^2 + 9y + 20 &= 0 \\ (y + 5)(y + 4) &= 0 \end{aligned}$$

$y = -5$	$y = -4$
$x = y + 9 = -5 + 9 = 4$	$x = y + 9 = -4 + 9 = 5$

Therefore the two numbers can be either $\{4, -5\}$ or $\{5, -4\}$.

EXERCISES 9-8

Solve each word problem by using a system of equations.

1. The sum of two numbers is 30. Their difference is 12. What are the numbers?
2. The sum of two angles is 180°. Their difference is 70°. Find the angles.

3. Beatrice has 15 coins having a total value of \$1.75. If these coins are nickels and quarters, how many of each kind are there?

4. Raul has 22 coins having a total value of \$5.00. If these coins are dimes and half-dollars, how many of each kind are there?

5. Find two numbers such that twice the smaller plus three times the larger is 34, and five times the smaller minus twice the larger is 9.

6. One-third the sum of two numbers is 12. Twice their difference is 12. Find the numbers.

7. Don spent \$3.70 for 22 stamps. If he bought only 15¢ and 20¢ stamps, how many of each kind did he buy?

8. Several families went to a movie together. They spent \$19.25 for 8 tickets. If adult tickets cost \$3.50 and children's tickets cost \$1.75 how many of each kind of ticket were bought?

9. A fraction has a value of two-thirds. If 10 is added to its numerator and 5 is subtracted from its denominator, the value of the fraction becomes 1. What was the original fraction?

10. A fraction has a value of one-half. If 8 is added to its numerator and 6 is added to its denominator, the value of the fraction becomes two-thirds. What was the original fraction?

11. The sum of the digits of a three-digit number is 20. The tens digit is 3 more than the units digit. The sum of the hundreds digit and tens digit is 15. Find the number.

12. The sum of the digits of a three-digit number is 21. The units digit is one less than the tens digit. Twice the hundreds digit plus the tens digit is 17. Find the number.

13. A pilot takes $2\frac{1}{2}$ hr to fly 1200 miles against the wind and only 2 hr to return with the wind. Find the average speed of the plane in still air and the average speed of the wind.

14. Jerry takes 6 hr to ride his bicycle 30 miles against the wind. He takes 2 hr to return with the wind. Find Jerry's average riding speed in still air and the average speed of the wind.

15. Albert, Bill, and Carlos working together can do a job in 2 hr. Bill and Carlos together can do the job in 3 hr. Albert and Bill together can do the job in 4 hr. How long would it take each man to do the entire job working alone?

16. A refinery tank has one fill pipe and two drain pipes. Pipe 1 can fill the tank in 3 hr. Pipe 2 takes 6 hr longer to drain the tank than pipe 3. If it takes 12 hr to fill the tank when all three pipes are open, how long does it take pipe 3 to drain the tank alone?

17. Al, Chet, and Muriel are two brothers and a sister. Ten years ago Al was twice as old as Chet was then. Three years ago Muriel was three-fourths Chet's age at that time. In 15 years Al will be 8 years older than Muriel is then. Find their ages now.

18. Mrs. Rice has two daughters, Alice and Shelley. At present, Mrs. Rice is 15 years older than the sum of her daughter's ages. Six years ago, Alice was twice as old as Shelley. In 10 years Mrs. Rice will be twice as old as Alice. Find their present ages.

19. The sum of two numbers is 4. Their product is -21. What are the numbers?

20. The difference of two numbers is 14. Their product is -48. What are the numbers?

21. One side of a rectangle is 2 less than its diagonal. If the area of the rectangle is 12, what are its dimensions?

22. The diagonal of a rectangle is 2 more than one of its sides. If the area of the rectangle is 48, what are its dimensions?

23. A tie and a pin cost \$1.10. The tie costs \$1.00 more than the pin. What is the cost of each?

24. A number of birds are resting on two limbs of a tree. One limb is above the other. A bird on the lower limb says to the birds on the upper limb, "If one of you will come down here, we will have an equal number on each limb." A bird from above replies, "If one of you will come up here, we will have twice as many up here as you have down there." How many birds are sitting on each limb?

25. One side of a rectangle is 1 less than its diagonal. If the area of the rectangle is 60, what are its dimensions?

26. *Brainteaser.* A farmer purchases 100 animals for \$100—sheep at \$5 each, pigs at \$3 each, and chickens at 50¢ each. How many of each did he buy?

27. *Brainteaser.* Fran gave Dan twice as much money as he already had. Dan said that was too much and gave her back one-fourth as much money as she had left. Pleased by his consideration, Fran gave Dan back as much money as he had left. This left Fran with \$75 and Dan wth \$100. How much did Fran and Dan each have in the beginning?

28. *Brainteaser.* Find the length of a fish whose head is 4 in. long, whose tail is equal to the length of the head plus one-half the length of the body, and whose body is equal to the length of the head plus the length of the tail.

9–9 Solving Systems of Inequalities by Graphing

So far we have discussed only systems of *equations*. In this section we discuss solving systems of linear *inequalities*. You may wish to refer to Section 7–4, where drawing the graph of a linear inequality was discussed in detail.

TO SOLVE A SYSTEM OF TWO LINEAR INEQUALITIES GRAPHICALLY

1. Graph the first inequality, shading the half-plane that represents its solution (Section 7–4).
2. Graph the second inequality on the same set of axes. Use a different type of shading for the half-plane that represents its solution.
3. The solution of the system is represented by the area with *both* types of shading (see Figure 9–9A).

Example 1 Solve the system $\begin{Bmatrix} (1) & 2x - 3y < 6 \\ (2) & 3x + 4y \leq -12 \end{Bmatrix}$ graphically

1. <u>*Graph inequality (1):*</u> $2x - 3y < 6$

Boundary $2x - 3y = 6$
is a dashed line because $=$ *not* included

x-intercept: Set $y = 0$. Then $2x - 3(0) = 6$
$2x = 6$
$x = 3$

y-intercept: Set $x = 0$. Then $2(0) - 3y = 6$
$-3y = 6$
$y = -2$

x	y
3	0
0	-2

Therefore the boundary line goes through (3,0) and (0,−2).
Substitute the coordinates of the origin (0,0) into inequality (1):

$$2x - 3y < 6$$
$$2(0) - 3(0) < 6$$
$$0 < 6 \quad \textit{True}$$

Therefore the solution of (1) is the half-plane containing (0,0) (see Figure 9–9A).

2. <u>*Graph inequality (2):*</u> $3x + 4y \leq -12$

Boundary $3x + 4y = -12$
is a solid line because $=$ *is* included

x-intercept: Set $y = 0$. Then $3x + 4(0) = -12$
$3x = -12$
$x = -4$

y-intercept: Set $x = 0$. Then $3(0) + 4y = -12$
$4y = -12$
$y = -3$

x	y
-4	0
0	-3

Substitute the coordinates of the origin (0,0) into inequality (2):

$$3x + 4y \leq -12$$
$$3(0) + 4(0) \leq -12$$
$$0 \leq -12 \quad \textit{False}$$

Therefore the solution of (2) is the half-plane *not* containing (0,0) (see Figure 9–9A).

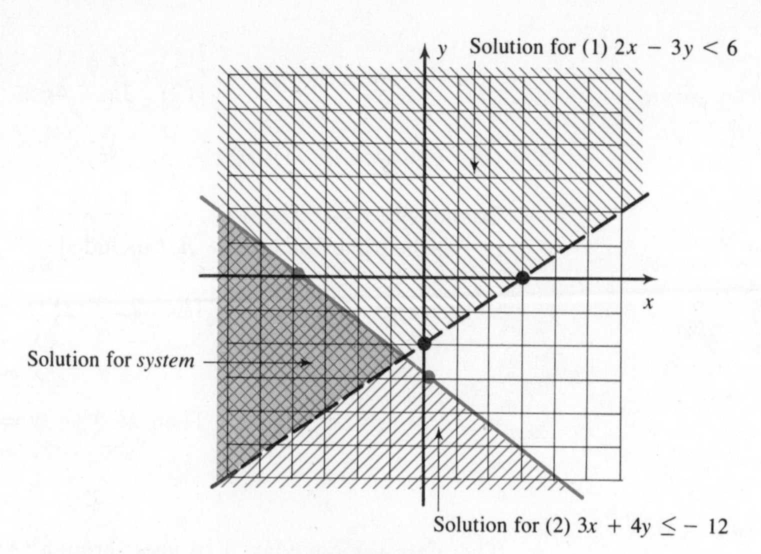

Figure 9–9A

Example 2 Solve the system $\begin{Bmatrix} (1) & x - 2y < -4 \\ (2) & x^2 + y^2 < 16 \end{Bmatrix}$

1. <u>*Graph inequality (1):*</u> $x - 2y < -4$

   Boundary $x - 2y = -4$
   is a dashed line because = *not* included

   x-intercept: $y = 0$: $x - 2(0) = -4$; $(-4,0)$

   y-intercept: $x = 0$: $0 - 2y = -4$

   $y = 2$; $(0,2)$

   Test Point (0,0): $x - 2y < -4$

   $0 - 2(0) < -4$ False

   Therefore the solution of (1) is the half-plane *not* containing (0,0) (Figure 9–9B).

2. <u>*Graph inequality (2):*</u> $x^2 + y^2 < 16$

   Boundary: $x^2 + y^2 = 16$

   x-intercepts: $y = 0$: $x^2 + 0^2 = 16$

   $x = \pm 4$

   y-intercepts: $x = 0$: $0^2 + y^2 = 16$

   $y = \pm 4$

Therefore the boundary is a circle of radius 4 with center at the origin. Circle is dashed because = *not* included

Test Point (0,0): $x^2 + y^2 < 16$

$0^2 + 0^2 < 16$ True

Therefore the solutions of (2) lie inside the circle, along with (0,0) (Figure 9–9B).

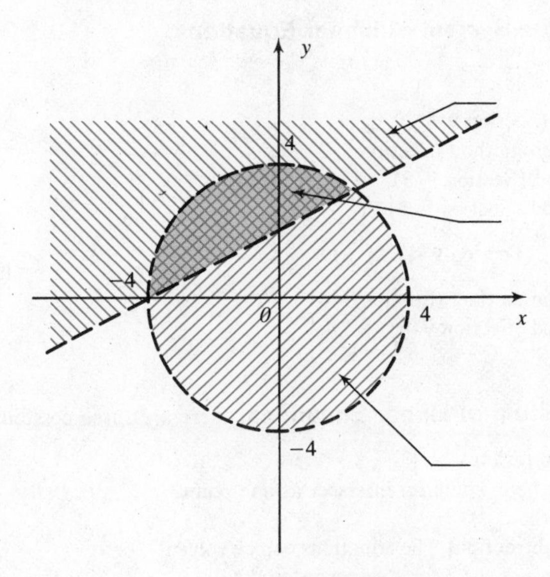

Figure 9–9B

EXERCISES 9–9

Solve each system of inequalities graphically.

1. $\begin{aligned} 4x - 3y &> -12 \\ y &> 2 \end{aligned}$

2. $\begin{aligned} x - y &> -2 \\ 2x + y &> 2 \end{aligned}$

3. $\begin{aligned} 2x - y &\le 2 \\ x + y &\ge 5 \end{aligned}$

4. $\begin{aligned} x - y &\le 4 \\ x + y &\ge -2 \end{aligned}$

5. $\begin{aligned} 2x + y &< 0 \\ x - y &\ge -3 \end{aligned}$

6. $\begin{aligned} 2x + 3y &< 6 \\ x - 3y &< 3 \end{aligned}$

7. $\begin{aligned} 3x &> 6 - y \\ y + 3x &\le 0 \end{aligned}$

8. $\begin{aligned} 2x &\ge y - 4 \\ y - 2x &> 0 \end{aligned}$

9. $\begin{aligned} x^2 &< 4y \\ x + 2y &< 4 \end{aligned}$

10. $\begin{aligned} x^2 + y^2 &< 25 \\ x - 3y &< -5 \end{aligned}$

11. $\begin{aligned} x^2 + 4y^2 &< 4 \\ x^2 - y^2 &\ge 1 \end{aligned}$

12. $\begin{aligned} 2x^2 + y^2 &\le 24 \\ x^2 + 5y^2 &\ge 45 \end{aligned}$

13. $\begin{aligned} 3x - 2y &< 6 \\ x + 2y &\le 4 \\ 6x + y &> -6 \end{aligned}$

14. $\begin{aligned} 3x + 4y &\le 15 \\ 2x + y &\le 2 \\ -4 < x &< 1 \\ -2 < y &< 6 \end{aligned}$

15. $\begin{aligned} y &> x^2 - 2x \\ x &\ge 0 \\ y &< 3 \end{aligned}$

16. $\begin{aligned} x^2 + y^2 &> 9 \\ -2 < y &\le 5 \\ -4 \le x &< 0 \end{aligned}$

Chapter Nine Summary

Systems of Equations. (Section 9–1)

Linear. Each equation is first degree.

Quadratic. The highest-degree equation is second degree.

A Solution of a System of Equations must make each equation of the system a true statement when substituted into it. (Section 9–1)

Methods of Solving a System of Linear Equations.

Second-Order System.

1. Graphical method (Section 9–1)
2. Addition-subtraction method (Section 9–2)
3. Substitution method (Section 9–3)
4. Determinant method (Section 9–5)

Higher-Order System. (Section 9–4)

1. Addition-subtraction method (Section 9–4)
2. Determinant method (Section 9–6)

When Solving a System of Linear Equations, there are three possibilities:

1. *There is only one solution.*
   (a) Graphical method: The lines intersect at one point.
   (b) Algebraic method:
   Addition-subtraction } The equations can be solved
   Substitution } for a single ordered pair.
   (c) Determinant method: The determinant of the coefficients $D \neq 0$.

2. *There is no solution.*
   (a) Graphical method: The lines are parallel.
   (b) Algebraic method:
   Addition-subtraction } Both letters drop out and
   Substitution } an *inequality* results.
   (c) Determinant method: $D = 0$; *and* $D_x, D_y, (D_z) \neq 0$.

3. *There are an infinite number of solutions.*
   (a) Graphical method: Both equations have the same line for a graph.
   (b) Algebraic method:
   Addition-subtraction } Both letters drop out and
   Substitution } an *equality* results.
   (c) Determinant method: $D = 0$; *and* $D_x, D_y, (D_z) = 0$.

Methods of Solving a Quadratic System of Equations. (Section 9–7)

1. *One quadratic and one linear equation:* Solve the linear equation for one letter, then substitute the expression obtained into the quadratic equation.

2. *Two quadratic equations* (with like terms): Use addition-subtraction and substitution when necessary.

Systems of Inequalities. (Section 9–9)

Linear. Each inequality is first degree.

Quadratic. The highest-degree inequality is second degree.

To Solve a System of Inequalities, graph each inequality. Use a different shading to represent the solution of each inequality. The solution of the system is represented by the shaded area common to all the solutions of the inequalities in the system. (Section 9–9)

To Solve a Word Problem Using a System of Equations. (Section 9–8)

1. Read the problem completely and determine how many unknown numbers there are.
2. Draw a diagram showing the relationships in the problem whenever possible.
3. Represent each unknown number by a different letter.
4. Use word statements to write a system of equations. There must be as many equations as unknown letters.
5. Solve the system of equations using one of the following:
   (a) Addition-subtraction method (Section 9–2)
   (b) Substitution method (Section 9–3)
   (c) Determinant method (Sections 9–5 and 9–6)
   (d) Graphical method (Section 9–1)

Chapter Nine Diagnostic Test or Review

Allow yourself about one hour to do these problems. Complete solutions for every problem, together with section references, are given at the end of the book.

---

1. Solve graphically: $\begin{Bmatrix} 3x + 2y = 4 \\ x - y = 3 \end{Bmatrix}$

2. Solve by addition-subtraction: $\begin{Bmatrix} 4x - 3y = 13 \\ 5x - 2y = 4 \end{Bmatrix}$

3. Solve by substitution: $\begin{Bmatrix} 5x + 4y = 23 \\ 3x + 2y = 9 \end{Bmatrix}$

4. Solve by any convenient method: $\begin{Bmatrix} 15x + 8y = -18 \\ 9x + 16y = -8 \end{Bmatrix}$

5. Solve by any convenient method: $\begin{Bmatrix} 35y - 10x = -18 \\ 4x - 14y = 8 \end{Bmatrix}$

6. The sum of two numbers is 12. Their difference is 34. What are the numbers?

7. Solve the system: $\begin{Bmatrix} x + y + z = 0 \\ 2x \qquad - 3z = 5 \\ \qquad 3y + 4z = 3 \end{Bmatrix}$

8. Evaluate each of the following determinants:

(a) $\begin{vmatrix} 8 & -9 \\ 5 & -3 \end{vmatrix}$ (b) $\begin{vmatrix} 2 & 0 & -1 \\ -3 & 1 & 4 \\ -1 & 5 & 6 \end{vmatrix}$

9. Given the system $\left\{\begin{matrix} x - y - z = 0 \\ x + 3y + z = 4 \\ 7x - 2y - 5z = 2 \end{matrix}\right\}$, express y as a ratio of two third-order determinants.

10. Solve the system $\left\{\begin{matrix} y^2 = 8x \\ 3x + y = 2 \end{matrix}\right\}$ algebraically.

11. Solve the system $\left\{\begin{matrix} 2x + 3y \leq 6 \\ y - 2x < 2 \end{matrix}\right\}$ graphically.

12. Solve the system $\left\{\begin{matrix} 4x^2 + y^2 < 16 \\ x + y \geq 1 \end{matrix}\right\}$ graphically.

10 LOGARITHMS

In this chapter we discuss logarithms and some of their applications. Logarithms are important because they have many applications in mathematics and science.

10–1 Basic Definitions

You know that $10^2 = 100$.

Exponent

$$10^2 = 100$$

Base — Power

This same relationship can be thought of in another way:

Logarithm

Logarithm

$$10^2 = 100$$

Base — Number

The **logarithm** is the exponent (2) to which the *base* (10) must be raised to give the *number* (100).

This can be written

$$\log_{10} 100 = 2$$

and is read "the logarithm of 100 to the base 10 equals 2."

Exponential and Logarithmic Forms

Exponential form		*Logarithmic form*
$10^2 = 100$	$\iff$	$\log_{10}100 = 2$

This symbol ($\iff$) means that the exponential form and the logarithmic form are *equivalent*

> **The logarithm of a positive* number** N is the exponent ℓ to which the base $b(b > 1)$† must be raised to give N.
>
> | *Exponential form* | | *Logarithmic form* |
> |---|---|---|
> | $b^\ell = N$ | $\iff$ | $\log_b N = \ell$ |

We use ℓ for "logarithm" because it is the first letter in that word. N is used for "number" because it is the first letter in that word. We use b for "base" because it is the first letter in that word.

Example 1

	Exponential form		*Logarithmic form*
(a)	$5^2 = 25$	$\iff$	$\log_5 25 = 2$
(b)	$2^3 = 8$	$\iff$	$\log_2 8 = 3$
(c)	$10^4 = 10{,}000$	$\iff$	$\log_{10}10{,}000 = 4$
(d)	$10^{-2} = \dfrac{1}{10^2} = \dfrac{1}{100} = 0.01$	$\iff$	$\log_{10}0.01 = -2$

Solving for the Logarithm

Example 2 Find $\log_{10}1000$

	Logarithmic form		*Exponential form*
Let	$x = \log_{10}1000$	$\iff$	$10^x = 1000 = 10^3$
			$x = 3$

Therefore $\log_{10}1000 = 3$.

Solving for the Base

Example 3 Solve for x: $\log_x 27 = 3$

Logarithmic form		*Exponential form*
$\log_x 27 = 3$	$\iff$	$x^3 = 27 = 3^3$
		$x = 3$

---

*The logarithm of a negative number exists, but it is a complex number. We will not discuss the logarithm of a negative number in this book.

†Positive bases less than 1 are not discussed in this book.

Solving for the Number

Example 4 Solve for x: $\log_2 x = 4$

Logarithmic form		*Exponential form*
$\log_2 x = 4$	$\Longleftrightarrow$	$2^4 = x$
		$16 = x$

Example 5 Find $\log_8 4$

	Logarithmic form		*Exponential form*	
Let	$x = \log_8 4$	$\Longleftrightarrow$	$8^x = 4$	
			$(2^3)^x = 2^2$	Wrote 8 and 4 as powers of 2
			$2^{3x} = 2^2$	
			$3x = 2$	
			$x = \dfrac{2}{3}$	

Therefore $\log_8 4 = \dfrac{2}{3}$.

Example 6 Find $\log_7\left(\dfrac{1}{49}\right)$

	Logarithmic form		*Exponential form*
Let	$x = \log_7\left(\dfrac{1}{49}\right)$	$\Longleftrightarrow$	$7^x = \dfrac{1}{49} = \dfrac{1}{7^2} = 7^{-2}$
			$x = -2$

Therefore $\log_7\left(\dfrac{1}{49}\right) = -2$.

Note. The logarithm of a positive number *can* be negative, but the logarithm of a negative number is *complex* (and is not discussed in this book). ■

Logarithm of the Base

Example 7 Find $\log_6 6$

	Logarithmic form		*Exponential form*
Let	$x = \log_6 6$	$\Longleftrightarrow$	$6^x = 6 = 6^1$
			$x = 1$

Therefore $\log_6 6 = 1$.

This result can be generalized as follows:

> **The logarithm of the base** equals 1.
>
> $$\log_b b = 1 \quad (b > 1)$$

Logarithm of 1

Example 8 Find $\log_5 1$

	Logarithmic form		*Exponential form*	
Let	$x = \log_5 1$	$\Longleftrightarrow$	$5^x = \boxed{1 = 5^0}$ $x = 0$	Rule 5, Section 1-6B

Therefore $\log_5 1 = 0$.

This result can be generalized as follows:

> **The logarithm of 1 to any base ($b > 1$) equals 0.**
>
> $$\log_b 1 = 0$$

Logarithm of 0

Example 9 Find $\log_{10} 0$.

	Logarithmic form		*Exponential form*
Let	$x = \log_{10} 0$	$\Longleftrightarrow$	$10^x = 0$ *No solution* because no value of x makes $10^x = 0$

Therefore $\log_{10} 0$ does not exist.

This result can be generalized as follows:

> **The logarithm of 0 to any base ($b > 1$) does not exist.**

EXERCISES 10-1

In Exercises 1–16 write each equation in logarithmic form.

1. $3^2 = 9$ **2.** $4^3 = 64$ **3.** $10^3 = 1000$ **4.** $10^5 = 100{,}000$

5. $2^4 = 16$ **6.** $4^2 = 16$ **7.** $3^{-2} = \frac{1}{9}$ **8.** $2^{-3} = \frac{1}{8}$

9. $12^0 = 1$ **10.** $8^0 = 1$ **11.** $16^{1/2} = 4$ **12.** $8^{1/3} = 2$

13. $2^3 = 8$ **14.** $10^4 = 10{,}000$ **15.** $4^{-2} = \frac{1}{16}$ **16.** $3^4 = 81$

In Exercises 17–30 write each equation in exponential form.

17. $\log_8 64 = 2$ **18.** $\log_2 32 = 5$ **19.** $\log_7 49 = 2$ **20.** $\log_4 64 = 3$

21. $\log_5 1 = 0$ **22.** $\log_6 1 = 0$ **23.** $\log_{10} 100 = 2$ **24.** $\log_{10} 1000 = 3$

25. $\log_{10}0.001 = -3$ **26.** $\log_{10}0.01 = -2$ **27.** $\log_{16}4 = \frac{1}{2}$ **28.** $\log_7 7 = 1$

29. $\log_{10}100{,}000 = 5$ **30.** $\log_b a = c$

In Exercises 31–42 find the value of each logarithm.

31. $\log_{10}10{,}000$ **32.** $\log_{10}100{,}000$ **33.** $\log_4 8$ **34.** $\log_{27}9$

35. $\log_3 3^4$ **36.** $\log_2 2^5$ **37.** $\log_{16}16$ **38.** $\log_{20}20$

39. $\log_8 1$ **40.** $\log_7 1$ **41.** $\log_{64}8$ **42.** $\log_{81}3$

In Exercises 43–68 find the value of the unknown b, N, or x.

43. $\log_5 N = 2$ **44.** $\log_2 N = 5$ **45.** $\log_3 9 = x$ **46.** $\log_2 16 = x$

47. $\log_b 27 = 3$ **48.** $\log_b 81 = 4$ **49.** $\log_9\left(\frac{1}{3}\right) = x$ **50.** $\log_8\left(\frac{1}{64}\right) = x$

51. $\log_{10}10^{-4} = x$ **52.** $\log_{10}10^{-3} = x$ **53.** $\log_{3/2}N = 2$ **54.** $\log_{5/3}N = 3$

55. $\log_5 125 = x$ **56.** $\log_3 81 = x$ **57.** $\log_b 8 = 1.5$ **58.** $\log_b 125 = 1.5$

59. $\log_2 N = -2$ **60.** $\log_{10}N = -2$ **61.** $\log_{25}N = 1.5$ **62.** $\log_b 9 = \frac{2}{3}$

63. $\log_{16}32 = x$ **64.** $\log_{16}N = \frac{5}{4}$ **65.** $\log_4 8 = x$ **66.** $\log_b 32 = \frac{5}{3}$

67. $\log_b\left(\frac{1}{2}\right) = -\frac{1}{3}$ **68.** $\log_b\left(\frac{27}{8}\right) = 3$

10–2 Exponential and Logarithmic Functions

The Exponential Function

An **exponential function** is a function of the form $y = f(x) = a^x$, where a is a positive real number.

Example 1 Graph the exponential function $y = 2^x$

Calculate the value of the function for integral values of x from -3 to $+3$ (Section 1–6).

$$y = 2^x$$
$$y = 2^{-3} = \frac{1}{2^3} = \frac{1}{8}$$
$$y = 2^{-2} = \frac{1}{2^2} = \frac{1}{4}$$
$$y = 2^{-1} = \frac{1}{2^1} = \frac{1}{2}$$
$$y = 2^0 = 1$$
$$y = 2^1 = 2$$
$$y = 2^2 = 4$$
$$y = 2^3 = 8$$

x	y
-3	$\frac{1}{8}$
-2	$\frac{1}{4}$
-1	$\frac{1}{2}$
0	1
1	2
2	4
3	8

Next, plot the points listed in the table of values and draw a smooth curve through them (joining the points in order from left to right). The graph of $y = 2^x$ is the curve drawn with the solid black line in Figure 10–2A.

Graphs of Exponential and Logarithmic Functions

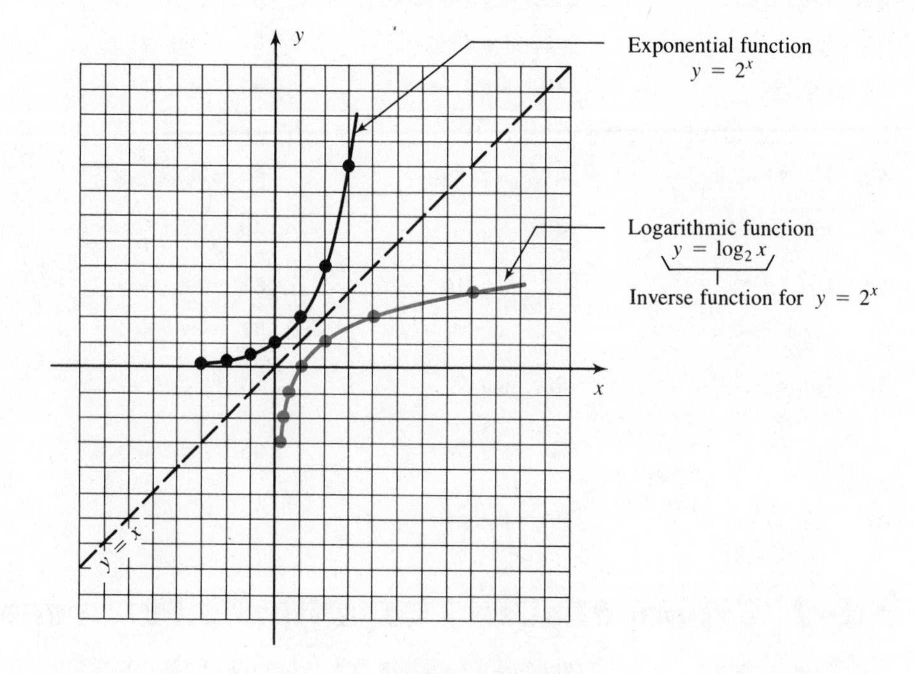

Figure 10–2A

For the exponential function $y = 2^x$, its inverse function is $x = 2^y$ (Section 7–6).

Exponential Function		*Logarithmic Function*
$x = 2^y$	$\Longleftrightarrow$	$y = \log_2 x$

$y = \log_2 x$ is the inverse function for $y = 2^x$

Example 2 Graph the logarithmic function $y = \log_2 x$

Since $y = \log_2 \Longleftrightarrow x = 2^y$, we can draw the graph of the logarithmic function by drawing the graph of its equivalent exponential function $x = 2^y$
Calculate the value of the function $x = 2^y$ for integral values of y from -3 to 3.

Next, plot the points listed in the table of values and draw a smooth curve through them (joining the points in order from left to right). The graph of $y = \log_2 x$ is the curve drawn with the solid color line in Figure 10–2A.

$$x = 2^y$$

$$x = 2^{-3} = \frac{1}{2^3} = \frac{1}{8}$$

$$x = 2^{-2} = \frac{1}{2^2} = \frac{1}{4}$$

$$x = 2^{-1} = \frac{1}{2^1} = \frac{1}{2}$$

$$x = 2^0 = 1$$

$$x = 2^1 = 2$$

$$x = 2^2 = 4$$

$$x = 2^3 = 8$$

x	y
$\frac{1}{8}$	-3
$\frac{1}{4}$	-2
$\frac{1}{2}$	-1
1	0
2	1
4	2
8	3

Mirror Image

Since $x = 2^y$ is the *inverse* function for $y = 2^x$, $y = \log_2 x$ is the **mirror image** of $y = 2^x$ with respect to the 45° line, $y = x$.

In general, the graph of any logarithmic function for which the base is greater than 1 has the appearance of the curve shown in Figure 10–2B.

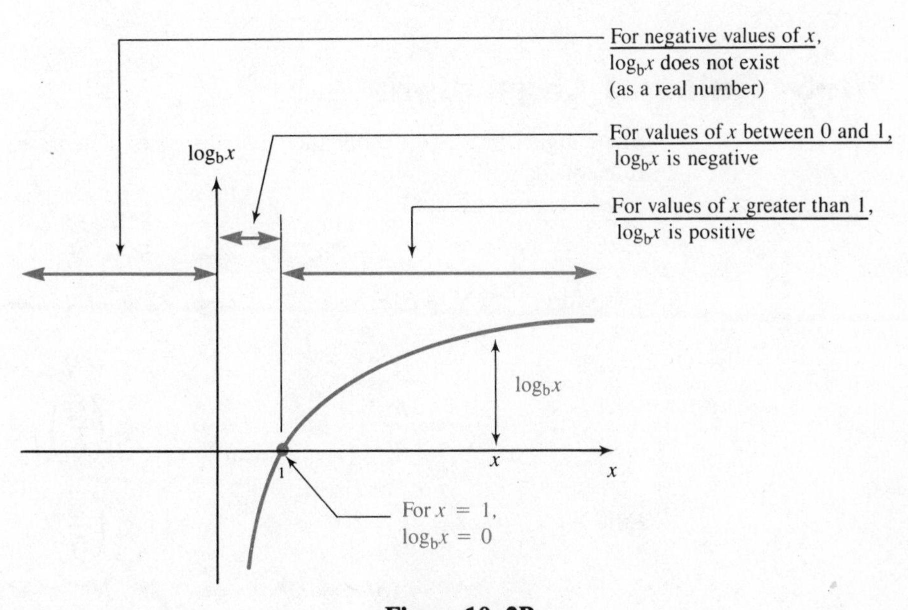

Figure 10–2B

It can also be seen in Figure 10–2B that:

1. Since $\log_b x$ is a *function, if two numbers are equal, then their logarithms must be equal.* This fact is used in solving logarithmic equations (Section 10–7).
2. The logarithmic function is an *increasing function.* This means: *as x gets larger, $\log_b x$ also gets larger.*
3. The $\log_b x$ does not exist (as a real number) if x is negative or zero.

EXERCISES 10-2

In Exercises 1–8 graph each function.

1. Graph $y = 4^x$

2. Graph $y = 2^{\frac{x}{2}}$

3. Graph $y = -2^x$

4. Graph $y = 2^{-x}$

5. Graph $y = \frac{1}{2}(2^x)$

6. Graph $y = 2(2^x)$

7. Graph $y = \left(\frac{3}{2}\right)^x$

8. Graph $y = \left(\frac{5}{2}\right)^x$

9. Graph $y = \log_2 x$ and $y = \log_{10} x$ on the same set of axes.

10. Graph $y = 2^x$ and $y = 10^x$ on the same set of axes.

11. Graph the exponential function $y = 3^x$ and its inverse function $y = \log_3 x$ on the same set of axes.

12. Graph $y = 5^x$ and its inverse function on the same set of axes.

10–3 Rules of Logarithms

Since logarithms are exponents, the rules of exponents enable us to derive the rules of logarithms.

Let
$$x = \log_b M \iff b^x = M$$
$$y = \log_b N \iff b^y = N$$

Therefore, $$MN = b^x b^y = b^{x+y} \iff \log_b MN = x + y$$

Rule 1 $$\log_b MN = \log_b M + \log_b N$$

$$\frac{M}{N} = \frac{b^x}{b^y} = b^{x-y} \iff \log_b \left(\frac{M}{N}\right) = x - y$$

Rule 2 $$\log_b \left(\frac{M}{N}\right) = \log_b M - \log_b N$$

$$N^p = (b^y)^p = b^{yp} \iff \log_b N^p = yp = py$$

Rule 3 $$\log_b N^p = p \log_b N$$

These rules of logarithms can be summarized as follows:

THE RULES OF LOGARITHMS

1. $\log_b MN = \log_b M + \log_b N$
2. $\log_b \left(\dfrac{M}{N}\right) = \log_b M - \log_b N$
3. $\log_b N^p = p \log_b N$

$$\text{where} \begin{cases} M \text{ and } N \text{ are positive real numbers} \\ b > 1 \\ p \text{ is any real number} \end{cases}$$

Example 1 Transforming logarithmic expressions by using the rules of logarithms

(a) $\log_{10}(56)(107) = \log_{10}56 + \log_{10}107$ — Rule 1

(b) $\log_{10}\left(\dfrac{275}{89}\right) = \log_{10}275 - \log_{10}89$ — Rule 2

(c) $\log_{10}(37)^2 = 2 \log_{10}37$ — Rule 3

(d) $\log_{10}\sqrt{5} = \log_{10}5^{1/2} = \dfrac{1}{2}\log_{10}5$ — Rule 3

(e) $\log_{10}\left[\dfrac{(57)(23)}{101}\right] = \log_{10}(57)(23) - \log_{10}101$ — Rule 2

$= \log_{10}57 + \log_{10}23 - \log_{10}101$ — Rule 1

(f) $\log_{10}\left[\dfrac{(49)(19)^3}{(1.04)^7}\right] = \log_{10}(49)(19)^3 - \log_{10}(1.04)^7$ — Rule 2

$= \log_{10}49 + \log_{10}(19)^3 - \log_{10}(1.04)^7$ — Rule 1

$= \log_{10}49 + 3 \log_{10}19 - 7 \log_{10}1.04$ — Rule 3

(g) $\log_{10}\left(\dfrac{\sqrt[5]{21.4}}{(3.5)^4}\right) = \log_{10}(21.4)^{1/5} - \log_{10}(3.5)^4$ — Rule 2

$= \dfrac{1}{5}\log_{10}21.4 - 4 \log_{10}3.5$ — Rule 3

Example 2 Given $\begin{Bmatrix} \log_{10}2 = 0.301 \\ \log_{10}3 = 0.477 \end{Bmatrix}$*, find:

(a) $\log_{10}6 = \log_{10}(2)(3) = \log_{10}2 + \log_{10}3$ — Rule 1

$= 0.301 + 0.477 = 0.778$

*The reason why $\log_{10}2 = 0.301$ and $\log_{10}3 = 0.477$ is explained in Section 10–4.

(b) $\log_{10}1.5 = \log_{10}\left(\frac{3}{2}\right) = \log_{10}3 - \log_{10}2$ Rule 2

$= 0.477 - 0.301 = 0.176$

(c) $\log_{10}8 = \log_{10}2^3 = 3\log_{10}2$ Rule 3

$= 3(0.301) = 0.903$

(d) $\log_{10}\sqrt{3} = \log_{10}3^{1/2} = \frac{1}{2}\log_{10}3$ Rule 3

$= \frac{1}{2}(0.477) = 0.2385 \doteq 0.239$

(e) $\log_{10}5 = \log_{10}\left(\frac{10}{2}\right) = \log_{10}10 - \log_{10}2$ Rule 2

$= \quad 1 \quad - 0.301 = 0.699$

↑ The logarithm of the base is 1

The rules of logarithms can be used to simplify some algebraic expressions.

Example 3

(a) $\log_b 5x + 2\log_b x = \log_b 5x + \log_b x^2$ Rule 3

$= \log_b(5x)(x^2)$ Rule 1

$= \log_b 5x^3$

(b) $\frac{1}{2}\log_b x - 4\log_b y = \log_{10} x^{1/2} - \log_b y^4$ Rule 3

$= \log_b\left(\frac{\sqrt{x}}{y^4}\right)$ Rule 2

A Word of Caution. A common mistake students make is to think that

$\log_b(m + n) = \log_b m + \log_b n$ This statement is incorrect

Since $m + n \neq mn$

then $\log_b(m + n) \neq \log_b mn = \log_b m + \log_b n$

Correct — Rule 1

Therefore $\log_b(m + n) \neq \log_b m + \log_b n$. ■

EXERCISES 10-3

In Exercises 1–14 transform each expression by using the rules of logarithms as was done in Example 1.

1. $\log_{10}(31)(7)$
2. $\log_{10}(17)(29)$
3. $\log_{10}\left(\frac{41}{13}\right)$
4. $\log_{10}\left(\frac{19}{23}\right)$
5. $\log_{10}(19)^3$
6. $\log_{10}(7)^4$
7. $\log_{10}\left[\frac{(17)(31)}{29}\right]$
8. $\log_{10}\left[\frac{(7)(11)}{13}\right]$

9. $\log_{10}\sqrt[5]{75}$ **10.** $\log_{10}\sqrt[4]{38}$ **11.** $\log_{10}\left[\frac{53}{(11)(19)^2}\right]$ **12.** $\log_{10}\left[\frac{29}{(31)^3(47)}\right]$

13. $\log_{10}\left[\frac{35\sqrt{73}}{(1.06)^8}\right]$ **14.** $\log_{10}\left[\frac{27\sqrt{31}}{(1.03)^{10}}\right]$

In Exercises 15–28 find the value of each logarithm, given that $\log_{10}2 = 0.301$, $\log_{10}3 = 0.477$, and $\log_{10}7 = 0.845$. (Remember, $\log_{10}10 = 1$.)

15. $\log_{10}14$ **16.** $\log_{10}21$ **17.** $\log_{10}\left(\frac{9}{7}\right)$ **18.** $\log_{10}\left(\frac{7}{4}\right)$

19. $\log_{10}\sqrt{27}$ **20.** $\log_{10}\sqrt{8}$ **21.** $\log_{10}90$ **22.** $\log_{10}7^3$

23. $\log_{10}(36)^2$ **24.** $\log_{10}(98)^2$ **25.** $\log_{10}\sqrt[4]{3}$ **26.** $\log_{10}\sqrt[5]{7}$

27. $\log_{10}6000$ **28.** $\log_{10}1400$

In Exercises 29–40 write each expression as a single logarithm and simplify.

29. $\log_b x + \log_b y$ **30.** $4 \log_b x + 2 \log_b y$

31. $2 \log_b x - 3 \log_b y$ **32.** $\frac{1}{2}\log_b x^4$

33. $3 (\log_b x - 2 \log_b y)$ **34.** $\frac{1}{3}\log_b y^3$

35. $3 \log_e v + \log_e v^2 - \log_e v$ **36.** $\log_b\left(\frac{6}{7}\right) - \log_b\left(\frac{27}{4}\right) + \log_b\left(\frac{21}{16}\right)$

37. $\log_a(x^2 - y^2) - 3 \log_a(x + y)$ **38.** $\log_e(x^2 - z^2) - \log_e(x - z)$

39. $2 \log_b 2xy - \log_b 3xy^2 + \log_b 3x$ **40.** $2 \log_a 3xy - \log_a 6x^2y^2 + \log_a 2y^2$

10–4 Common Logarithms

There are two systems of logarithms in widespread use:

1. Common logarithms (base 10)
2. Natural logarithms (base $e = 2.71828\ldots$)

In this section we discuss common logarithms and the use of the table of common logarithms (Table II, inside back cover). Natural logarithms are discussed briefly in Section 10–8.

First, we consider logarithms of powers of 10. *The logarithm of a number is the exponent to which the base must be raised to give that number.*

$\log_{10}10^3 = 3$ The base 10 must be raised to the exponent 3 to give the number 10^3

$\log_{10}10^2 = 2$ The base 10 must be raised to the exponent 2 to give the number 10^2

$\log_{10}10^1 = 1$

$\log_{10}10^0 = 0$

$\log_{10}10^{-1} = -1$

$\log_{10}10^{-2} = -2$

$\vdots$

RULE 4

In general, $\log_{10} 10^k = k$

Scientific Notation Any decimal number can be written in **scientific notation.** To write a number in scientific notation, place one nonzero digit to the left of the decimal point and then multiply this number by the appropriate power of 10.

TO WRITE A NUMBER IN SCIENTIFIC NOTATION

1. Place a caret ($\wedge$) to the right of the first nonzero digit.
2. Draw an arrow *from the caret to the actual decimal point.*
3. *The absolute value of the exponent of 10* is the number of digits separating the caret and the actual decimal point.
4. *The sign of the exponent of 10*
   is $\begin{cases} \text{positive if the arrow points right } \rightarrow \\ \text{negative if the arrow points left } \leftarrow \end{cases}$
5. Write the number with the decimal point after the first nonzero digit and multiply by the power of 10 found in Steps 3 and 4.

Example 1 Writing decimal numbers in scientific notation

	Decimal notation		*Scientific notation*
(a)	245 =	$2_\wedge 4\,5.$ (arrow →) =	2.45×10^2
(b)	2.45 =	$2_\wedge.\,4\,5$ =	2.45×10^0
(c)	0.0245 =	$0.\,0\,2_\wedge 4\,5$ (arrow ←) =	2.45×10^{-2}
(d)	92,900,000 =	$9_\wedge 2\,9\,0\,0\,0\,0\,0.$ (arrow →) =	9.29×10^7
(e)	0.0056 =	$0.\,0\,0\,5_\wedge 6$ (arrow ←) =	5.6×10^{-3}
(f)	684.5 =	$6_\wedge 8\,4.\,5$ (arrow →) =	6.845×10^2

Number between 1 and 10 (6.845) — Power of 10 (10^2)

Note. A number written in scientific notation is the *product of a number between 1 and 10, and a power of 10.* ■

The Two Parts of a Common Logarithm

The logarithm of a number is made of two parts:

1. *An integer part* called the **characteristic.** The characteristic is the exponent of 10 when the number is written in scientific notation.
2. *A decimal part* called the **mantissa** (found in Table II, inside the back cover). The mantissa is the logarithm of the number between 1 and 10 when the number is written in scientific notation.

Example 2 Find $\log_{10}727$

$\log_{10}7_\wedge 27. = \log_{10}7.27 \times 10^2$ — Because $727 = 7.27 \times 10^2$ in scientific notation

$= \log_{10}7.27 + \log_{10}10^2$ — Rule 1

$= \boxed{\log_{10}7.27} + 2$ — Rule 4

This can be found from Table II (inside the back cover)

$= .8615 + 2 = 2.8615$

Mantissa: .8615; Characteristic: 2

Example 3 Find $\log_{10}0.0438$

$\log_{10}0.04_\wedge 38 = \log_{10}4.38 \times 10^{-2}$

$= \log_{10}4.38 + \log_{10}10^{-2}$

$= \log_{10}4.38 + (-2)$

$= .6415 + (-2)$ — For computational purposes, it is better to write negative characteristics as the difference of two numbers: $-2 = 8 - 10$

$= .6415 + (8 - 10)$

$= (.6415 + 8) - 10$

$= 8.6415 - 10$

Characteristic: 8 … − 10

10-4A FINDING THE CHARACTERISTIC

Example 4 Writing the characteristic of the logarithm of a number

	Number	Characteristic	Written
(a)	$2_\wedge 45,000.$	5	5.
(b)	$7_\wedge 6.3$	1	1.
(c)	$0.5_\wedge 06$	−1	9. ______ − 10
(d)	$0.008_\wedge 21$	−3	7. ______ − 10

	Number	*Characteristic*	*Written*
(e)	0.000004‸79	−6	4. ____ − 10
(f)	1.83×10^4	4	4. ____
(g)	2.36×10^{-7}	−7	3. ____ − 10

↑ Mantissa goes here

EXERCISES 10–4A

In Exercises 1–6 write each decimal number in scientific notation.

1. 28.56 **2.** 375.4 **3.** 0.06184 **4.** 0.003056
5. 3700.5 **6.** 0.000904

In Exercises 7–22 write the characteristic of the logarithm of each number.

7. 386 **8.** 27 **9.** 5.67 **10.** 30.4
11. 0.516 **12.** 0.089 **13.** 93,000,000 **14.** 186,000
15. 0.0000806 **16.** 0.000777 **17.** 78,000 **18.** 1400
19. 2.06×10^5 **20.** 3.55×10^4 **21.** 7.14×10^{-3} **22.** 8.96×10^{-5}

10–4B FINDING THE MANTISSA

Hereafter, *when the logarithm of a number is written without giving the base, the base is understood to be 10.* We write log N instead of $\log_{10} N$.

Figure 10–4 is part of Table II (inside the back cover).

Example 5 Find the mantissa for log 5.74

Third digit of 5.74

N	0	1	2	3	4	5	6	7	8	9
5‸5	.7404	.7412	.7419	.7427	.7435	.7443	.7451	.7459	.7466	.7474
5‸6	.7482	.7490	.7497	.7505	.7513	.7520	.7528	.7536	.7543	.7551
5‸7	.7559	.7566	.7574	.7582	.7589	.7597	.7604	.7612	.7619	.7627
5‸8	.7634	.7642	.7649	.7657	.7664	.7672	.7679	.7686	.7694	.7701
5‸9	.7709	.7716	.7723	.7731	.7738	.7745	.7752	.7760	.7767	.7774
6‸0	.7782	.7789	.7796	.7803	.7810	.7818	.7825	.7832	.7839	.7846
6‸1	.7853	.7860	.7868	.7875	.7882	.7889	.7896	.7903	.7910	.7917

First two digits of 5.74

.7589 is the *mantissa* of log 5.74

Figure 10–4

Example 6 Find log 8360

$$\log 8{,}360. \Rightarrow \text{characteristic} = 3$$

N	0	1	2	3	4	5	6	7	8	9
7‸8	.8921	.8927	.8932	.8938	.8943	.8949	.8954	.8960	.8965	.8971
7‸9	.8976	.8982	.8987	.8993	.8998	.9004	.9009	.9015	.9020	.9025
8‸0	.9031	.9036	.9042	.9047	.9053	.9058	.9063	.9069	.9074	.9079
8‸1	.9085	.9090	.9096	.9101	.9106	.9112	.9117	.9122	.9128	.9133
8‸2	.9138	.9143	.9149	.9154	.9159	.9165	.9170	.9175	.9180	.9186
8‸3	.9191	.9196	.9201	.9206	.9212	.9217	.9222	.9227	.9232	.9238
8‸4	.9243	.9248	.9253	.9258	.9263	.9269	.9274	.9279	.9284	.9289

—Mantissa

Therefore, log 8‸360 = 3.9222

Characteristic

The work of finding the logarithm of a number can be conveniently arranged as shown in Example 7.

Example 7 Find log 0.0429

1. *Find characteristic:* log 0.04‸29 = 8 − 10
2. *Find mantissa:* = 8.6325 − 10

Mantissa from Table II for 4‸29

EXERCISES 10-4B

Find each logarithm.

1. log 754 **2.** log 186 **3.** log 17 **4.** log 29

5. log 3350 **6.** log 4610 **7.** log 7000 **8.** log 200

9. log 0.0604 **10.** log 0.0186 **11.** log 0.905 **12.** log 0.306

13. log 58.9 **14.** log 36.7 **15.** log (5.77×10^{-4}) **16.** log (3.96×10^{3})

17. log (5.64×10^{3}) **18.** log (2.14×10^{-4})

10-4C INTERPOLATION

Sometimes we want the logarithm of a number that lies *between* two consecutive numbers in Table II. The process of finding such a number is called **interpolation.**

Example 8 Find log 29.38

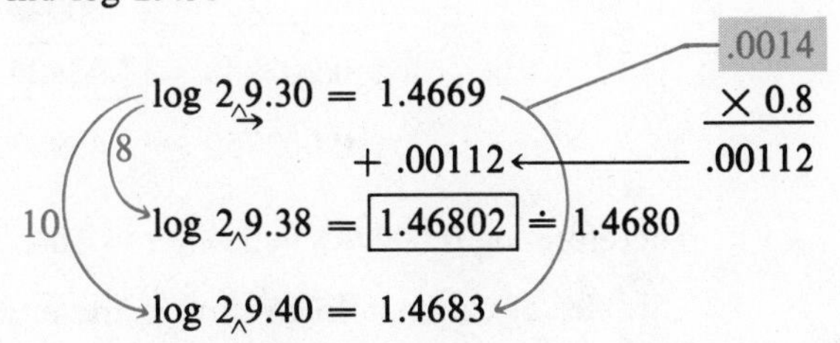

2‸938 lies *between* 2‸93 and 2‸94

N	0	1	2	3	4	5	6	7	8	9
2‸8	.4472	.4487	.4502	.4518	.4533	.4548	.4564	.4579	.4594	.4609
2‸9	.4624	.4639	.4654	.4669	.4683	.4698	.4713	.4728	.4742	.4757
3‸0	.4771	.4786	.4800	.4814	.4829	.4843	.4857	.4871	.4886	.4900
3‸1	.4914	.4928	.4942	.4955	.4969	.4983	.4997	.5011	.5024	.5038
3‸2	.5051	.5065	.5079	.5092	.5105	.5119	.5132	.5145	.5159	.5172
3‸3	.5185	.5198	.5211	.5224	.5237	.5250	.5263	.5276	.5289	.5307
3‸4	.5315	.5328	.5340	.5353	.5366	.5378	.5391	.5403	.5416	.5428

Mantissa for 2‸938 lies *between* .4669 and .4683

Since 29.38 is "eight-tenths of the way" from 29.30 to 29.40, we assume that log 29.38 is "eight-tenths of the way" from 1.4669 to 1.4683. Therefore we add 0.8 × 14 = 11.2 to 1.4669, and get log 29.38 = 1.46802 ≐ 1.4680.

When a four-place log table is used, the mantissa is rounded off to four decimal places.

Example 9 Find log 0.002743

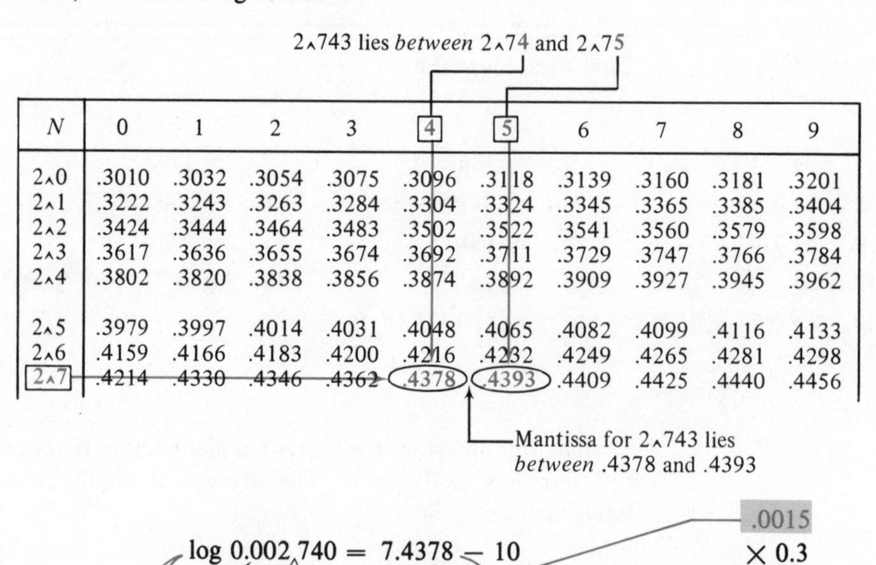

2‸743 lies *between* 2‸74 and 2‸75

N	0	1	2	3	4	5	6	7	8	9
2‸0	.3010	.3032	.3054	.3075	.3096	.3118	.3139	.3160	.3181	.3201
2‸1	.3222	.3243	.3263	.3284	.3304	.3324	.3345	.3365	.3385	.3404
2‸2	.3424	.3444	.3464	.3483	.3502	.3522	.3541	.3560	.3579	.3598
2‸3	.3617	.3636	.3655	.3674	.3692	.3711	.3729	.3747	.3766	.3784
2‸4	.3802	.3820	.3838	.3856	.3874	.3892	.3909	.3927	.3945	.3962
2‸5	.3979	.3997	.4014	.4031	.4048	.4065	.4082	.4099	.4116	.4133
2‸6	.4159	.4166	.4183	.4200	.4216	.4232	.4249	.4265	.4281	.4298
2‸7	.4214	.4330	.4346	.4362	.4378	.4393	.4409	.4425	.4440	.4456

Mantissa for 2‸743 lies *between* .4378 and .4393

log 0.002‸740 = 7.4378 − 10

+ .00045

log 0.002743 = 7.43825 − 10 ≐ 7.4382 − 10

log 0.002750 = 7.4393 − 10

3

10

.0015
× 0.3
.00045

Therefore log 0.002743 = 7.4383 − 10.

Tables of common logarithms come in different accuracies. The mantissa of the logarithm of a number is usually a never-ending decimal. Some tables have the mantissas rounded off to four places, others to five places, etc. Our Table II is a four-place table. If you use other than a four-place table to solve the problems in this chapter, you may get slightly different answers.

We will follow the customary practice of using the equal sign (=) instead of the approximately equal sign ($\doteq$) when writing logarithms of numbers.

The logarithms of numbers can be found by means of a calculator (with a LOG key) as well as by using tables. If you have a calculator with a LOG key, we suggest that you use your calculator to check logarithms found by using the tables. One difference between logarithms obtained by calculator and those obtained from Table II occurs when the characteristic of the logarithm is negative. In Example 9,

$$\log 0.002743 = 7.43825 - 10$$

was found by using Table II. If this same logarithm were found by depressing the LOG key of a calculator, we would get

$$\log 0.002743 = -2.56175$$

because

$$7.43825 - 10 = -2.56175$$

EXERCISES 10-4C

Find each logarithm.

1. log 23.35 **2.** log 27.85 **3.** log 3.062 **4.** log 4.098

5. log 0.06644 **6.** log 0.5839 **7.** log 150.7 **8.** log 20.88

9. log 186,300 **10.** log 92,840,000 **11.** log 0.8006 **12.** log 0.07093

13. log 0.004003 **14.** $\log (1.756 \times 10^{-6})$ **15.** $\log (8.375 \times 10^{6})$ **16.** $\log (3.875 \times 10^{-5})$

17. log 324.38 **18.** log 75.062

10–5 Common Antilogarithms

In Section 10–4 we discussed finding the logarithm of a given number.

Number ⟶ Logarithm

Given N, *find* $\log N$

In this section we discuss the inverse operation: finding the number when you know its logarithm.

Logarithm ⟶ Number

Given $\log N$, *find* N

Example 1 If $\log N = 3.6263$, find N

1. Locate the mantissa .6263 in the body of Table II, and find the number $4_{\wedge}23$ by the method shown in the figure below.

2. Use the characteristic 3 to locate the actual decimal point. Therefore

$$N = 4_\wedge 230. = 4230$$

3

N	0	1	2	3	4	5	6	7	8	9
4∧0	.6021	.6031	.6042	.6053	.6064	.6075	.6085	.6096	.6107	.6117
4∧1	.6128	.6138	.6149	.6160	.6170	.6180	.6191	.6201	.6212	.6222
4∧2	.6232	.6243	.6253	.6263	.6274	.6284	.6294	.6304	.6314	.6325
4∧3	.6335	.6345	.6355	.6365	.6375	.6385	.6395	.6405	.6415	.6425
4∧4	.6435	.6444	.6454	.6464	.6474	.6484	.6493	.6503	.6513	.6522

Mantissa

Antilogarithm

If the logarithm of N is L, then N is the **antilogarithm** of L:

$$\log N = L \iff N = \text{antilog } L$$

Example 2 If $N = \text{antilog } (7.8675 - 10)$, find N

$$\log N = 7.8675 - 10$$

1. Locate the mantissa .8675 in the body of Table II, and find the number $7_\wedge 37$ by the method shown in the figure below.
2. Use the characteristic $7 - 10 = -3$ to locate the actual decimal point. Therefore

$$N = .007_\wedge 37 = 0.00737$$

N	0	1	2	3	4	5	6	7	8	9
7∧0	.8451	.8457	.8463	.8470	.8476	.8482	.8488	.8494	.8500	.8506
7∧1	.8513	.8519	.8525	.8531	.8537	.8543	.8549	.8555	.8561	.8567
7∧2	.8573	.8579	.8585	.8591	.8597	.8603	.8609	.8615	.8621	.8627
7∧3	.8633	.8639	.8645	.8651	.8657	.8663	.8669	.8675	.8681	.8686
7∧4	.8692	.8698	.8704	.8710	.8716	.8722	.8727	.8733	.8739	.8745

Mantissa

When the mantissa falls between two consecutive numbers in Table II, we must *interpolate*.

Example 3 Find antilog 4.7129

Let $N = \text{antilog } 4.7129$
then $\log N = 4.7129$

N	0	1	2	3	4	5	6	7	8	9
$5_\wedge 0$	.6990	.6998	.7007	.7016	.7024	.7033	.7042	.7050	.7059	.7067
$5_\wedge 1$	.7076	.7084	.7093	.7101	.7110	.7118	.7126	.7135	.7143	.7152
$5_\wedge 2$	.7160	.7168	.7177	.7185	.7193	.7202	.7210	.7218	.7226	.7235
$5_\wedge 3$	.7243	.7251	.7259	.7267	.7275	.7284	.7292	.7300	.7308	.7316
$5_\wedge 4$	.7324	.7332	.7340	.7348	.7356	.7364	.7372	.7380	.7388	.7396

Mantissa .7129 lies *between* .7126 and .7135

Locate the mantissa .7129 *between* .7126 and .7135. The mantissa .7126 has antilog $5_\wedge 16$, and mantissa .7135 has antilog $5_\wedge 17$.

$$\log\ 5_\wedge 160 = .7126$$

$$\frac{3}{9} \times 10 \doteq 3 \longrightarrow +3 \qquad 3$$

$$\log\ \boxed{5_\wedge 163} = .7129$$

$$10 \qquad\qquad 9$$

$$\log\ 5_\wedge 170 = .7135$$

Since mantissa .7129 is "three-ninths of the way" from .7126 to .7135, we assume that N is "three-ninths of the way from $5_\wedge 160$ to $5_\wedge 170$." Therefore we add $\frac{3}{9} \times 10 \doteq 3$ to the last place of $5_\wedge 160$ and get $5_\wedge 163$.

Then $$N = 5_\wedge 1630. = 51{,}630$$

4 ← Characteristic

Therefore antilog 4.7129 = 51,630.

Example 4 Find antilog (8.7385 − 10)

$$\log N = 8.7385 - 10$$

Locate the mantissa .7385 between .7380 and .7388. The mantissa .7380 has antilog $5_\wedge 47$, and mantissa .7388 has antilog $5_\wedge 48$.

$$\frac{5}{8} \times 10 = 6.25 \qquad \log\ 5_\wedge 470 = .7380$$

$$\doteq 6 \longrightarrow +6 \qquad 5$$

$$\log\ \boxed{5_\wedge 476} = .7385$$

$$10 \qquad\qquad 8$$

$$\log\ 5_\wedge 480 = .7388$$

Therefore

$$N = .05_\wedge 476 = 0.05476$$

−2 ← Characteristic = 8 − 10 = −2

EXERCISES 10-5

In Exercises 1–10 find the antilogarithms.

1. antilog 3.5478
2. antilog 2.4409
3. antilog 0.9605
4. antilog 0.8848
5. antilog (9.2529 − 10)
6. antilog (8.1271 − 10)
7. antilog 3.5051
8. antilog 2.6335
9. antilog (7.6117 − 10)
10. antilog (6.6010 − 10)

In Exercises 11–18 find N.

11. $\log N = 4.0588$
12. $\log N = 3.0846$
13. $\log N = 6.9900 - 10$
14. $\log N = 7.9596 - 10$
15. $\log N = 1.1685$
16. $\log N = 2.5470$
17. $\log N = 8.6908 - 10$
18. $\log N = 9.7995 - 10$

10–6 Calculating with Logarithms

Logarithms can be used to perform arithmetic calculations.

Exact Numbers

In this chapter we assume all numbers given in examples and exercises are **exact numbers.** Answers are rounded off to the accuracy of Table II (four decimal places).

Example 1 Multiply (37.5)(0.00842)

Let $N = (37.5)(0.00842)$

then $\log N = \log(37.5)(0.00842)$

$= \log 37.5 + \log 0.00842$ Rule 1

$= 1.5740 + 7.9253 - 10$

$\log N = 9.4993 - 10$

$\therefore$* $N = .3_\wedge157 = 0.3157$

$$\begin{array}{r} 1.5740 \\ 7.9253 - 10 \\ \hline 9.4993 - 10 \end{array}$$

TO CALCULATE WITH LOGARITHMS

1. Analyze the problem and make a blank outline.
2. Write all the characteristics in the blank outline.
3. Use Table II to find all the mantissas, and then write them in the blank outline.
4. Carry out the calculations indicated in the outline.

*The symbol $\therefore$ means "therefore."

Example 2 Divide: $\dfrac{6.74}{0.0391}$

Let $N = \dfrac{6.74}{0.0391}$

then $\log N = \log 6.74 - \log 0.0391$ Rule 2

The characteristic 0 is written as 10 − 10 to help with the subtraction

Blank Outline of Procedure	*Actual Calculations*
$\log 6.74 =$	$\log 6.74 = 10.8287 - 10$
$\log 0.0391 =$ ______ (−)	$\log 0.0391 = 8.5922 - 10$ (−)
$\log N =$	$\log N = 2.2365$
$\therefore N =$	$\therefore N = 1_{\wedge}72.4 = 172.4$

Example 3 Find $(1.05)^{10}$

Let $N = (1.05)^{10}$

then $\log N = 10 \log 1.05$ Rule 3

Blank Outline	*Actual Calculations*
$\log 1.05 =$	$\log 1.05 = 0.0212$
$\log N = 10 \log 1.05 =$	$\log N = 10 \log 1.05 = 0.2120$
$N =$	$N = 1_{\wedge}.629$
	$= 1.629$

Example 4 Find $\sqrt[3]{0.506}$

Let $N = \sqrt[3]{0.506} = (0.506)^{1/3}$

then $\log N = \dfrac{1}{3} \log 0.506$ Rule 3

Blank Outline	*Actual Calculations*
$\log 0.506 =$	$\log 0.506 = 9.7042 - 10$
$\log N = \dfrac{1}{3} \log 0.506 =$	$= 29.7042 - 30$
$\therefore N =$	$\log N = \dfrac{1}{3} \log 0.506 = 9.9014 - 10$
	$\therefore N = .7\,968$
	$= 0.7968$

The characteristic 9 − 10 is written as 29 − 30 so that the second term, −30, is exactly divisible by 3

Example 5 Find $N = \dfrac{(1.16)^5(31.7)}{\sqrt{481}\,(0.629)}$

Blank Outline

$\log 1.16 =$
$5 \log 1.16 =$ ⎤
$\log 31.7 =$ ____ ⎦ +
☐ (subtract the box below from this one)

$\log 481 =$
$\frac{1}{2} \log 481 =$ ⎤
$\log 0.629 =$ ____ ⎦ +
☐ (−)

$\log N =$
$N =$

Actual Calculations

$\log 1.16 = 0.0645$
$5 \log 1.16 = 0.3225$ ⎤
$\log 31.7 = \underline{1.5011}$ ⎦ +
1.8236 (subtract the result below from this one)

$\log 481 = 2.6821$
$\frac{1}{2} \log 481 = 1.3411$ ⎤
$\log 0.629 = \underline{9.7987 - 10}$ ⎦ +
$11.1398 - 10$
$= \underline{1.1398}$ (−)

$\log N = 0.6838$
$N = 4.829 = 4.829$

Note. Today the use of calculators and computers makes calculating with logarithms less important than it used to be. However, because logarithms enter into the scientific analyses of many natural phenomena, an understanding of logarithms and their rules continues to be important. ■

EXERCISES 10-6

Use logarithms to perform the calculations. Assume that all the given numbers are *exact* numbers. Round off answers to the accuracy of Table II (four decimal places). We suggest that you use a calculator to verify the results obtained by the logarithmic calculations. There may be a slight difference in the answers obtained by using a calculator and those found by using the table because of their different levels of accuracy.

1. 74.3×0.618

2. 0.314×14.9

3. $\dfrac{562}{21.4}$

4. $\dfrac{651}{30.6}$

5. $(1.09)^5$

6. $(3.4)^4$

7. $\sqrt[3]{0.444}$

8. $\sqrt[4]{0.897}$

9. $\dfrac{(4.92)(25.7)}{388}$

10. $\dfrac{(2.04)^5}{(5.9)(0.66)}$

11. $2.863 + \log 38.46$

12. $\log 786.4 + 3.154$

13. $\dfrac{\log 7.86}{\log 38.4}$

14. $\dfrac{\log 58.4}{\log 2.50}$

15. $\dfrac{(85.3)^3(0.0409)}{\sqrt[5]{7.65}}$

16. $\dfrac{(5.65)\sqrt[6]{175}}{(2.4)^4}$

17. $\sqrt[5]{\dfrac{(5.86)(17.4)}{\sqrt{450}}}$

18. $\sqrt[4]{\dfrac{(39.4)(7.86)}{\sqrt[3]{704}}}$

19. The formula for finding the monthly payment on a homeowner's mortgage is

$$R = \frac{Ai(1+i)^n}{(1+i)^n - 1}$$

where R = Monthly payment
i = Interest rate per month expressed as a decimal
n = Number of months
A = Original amount of mortgage

Find the monthly payment on a 25-year, \$40,000 loan at 9% interest.

20. Find the monthly payment on a 30-year, \$30,000 loan at 9% interest, using the formula in Exercise 19.

10–7 Exponential and Logarithmic Equations

In this section we discuss the application of logarithms and the rules of logarithms in the solution of exponential and logarithmic equations. Although calculating with logarithms is being replaced by the use of hand calculators and computers, the use of logarithms and the rules of logarithms will continue to be important in the solution of exponential and logarithmic equations.

Exponential Equations

An exponential equation is an equation in which the unknown appears in one or more *exponents.*

Example 1 Exponential equations

(b) $(5.26)^{x+1} = 75.4$

Solving Exponential Equations

It is possible to solve some exponential equations if each side can be expressed as powers of the *same* base.

Example 2 Solve: $16^x = \frac{1}{4}$

$$\left.\begin{aligned} (2^4)^x &= \frac{1}{2^2} \\ 2^{4x} &= 2^{-2} \end{aligned}\right\}$$

Expressed both sides as powers of the same base, 2

Therefore $4x = -2$

$$x = -\frac{2}{4} = -\frac{1}{2}$$

When both sides of an exponential equation *cannot* be expressed as powers of the same base, we solve the equation by taking logarithms of both sides. Most exponential equations are of this type.

Example 3 Solve:

$$3^x = 17$$

$$\log 3^x = \log 17$$

$$x \log 3 = \log 17 \quad \text{Rule 3}$$

$$x = \frac{\log 17}{\log 3} = \boxed{\frac{1.2304}{0.4771}} \doteq 2.579$$

This division could be done by logarithms

A Word of Caution. A common mistake students make is to think that

$$\frac{\log 17}{\log 3} = \log\left(\frac{17}{3}\right)$$ This statement is *incorrect.*

$\frac{\log 17}{\log 3}$ *cannot* be evaluated as $\log 17 - \log 3$

Actually $\frac{\log 17}{\log 3} = \frac{1.2304}{0.4771} \doteq 2.579$ ← unequal

whereas $\log 17 - \log 3 = 1.2304 - 0.4771 = 0.7533$ ■

Argument of Log Function

Example 4 **Argument of the Log Function**

(a) $\log(2x + 1)$ $(2x + 1)$ is the *argument* of this log function

(b) $\log 5$ 5 is the *argument* of this log function

Logarithmic Equations

A logarithmic equation is an equation in which the unknown appears in the *argument* of a *logarithm.*

Example 5 Logarithmic equations

(a) $\log(2x + 1) + \log 5 = \log(x + 6)$

(b) $\log x + \log(11 - x) = 1$

Solving Logarithmic Equations

Logarithmic equations can be solved by using the rules of logarithms to simplify both sides of the equation.

Example 6 Solve: $\log(2x + 1) + \log 5 = \log(x + 6)$

$$\log(2x + 1)5 = \log(x + 6) \quad \text{Rule 1}$$

Therefore

$$(2x + 1)5 = (x + 6)$$

$$10x + 5 = x + 6$$

$$9x = 1$$

$$x = \frac{1}{9}$$

Apparent solutions to logarithmic equations must be checked, because it is possible to obtain extraneous roots.

Check for $x = \frac{1}{9}$:

$$\log(2x + 1) + \log 5 = \log(x + 6)$$

$$\log\left[2\left(\frac{1}{9}\right) + 1\right] + \log 5 \stackrel{?}{=} \log\left(\frac{1}{9} + 6\right)$$

$$\log\left(\frac{11}{9}\right) + \log 5 \stackrel{?}{=} \log\left(\frac{55}{9}\right)$$

$$\log\left(\frac{11}{9}\right)5 \stackrel{?}{=} \log\left(\frac{55}{9}\right)$$

$$\log\left(\frac{55}{9}\right) = \log\left(\frac{55}{9}\right) \quad \text{True}$$

Therefore $x = \frac{1}{9}$ is a solution.

Example 7 Solve: $\log(x - 1) + \log(x + 2) = \log 4$

$$\log(x - 1)(x + 2) = \log 4$$

Therefore

$$(x - 1)(x + 2) = 4$$

$$x^2 + x - 2 = 4$$

$$x^2 + x - 6 = 0$$

$$(x - 2)(x + 3) = 0$$

$$x - 2 = 0 \quad | \quad x + 3 = 0$$

$$x = 2 \quad | \quad x = -3$$

Check for $x = 2$:

$$\log(x - 1) + \log(x + 2) = \log 4$$

$$\log(2 - 1) + \log(2 + 2) \stackrel{?}{=} \log 4$$

$$\log 1 + \log 4 \stackrel{?}{=} \log 4$$

$$0 + \log 4 \stackrel{?}{=} \log 4$$

$$\log 4 = \log 4 \quad \text{True}$$

Therefore $x = 2$ *is* a solution.

Check for $x = -3$:

$$\log(x - 1) + \log(x + 2) = \log 4$$

$$\log(-3 - 1) + \log(-3 + 2) \stackrel{?}{=} \log 4$$

$$\log(-4) + \log(-1) \neq \log 4$$

Logarithms of negative numbers are not real numbers

Therefore $x = -3$ *is not* a solution (-3 is an *extraneous root*).

Example 8 Solve: $\log(2x + 3) - \log(x - 1) = 0.73$

$$\log\left(\frac{2x+3}{x-1}\right) = \boxed{\log N = 0.7300}$$

Finding antilog 0.7300 ⟶ $N = 5{,}370 = 5.370$

$$\log\left(\frac{2x+3}{x-1}\right) = \log 5.370$$

Therefore

$$\frac{2x+3}{x-1} = 5.370$$

$$2x + 3 = 5.370(x - 1)$$

$$2x + 3 = 5.370x - 5.370$$

$$8.370 = 3.370x$$

$$\frac{8.370}{3.370} = x$$

$$x \doteq 2.484$$

Check for $x \doteq 2.484$:

$$\log(2x + 3) - \log(x - 1) = 0.73$$

$$\log[2(2.484) + 3] - \log[2.484 - 1] \stackrel{?}{=} 0.73$$

$$\log 7.968 - \log 1.484 \stackrel{?}{=} 0.73$$

$$0.9014 - 0.1715 \stackrel{?}{=} 0.73$$

$$0.7299 \doteq 0.73 \quad \textit{True}$$

Therefore $x \doteq 2.484$ *is* a solution.

EXERCISES 10-7

In Exercises 1–8 solve by using the rules of exponents to express each side as a power of the same base.

1. $27^x = \frac{1}{9}$
2. $8^x = \frac{1}{16}$
3. $4^x = \frac{1}{8}$
4. $9^x = \frac{1}{3^{-2}}$
5. $2^{-3x} = \frac{1}{8}$
6. $5^{3x-2} = 25^x$
7. $25^{2x+3} = 5^{x-1}$
8. $27^{3x-1} = 9^{x+2}$

In Exercises 9–16 solve by taking the logarithms of both sides of the equation. Assume the given numbers to be exact and carry out the work as far as possible using Table II.

9. $2^x = 3$
10. $5^x = 4$
11. $(7.43)^{x+1} = 9.55$
12. $(5.14)^{x-1} = 7.08$
13. $(4.6)^{x+1} = 100$
14. $(34.7)^{2x} = (12.5)^{3x-2}$
15. $(8.71)^{2x+1} = 8.57$
16. $(9.55)^{3x-1} = 3.09$

In Exercises 17–26 solve by using the rules of logarithms. Check your answers.

17. $\log(3x - 1) + \log 4 = \log(9x + 2)$

18. $\log(2x - 1) + \log 3 = \log(4x + 1)$

19. $\log(x + 4) - \log 3 = \log(x - 2)$

20. $\log(2x + 1) - \log 5 = \log(x - 1)$

21. $\log(x + 4) - \log 10 = \log 6 - \log x$

22. $\log(2x + 1) = \log 1 + \log(x + 2)$

23. $\log x + \log(7 - x) = \log 10$

24. $\log x + \log(11 - x) = \log 10$

25. $\log x + \log(x - 3) = \log 4$

26. $2 \log(x + 3) = \log(7x + 1) + \log 2$

In Exercises 27–32 solve by using Table II.

27. $\log(5x + 2) - \log(x - 1) = 0.7782$

28. $\log(8x + 11) - \log(x + 1) = 0.9542$

29. $6.73 \log(x - 1) = 23.2$

30. $27.5 \log(3x + 2)^2 = 5.84$

10–8 Change of Base

There are two systems of logarithms in widespread use for which tables are available:

1. *Common logarithms* (base 10).
2. *Natural logarithms* (base $e \doteq 2.7183$).

Natural logarithms are used in more advanced courses in mathematics. No tables for natural logarithms are provided in this book. However, in this section we show how natural logarithms can be found by using common logarithms.

In some applications, logarithms to a base different from 10 are needed. To find such logarithms, we must be able to **change the base** of a logarithm from 10 to some other number.

Let $x = \log_b N \iff b^x = N$

then $\log_a b^x = \log_a N$ — Took logarithms of both sides to the base a

$x \log_a b = \log_a N$ — Rule 3

$x = \dfrac{\log_a N}{\log_a b}$ — Divided both sides by $\log_a b$

Therefore $\log_b N = \dfrac{\log_a N}{\log_a b}$ — Because $x = \log_b N$

RULE 5

To change a logarithm from base a to base b

$$\log_b N = \frac{\log_a N}{\log_a b}$$

Rule 5 makes it possible to use any table of logarithms available to find the logarithm of a number to a different base.

Example 1 Find $\log_5 51.7$

$$\log_b N = \frac{\log_a N}{\log_a b}$$ Rule 5

$$\log_5 51.7 = \frac{\log_{10} 51.7}{\log_{10} 5}$$ By setting $\begin{cases} b = 5 \\ a = 10 \\ N = 51.7 \end{cases}$

$$\log_5 51.7 = \frac{1.7135}{0.6990} \doteq 2.451$$ Found logarithms in Table II

Example 2 Find $\log_e 51.7$ ($e \doteq 2.7183$)

$$\log_b N = \frac{\log_a N}{\log_a b}$$ Rule 5

$$\log_e 51.7 = \frac{\log_{10} 51.7}{\log_{10} 2.718} = \frac{1.7135}{0.4343} \doteq 3.945$$

EXERCISES 10-8 Find each logarithm.

1. $\log_2 156$
2. $\log_3 231$
3. $\log_5 29.8$
4. $\log_{14} 0.842$
5. $\log_e 3.04$
6. $\log_e 4.08$
7. $\log_e 53.7$
8. $\log_e 0.926$
9. $\log_{6.8} 0.00507$
10. $\log_{8.3} 0.00304$

Chapter Ten Summary

The logarithm of a number N is the exponent l to which the base must be raised to give N (Section 10–1).

Logarithmic form		*Exponential form*
$\log_b N = \ell$	$\Longleftrightarrow$	$b^\ell = N$

The logarithm of a number less than 0 does not exist (as a real number).

The logarithm of a number between 0 and 1 is negative.

The logarithm of a number greater than 1 is positive.

The logarithm of 1 to any base equals 0.

The logarithm of 0 to any base does not exist.

The logarithm of the base equals 1.

The Rules of Logarithms. (Section 10–3)

1. $\log_b MN = \log_b M + \log_b N$

2. $\log_b\left(\frac{M}{N}\right) = \log_b M - \log_b N$

3. $\log_b N^p = p \log_b N$

4. $\log_{10} 10^k = k$ (Section 10–4)

5. $\log_b N = \frac{\log_a N}{\log_a b}$ *Change of base* (Section 10–8).

The common logarithm (base 10) of a number is made up of two parts (Section 10–4):

1. An integer part called the *characteristic*. The characteristic is the exponent of 10 when the number is written in scientific notation.

2. A decimal part called the *mantissa,* which is found in Table II. The mantissa is the logarithm of the number between 1 and 10 when the number is written in scientific notation.

To Calculate with Logarithms. (Section 10–6)

1. Analyze the problem and make a blank outline.

2. Write all the characteristics in the blank outline.

3. Use Table II to find all the mantissas and then write them in the blank outline.

4. Carry out the calculations indicated in the outline.

An exponential equation is an equation in which the unknown appears in one or more exponents (Section 10–7).

To Solve an Exponential Equation. (Section 10–7)

Method I:
1. Express both sides as powers of the same base (if possible).
2. Equate the exponents.
3. Solve the resulting equation for the unknown.

Method II:
1. Take the logarithm of both sides.
2. Use Rule 3 to get the unknown from the exponent.
3. Solve the resulting equation for the unknown.

A logarithmic equation is an equation in which the unknown appears in the argument of a logarithm (Section 10–7).

To Solve a Logarithmic Equation. (Section 10–7)

1. Use the rules of logarithms to write each as the logarithm of a single expresion.

2. Equate the expressions found in Step 1.

3. Solve the resulting equation for the unknown.

4. Check apparent solutions in the given logarithmic equation.

To Change a Base 10 Logarithm to a Base *b* Logarithm. (Section 10–8)

$$\log_b N = \frac{\log_{10} N}{\log_{10} b} \quad \text{where } b \text{ is the base desired}$$

Chapter Ten Diagnostic Test or Review

Allow yourself about one hour to do these problems. Complete solutions for every problem, together with section references, are given in the answer section at the end of the book.

---

1. Write $2^4 = 16$ in logarithmic form.

2. Write $\log_{2.5} 6.25 = 2$ in exponential form.

In Problems 3–7 find the value of the unknown b, N, or x.

3. $\log_4 N = 3$

4. $\log_{10} 10^{-2} = x$

5. $\log_b 6 = 1$

6. $\log_5 1 = x$

7. $\log_{0.5} N = -2$

In Problems 8–10 write each expression as a single logarithm.

8. $\log x + \log y - \log z$

9. $\frac{1}{2} \log x^4 + 2 \log x$

10. $\log (x^2 - 9) - \log(x - 3)$

11. Use the rules of logarithms to solve:

$$\log(3x + 5) - \log 7 = \log(x - 1)$$

12. Use logarithms to solve: $x^{1.24} = 302$

In Problems 13 and 14 use logarithms to perform the calculations.

13. $(1.71)^4$

14. $\dfrac{\sqrt[3]{8.84}}{30.3}$

15. Find $\log_2 718$.

16. Graph $y = 7^x$ and its inverse logarithmic function on the same set of axes.

11 SEQUENCES AND SERIES

In this chapter we introduce *sequences* and *series*. There are many applications of sequences and series in the sciences and in the mathematics of finance. For example, formulas for the calculation of interest, annuities, and mortgage loans are derived using series. The numbers in the logarithm and square root tables in this book were calculated using series. This chapter is just a brief introduction to a very extensive and important part of mathematics.

11–1 Basic Definitions

SEQUENCES

When you look at the following set of numbers, do you know what number comes next?

$$30, 40, 50, 60, 70, \ldots$$

What number comes next in the following set of numbers?

$$7, 10, 13, 16, \ldots$$

These sets of numbers are examples of *sequences* of numbers. A **sequence** of numbers is a set of numbers arranged in a definite order. The numbers that make up the sequence are called the **terms** of the sequence.

A sequence is usually written

$$a_1, a_2, a_3, \ldots, a_n, \ldots$$

where a_1 = First term
a_2 = Second term
a_3 = Third term
⋮

General Term

a_n = ***n*th term** (also called the **general term** of the sequence)
⋮

The *subscript* of each term represents the *term number*.

Finite and Infinite Sequences

If in counting the terms of a sequence, the counting comes to an end, the sequence is called a **finite sequence.** The last term of a finite sequence is represented by the symbol a_n or ℓ.

If in counting the terms of a sequence, the counting never comes to an end, the sequence is called an **infinite sequence.**

The symbol a_n represents the nth term of *any* sequence.

Example 1 Finite and infinite sequences

(a) 15, 10, 5, 0, −5, −10.
↑ The period indicates that the sequence ends here

This is a finite sequence. Each term after the first is found by adding −5 to the preceding term.

(b) 0, 1, 2, 3, . . .
↑ The 3 dots indicate that the sequence never ends

The set of whole numbers is an *infinite* sequence. Each term except the first term is found by adding 1 to the preceding term.

Sequences in Which $a_n = f(n)$

In the sequences discussed so far it is possible to discover each succeeding term *by inspection.* For many sequences this is not possible.

Sometimes each term of a sequence is a function of n, where n is the term number; n is a natural number.

$$a_n = f(n)$$

This means that a sequence can be thought of as a *function* with domain the set of natural numbers.

Example 2 Sequences in which each term is a function of n

(a) If $a_n = f(n) = \frac{n}{2}$,

then $a_1 = \frac{n}{2} = \frac{(1)}{2} = \frac{1}{2}$

$a_2 = \frac{(2)}{2} = 1$

$a_3 = \frac{(3)}{2} = \frac{3}{2}$

$\vdots$

(b) $a_n =$

$a_1 = \frac{(1)+1}{4} = \frac{2}{4} = \frac{1}{2}$

$a_2 = \frac{(2)+1}{4} = \frac{3}{4}$

$\vdots$

$a_{11} = \frac{(11)+1}{4} = \frac{12}{4} = 3$

$\vdots$

SERIES

Finite and Infinite Series

A **series** is the indicated sum of a finite or infinite sequence of terms. It is a **finite** or an **infinite series** according to whether the number of terms is finite or infinite.

An infinite series is usually written

$$a_1 + a_2 + a_3 + \cdots + a_n + \cdots$$

Partial Sum

A **partial sum** of a series is the sum of a finite number of consecutive terms of the series, beginning with the first term.

$S_1 = a_1$ — *First* partial sum

$S_2 = a_1 + a_2$ — *Second* partial sum

$S_3 = a_1 + a_2 + a_3$ — *Third* partial sum

$\vdots$

$S_n = a_1 + a_2 + \cdots + a_n$ — nth partial sum

$\vdots$

Example 3 Given the infinite *sequence*, $f(n) = \frac{1}{n}$:

The *sequence* is: $\frac{1}{1}, \frac{1}{2}, \frac{1}{3}, \ldots$

The *series for this sequence* is: $1 + \frac{1}{2} + \frac{1}{3} + \cdots$

The *partial sums* are:

$$\begin{aligned} S_1 &= 1 &&= 1 \\ S_2 &= 1 + \frac{1}{2} &&= \frac{3}{2} \\ S_3 &= 1 + \frac{1}{2} + \frac{1}{3} &&= \frac{11}{6} \\ &\vdots \end{aligned}$$

EXERCISES 11-1

In Exercises 1–6 for each given sequence, determine the next three terms by inspection.

1. 10, 15, 20, __, __, __, . . .

2. 8, 11, 14, __, __, __, . . .

3. 15, 13, 11, __, __, __, . . .

4. $1, \frac{4}{3}, \frac{5}{3}$, __, __, __, . . .

5. 20, 16, 12, __, __, __, . . .

6. $\frac{1}{2}, \frac{3}{4}, 1$, __, __, __, . . .

In Exercises 7–14 use the given general term to write the terms specified.

7. $a_n = f(n) = n + 4$; first three terms

8. $a_n = f(n) = 1 - 2n$; first four terms

9. $a_n = \frac{1-n}{n}$; first four terms

10. $a_n = \frac{n(n-1)}{2}$; first three terms

11. $a_n = \frac{n}{1-2n}$; first three terms

12. $a_n = \frac{2-n}{1+n^2}$; first three terms

13. $a_n = n^2 - 1$; first three terms

14. $a_n = n^3 + 1$; first four terms

In Exercises 15–20 find the indicated partial sum by using the given general term.

15. Given: $a_n = 2n - 3$
Find: S_4

16. Given: $a_n = 4 - n$
Find: S_5

17. Given: $a_n = \frac{n-1}{n+1}$
Find: S_3

18. Given: $a_n = \frac{2n-1}{5-n}$
Find: S_4

19. Given: $a_n = 3^n + 2$
Find: S_3

20. Given: $a_n = \frac{1-2^n}{3}$
Find: S_4

11–2 Arithmetic Progressions

ARITHMETIC SEQUENCES

Common Difference

An **arithmetic progression** (*arithmetic sequence*) is a sequence in which each term after the first is found by *adding* the same fixed number to the preceding term. The fixed number added is called the **common difference, *d***. *Arithmetic progression is abbreviated* **AP.**

Example 1 Write a six-term arithmetic progression having first term $a_1 = 15$ and common difference $d = -7$

$$a_1 = 15$$
$$a_2 = 15 + (-7) = 8$$
$$a_3 = 8 + (-7) = 1$$
$$a_4 = 1 + (-7) = -6$$
$$a_5 = -6 + (-7) = -13$$
$$a_6 = -13 + (-7) = -20$$

Therefore the AP is 15, 8, 1, -6, -13, -20.

Example 2 Determine which of the sequences are arithmetic progressions

Method. Subtract each term from the following term. If every difference found this way is the same, the sequence is an AP.

(a) $-8, -3, 2, 7, 12$

$$12 - 7 = 5$$
$$7 - 2 = 5$$
$$2 - (-3) = 5$$
$$-3 - (-8) = 5$$

Since all these differences are the same 5, the sequence *is* an AP

(b) $1, 5, 9, 12, 16$

$$16 - 12 = 4$$
$$12 - 9 = 3$$

Since these differences are *not* the same, the sequence is *not* an AP

(c) $(5x - 3), (7x - 4), (9x - 5), \cdots$

$$(9x - 5) - (7x - 4) = 2x - 1$$
$$(7x - 4) - (5x - 3) = 2x - 1$$

AP because differences are the same $(2x - 1)$

In general, an arithmetic progression with first term a_1 and common difference d has the following terms:

$$a_1 = a_1$$
$$a_2 = a_1 + 1d$$
$$a_3 = (a_1 + d) + d = a_1 + 2d$$
$$a_4 = (a_1 + 2d) + d = a_1 + 3d$$
$$\vdots$$
$$a_n = ?$$

Finding the *n*th Term

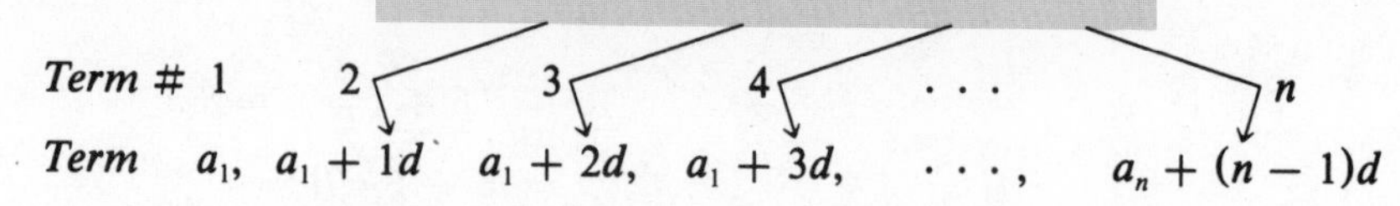

Term #	1	2	3	4	. . .	n
Term	a_1,	$a_1 + 1d$	$a_1 + 2d$,	$a_1 + 3d$,	. . . ,	$a_n + (n - 1)d$

a_n

Therefore:

THE *n*TH TERM OF AN AP

$$a_n = a_1 + (n - 1)d$$

Example 3 Find the twenty-first term of the AP having $a_1 = 23$ and $d = -2$

$$a_n = a_1 + (n - 1)d$$

$$a_{21} = 23 + (21 - 1)(-2)$$

$$a_{21} = 23 + (-40) = -17$$

Example 4 Given an AP with $a_7 = -10$ and $a_{12} = 5$, find a_1 and d

$$a_n = a_1 + (n - 1)d$$

(1) $a_7 = a_1 + (7 - 1)d = -10 \Rightarrow a_1 + 6d = -10$

(2) $a_{12} = a_1 + (12 - 1)d = 5 \Rightarrow a_1 + 11d = 5$

$$5d = 15$$

$$d = 3$$

Substitute $d = 3$ into Equation (1): $a_1 + 6d = -10$

$$a_1 + 6(3) = -10$$

$$a_1 = -28$$

ARITHMETIC SERIES

An **arithmetic series** is the *sum of the terms of an arithmetic progression.* An infinite arithmetic series can be written

$$a_1 + (a_1 + d) + (a_1 + 2d) + \cdots + a_n + \cdots$$

For a finite arithmetic series of n terms, the sum of those terms, S_n, is:

$$S_n = a_1 + (a_1 + d) + (a_1 + 2d) + \cdots + (a_n - 2d) + (a_n - d) + a_n$$

Subtract d from the last term to give the preceding term

Finding the Sum of n Terms

A formula for S_n can be found by adding the reverse of S_n to itself:

$$
\begin{array}{llcccccc}
(1) & S_n & = & a_1 & + (a_1 + d) & + \cdots + & & + \\
(2) & S_n & = & a_n & + (a_n - d) & + \cdots + & (a_1 + d) + & a_1 \\
\hline
& 2S_n & = & (a_1 + a_n) & + (a_1 + a_n) & + \cdots + & (a_1 + a_n) + & (a_1 + a_n)
\end{array}
$$

The right side of the last equation has n terms of $(a_1 + a_n)$, so

$$2S_n = n(a_1 + a_n)$$

$$S_n = \frac{n(a_1 + a_n)}{2}$$

S_n

Therefore:

THE SUM OF n TERMS OF AN AP

$$S_n = \frac{n(a_1 + a_n)}{2}$$

Example 5 Find the sum of the first 100 natural numbers

$$
\left.\begin{array}{l}
a_1 = 1 \\
a_n = 100 \\
n = 100
\end{array}\right\}
\quad
\begin{aligned}
S_n &= \frac{n(a_1 + a_n)}{2} \\
S_{100} &= \frac{\overset{50}{\cancel{100}}(1 + 100)}{\cancel{2}} = 50(101) = 5050
\end{aligned}
$$

The story is told that the famous German mathematician, Carl Friedrich Gauss, at the age of ten very quickly solved a problem like this when it was presented in his first arithmetic class.

Example 6 Given an AP having $a_1 = -8$, $a_n = 20$, and $S_n = 30$, find d and n

(1) $a_n = a_1 + (n - 1)d \Rightarrow 20 = -8 + (n - 1)d$

(2) $S_n = \dfrac{n(a_1 + a_n)}{2} \Rightarrow 30 = \dfrac{n(-8 + 20)}{2} \Rightarrow 30 = 6n$

$$5 = n$$

Substitute $n = 5$ into Equation (1):

$$
\begin{aligned}
20 &= -8 + (n - 1)d \\
20 &= -8 + (5 - 1)d \\
20 &= -8 + 4d \\
28 &= 4d \\
7 &= d
\end{aligned}
$$

EXERCISES 11-2

In Exercises 1–10 determine whether each sequence is an AP. If it is, find the common difference.

1. $3, 8, 13, 18.$
2. $7, 11, 15, 19.$
3. $7, 4, 1, -2, \ldots$
4. $9, 4, -1, -6, \ldots$
5. $4, 5\frac{1}{2}, 7, 9.$
6. $3, 4\frac{1}{4}, 5\frac{1}{2}, 6\frac{1}{2}.$
7. $-1, \frac{1}{2}, 1, 2\frac{1}{2}.$
8. $-2 + x, -1 - x, -3x, 1 - 5x, \ldots$
9. $2x - 1, x, 1, -x + 2.$
10. $3 - 2x, 2 - x, 1, x.$
11. Write the first four terms of the AP for which $a_1 = 5$ and $d = -7$.
12. Write the first five terms of the AP for which $a_1 = 4$ and $d = -5$.
13. Write the first four terms of the AP for which $a_1 = 19$ and $d = -24$.
14. Write the AP with five terms for which $a_1 = 7$ and $a_7 = -11$.
15. Write the AP with five terms for which $a_1 = 7$ and $a_5 = 31$.
16. Write the AP with six terms for which $a_1 = 6$ and $a_6 = 51$.
17. Write the thirty-first term of the AP: $-8, -2, 4, \ldots$
18. Write the forty-first term of the AP: $-5, -1, 3, \ldots$
19. Write the twenty-ninth term of the AP: $-26, -22, -18, \ldots$
20. Write the seventh term of the AP: $-x, x + 3, 3x + 6, \ldots$
21. Write the eleventh term of the AP: $x, 2x + 1, 3x + 2, \ldots$
22. Write the ninth term of the AP: $2z + 1, 3z, 4z - 1, \ldots$
23. Find the sum of the even integers from 2 to 100, inclusive.
24. Find the sum of the odd integers from 1 to 99, inclusive.
25. Find the sum of the even integers from 100 to 200, inclusive.
26. Given an AP with $a_7 = 11$ and $a_{13} = 29$, find a_1 and d.
27. Given an AP with $a_6 = 15$ and $a_{12} = 39$, find a_1 and d.
28. Given an AP with $a_5 = 12$ and $a_{14} = 57$, find a_1 and d.

In Exercises 29–37 certain elements of an arithmetic progression are given. Solve for the indicated elements.

29. $a_1 = -5, \quad d = 3, \quad a_n = 16$; find n and S_n.
30. $a_1 = -7, \quad d = 4, \quad a_n = 25$; find n and S_n.
31. $a_1 = 17, \quad d = -12, \quad a_n = -103$; find n and S_n.
32. $a_1 = 7, \quad a_n = -83, \quad S_n = -722$; find d and n.
33. $a_1 = 5, \quad a_n = 17, \quad S_n = 44$; find d and n.
34. $a_1 = 3, \quad a_n = 42, \quad S_n = 180$; find d and n.
35. $d = \frac{3}{2}, n = 9, S_n = -\frac{9}{4}$; find a_1 and a_n.
36. $d = \frac{3}{4}, n = 7, S_n = \frac{21}{4}$; find a_1 and a_n.
37. $d = \frac{2}{3}, n = 15, S_n = 35$; find a_1 and a_n.

38. If we put one penny on the first square of a chessboard, three pennies on the second square, five pennies on the third square, and continue in this way until all the squares are covered, how much money will there be on the board? A chessboard has 64 squares.

39. A rock dislodged by a mountain climber falls approximately 16 ft during the first second, 48 ft during the second second, 80 ft during the third second, and so on. Find the distance it falls during the tenth second, and the total distance it falls during the first 12 seconds.

40. A college student's young son saves 10¢ on the first day of May, 12¢ on the second, 14¢ on the third, etc. If he continues saving in this manner, how much does he save during the month of May?

11-3 Geometric Progressions

GEOMETRIC SEQUENCES

Common Ratio

A **geometric progression** (*geometric sequence*) is a sequence in which each term after the first is found by *multiplying* the preceding term by the same fixed number, called the **common ratio, r**. *Geometric progression is abbreviated* **GP.**

Example 1 Write the first four terms of the geometric progression having first term $a_1 = 5$ and common ratio $r = 2$

$a_1 = 5$

$a_2 = 5(2) = 10$ Multiplied first term (5) by common ratio (2)

$a_3 = 10(2) = 20$ Multiplied second term (10) by common ratio (2)

$a_4 = 20(2) = 40$ Etc.

Therefore the GP is 5, 10, 20, 40.

Example 2 Determine which of the sequences are geometric progressions

Method. Divide each term by the preceding term. If every ratio found this way is the same, the sequence is a GP.

(a) $24, -12, 6, -3$

$$\frac{-3}{6} = -\frac{1}{2}$$

$$\frac{6}{-12} = -\frac{1}{2}$$

$$\frac{-12}{24} = -\frac{1}{2}$$

Since all these ratios are the same $(-\frac{1}{2})$, the sequence *is* a GP

(b) $36, 9, 3, 1$

$$\frac{1}{3} = \frac{1}{3}$$

$$\frac{3}{9} = \frac{1}{3}$$

$$\frac{9}{36} = \frac{1}{4}$$

Since these ratios are *not* the same, the sequence is *not* a GP

(c) $-\dfrac{3x}{y^2}, \dfrac{9x^2}{y}, -27x^3, \ldots$

$$-27x^3 \div \frac{9x^2}{y} = \frac{-27x^3}{1} \cdot \frac{y}{9x^2} = -3xy$$

$$\frac{9x^2}{y} \div \left(-\frac{3x}{y^2}\right) = \frac{9x^2}{y} \cdot \left(-\frac{y^2}{3x}\right) = -3xy$$

Equal

Therefore this *is* a GP

In general, a geometric progression with first term a_1 and common ratio r has the following terms:

$$a_1 = a_1$$
$$a_2 = a_1 r$$
$$a_3 = (a_1 r)r = a_1 r^2$$
$$a_4 = (a_1 r^2)r = a_1 r^3$$
$$\vdots$$
$$a_n = ?$$

Finding the *n*th Term

Exponent of r is always 1 less than term number

Term #	1	2	3	4	$\cdots$	n
Term	a_1,	$a_1 r^1$,	$a_1 r^2$,	$a_1 r^3$,	$\cdots$,	$a_1 r^{n-1}$

a_n

Therefore:

THE *n*TH TERM OF A GP

$$a_n = a_1 r^{n-1}$$

Example 3 Find the fifth term of the GP having $a_1 = 18$ and $r = -\dfrac{1}{3}$

$$a_n = a_1 r^{n-1}$$

$$a_5 = 18\left(-\frac{1}{3}\right)^4$$

$$a_5 = \frac{18}{1} \cdot \frac{1}{81} = \frac{2}{9}$$

Example 4 Given a GP with $a_2 = 12$ and $a_5 = 96$, find a_1 and r

$$a_n = a_1 r^{n-1}$$

$$\left.\begin{array}{ll}(1) & a_5 = 96 = a_1 r^4 \\ (2) & a_2 = 12 = a_1 r^1\end{array}\right\} \text{Therefore} \quad \frac{96}{12} = \frac{a_1 r^4}{a_1 r} = r^3$$

$$8 = r^3$$

$$2 = r$$

Substitute $r = 2$ into Equation (2): $12 = a_1 r^1$

$$12 = a_1(2)$$

$$6 = a_1$$

GEOMETRIC SERIES

A **geometric series** is the *sum of the terms of a geometric progression.* An infinite geometric series can be written

$$a_1 + a_1 r^1 + a_1 r^2 + \cdots + a_1 r^{n-1} + \cdots$$

Finding the Sum of n Terms

For a finite geometric series of n terms, the sum of those terms, S_n, is:

$$(1) \qquad S_n = a_1 + a_1 r^1 + a_1 r^2 + \cdots + a_1 r^{n-1}$$

Another formula for S_n can be found as follows: Multiply both sides of Equation (1) by r; then subtract the resulting equation from Equation (1).

$$\begin{aligned} S_n &= a_1 + a_1 r^1 + a_1 r^2 + \cdots + a_1 r^{n-1} \\ rS_n &= \phantom{a_1 +{}} a_1 r^1 + a_1 r^2 + \cdots + a_1 r^{n-1} + a_1 r^n \\ \hline S_n - rS_n &= a_1 \phantom{+ a_1 r^1 + a_1 r^2 + \cdots + a_1 r^{n-1}} - a_1 r^n \end{aligned}$$

$$(1 - r)S_n = a_1(1 - r^n) \qquad \text{Factored both sides}$$

$$S_n = \frac{a_1(1 - r^n)}{1 - r} \qquad \text{Divided both sides by } 1 - r$$

S_n

Therefore:

THE SUM OF n TERMS OF A GP

$$S_n = \frac{a_1(1 - r^n)}{1 - r}, \quad r \neq 1$$

Example 5 Find the sum of the first six terms of a GP having $a_1 = 24$ and $r = \frac{1}{2}$

$$S_n = \frac{a_1(1 - r^n)}{1 - r}$$

$$S_6 = \frac{24\left[1 - \left(\frac{1}{2}\right)^6\right]}{1 - \frac{1}{2}} = \frac{24\left[1 - \frac{1}{64}\right]}{\frac{1}{2}} = \frac{24\left[\frac{63}{64}\right]}{\frac{1}{2}} = \frac{189}{4}$$

Example 6 Given a GP having $r = -2$, $a_n = 80$, and $S_n = 55$, find a_1 and n

$$a_n = 80 = a_1(-2)^{n-1}$$

(1) $\quad -160 = a_1(-2)^n \qquad$ Multiplied both sides by -2

$$S_n = 55 = \frac{a_1(1 - r^n)}{1 - r} = \frac{a_1[1 - (-2)^n]}{1 - (-2)}$$

$$55 = \frac{a_1 - a_1(-2)^n}{3}$$

$165 = a_1 - a_1(-2)^n \qquad$ Multiplied both sides by 3

$165 = a_1 - (-160) \qquad$ Since $a_1(-2)^n = -160$ from Equation (1)

$$a_1 = 5$$

Substitute $a_1 = 5$ into Equation (1): $\quad -160 = a_1(-2)^n$

$$-160 = 5(-2)^n$$

$$-32 = (-2)^n$$

$(-2)^5 = (-2)^n \qquad$ Exponents must be equal

$$5 = n$$

EXERCISES 11-3

In Exercises 1–10 determine whether each sequence is a GP. If it is, find the common ratio.

1. 4, 12, 36, 108.

2. 7, 14, 28, 56.

3. $-5, 15, -45, 135, \ldots$

4. $-6, 24, -96, 384, \ldots$

5. $2, \frac{1}{2}, \frac{1}{8}, \frac{1}{16}.$

6. $3, \frac{1}{2}, \frac{1}{8}, \frac{1}{48}.$

7. $1, -\frac{1}{2}, -\frac{1}{4}, -\frac{1}{8}.$

8. $20ab, -5a^3b, \frac{5}{4}a^5b, -\frac{5}{16}a^7b, \ldots$

9. $5x, 10xy, 20xy^2, 40xy^3, \ldots$

10. $4xz, 12xz^2, 36xz^3, 108xz^4, \ldots$

11. Write the first five terms of the GP for which $a_1 = 12$ and $r = 1/3$.

12. Write the first four terms of the GP for which $a_1 = 8$ and $r = 3/2$.

13. Write the first four terms of the GP for which $a_1 = -9$ and $r = -1/3$.

14. Write the GP with five terms for which $a_1 = -3/4$ and $a_5 = -12$.

15. Write the GP with five terms for which $a_1 = 2/3$ and $a_5 = 54$ (two answers).

16. Write the GP with five terms for which $a_1 = 3/25$ and $a_5 = 75$ (two answers).

17. Write the seventh term of the GP: $-9, -6, -4, \ldots$

18. Write the eighth term of the GP: $-12, -18, -27, \ldots$

19. Write the seventh term of the GP: $\frac{-25}{54}, \frac{5}{18}, \frac{-1}{6}, \ldots$

20. Write the eighth term of the GP: $\frac{24}{hk}, -\frac{12k}{h}, \frac{6k^3}{h}, \ldots$

21. Write the eighth term of the GP: $16x, 8xy, 4xy^2, \ldots$

22. Write the seventh term of the GP: $27y, 9x^2y, 3x^4y, \ldots$

23. Given a GP with $a_5 = 80$ and $r = 2/3$, find a_1 and S_5.

24. Given a GP with $a_7 = 320$ and $r = 2$, find a_1 and S_5.

25. Given a GP with $a_5 = 40$ and $r = -2/3$, find a_1 and S_5.

26. Given a GP with $a_3 = 16$ and $a_5 = 9$, find a_1, r, and S_5.

27. Given a GP with $a_3 = 28$ and $a_5 = 112/9$, find a_1, r, and S_5 (two answers).

28. Given a GP with $a_2 = 384$ and $a_4 = 24$, find a_1, r, and S_4 (two answers).

29. Given a GP having $r = 1/2$, $a_n = 3$, and $S_n = 189$, find a_1 and n.

30. Given a GP having $r = -1/2$, $a_n = 5$, and $S_n = 55$, find a_1 and n.

31. A woman invested a certain amoung of money which earned $1\frac{1}{5}$ times as much in the second year as in the first year, and $1\frac{1}{5}$ times as much in the third year as in the second year, and so on. If the investment earned her \$22,750 in the first 3 years, how much would it earn her in the fifth year?

32. A man invested a certain amoung of money which earned $1\frac{1}{4}$ times as much in the second year as in the first year, and $1\frac{1}{4}$ times as much in the third year as in the second, and so on. If the investment earned him \$9760 in the first 3 years, how much would it earn him in the fifth year?

33. If it takes 1 sec for a certain type of microbe to split into two microbes, how long will it take a colony of 1500 such microbes to exceed 6 million?

34. If we put one penny on the first square of a chessboard, two pennies on the second square, four pennies on the third square, and continue in this way until all squares are covered, how much money will there be on the board? A chessboard has 64 squares. Compare your answer to the answer to Exercise 38 in Exercises 11–2.

35. Suppose you took a job that pays 1¢ the first day, 2¢ the second day, and 4¢ the third day, with the pay continuing to increase in this manner for a month of 31 days.
(a) How much would you make on the tenth day?
Use logarithms as an aid in working parts (b) and (c).
(b) How much would you make on the thirty-first day?
(c) What would be your total earnings for the month?

36. Suppose you took a job that pays 1¢ the first day, 3¢ the second day, and 9¢ the third day, with the pay continuing to increase in this manner for a month of 31 days.
(a) How much would you make on the eighth day?
Use logarithms as an aid in working parts (b) and (c).
(b) How much would you make on the thirty-first day?
(c) What would be your total earnings for the month?

11–4 Infinite Geometric Series

Consider the formula derived in Section 11–3 for the sum of n terms of a geometric series:

$$S_n = \frac{a_1(1 - r^n)}{1 - r}, \quad r \neq 1$$

$|r| < 1$

If $|r| < 1$, then r^n gets smaller and smaller as n gets larger. Suppose $r = \frac{1}{2}$ and n becomes larger and larger. Then:

$$\left(\frac{1}{2}\right)^1 = \frac{1}{2} = 0.5$$

$$\left(\frac{1}{2}\right)^2 = \frac{1}{4} = 0.25$$

$$\left(\frac{1}{2}\right)^3 = \frac{1}{8} = 0.125$$

$$\vdots$$

$$\left(\frac{1}{2}\right)^{10} = \frac{1}{1024} \doteq 0.001$$

$$\vdots$$

$$\left(\frac{1}{2}\right)^{20} \doteq 0.000001$$

$$\vdots$$

$$\left(\frac{1}{2}\right)^{100} \doteq 8 \times 10^{-31}$$

From this you can see that when a large number of terms are taken, r^n contributes essentially nothing to the sum S_n when $|r| < 1$. Therefore the formula for S_n becomes

This term contributes essentially nothing when n becomes infinitely large

$$S_n = \frac{a_1(1 - r^n)}{1 - r} \doteq \frac{a_1}{1 - r}$$

The symbol S_∞ represents S_n when n becomes infinitely large.

Sum of Infinitely Many Terms

Therefore:

$$S_\infty = \frac{a_1}{1 - r}, \quad |r| < 1$$

Example 1 Evaluate the GP: $6 - 4 + \frac{8}{3} - + \cdots$

$$6, -4, \frac{8}{3}, \ldots$$

$$\frac{-4}{6} = -\frac{2}{3} = r$$

$$S_\infty = \frac{a_1}{1 - r} = \frac{(6)}{1 - \left(-\frac{2}{3}\right)} = \frac{6}{\frac{5}{3}} = \frac{18}{5}$$

Example 2 Write the repeating decimal 0.252525 . . . as a fraction

$$0.252525\ldots = 0.25 + 0.0025 + 0.000025 + \cdots$$

$$\frac{0.0025}{0.25} = 0.01 = r$$

$$S_\infty = \frac{a_1}{1 - r} = \frac{(0.25)}{1 - (0.01)} = \frac{0.25}{0.99} = \frac{25}{99}$$

$|r| > 1$

The reason $|r|$ is restricted to less than 1 is that **if $|r| > 1$,** the absolute value of succeeding terms in the geometric series becomes larger and larger. Therefore S_∞ is infinitely large.

$r = 1$

If $r = 1$, then every term of the series is the same as the first term, so that S_∞ is infinitely large.

$r = -1$

If $r = -1$, then the series becomes:

$$a_1 - a_1 + a_1 - + \cdots$$

In this case as n becomes infinitely large:

$S_n = 0$ if n is *even*

and $S_n = a_1$ if n is *odd*.

EXERCISES 11-4

In Exercises 1–8 find the sum of each geometric series.

1. $3 + 1 + \frac{1}{3} + \cdots$

2. $9 - 1 + \frac{1}{9} - + \cdots$

3. $\frac{4}{3} - 1 + \frac{3}{4}$

4. $\frac{6}{5} + 1 + \frac{5}{6} + \cdots$

5. $10^{-1} + 10^{-2} + 10^{-3} + \cdots$

6. $10^{-2} + 10^{-4} + 10^{-6} + \cdots$

7. $-6 - 4 - \frac{8}{3} - \cdots$

8. $-49 - 35 - 25 - \cdots$

In Exercises 9–15 write each repeating decimal as a fraction.

9. 0.2222 . . .
10. 0.2121 . . .
11. 0.05454 . . .
12. 0.03939 . . .
13. 8.6444 . . .
14. 5.2666 . . .
15. 3.7656565 . . .
16. 0.00136136136. . .

17. A rubber ball is dropped from a height of 9 ft. Each time it strikes the floor it rebounds to a height that is two-thirds the height from which it last fell. Find the total distance the ball travels before coming to rest.

18. A ball bearing is dropped from a height of 10 ft. Each time it strikes the metal floor it rebounds to a height that is three-fifths the height from which it last fell. Find the total distance the bearing travels before coming to rest.

19. The first swing of a pendulum is 12 in. Each succeeding swing is nine-tenths as long as the preceding one. Find the total distance traveled by the pendulum before it comes to rest.

20. The first swing of a pendulum is 10 in. Each succeeding swing is eight-ninths as long as the preceding one. Find the total distance traveled by the pendulum before it comes to rest.

21. When Bob stops his car suddenly, the 3-ft radio antenna oscillates back and forth. Each swing is three-fourths as great as the previous one. If the initial travel of the antenna tip is 4 in., how far will the tip travel before it comes to rest?

22. If a rabbit moves 10 yd in the first second, 5 yd in the second second, and continues to move one-half as far in each succeeding second as he did in the preceding second:
(a) How many yards does the rabbit travel before he comes to rest?
(b) What is the total time the rabbit is moving?

23. Suppose a sheet of paper 0.001 in. thick were cut in half, the pieces stacked with edges aligned, and the stack cut in half. If this procedure were repeated over and over, how high would the stack be after 50 cuts?

24. If the population of a city increases by 10% each year, how long will it take for its population to double?

11–5 The Binomial Expansion

When we worked with special products and factoring, the forms $(a + b)^2$ and $(a - b)^2$ occurred so often that formulas were derived for their expansions:

$$(a + b)^2 = a^2 + 2ab + b^2$$

$$(a - b)^2 = a^2 - 2ab + b^2$$

Other powers of binomials occur so frequently that it is convenient to have a method for expanding *any* power of *any* binomial.

Consider the following powers of $(a + b)$, which are found either by inspection or by actual multiplication:

$$(a + b)^1 = a + b$$

$$(a + b)^2 = a^2 + 2ab + b^2$$

$$(a + b)^3 = a^3 + 3a^2b + 3ab^2 + b^3$$

$$(a + b)^4 = a^4 + 4a^3b + 6a^2b^2 + 4ab^3 + b^4$$

$$\vdots$$

A careful examination of the above binomial expansions shows that they are always polynomials in a and b. Each term has the form Ca^rb^s (Section 1–10).

The degree of each term is n. The number of terms in each expansion is one more than n.

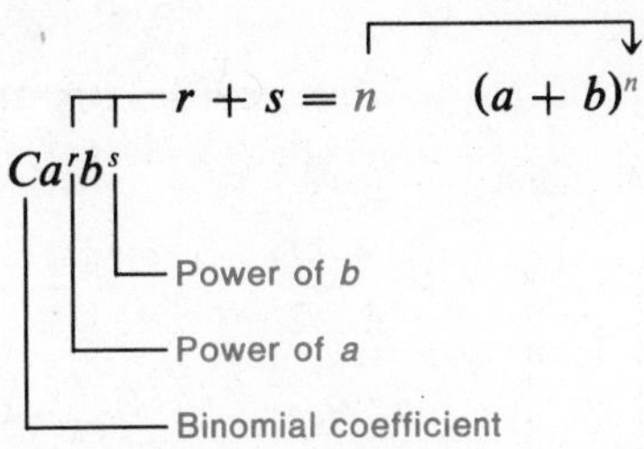

Look at one particular expansion in detail:

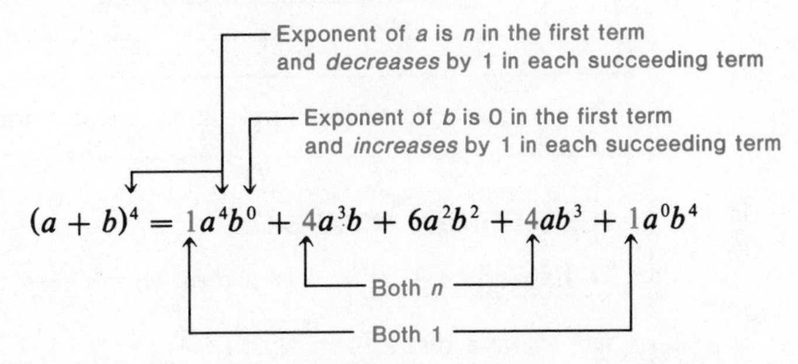

Since the binomial coefficients are *symmetrical,* we need to find only half of them. The remaining coefficients are obtained by setting them equal to their corresponding symmetrical coefficients. The binomial coefficients are written according to the following pattern:

$$1,\ n,\ \frac{n(n-1)}{1\cdot 2},\ \frac{n(n-1)(n-2)}{1\cdot 2\cdot 3},\ \ldots$$

Coefficient of second term is n

Coefficient of first term is 1

Finding the Binomial Coefficients

Example 1 Finding the binomial coefficients

(a) Coefficients for expansion of $(a + b)^5$

$$n + 1 = 5 + 1 = 6 \text{ terms}$$

Find the first three coefficients, then use symmetry:

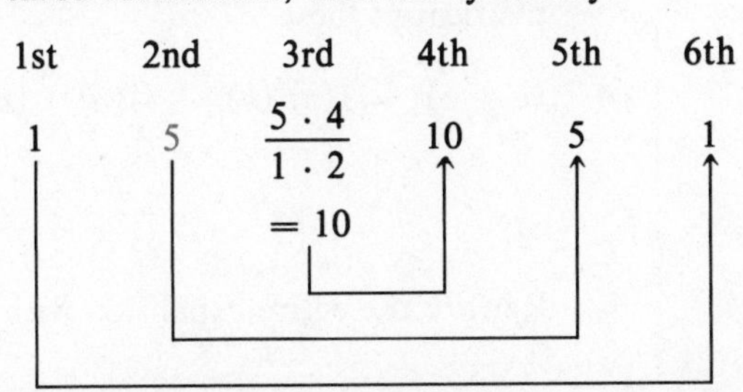

(b) Coefficients for expansion of $(a + b)^8$

$$n + 1 = 8 + 1 = 9 \text{ terms}$$

Find the first five coefficients; then use symmetry.

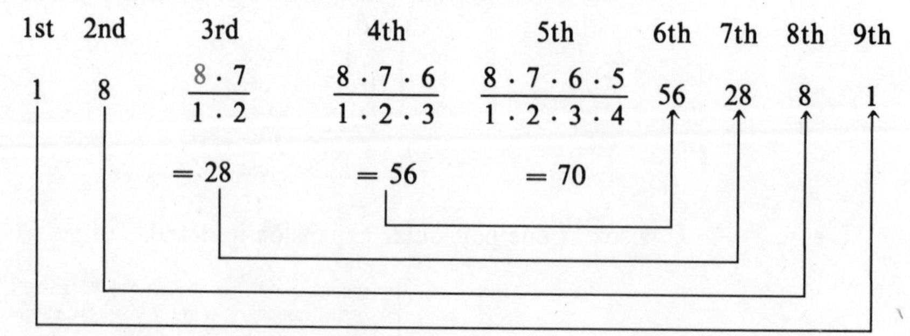

The procedure for expanding a binomial is summarized as follows:

THE BINOMIAL EXPANSION

To Expand $(a + n)^n$ (n a positive integer)

1. Make a blank form with $(n + 1)$ terms.

$$(\ )(\ ) + (\ )(\ ) + (\ )(\ ) + (\ )(\ ) + \cdots$$

2. Fill in the powers of a and b.

Exponent of a is n in first term and *decreases* by 1 in each succeeding term

Exponent of b is 0 in first term and *increases* by 1 in each succeeding term

$$(a)^n(b)^0 + (a)^{n-1}(b)^1 + (a)^{n-2}(b)^2 + (a)^{n-3}(b)^3 + \cdots$$

3. Write the coefficients as follows:

$$1, \ \frac{n}{1}, \ \frac{n(n-1)}{1 \cdot 2}, \ \frac{n(n-1)(n-2)}{1 \cdot 2 \cdot 3}, \ \ldots$$

Coefficient of second term is n

Coefficient of first term is 1

Calculate *half* the coefficients. The remaining coefficients are symmetrical to these.

4. $$(a + b)^n = 1(a)^n(b)^0 + \frac{n}{1}(a)^{n-1}(b)^1 + \frac{n(n-1)}{1 \cdot 2}(a)^{n-2}(b)^2 + \frac{n(n-1)(n-2)}{1 \cdot 2 \cdot 3}(a)^{n-3}(b)^3 + \cdots$$

5. Rewrite the entire expansion with each term simplified.

Example 2 Expand: $(a + b)^5$

Step 1. Make a blank form with $n + 1 = 6$ terms.

$$(\ )(\ ) + (\ )(\ ) + (\ )(\ ) + (\ )(\ ) + (\ )(\ ) + (\ )(\ )$$

Step 2. Fill in the powers of a and b.

$$(a)^5(b)^0 + (a)^4(b)^1 + (a)^3(b)^2 + (a)^2(b)^3 + (a)^1(b)^4 + (a)^0(b)^5$$

Steps 3 and 4. Write in the coefficients.

$$1(a)^5(b)^0 + 5(a)^4(b)^1 + \frac{5 \cdot 4}{1 \cdot 2}(a)^3(b)^2 + 10(a)^2(b)^3 + 5(a)^1(b)^4 + 1(a)^0(b)^5$$

$$= 10$$

Step 5. Simplify the expansion.

$$(a + b)^5 = a^5 + 5a^4b + 10a^3b^2 + 10a^2b^3 + 5ab^4 + b^5$$

Example 3 Identify a and b in each binomial expansion

(a) $(3x^2 + 5y)^4$ In this binomial $\begin{cases} a = 3x^2 \\ b = 5y \end{cases}$

(b) $(2e^3 - 7f^2)^5$ In this binomial $\begin{cases} a = 2e^3 \\ b = -7f^2 \end{cases}$

(c) $\left(\frac{h}{4} - 6k^2\right)^7$ In this binomial $\begin{cases} a = \frac{h}{4} \\ b = -6k^2 \end{cases}$

Example 4 Expand: $\left(\frac{x}{2} - 4\right)^4$

Steps 1–4

$$1\left(\frac{x}{2}\right)^4(-4)^0 + 4\left(\frac{x}{2}\right)^3(-4)^1 + \frac{4 \cdot 3}{1 \cdot 2}\left(\frac{x}{2}\right)^2(-4)^2 + 4\left(\frac{x}{2}\right)^1(-4)^3 + 1\left(\frac{x}{2}\right)^0(-4)^4$$

Step 5 $\left(\frac{x}{2} - 4\right)^4$

$$= \frac{x^4}{16} - 2x^3 + 24x^2 - 128x + 256$$

Example 5 Write the first five terms of the expansion of $(x + \sqrt{y})^{12}$

Steps 1–4

$$1(x)^{12}(\sqrt{y})^0 + 12(x)^{11}(\sqrt{y})^1 + \frac{12 \cdot 11}{1 \cdot 2}(x)^{10}(\sqrt{y})^2 + \frac{12 \cdot 11 \cdot 10}{1 \cdot 2 \cdot 3}(x)^9(\sqrt{y})^3$$
$$+ \frac{12 \cdot 11 \cdot 10 \cdot 9}{1 \cdot 2 \cdot 3 \cdot 4}(x)^8(\sqrt{y})^4 + \cdots$$

Step 5 $(x + \sqrt{y})^{12}$

$$= x^{12} + 12x^{11}\sqrt{y} + 66x^{10}y + 220x^9y\sqrt{y} + 495x^8y^2 + \cdots$$

In this section we discussed the binomial expansion for *natural-number powers* only. Other numbers can be used for powers of binomials, but binomial expansions using such numbers as exponents are not discussed in this book.

Pascal's Triangle

If the binomial expansion for the first five values of n is written, we have:

$$(a + b)^0 = 1$$
$$(a + b)^1 = 1a + 1b$$
$$(a + b)^2 = 1a^2 + 2ab + 1b^2$$
$$(a + b)^3 = 1a^3 + 3a^2b + 3ab^2 + 1b^3$$
$$(a + b)^4 = 1a^4 + 4a^3b + 6a^2b^2 + 4ab^3 + 1b^4$$
$$\vdots$$

If we omit everything in the above display except the numerical coefficients, we get a triangular array of numbers known as **Pascal's Triangle:**

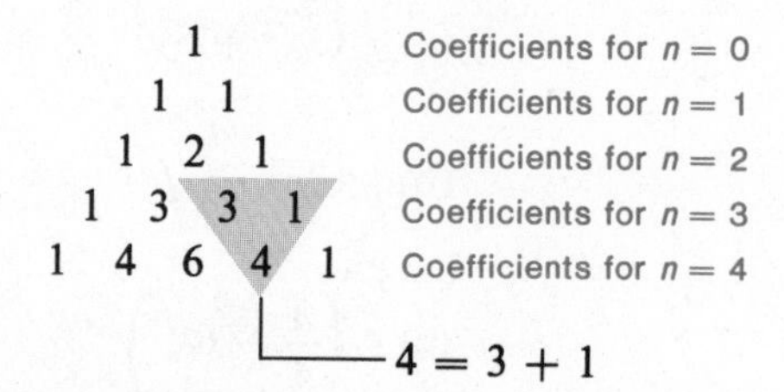

A close examination of the numbers in Pascal's triangle reveals:

1. The first and last numbers in any row are 1.
2. Any other number in Pascal's triangle is the sum of the two closest numbers in the row above it (see shaded triangle).

Pascal's triangle can be used to find the coefficients in a binomial expansion. The triangle we have shown can be extended to any size by using the two rules given above.

EXERCISES 11-5

In Exercises 1–14 expand each binomial.

1. $(x + 2)^4$ **2.** $(x - y)^6$ **3.** $(3r + s)^6$ **4.** $(x + y^2)^4$

5. $\left(2x - \frac{1}{2}\right)^5$ **6.** $(4x^2 - 3y^2)^5$ **7.** $\left(\frac{1}{3}x + \frac{3}{2}\right)^4$ **8.** $\left(\frac{4}{3} + \frac{3}{2}a\right)^6$

9. $(x + x^{-1})^4$ **10.** $(y^{-2} - 2y)^4$ **11.** $(x^{1/2} + y^{1/2})^4$ **12.** $(u^{1/3} - v^{1/2})^4$

13. $(1 - \sqrt[3]{x})^3$ **14.** $(1 - \sqrt[5]{y})^5$

15. Write the first four terms of the expansion of $(x + 2y^2)^{10}$.

16. Write the first four terms of the expansion of $(x - 3y^2)^{10}$.

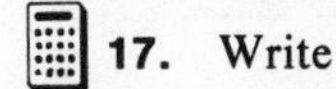

17. Write the first four terms of $(2h - k^2)^{11}$.

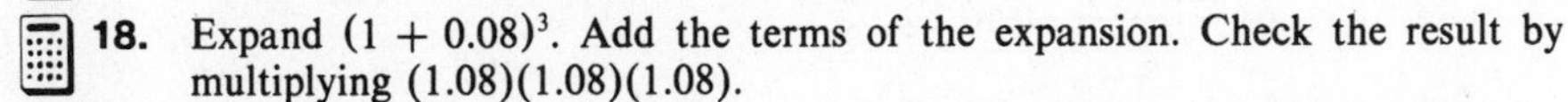

18. Expand $(1 + 0.08)^3$. Add the terms of the expansion. Check the result by multiplying $(1.08)(1.08)(1.08)$.

19. Expand $(1 + 0.02)^3$. Add the terms of the expansion. Check the result by multiplying $(1.02)(1.02)(1.02)$.

20. Expand $(2 + 0.1)^4$. Add the terms of the expansion. Check the result by multiplying $(2.1)(2.1)(2.1)(2.1)$.

Chapter Eleven Summary

Sequences. (Section 11–1) A *sequence* of numbers is a set of numbers arranged in a definite order. The numbers that make up the sequence are called the *terms* of the sequence. If in counting the terms of a sequence, the counting comes to an end, the sequence is a *finite sequence;* if the counting never ends, the sequence is an *infinite sequence.*

Series. (Section 11–1) A *series* is the indicated sum of a finite or infinite sequence of terms. It is a finite or an infinite series according to whether the number of terms is finite or infinite. The *partial sum of a series* is the sum of a finite number of consecutive terms of the series, beginning with the first term.

Arithmetic Progressions. (Section 11–2) An *AP* is a sequence in which each term after the first is found by *adding* the same fixed number to the preceding term. The fixed number added is called the *common difference, d; d* can be found by subtracting any term from the term that follows it.

The *n*th Term of an AP: $a_n = a_1 + (n - 1)d$

The Sum of *n* Terms of an AP: $S_n = \dfrac{n(a_1 + a_n)}{2}$

Geometric Progressions. (Section 11–3) A *GP* is a sequence in which each term after the first is found by *multiplying* the preceding term by the same fixed number, called the *common ratio, r*. The common ratio can be found by dividing any term by the term that precedes it.

The *n*th Term of a GP: $a_n = a_1r^{n-1}$

The Sum of *n* Terms of a GP: $S_n = \dfrac{a_1(1 - r^n)}{1 - r}, \quad r \neq 1$

The Sum of an Infinite Geometric Series (Section 11–4):

$$S_\infty = \frac{a_1}{1 - r}, \quad |r| < 1$$

The Binomial Expansion (Section 11–5):

$$(a + b)^n = 1(a)^n(b)^0 + \frac{n}{1}(a)^{n-1}(b)^1 + \frac{n(n-1)}{1 \cdot 2}(a)^{n-2}(b)^2 + \frac{n(n-1)(n-2)}{1 \cdot 2 \cdot 3}(a)^{n-3}(b)^3 + \cdots$$

Chapter Eleven Diagnostic Test or Review

Allow yourself about 1 hour and 10 minutes to do these problems. Complete solutions for every problem, together with section references, are given in the answer section at the end of the book.

---

1. Given $a_n = \dfrac{2n - 1}{n}$, find S_4.

2. Determine which of the following sequences are an AP, a GP, or neither. If AP, give d. If a GP, give r.

   (a) $8, -20, 50, - + \ldots$ (b) $\dfrac{1}{2}, \dfrac{3}{4}, 1, \dfrac{5}{4}, \dfrac{3}{2}, \ldots$

   (c) $2x - 1, 3x, 4x + 2, \ldots$ (d) $\dfrac{c^4}{16}, -\dfrac{c^3}{8}, \dfrac{c^2}{4}, - + \ldots$

3. Write the first five terms of the AP for which $a_1 = x + 1$ and $d = x - 1$.

4. Write the first five terms of the AP having $a_1 = 2$ and $a_5 = -2$.

5. Write the fifteenth term of the AP: $1 - 6h, 2 - 4h, 3 - 2h, \ldots$

6. Given an AP with $a_3 = 1$ and $a_7 = 2$, find a_1 and d.

7. Given an AP with $a_1 = 10$, $a_n = -8$, and $S_n = 7$, find d and n.

8. Write the first five terms of the GP for which $a_1 = \dfrac{C^4}{16}$ and $r = -\dfrac{2}{c}$.

9. Write the seventh term of the GP: $8, -20, 50, - + \cdots$

10. Given a GP with $a_2 = -18$ and $a_4 = -8$, find a_1, r, and S_4.

11. Given a GP with $r = -\frac{2}{3}$, $a_n = -16$, and $S_n = 26$, find a_1 and n.

12. Find the sum of: $8, -4, 2, -1, \ldots$

13. Write $3.0333\ldots$ as a fraction.

14. A ball bearing dropped from a height of 6 ft rebounds to a height that is two-thirds of the height from which it fell. Find the total distance the ball bearing travels vertically before it comes to rest.

15. A ball rolling down a hill travels 9 ft in the first second, 27 ft in the second second, 45 ft in the third second, etc. Find the distance traveled during the tenth second and the total distance traveled during the first 12 sec.

16. Expand the binomial: $\left(\dfrac{1}{5} - 5y\right)^4$

17. Write the first three terms of the expansion $\left(2x - \dfrac{1}{2}y\right)^8$.

ANSWERS

Exercises 1–1 (page 5)

1. Yes

3. (a) True
(b) False
(c) True

5. (a) {0, 2, 4, 6, 8}
(b) {4}

7. (a) {5, 11}
(b) {2, 5, 6, 11, 0, 3, 4}
(c) Y and Z are disjoint.

9. $R \cap S$

Exercises 1–2 (page 11)

1. (a) R, D, N, W, J, Q
(b) R, J, Q
(c) R, N, W, J, Q
(d) R, Q
(e) R, H
(f) R, Q

3. $\{-7, -6\}$

5. (a) 5
(b) 3
(c) -14

7. (a) $\subseteq$
(b) $\in$
(c) $\subseteq$

9. (a) $\subseteq$
(b) $\subseteq$
(c) $\in$

Exercises 1–3A (page 16)

1. -14 **3.** 5 **5.** -39 **7.** -17 **9.** -21.56

11. 57.389 **13.** -26 **15.** 253 **17.** -60 **19.** -18

21. -5.20882 **23.** -260 **25.** 77 **27.** 0 **29.** -2625

31. $-25{,}600$ **33.** -15 **35.** -3 **37.** 0 **39.** $2\frac{1}{2}$

41. Not possible **43.** Cannot be determined **45.** -15 **47.** 15.625

49. 19,777 ft **51.** 18°F **53.** -209 ft **55.** 65,226 ft **57.** 6,064 ft

Exercises 1–3B (page 19)

1. True, because of the commutative property of addition (order of numbers changed).

3. True, commutative property of addition and commutative property of multiplication.

5. False, commutative property does not hold for division.

7. True, commutative property holds for both addition and multiplication.

9. False, commutative property does not hold for subtraction.

11. True, commutative property holds for addition.

Exercises 1–3C (page 23)

1. $(-3)^4 = (-3)(-3)(-3)(-3) = 81$ **3.** $-2^4 = -(2 \cdot 2 \cdot 2 \cdot 2) = -16$ **5.** 0

7. $(-1)^{49} = -1$; an odd power of a negative number is negative. **9.** 299.29

11. $(-1.5)^5 = (-1.5)(-1.5)(-1.5)(-1.5)(-1.5) = -7.59375$ **13.** 6, −6 **15.** 12, −12 **17.** $\sqrt{25} = 5$

19. $-\sqrt{100} = -(\sqrt{100}) = -(10) = -10$ **21.** 9 **23.** −13 **25.** 15 **27.** 3.464 **29.** 5.292

31. 6.708 **33.** 13.565 **35.** 30.62678566 **37.** 13.14534138 **39.** 2

41. $-\sqrt[3]{27} = -(\sqrt[3]{27}) = -(3) = -3$ **43.** −4 because $(-4)^3 = -64$

45. $-\sqrt[5]{32} = -(\sqrt[5]{32}) = -(2) = -2$ **47.** 0 **49.** −2 because $(-2)^5 = -32$ **51.** 1

53. −10 because $(-10)^3 = -1000$ **55.** −3

Exercises 1–4A (page 26)

1. 10^6 **3.** x^7 **5.** a^5 **7.** 2^{x+y} **9.** $5^{3z} = 125^z$

11. 10^6 **13.** 3^{ab} **15.** 3^{x^2} **17.** a^3b^3 **19.** $27x^3$

21. 2^ky^k **23.** x^2 **25.** 10^{b-c} **27.** $\frac{a^{5x}}{a^{2x}} = a^{5x-2x} = a^{3x}$

29. 5^{2x} **31.** $\frac{x^4}{y^2}$ cannot be reduced because the bases are different. **33.** $\frac{w^2}{z^2}$ **35.** $\frac{x^3}{125}$

37. $\frac{2^n}{u^n}$

Exercises 1–4B (page 30)

1. $\frac{1}{a^3}$ **3.** $\frac{1}{10^3}$ **5.** $\frac{y^2}{x^3}$ **7.** $\frac{x}{y^2z^3}$ **9.** a^3b^4 **11.** $\frac{y^2}{x^3}$ **13.** $\frac{x}{y^2}$

15. $\frac{1}{a^6}$ **17.** x^2 **19.** e^7 **21.** $\frac{1}{x^{6a}}$ **23.** $\frac{x^{3a}}{x^{-a}} = x^{3a-(-a)} = x^{4a}$

25. $(3^x)^0 = 1$, any expression other than 0 to the 0 power is 1. **27.** $\frac{6x^2}{2x^{-3}} = \frac{6}{2} \cdot x^{2-(-3)} = 3x^5$ **29.** $\frac{1}{x^3y^3}$

31. $\frac{1}{9x^2}$ **33.** $\frac{f}{e}$ **35.** $\frac{25}{u^2}$ **37.** $\frac{15m^0n^{-3}}{5m^{-2}n^2} = \frac{15}{5} \cdot \frac{m^0}{m^{-2}} \cdot \frac{n^{-3}}{n^2} = 3m^{0-(-2)}n^{-3-(2)} = 3m^2n^{-5} = \frac{3m^2}{n^5}$

39. $\frac{x^{-1}+y}{y} = \frac{\frac{1}{x}+y}{y}$ **41.** 1000 **43.** 81 **45.** 10,000 **47.** $(10^0)^5 = 1^5 = 1$

49. $x^{-3}y$ **51.** a^4x^3 **53.** $x^4y^{-3}z^2$

Exercises 1–4C (page 32)

1. $(a^2b^3)^2 = a^{2\cdot 2}b^{3\cdot 2} = a^4b^6$ **3.** $(m^{-2}n)^4 = m^{(-2)4}n^{1\cdot 4} = m^{-8}n^4 = \frac{n^4}{m^8}$ **5.** $\frac{16x^{12}}{y^8}$

7. $(x^{-2}y^3)^{-4} = x^{(-2)(-4)}y^{3(-4)} = x^8y^{-12} = \frac{x^8}{y^{12}}$ **9.** $\frac{a^3}{125b^6}$ **11.** $(10^0k^{-4})^{-2} = k^{(-4)(-2)} = k^8$

13. $\left(\frac{xy^4}{z^2}\right)^2 = \frac{x^{1\cdot 2}y^{4\cdot 2}}{z^{2\cdot 2}} = \frac{x^2y^8}{z^4}$ **15.** $\left(\frac{M^{-2}}{N^3}\right)^4 = \frac{M^{(-2)4}}{N^{3\cdot 4}} = \frac{M^{-8}}{N^{12}} = \frac{1}{M^8N^{12}}$

17. $\left(\dfrac{-x^{-5}}{y^4z^{-3}}\right)^{-2} = \dfrac{x^{(-5)(-2)}}{y^{4(-2)}z^{(-3)(-2)}} = \dfrac{x^{10}}{y^{-8}z^6} = \dfrac{x^{10}y^8}{z^6}$ **19.** $\left(\dfrac{r^7s^8}{r^9s^6}\right)^0 = 1$ **21.** $\dfrac{x^3y^6}{8}$ **23.** $\dfrac{a^3b^6}{64}$ **25.** $u^{10}v^6$

27. $\left(\dfrac{x^{3n}x^{2n}}{x^{4n}}\right)^2 = (x^{3n+2n-4n})^2 = (x^n)^2 = x^{2n}$ **29.** $x^{9n-2}y^{6n-5}$

Exercises 1–5A (page 34)

1. $10 \div 2 \cdot 5 = 5 \cdot 5 = 25$ **3.** $12 \div 6 \div 2 = 2 \div 2 = 1$ **5.** $14 \cdot 2 \div 4 = 28 \div 4 = 7$

7. $3(2^4) = 3(16) = 48$ **9.** $4 \cdot 3 + 15 \div 5 = 12 + 3 = 15$ **11.** $10 - 3 \cdot 2 = 10 - 6 = 4$

13. $10 \cdot 15^2 - 4^3 = 10 \cdot 225 - 64$
$= 2250 - 64 = 2186$ **15.** $(785)^3(0) + 1^5 = 0 + 1 = 1$

17. $(10^2)\sqrt{16} \cdot 5 = 100 \cdot 4 \cdot 5$
$= 400 \cdot 5 = 2000$ **19.** $28 \div 4 \cdot 2(6) = 7 \cdot 2(6)$
$= 14(6) = 84$ **21.** $(8 - 2) \cdot 6 = (6) \cdot 6 = 36$

23. $2\sqrt{9}(2^3 - 5) = 2 \cdot 3(8 - 5)$
$= 6(3) = 18$ **25.** $(3 \cdot 5^2 - 15 \div 3) \div (-7)$
$= (3 \cdot 25 - 15 \div 3) \div (-7)$
$= (75 - 5) \div (-7)$
$= (70) \div (-7) = -10$ **27.** $(-10) \div 2 - (-3)$
$= -5 - (-3) = -5 + 3$
$= -5 - (-3) = -5 + 3$
$= -2$

29. $(-10)^3 - 5(10^2)\sqrt{100}$
$= -1000 - 5(100) \cdot 10$
$= -1000 - 500 \cdot 10$
$= -1000 - 5000 = -6000$ **31.** $-18 \div (-\sqrt{81})$
$= -18 \div (-9) = 2$ **33.** $\dfrac{7 + (-12)}{8 - 3} = \dfrac{-5}{5} = -1$

35. $2\frac{2}{3}$ **37.** $8 - [5(-2)^3 - \sqrt{16}]$
$= 8 - [5(-8) - 4]$
$= 8 - [-40 - 4]$
$= 8 - [-44] = 52$ **39.** $(667.5) \div (25.8) \cdot (2.86)$
$\doteq (25.8721)(2.86) \doteq 73.99$

Exercises 1–5B (page 36)

1. -28 **3.** 19 **5.** 37 **7.** -1 **9.** $\dfrac{\sqrt{5.6F - 1.7G}}{0.78H^2} = \dfrac{\sqrt{5.6(3) - 1.7(-5)}}{0.78(-4)^2} \doteq 0.4030$ **11.** 70

13. $46\frac{2}{3}$ **15.** $C = \frac{5}{9}(F - 32) = \frac{5}{9}(-10 - 32) = \frac{5}{9}(-42) = -23\frac{1}{3}$ **17.** 5 **19.** 314 **21.** $\doteq 19.15$

Exercises 1–6A (page 40)

1. Three terms Second term: $\dfrac{2x + y}{3xy}$ **3.** Two terms Second term: $-2(x + y)$ **5.** 2, RT

7. $x^2y - 3x$ **9.** $-4x + 20$ **11.** $-3x + 6y - 6$ **13.** $(3x^3 - 2x^2y + y^3)(-2xy)$
$= (3x^3)(-2xy) + (-2x^2y)(-2xy) + (y^3)(-2xy)$
$= -6x^4y + 4x^3y^2 - 2xy^4$

15. $-36a^4b^3c^3$ **17.** $x - y + 10$ **19.** $7 + 4R - S$ **21.** $6 - 2a + 6b$

23. $-2x^2 + 8xy + 3$ **25.** $-x + y + 2 - a$ **27.** $5 - 3[a - 4(2x - y)] = 5 - 3[a - 8x + 4y]$
$= 5 - 3a + 24x - 12y$

29. $6 - 8x + 12a - 24b$

Exercises 1–6B (page 42)

1. $-a$ **3.** $6h^3 - 3hk + 3k^4$ **5.** $5x^3y^3 - 9x^2y^2$ **7.** $13xy^2 - 11x^2$

9. $2x(3x^2 - 5x + 1) - 4x(2x^2 - 3x - 5)$
$= \underline{\underline{6x^3}} - \underline{10x^2} + 2x - \underline{\underline{8x^3}} + \underline{12x^2} + 20x$
$= -2x^3 + 2x^2 + 22x$ **11.** $u^3 - 8$ **13.** $2a - 2x$ **15.** $3x + 17$

17. $56x - 270$ **19.** $14g - 45$ **21.** $(2x - y) - \{[3x - (7 - y)] - 10\}$
$= 2x - y - \{[3x - 7 + y] - 10\}$
$= 2x - y - \{3x - 7 + y - 10\}$
$= 2x - y - 3x + 7 - y + 10$
$= -x - 2y + 17$
23. $6x^2y^2$ **25.** $2z + 96$

27. $60x - 2\{-3[-5(-5 - x) - 6x]\}$
$= 60x - 2\{-3[25 + 5x - 6x]\}$
$= 60x - 2\{-3[25 - x]\}$
$= 60x - 2\{-75 + 3x\}$
$= 60x + 150 - 6x = 54x + 150$
29. $120 + 18z$

Exercises 1–7A (page 45)

1. (a) 2nd degree (b) Not a polynomial because the exponents are not positive integers. (c) 0 degrees (d) 6th degree (e) Not a polynomial because the term is not of the form ax^n.

3. $5m^2 - 1$ **5.** $-8y^3 + y^2 + 12$ **7.** $-a^2 + 3a + 12$ **9.** $-7x^4 - 4x^3 + 9$ **11.** $v^3 + v^2 + 8v + 13$

13. $(4x^3 + 6 + x) - (6 + 3x^5 - 4x^2)$
$= 4x^3 + 6 + x - 6 - 3x^5 + 4x^2$
$= -3x^5 + 4x^3 + 4x^2 + x$

15.
$$\begin{array}{r} 4x^3 + 7x^2 - 5x + 4 \\ 2x^3 - 5x^2 + 5x - 6 \\ \hline 6x^3 + 2x^2 \qquad\quad - 2 \end{array}$$

17. $-x^4 + 2x^3 + 13$

19. $(6r^3t + 14r^2t - 11) - (19 - 8r^2t + r^3t) + (8 - 6r^2t)$
$= 6r^3t + 14r^2t - 11 - 19 + 8r^2t - r^3t + 8 - 6r^2t$
$= 5r^3t + 16r^2t - 22$

21. $(6m^2n^2 - 8mn + 9) + (-10m^2n^2 + 18mn - 11) - (-3m^2n^2 + 2mn - 7)$
$= 6m^2n^2 - 8mn + 9 - 10m^2n^2 + 18mn - 11 + 3m^2n^2 - 2mn + 7$
$= -m^2n^2 + 8mn + 5$

23. $4x^2 + 3x + 8$

25. $[(5 + xy^2 + x^3y) + (-6 - 3xy^2 + 4x^3y)] - [(x^3y + 3xy^2 - 4) + (2x^3y - xy^2 + 5)]$
$= [5 + xy^2 + x^3y - 6 - 3xy^2 + 4x^3y] - [x^3y + 3xy^2 - 4 + 2x^3y - xy^2 + 5]$
$= [-1 - 2xy^2 + 5x^3y] - [3x^3y + 2xy^2 + 1] = -1 - 2xy^2 + 5x^3y - 3x^3y - 2xy^2 - 1$
$= 2x^3y - 4xy^2 - 2$

27. $2.726x^2 + 32.42x + 14.444$

Exercises 1–7B (page 48)

1. $30x^3 - 18x^2 - 42x$ **3.** $-8m^5 - 4m^4 + 24m^3$ **5.** $-15z^5 + 12z^4 - 6z^3 + 24z^2$

7. $(4y^4 - 7y^2 - y + 12)(-5y^3)$
$= (4y^4)(-5y^3) + (-7y^2)(-5y^3) + (-y)(-5y^3) + (12)(-5y^3)$
$= -20y^7 + 35y^5 + 5y^4 - 60y^3$

9. $8h^3 - 22h^2 + 29h - 21$

11. $a^5 + 3a^4 + 3a^3 + 7a^2 - 2a + 12$ **13.** $3z^4 - 13z^3 + 9z^2 - 24z + 16$

15.
$$\begin{array}{r} 3u^2 - u + 5 \\ 2u^2 + 4u - 1 \\ \hline -3u^2 + u - 5 \\ 12u^3 - 4u^2 + 20u \phantom{{}-5} \\ 6u^4 - 2u^3 + 10u^2 \phantom{{}+20u-5} \\ \hline 6u^4 + 10u^3 + 3u^2 + 21u - 5 \end{array}$$

17. $6x^3y^2 - 2x^2y^3 + 8xy^4$ **19.** $5a^5b - 15a^4b^2 + 15a^3b^3 - 5a^2b^4$ **21.** $x^4 + 4x^3 + 10x^2 + 12x + 9$

23. $x^6 - y^6$ **25.** $\doteq 7.81x^3 - 39.1x^2 + 59.5x - 39.1$

Exercises 1–7C (page 50)

1. $3x^3 - 4x^2 - 2x$ **3.** $-5a^3b^2 + 3b$ **5.** $3xyz + 6$

7. $\dfrac{5x^3 - 4x^2 + 10}{-5x^2} = \dfrac{5x^3}{-5x^2} + \dfrac{-4x^2}{-5x^2} + \dfrac{10}{-5x^2} = -x + \dfrac{4}{5} - \dfrac{2}{x^2}$ **9.** $x - \dfrac{2y}{x} + \dfrac{3}{xy}$

11. $\dfrac{6a^2bc^2 - 4ab^2c^2 + 12bc}{6abc} = \dfrac{6a^2bc^2}{6abc} + \dfrac{-4ab^2c^2}{6abc} + \dfrac{12bc}{6abc} = ac - \dfrac{2bc}{3} + \dfrac{2}{a}$ **13.** $2x + 3$

15.

$$\begin{array}{r} 3v + 8 \quad \text{R } 66 \\ 5v - 7\,\overline{)\,15v^2 + 19v + 10} \\ \underline{15v^2 - 21v} \\ 40v + 10 \\ \underline{40v - 56} \\ 66 \end{array}$$

Check

$$\begin{array}{rl} 3v + 8 & \text{Quotient} \\ \underline{5v - 7} & \text{Divisor} \\ -21v - 56 & \\ \underline{15v^2 + 40v} & \\ 15v^2 + 19v - 56 & \\ \underline{+66} & \text{Remainder} \\ 15v^2 + 19v + 10 & \text{Dividend} \end{array}$$

17. $3x^2 - x - 4$ **19.** $4x^2 + 8x + 8$ R -6 **21.** $4a^2 - ab + 5b^2$ **23.** $x^2 + x - 1$

25.

$$\begin{array}{r} 2u^2 + 3uv - v^2 \quad \text{R } -4v^4 \\ u^2 - uv - 2v^2\,\overline{)\,2u^4 + u^3v - 8u^2v^2 - 5uv^3 - 2v^4} \\ \underline{2u^4 - 2u^3v - 4u^2v^2} \\ 3u^3v - 4u^2v^2 - 5uv^3 \\ \underline{3u^3v - 3u^2v^2 - 6uv^3} \\ -u^2v^2 + uv^3 - 2v^4 \\ \underline{-u^2v^2 + uv^3 + 2v^4} \\ -4v^4 \end{array}$$

27.

$$\begin{array}{r} 6.15x + 53.669 \quad \text{R } 504.81494 \\ x - 9.26\,\overline{)\,6.15x^2 - 3.28x + 7.84} \\ \underline{6.15x^2 - 56.949x} \\ 53.669x + 7.84 \\ \underline{53.669x - 496.97694} \\ 504.81494 \end{array}$$

Chapter One Diagnostic Test or Review (page 53)

Following each problem number is the textbook section number (in parentheses) where that kind of problem is discussed.

1. (1–1) Yes, they have the same elements. **2.** (1–1) $\{7, 8, 9\}$

3. (1–1, 1–2) (a) Infinite; $\{\ldots, -3, -2, -1, 0, 1, 2, 3, \ldots\}$
(b) Finite; $\{2, 3, 4\}$
(c) Infinite; there are an infinite number of real numbers between 1 and 5.

4. (1–1) (a) $A \cup C = \{x, z, w, y, r, s\}$; all elements of A as well as all elements of C
(b) $B \cap C = \{y\}$; y is the only element in *both* B and C
(c) $A \cap B = \{x, w\}$; x and w are the only elements in *both* A and B
(d) $B \cup C = \{x, y, w, r, s\}$; all elements of B as well as all elements of C

5. (1–2) (a) $-3, 2.4, 0, \sqrt{3}, 5, \frac{1}{2}$
(b) $-3, 0, 5$
(c) 5
(d) $\sqrt{3}$
(e) $-3, 2.4, 0, 5, \frac{1}{2}$

6. (1–2) $|-17| = 17$ **7.** (1–3A) $(-11) + (15) = 4$

8. (1–3A) $(14)(-2) = -28$ **9.** (1–3C) $(-5)^2 = (-5)(-5) = 25$ **10.** (1–3A) $(30) \div (-5) = -6$

11. (1–2) $|0| = 0$ **12.** (1–3A) $\dfrac{9}{0}$ is not possible **13.** (1–3A) $(-35) - (2) = (-35) + (-2) = -37$

14. (1–3A) $(-27) - (-17) = (-27) + (17) = -10$ **15.** (1–3A) $(-9)(-8) = 72$ **16.** (1–2) $|-3| = 3$

17. (1–3A) $(-19)(0) = 0$ **18.** (1–2) $|5| = 5$ **19.** (1–3A) $\dfrac{-40}{-8} = 5$

20. (1–3A) $(-9) + (-13) = -22$ **21.** (1–3C) $0^4 = 0 \cdot 0 \cdot 0 \cdot 0 = 0$

22. (1–3C) $-6^2 = -(6^2) = -36$ **23.** (1–4B) $(-2)^0 = 1$ **24.** (1–3A) $\dfrac{0}{-5} = 0$

25. (1–2) $-|-3| = -(3) = -3$ **26.** (1–3C) $\sqrt[3]{-27} = -3$ **27.** (1–3C) $\sqrt[7]{-1} = -1$

28. (1–3C) $\sqrt[4]{16} = 2$ **29.** (1–3C) $\sqrt{81} = 9$ **30.** (1–4B) $(3^{-2})^{-1} = 3^{(-2)(-1)} = 3^2 = 9$

31. (1–4B) $10^{-3} \cdot 10^5 = 10^{-3+5} = 10^2 = 100$ **32.** (1–4B) $\dfrac{2^{-4}}{2^{-7}} = 2^{-4-(-7)} = 2^{-4+7} = 2^3 = 8$

33. (1–4B) $\left(\dfrac{1}{10^{-3}}\right)^2 = (10^3)^2 = 10^6 = 1{,}000{,}000$

34. (1–5A) (a) $16 \div 4 \cdot 2 = 4 \cdot 2 = 8$
(b) $3 + 2 \cdot 5 = 3 + 10 = 13$
(c) $2\sqrt{9} - 5 = 2 \cdot 3 - 5 = 6 - 5 = 1$
(d) $16 \div (-2)^2 - \dfrac{7-1}{2} = 16 \div 4 - \dfrac{7-1}{2}$
$= 4 - 3 = 1$

35. (1–4B) $x^2 \cdot x^{-5} = x^{2-5} = x^{-3} = \dfrac{1}{x^3}$ **36.** (1–4A) $(N^2)^4 = N^{2\cdot 4} = N^8$ **37.** (1–4C) $\left(\dfrac{2x^3}{y}\right)^2 = \dfrac{2^2x^6}{y^2} = \dfrac{4x^6}{y^2}$

38. (1–4C) $\left(\dfrac{xy^{-2}}{y^{-3}}\right)^{-1} = (xy)^{-1} = \dfrac{1}{xy}$ **39.** (1–4B) $\dfrac{1}{a^{-3}} = a^3$ **40.** (1–4B) $\dfrac{x^2y^{-1}}{y^3z^4} = x^2y^{-4}z^{-4}$

41. (1–4B) $\dfrac{x^{4a}}{x^{7a}} = x^{4a-7a} = x^{-3a}$ **42.** (1–4C) $\left(\dfrac{x^{-a}}{2}\right)^{-3} = \dfrac{x^{3a}}{2^{-3}} = 8x^{3a}$

43. (1–5B) $C = \dfrac{5}{9}(F - 32)$
$= \dfrac{5}{9}(-4 - 32) = \dfrac{5}{9}(-36) = -20$

44. (1–6B) $7x - 2(5 - x) = 7x + (-2)(5) + (-2)(-x)$
$= 7x - 10 + 2x$
$= 9x - 10$

45. (1–6B) $y - [2(x - y) - 3(1 - y)]$
$= y - [2x - 2y - 3 + 3y]$
$= y - 2x + 2y + 3 - 3y = -2x + 3$

46. (1–6B) $6x(2xy^2 - 3x^3) - 3x^2(2y^2 - 6x^2)$
$= \underline{\underline{12x^2y^2}} - \underline{18x^4} - \underline{\underline{6x^2y^2}} + \underline{18x^4}$
$= 6x^2y^2$

47. (1–6B) $7x - 2\{6 - 3[8 - 2(x - 3) - 2(6 - x)]\}$
$= 7x - 2\{6 - 3[8 - 2x + 6 - 12 + 2x]\}$
$= 7x - 2\{6 - 3[2]\} = 7x - 2\{6 - 6\}$
$= 7x - 2\{0\}$
$= 7x - 0 = 7x$

48. (1–7C) $\dfrac{42u^4 - 7u^2 + 28}{-14u^2}$
$= \dfrac{42u^4}{-14u^2} + \dfrac{-7u^2}{-14u^2} + \dfrac{28}{-14u^2}$
$= -3u^2 + \dfrac{1}{2} - \dfrac{2}{u^2}$

49. (1–7B)

$$\begin{array}{r} 2z^4 + 0z^3 - z^2 - 9z + 8 \\ z - 4 \\ \hline -8z^4 - 0z^3 + 4z^2 + 36z - 32 \\ 2z^5 + 0z^4 - 0z^3 - 9z^2 + 8z \\ \hline 2z^5 - 8z^4 - z^3 - 5z^2 + 44z - 32 \end{array}$$

50. (1–7C)

$$\begin{array}{r} 4a^2 - 3a - 7 \\ 5a - 2\,\overline{)\,20a^3 - 23a^2 - 29a + 14} \\ 20a^3 - 8a^2 \\ \hline -15a^2 - 29a \\ -15a^2 + 6a \\ \hline -35a + 14 \\ -35a + 14 \\ \hline \end{array}$$

Exercises 2–1A (page 61)

1. $\{-3\}$

−3 0

3. $\{-5\}$

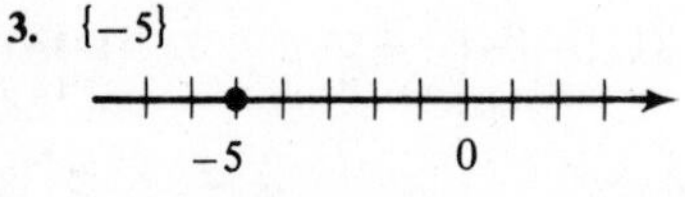

5. $\{4\}$

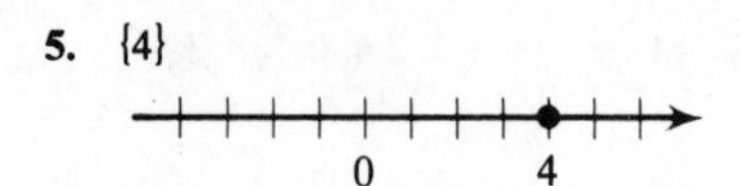

7.
$2(3x - 6) - 3(5x + 4) = 5(7x - 8)$
$6x - 12 - 15x - 12 = 35x - 40$
$-9x - 24 = 35x - 40$
$16 = 44x$
$\dfrac{16}{44} = x$
$\dfrac{4}{11} = x$

$\frac{4}{11}$

0 1

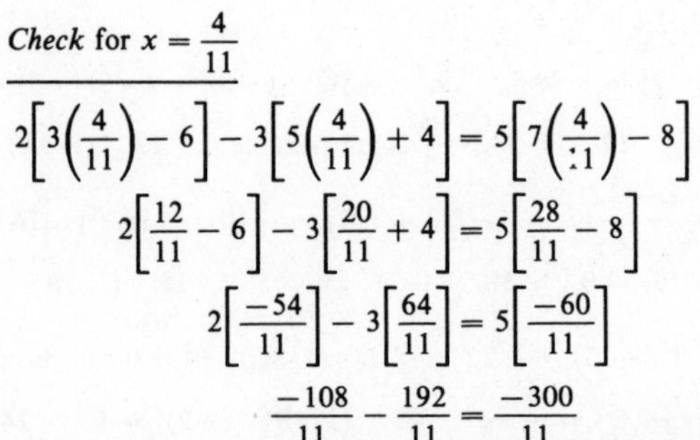

Check for $x = \dfrac{4}{11}$

$2\left[3\left(\dfrac{4}{11}\right) - 6\right] - 3\left[5\left(\dfrac{4}{11}\right) + 4\right] = 5\left[7\left(\dfrac{4}{11}\right) - 8\right]$

$2\left[\dfrac{12}{11} - 6\right] - 3\left[\dfrac{20}{11} + 4\right] = 5\left[\dfrac{28}{11} - 8\right]$

$2\left[\dfrac{-54}{11}\right] - 3\left[\dfrac{64}{11}\right] = 5\left[\dfrac{-60}{11}\right]$

$\dfrac{-108}{11} - \dfrac{192}{11} = \dfrac{-300}{11}$

9. $\{1\}$

1

0

11. $\{20\}$ **13.** $\{12\}$ **15.** $\left\{\frac{9}{8}\right\}$ **17.** $\left\{\frac{14}{17}\right\}$ **19.** $\{1\}$ **21.** $\{7\frac{1}{5}\}$

23. $\left\{\frac{7}{10}\right\}$ **25.** LCD = 10

$$\frac{\overset{2}{\cancel{10}}}{1}\left[\frac{2(m-3)}{\cancel{5}}\right]+\frac{\overset{5}{\cancel{10}}}{1}\left[\frac{(-3)(m+2)}{\cancel{2}}\right]=\frac{10}{1}\left(\frac{7}{\cancel{10}}\right)$$

$$\begin{aligned}4(m-3)-15(m+2)&=7\\4m-12-15m-30&\doteq 7\\-11m&=49\\m&=-4\tfrac{5}{11}\end{aligned}$$

27.

$$\begin{aligned}5(3-2x)-10&=4x+[-(2x-5)+15]\\15-10x-10&=4x+[-2x+5+15]\\15-10x-10&=4x-2x+20\\-15&=12x\\-\frac{15}{12}&=x\\-1\tfrac{1}{4}&=x\end{aligned}$$

29. $\{0\}$ **31.** $\{\doteq 1.57\}$ **33.** $\{\doteq 1.04\}$

Exercises 2–1B (page 63)

1.
$$\begin{aligned}5x-2(4-x)&=6\\5x-8+2x&=6\\7x&=14\\x&=2\end{aligned}$$
Conditional

3.
$$\begin{aligned}6x-3(5+2x)&=-15\\6x-15-6x&=-15\\0&=0\end{aligned}$$
Identity

5.
$$\begin{aligned}4x-2(6+2x)&=-15\\4x-12-4x&=-15\\0&\neq -3\end{aligned}$$
No solution

7. Identity; solution set = R **9.** $\left\{\frac{34}{41}\right\}$ Conditional **11.** Conditional; $\{6\}$

Exercises 2–2 (page 69)

1. $x<4$ **3.** $x\leq 13$ **5.** $x\geq -7$ **7.** $y<-11$ **9.** $z>10$ **11.** $a<-1$

13.
$$\begin{aligned}9(2-5m)-4&\geq 13m+8(3-7m)\\18-45m-4&\geq 13m+24-56m\\14-45m&\geq 24-43m\\+45m\quad&\quad+45m\\\hline 14&\geq 24+2m\\-10&\geq 2m\\-5&\geq m\end{aligned}$$

15.
$$\begin{aligned}10-5x&>2[3-5(x-4)]\\10-5x&>2[3-5x+20]\\10-5x&>2[23-5x]\\10-5x&>46-10x\\10+5x&>46\\5x&>36\\x&>\frac{36}{5}\text{ or }x>7\tfrac{1}{5}\end{aligned}$$

17. $k\leq 3\frac{1}{8}$

19. $z>12$

−4 0 4 8 12 16 20 24

21. LCD = 15

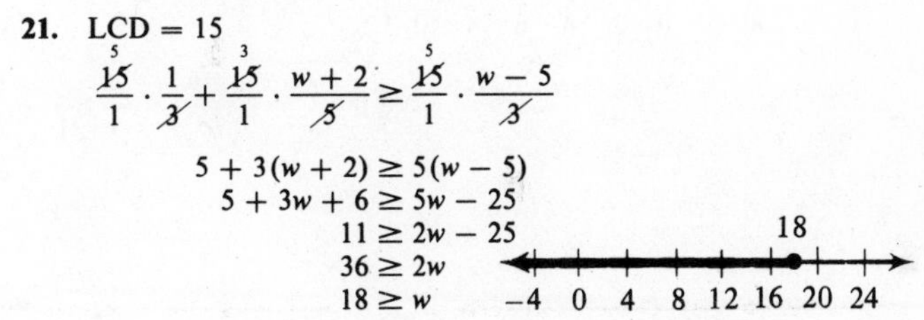

$$\frac{\overset{5}{\cancel{15}}}{1}\cdot\frac{1}{\cancel{3}}+\frac{\overset{3}{\cancel{15}}}{1}\cdot\frac{w+2}{\cancel{5}}\geq\frac{\overset{5}{\cancel{15}}}{1}\cdot\frac{w-5}{\cancel{3}}$$

$$\begin{aligned}5+3(w+2)&\geq 5(w-5)\\5+3w+6&\geq 5w-25\\11&\geq 2w-25\\36&\geq 2w\\18&\geq w\end{aligned}$$

23. $w>24$

0 4 8 12 16 20 24 28

25. $\{x\mid x>-2\}$

−4 −3 −2 −1 0 1 2 3 4 5

27. $\{x\mid -4\leq x<1\}$

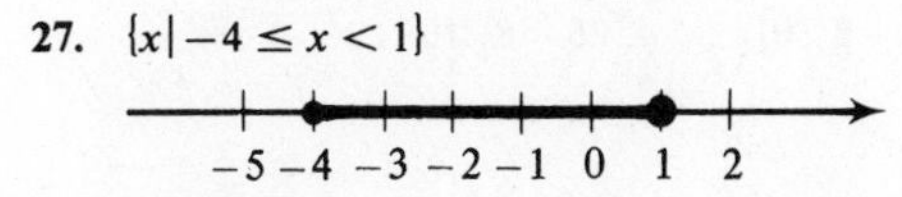

29.
$$\begin{aligned}4\geq x-3&\geq -5\\+3\quad\;+3\;&\quad+3\\\hline 7\geq x\quad&\geq -2,\ x\in N\end{aligned}$$
$=\{1, 2, 3, 4, 5, 6, 7\}$

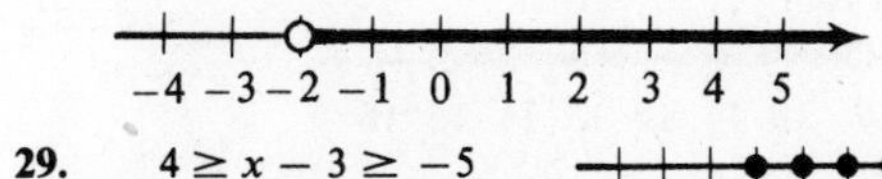

31. $\{x\mid -3<x<2,\ x\in J\}$

−5 −4 −3 −2 −1 0 1 2 3 4

33. $x\leq -0.300$ (approx.) **35.** $x\geq .156$ (approx.)

Exercises 2–3 (page 75)

1. $\{3, -3\}$

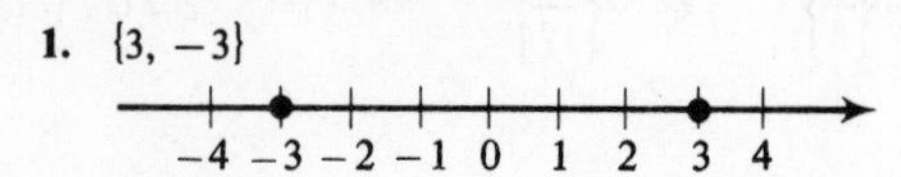

3. $\{2, -2\}$

5. $\{x \mid -2 < x < 2\}$

7. $\{x \mid -3 \le x \le 3\}$

9. $\{7, -1\}$

11. $\{x \mid x \le 1 \text{ or } x \ge 9\}$

13. $\{2, -3\}$

15. $\{\ \}$ The absolute value of a number can never be negative.

17. $\{7, 1\}$

19. $\left\{x \mid -3 < x < \frac{11}{3}\right\}$

21. $\left|\frac{1-x}{2}\right| = 6$

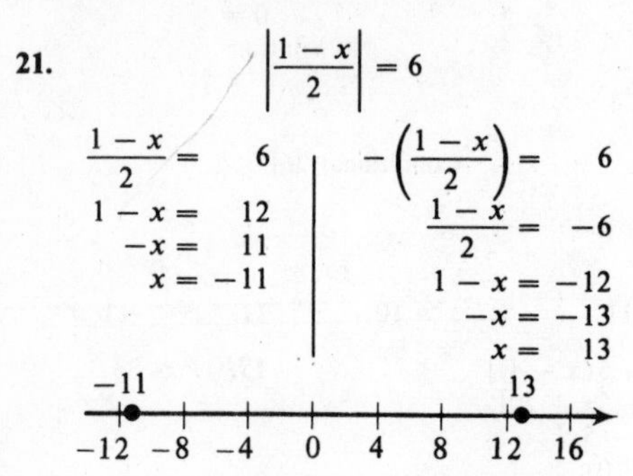

$$\frac{1-x}{2} = 6 \qquad\qquad -\left(\frac{1-x}{2}\right) = 6$$
$$1 - x = 12 \qquad\qquad \frac{1-x}{2} = -6$$
$$-x = 11 \qquad\qquad 1 - x = -12$$
$$x = -11 \qquad\qquad -x = -13$$
$$x = 13$$

23. $\left|\frac{3x-5}{2}\right| \le 10$

$$-10 \le \frac{3x-5}{2} \le 10$$
$$\frac{2}{1}\cdot\frac{-10}{1} \le \frac{2}{1}\cdot\frac{3x-5}{2} \le \frac{2}{1}\cdot\frac{10}{1}$$
$$-20 \le 3x - 5 \le 20$$
$$-15 \le 3x \le 25$$
$$-5 \le x \le \frac{25}{3}$$

25. $\{x \mid x > 10 \text{ or } x < -2\}$

27. $\left|3 - \frac{x}{2}\right| > 4$

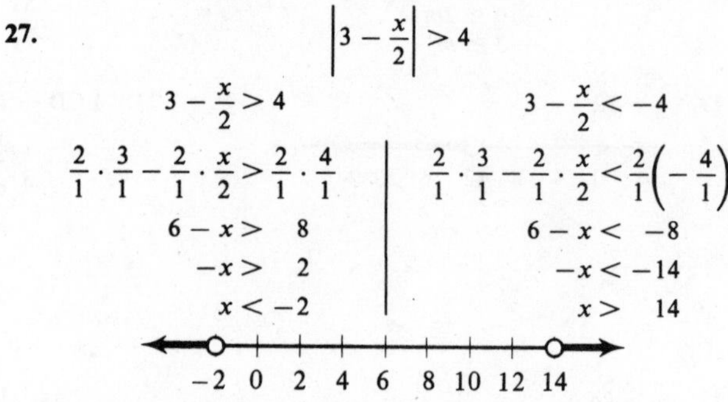

$$3 - \frac{x}{2} > 4 \qquad\qquad 3 - \frac{x}{2} < -4$$
$$\frac{2}{1}\cdot\frac{3}{1} - \frac{2}{1}\cdot\frac{x}{2} > \frac{2}{1}\cdot\frac{4}{1} \qquad\qquad \frac{2}{1}\cdot\frac{3}{1} - \frac{2}{1}\cdot\frac{x}{2} < \frac{2}{1}\left(-\frac{4}{1}\right)$$
$$6 - x > 8 \qquad\qquad 6 - x < -8$$
$$-x > 2 \qquad\qquad -x < -14$$
$$x < -2 \qquad\qquad x > 14$$

29. $\{x \mid 9 < x < 15\}$

31. $\{x \mid x > 10.49 \text{ or } x < -4.99\}$

Chapter Two Diagnostic Test or Review (page 76)

Following each problem number is the textbook section number (in parentheses) where that kind of problem is discussed.

1. (2–1A)

$$8x - 4(2 + 3x) = 12$$
$$8x - 8 - 12x = 12$$
$$-4x = 20$$
$$x = -5$$

Check

$$8x - 4(2 + 3x) = 12$$
$$8(-5) - 4[2 + 3(-5)] \stackrel{?}{=} 12$$
$$-40 - 4[2 - 15] \stackrel{?}{=} 12$$
$$-40 - 4[-13] \stackrel{?}{=} 12$$
$$-40 + 52 \stackrel{?}{=} 12$$
$$12 = 12$$

2. (2–1B)
$$\begin{aligned} 3(x-6) &= 5(1+2x)-7(x-4) \\ 3x-18 &= 5+10x-7x+28 \\ 3x-18 &= 3x+33 \\ -18 &\neq 33 \quad \text{No solution} \end{aligned}$$

3. (2–1A)
$$\begin{aligned} 4(3x-8)-2(-13-7x) &= 5(4x+6) \\ 12x-32+26+14x &= 20x+30 \\ 26x-6 &= 20x+30 \\ 6x &= 36 \\ x &= 6 \end{aligned}$$

Check
$$\begin{aligned} 4(3x-8)-2(-13-7x) &= 5(4x+6) \\ 4[3(6)-8]-2[-13-7(6)] &\stackrel{?}{=} 5[4(6)+6] \\ 4[18-8]-2[-13-42] &\stackrel{?}{=} 5[24+6] \\ 4[10]-2[-55] &\stackrel{?}{=} 5[30] \\ 40+110 &\stackrel{?}{=} 150 \\ 150 &= 150 \end{aligned}$$

4. (2–1B)
$$\begin{aligned} 2[7x-4(1+3x)] &= 5(3-2x)-23 \\ 2[7x-4-12x] &= 15-10x-23 \\ 2[-4-5x] &= -10x-8 \\ -8-10x &= -8-10x \\ 0 &= 0 \quad \text{Identity} \end{aligned}$$

5. (2–2)
$$\begin{aligned} 5w+2 &\le 10-w \\ 6w &\le 8 \\ w &\le \frac{8}{6} \\ w &\le \frac{4}{3} \end{aligned}$$

6. (2–2)
$$\begin{aligned} 13h-4(2+3h) &\ge 0 \\ 13h-8-12h &\ge 0 \\ h &\ge 8 \end{aligned}$$

7. (2–2)
$$\begin{aligned} 2[-5y-6(y-7)] &< 6+4y \\ 2[-5y-6y+42] &< 6+4y \\ 2[-11y+42] &< 6+4y \\ -22y+84 &< 6+4y \\ 78 &< 26y \\ 3 < y &\text{ or } y > 3 \end{aligned}$$

8. (2–3)
$$\begin{aligned} \{x \mid 4 \ge 3x+7 &> -2,\ x \in R\} \\ -3 \ge 3x &> -9 \\ -1 \ge x &> -3,\ x \in R \end{aligned}$$

Solution set $= \{x \mid -1 \ge x > -3,\ x \in R\}$

9. (2–2)
$$\left\{x \,\middle|\, \frac{5(x-2)}{3} + \frac{x}{4} \le 12,\ x \in R\right\}$$
$$\frac{\overset{4}{\cancel{12}}}{1} \cdot \frac{5(x-2)}{\cancel{3}} + \frac{\overset{3}{\cancel{12}}}{1} \cdot \frac{x}{\cancel{4}} \le \frac{12}{1} \cdot \frac{12}{1}$$
$$\begin{aligned} 20(x-2)+3x &\le 144 \\ 20x-40+3x &\le 144 \\ 23x &\le 184 \\ x &\le 8,\ x \in R \end{aligned}$$
$\{x \mid x \le 8,\ x \in R\}$

10. (2–4)
$$|7-3x| \ge 6$$
$$\begin{aligned} 7-3x &\ge 6 \\ 7 &\ge 3x+6 \\ 1 &\ge 3x \\ \frac{1}{3} &\ge x \end{aligned} \qquad \begin{aligned} 7-3x &\le -6 \\ 7 &\le 3x-6 \\ 13 &\le 3x \\ \frac{13}{3} &\le x \end{aligned}$$

Solution set $= \left\{x \mid x \ge 4\frac{1}{3}\right\} \cup \left\{x \mid x \le \frac{1}{3}\right\}$

11. (2–3)
$$\begin{aligned} \{x \mid |2x-5| < 11,\ x \in J\} \\ -11 < 2x-5 < 11 \\ -6 < \ 2x \ < 16 \\ -3 < \ x \ < 8 \end{aligned}$$

Solution set $= \{x \mid -3 < x < 8,\ x \in J\}$

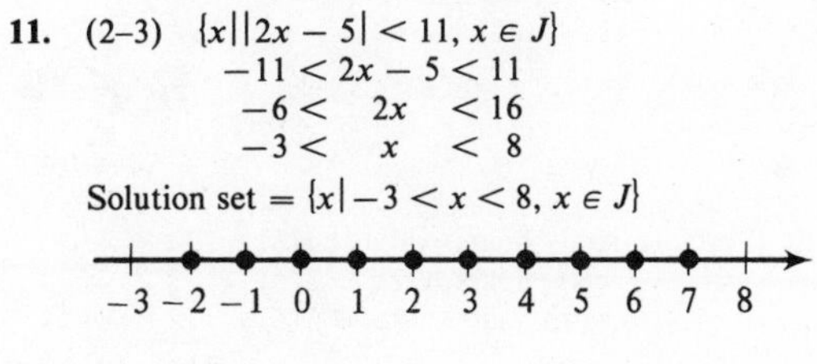

12. (2–4) $\left\{x \,\middle|\, \left|\frac{2x+3}{5}\right| = 1,\ x \in J\right\}$

$$\begin{aligned} \frac{2x+3}{5} &= 1 \\ 2x+3 &= 5 \\ 2x &= 2 \\ x &= 1 \end{aligned} \qquad \begin{aligned} -\left(\frac{2x+3}{5}\right) &= 1 \\ \frac{2x+3}{5} &= -1 \\ 2x+3 &= -5 \\ 2x &= -8 \\ x &= -4 \end{aligned}$$

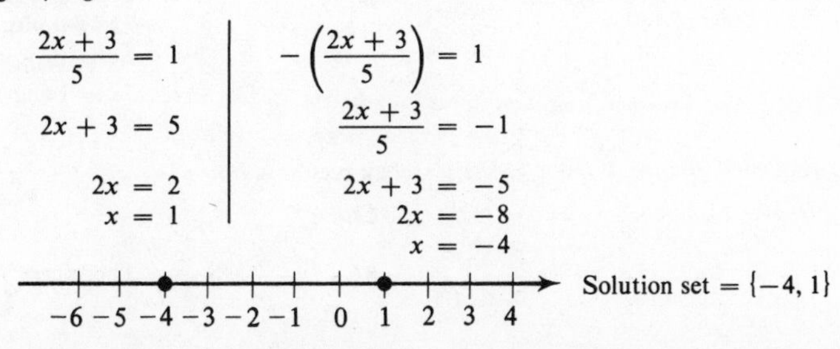

Solution set $= \{-4, 1\}$

Exercises 3–1 (page 82)

1. 8 **3.** 6 **5.** 8 **7.** 6 **9.** 11 **11.** 3 **13.** One-sixth of an unknown number is three.

$$\frac{1}{6}x = 3$$
$$6\left(\frac{1}{6}x\right) = 6(3)$$
$$x = 18$$

15. 56 **17.** When an unknown number is decreased by seven | the difference is | half the unknown number.

$x - 7 \quad = \quad \frac{1}{2}x$

$$2(x - 7) = 2(\tfrac{1}{2}x)$$
$$2x - 14 = x$$
$$x = 14$$

19. 7

21. 20 **23.** Three times the sum of eight and an unknown number | is equal to | twice the sum of the unknown number and seven.

$3(8 + x) \quad = \quad 2(x + 7)$

$$24 + 3x = 2x + 14$$
$$x = -10$$

25. When twice the sum of five and an unknown number is subtracted from | eight times the unknown number | the result is equal to | four times the sum of eight and twice the unknown number.

$8x \quad - \quad 2(5 + x) \quad = \quad 4(8 + 2x)$

$$8x - 10 - 2x = 32 + 8x$$
$$-42 = 2x$$
$$-21 = x$$

27. −8

Exercises 3–2A (page 85)

1. Smaller number = 36
Larger number = 45

3. 22, 77

5. First side = 27
Second side = 36
Third side = 45

7. 4 :2 :3 the ratio
$4x{:}2x{:}3x$ multiplied each term of the ratio by x
$3x$ = work hours
$2x$ = class hours
$4x$ = study hours

54 hours | of a student's week | are spent in | study, | class, | and | work.

$54 \quad = \quad 4x \quad + \quad 2x \quad + \quad 3x$

$$54 = 9x$$
$$6 = x$$

Therefore, study hours = $4x = 4(6) = 24$
class hours = $2x = 2(6) = 12$
work hours = $3x = 3(6) = 18$
54 *Check*

9. 10.5 liters of blue
3.5 liters of yellow

11. 7 :6 the ratio
$7x{:}6x$ multiplied each term of the ratio by x
$6x$ = Width (W)
$7x$ = Length (L)
Perimeter is 78.

$$2(W) + 2(L) = 78$$
$$2(6x) + 2(7x) = 78$$
$$12x + 14x = 78$$
$$26x = 78$$
$$x = 3$$

Therefore, Width $= 6x = 6(3) = 18$
Length $= 7x = 7(3) = 21$

13. \$12,000 \$10,000 \$6,000 **15.** 54 in., 24 in. **17.** 98

Exercises 3–2B (page 88)

1. {4.5} **3.** {14.4} **5.** {3.5}

7. $\frac{P}{3} = \frac{\frac{5}{6}}{5}$

$5P = (3)\left(\frac{5}{6}\right) = 2.5$

$P = \frac{2.5}{5} = 0.5$

Check $\frac{0.5}{3} = \frac{\frac{5}{6}}{5}$

$5(0.5) = (3)\left(\frac{5}{6}\right)$

$2.5 = 2.5$

9. $\frac{2x+7}{3x+10} = \frac{3}{4}$

$4(2x+7) = 3(3x+10)$

$8x + 28 = 9x + 30$

$-2 = x$

Check $\frac{2(-2)+7}{3(-2)+10} = \frac{3}{4}$

$\frac{-4+7}{-6+10} = \frac{3}{4}$

$\frac{3}{4} = \frac{3}{4}$

11. {5} **13.** $\left\{\frac{3}{13}\right\}$ **15.** {−2} **17.** {≐ 12.49}

Exercises 3–2C (page 90)

1. 21 games **3.** 45 ft **5.** \$8750

7. $\frac{1}{8} = \frac{1.5}{W}$

$1 \cdot W = (8)(1.5) = 12$ ft

$\frac{1}{8} = \frac{2.5}{L}$

$1 \cdot L = (8)(2.5) = 20$ ft

9. $\frac{\text{Wt on earth}}{\text{Wt on moon}} = \frac{6}{1}$

$\frac{126}{x} = \frac{6}{1}$

$6x = 1(126)$

$x = \frac{126}{6} = 21$ lb

11. 1050 **13.** 8¢

15. $\frac{\text{Wt of aluminum}}{\text{Wt of iron}} = \frac{9}{25}$

$\frac{45}{x} = \frac{9}{25}$

$9x = (45)(25)$

$x = \frac{(45)(25)}{9} = 125$ lb

17. 7.5 ft

Exercises 3–3 (page 94)

1. \$9.36

3. (12 / A) is (what percent / P) of (60 / B)?

$\frac{A}{B} = \frac{P}{100}$

$\frac{\overset{1}{\cancel{12}}}{\underset{5}{\cancel{60}}} = \frac{P}{100}$

$5P = 100$

$P = 20$

Therefore, 12 is 20% of 60.

5. \$624

7. (8% / P) of (some number / B) is (\$328.56 / A)

$\frac{A}{B} = \frac{P}{100}$

$\frac{328.56}{B} = \frac{8}{100}$

$8B = (100)(328.56)$

$B = \frac{(100)(328.56)}{8} = 4107$

Therefore, her gross sales for the week were \$4107.

9. \$6162

11. (Markup / A) = (60% / P) of (\$80 cost / B)

$\frac{A}{B} = \frac{P}{100}$

$\frac{A}{80} = \frac{\overset{3}{\cancel{60}}}{\underset{5}{\cancel{100}}}$

$A = \frac{(3)(80)}{5} = 48$

Selling cost = Cost + Markup
= \$80 + \$48 = \$128

13. \$105 **15.** \$20.57 **17.** \$9600 **19.** 93%

Exercises 3–4 (page 99)

1. 21.65 **3.** $\doteq$ 42 miles **5.** When $F = 3$ and $d = 4$, the formula $F = \frac{k}{d^2}$ becomes $3 = \frac{k}{4^2}$. Therefore, $k = 3(4^2) = 48$.

7. Condition 1 $\begin{cases} F_1 = 36 \\ d_1 = 4 \end{cases}$ Condition 2 $\begin{cases} F_2 = ? \\ d_2 = 80 \end{cases}$

$$\frac{F_1}{F_2} = \left(\frac{d_2}{d_1}\right)^2$$
$$\frac{36}{F_2} = \left(\frac{80}{4}\right)^2$$
$$\frac{36}{F_2} = 400$$
$$400\,F_2 = 36$$
$$F_2 = \frac{36}{400} = \frac{9}{100} = 0.09$$

9. $\doteq$ 2475 **11.** 1996.5 **13.** 8

15. Condition 1 $\begin{cases} e_1 = 3 \\ L_1 = 45 \\ P_1 = 2.4 \\ A_1 = 0.9 \end{cases}$ Condition 2 $\begin{cases} e_2 = ? \\ L_2 = 40 \\ P_2 = 1.5 \\ A_2 = 0.75 \end{cases}$

$$\frac{e_1}{e_2} = \frac{P_1}{P_2} \cdot \frac{L_1}{L_2} \cdot \frac{A_2}{A_1}$$
$$\frac{3}{e_2} = \frac{2.4}{1.5} \cdot \frac{45}{40} \cdot \frac{0.75}{0.9}$$
$$e_2 = \frac{3(1.5)(40)(0.9)}{(2.4)(45)(0.75)} = 2$$

17. $\doteq$ 4161 **19.** 847 lb.

21. (50-cent piece) + (Dime) = 60¢

This coin is not a dime. (pointing to the 50-cent piece)

Exercises 3–5A (page 103)

1. (a) Malone car 45 mph
King car 54 mph
(b) 270 miles

3. (a) 5 hr.
(b) 10 miles

5. Let x = Fran's speed
$\frac{4}{5}x$ = Ron's speed

	d =	r ·	t
Fran	$3x$	x	3
Ron	$3(\frac{4}{5}x)$	$\frac{4}{5}x$	3

$d = rt$
$d_1 = (x)3 = 3x$
$d = rt$
$d_2 = (\frac{4}{5}x)3 = 3(\frac{4}{5}x)$

Distance traveled by Fran + Distance traveled by Ron = 54 miles
$3x + 3(\frac{4}{5}x) = 54$

LCD = 5 $\quad \frac{5}{1}\left(\frac{3}{1}x\right) + \frac{5}{1}\left(\frac{3}{1}\right)\left(\frac{4}{5}x\right) = \frac{5}{1}\left(\frac{54}{1}\right)$

$$15x + 12x = 270$$
$$27x = 270$$
$$x = 10 \text{ mph} \quad \text{Fran's speed}$$
$$\frac{4}{5}x = \frac{4}{5}(10)$$
$$= 8 \text{ mph} \quad \text{Ron's speed}$$

7. Let $x =$ Speed of boat in still water
$x + 2 =$ Speed of boat downstream
$x - 2 =$ Speed of boat upstream

	d =	r ·	t
Upstream	$5(x-2)$	$x-2$	5
Downstream	$3(x+2)$	$x+2$	3

$d = rt$
$d_1 = (x-2)5$

$d = rt$
$d_2 = (x+2)3$

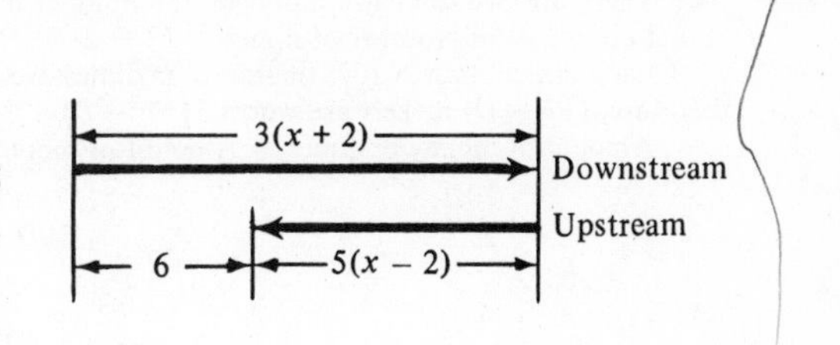

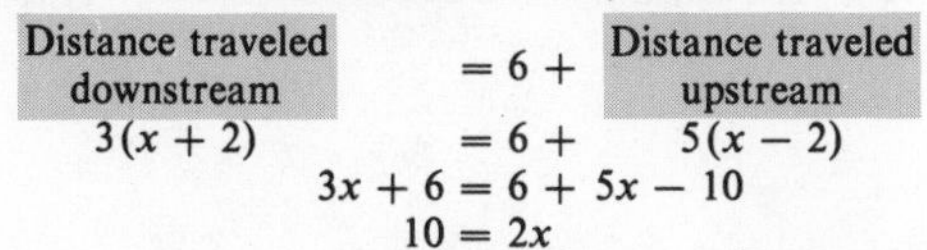

Distance traveled downstream $= 6 +$ Distance traveled upstream

$$3(x+2) = 6 + 5(x-2)$$
$$3x + 6 = 6 + 5x - 10$$
$$10 = 2x$$
$$x = 5 \text{ mph} \quad \text{Speed in still water}$$
$$3(x+2) = 3(5+2) = 3(7) = 21 \text{ miles} \quad \text{Traveled downstream}$$

9. 480 miles

11. 2 minutes. The last car of the train is a kilometer from the tunnel when the locomotive enters, so it must travel 2 kilometers.

Exercises 3–5B (page 105)

1. 30 lb Colombian coffee, 70 lb Brazilian coffee

3. 10 lb

5.

	Ingredient A	Ingredient B	Mixture
Unit cost	0.30	0.20	
Amount	x	$50 - x$	50
Total cost	$0.30x$	$0.20(50 - x)$	14.50

Cost of ingredient A + Cost of ingredient B = Cost of mixture

$$0.30x + 0.20(50 - x) = 14.50$$
$$30x + 20(50 - x) = 1450$$
$$30x + 1000 - 20x = 1450$$
$$10x = 450$$
$$x = 45 \text{ lb Delicious}$$
$$50 - x = 50 - 45 = 5 \text{ lb Jonathan}$$

7. \$3.50

Exercises 3–5C (page 106)

1.

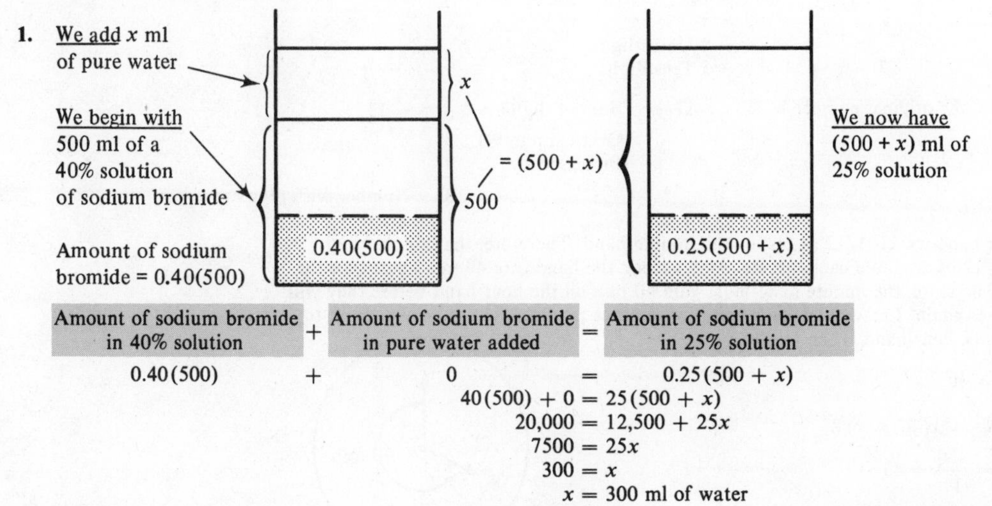

Amount of sodium bromide in 40% solution + Amount of sodium bromide in pure water added = Amount of sodium bromide in 25% solution

$$0.40(500) + 0 = 0.25(500 + x)$$
$$40(500) + 0 = 25(500 + x)$$
$$20{,}000 = 12{,}500 + 25x$$
$$7500 = 25x$$
$$300 = x$$
$$x = 300 \text{ ml of water}$$

3. 6 liters

5. 500 cc

7. 25 gal (30% solution)
75 gal (90% solution)

9. 8 liters (40% solution)
2 liters (90% solution)

Exercises 3–5D (page 107)

1. There are two unknown numbers: the number of nickels and the number of dimes.
Let $D =$ Number of dimes $17 - D =$ Number of nickels
Each dime is worth 10¢; therefore, D dimes are worth $10D$¢. Each nickel is worth 5¢; therefore; $(17 - D)$ nickels are worth $5(17 - D)$¢.

Amount of money in dimes + Amount of money in nickels = 115¢

$$\begin{aligned} 10D + 5(17 - D) &= 115 \\ 10D + 85 - 5D &= 115 \\ 5D &= 30 \\ D &= 6 \text{ dimes} \\ 17 - D = 17 - 6 &= 11 \text{ nickels} \end{aligned}$$

Check
$$\begin{aligned} 6 \text{ dimes} &= 6(10) = 60¢ \\ 11 \text{ nickels} &= 11(5) = 55¢ \\ 115¢ &= \$1.15 \end{aligned}$$

3. Let $x =$ First integer
$x + 1 =$ Next consecutive integer

The sum of two consecutive integers is thirty-three.

$$\begin{aligned} x + (x + 1) &= 33 \\ x + x + 1 &= 33 \\ 2x &= 32 \\ x &= 16 \\ x + 1 &= 17 \end{aligned}$$

Therefore, the consecutive integers are 16 and 17.

5. Let $x =$ Units digit
$13 - x =$ Tens digit
13 (Sum of the digits is 13)

Original number	*Number with digits reversed*
Tens digit, Units digit	Tens digit, Units digit
Written: $(13 - x)x$	Written: $x(13 - x)$
Value: $10(13 - x) + x$	Value: $10x + (13 - x)$

Value of the original number + 27 = Value of the number with digits reversed

$$\begin{aligned} 10(13 - x) + x + 27 &= 10x + (13 - x) \\ 130 - 10x + x + 27 &= 10x + 13 - x \\ -9x + 157 &= 9x + 13 \\ 144 &= 18x \\ 8 &= x \\ x &= 8 \text{ Units digit} \\ 13 - x = 13 - 8 &= 5 \text{ Tens digit} \end{aligned}$$

Therefore, the original number is 58.

Check Sum of digits $= 5 + 8 = 13$
Original number 58
Increased by 27
85 Number with digits reversed

7. The hour hand travels 1/12 as fast as the minute hand. Therefore, the minute hand gains 11/12 of a minute each minute. At 8 o'clock the hands are 40 min apart (see the figure). Therefore, the minute hand must gain 40 min on the hour hand before they will be together again. Let x equal the number of minutes required for the minute hand to overtake the hour hand. Then

$$x\left(\frac{11}{12}\right) = 40$$
$$11x = (12)(40) = 480$$
$$x = \frac{480}{11} = 43\frac{7}{11}$$

Therefore, the hands will be together at $8{:}43\frac{7}{11}$

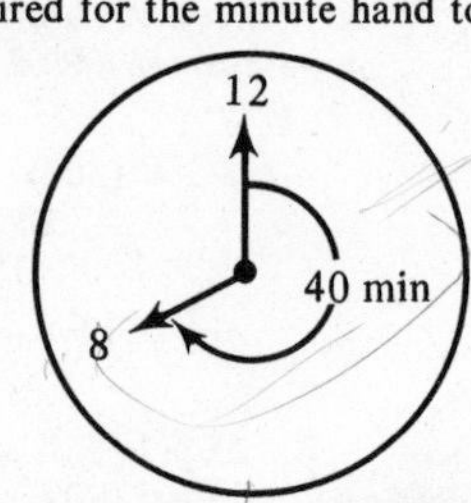

9. \$60, \$108 **11.** 584 days

Chapter Three Diagnostic Test or Review (page 108)

Following each problem number is the textbook section number (in parentheses) where that kind of problem is discussed.

1. (3–1) Let x = Unknown number

When 23	is added to	four times an unknown number	the sum is	31.
23	+	$4 \cdot x$	=	31

$$23 + 4x = 31$$
$$4x = 8$$
$$x = 2$$

2. (3–2)

Man / *Building*

$$\frac{6 \text{ ft height}}{4 \text{ ft shadow}} = \frac{x \text{ ft height}}{24 \text{ ft shadow}}$$
$$\frac{6}{4} = \frac{x}{24}$$
$$4 \cdot x = 6(24)$$
$$x = \frac{6(\overset{6}{\cancel{24}})}{\cancel{4}} = 36 \text{ ft}$$

3. (3–2A) Let $5x$ = Length
$4x$ = Width

Length	$\cdot$	Width	=	Area
$5x$	$\cdot$	$4x$	=	180

$$(5x)(4x) = 180$$
$$20x^2 = 180$$
$$x^2 = 9$$
$$x = 3$$

Therefore,
Length $= 5x = 5(3) = 15$
Width $= 4x = 4(3) = 12$

4. (3–5D) Let x = Number of dimes $\rightarrow$ Value of dimes $= 10x$
$14 - x$ = Number of quarters $\rightarrow$ Value of quarters $= 25(14 - x)$

Value of dimes	+	Value of quarters	=	Total value \$2.15
$10x$	+	$25(14 - x)$	=	215

$$10x + 350 - 25x = 215$$
$$-15x = -135$$

x = 9 dimes
$14 - x$ = 5 quarters

5. (3–5B) Let x = Pounds of cashews $\Rightarrow$ Cost of cashews $= 100x$
$60 - x$ = Pounds of peanuts $\Rightarrow$ Cost of peanuts $= 80(60 - x)$
Cost of mixture $= 85(60)$

Cost of cashews	+	Cost of peanuts	=	Total cost of mixture
$100x$	+	$80(60 - x)$	=	$85(60)$

$$100x + 4800 - 80x = 5100$$
$$20x = 300$$
x = 15 pounds of cashews
$60 - x$ = 45 pounds of peanuts

6. (3–5C) Let x = cc of water added
$600 + x$ = cc of 15% solution

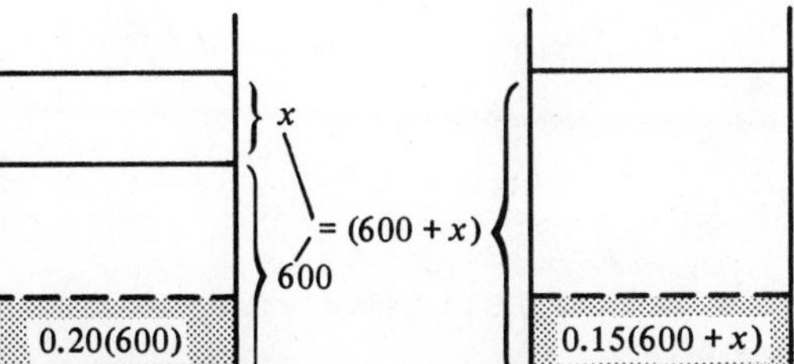

Amount of potassium chloride in 20% solution	=	Amount of potassium chloride in 15% solution
$0.20(600)$	=	$0.15(600 + x)$

$$20(600) = 15(600 + x)$$
$$12{,}000 = 9000 + 15x$$
$$3000 = 15x$$
x = 200 cc of water

7. (3–3)

Salary = \$240.00
Deductions = 57.60
Take-home pay = \$182.40

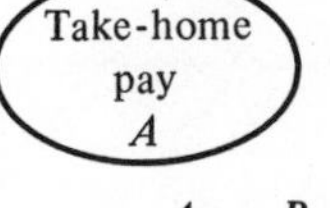

is

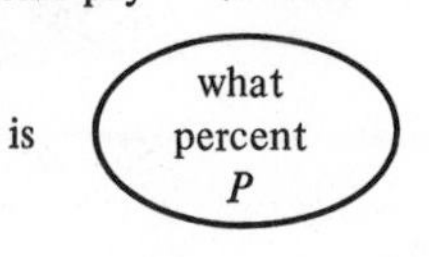

of

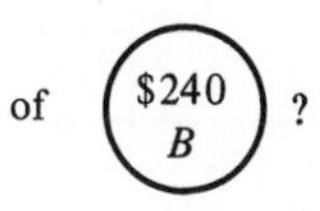

?

$$\frac{A}{B} = \frac{P}{100}$$
$$\frac{182.40}{240} = \frac{P}{100}$$
$$240P = (182.40)(100) = 18{,}240$$
$$P = \frac{18{,}240}{240} = 76$$

Therefore, the take-home pay is 76% of the salary.

8. (3–5D) Let x = Tens digit
$2x$ = Units digit

Value of original number + 27 = Value of number with digits reversed

$$10(x) + 1(2x) + 27 = 10(2x) + 1(x)$$
$$12x + 27 = 21x$$
$$27 = 9x$$
$x = 3$ Tens digit
$2x = 6$ Units digit

Therefore, the original number is 36.

9. (3–4) Condition 1 $\begin{cases} x_1 = 8 \\ y_1 = 6 \\ z_1 = 12 \\ w_1 = 20 \end{cases}$

$$\frac{w_1}{w_2} = \frac{x_1}{x_2} \cdot \frac{y_1}{y_2} \cdot \frac{(z_2)^2}{(z_1)^2}$$

$$\frac{20}{w_2} = \frac{8}{6} \cdot \frac{6}{10} \cdot \frac{(5)(5)}{(12)(12)} = \frac{5}{36}$$

Condition 2 $\begin{cases} x_2 = 6 \\ y_2 = 10 \\ z_2 = 5 \\ w_2 = ? \end{cases}$

$$5w_2 = (20)(36)$$
$$w_2 = 144$$

10. (3–5A) Let t = Time to hike from lake
$t + 2$ = Time to hike to lake

	d	=	r	·	t
To lake	$3(t + 2)$		3		$t + 2$
From lake	$5t$		5		t

Distance to lake = Distance from lake

$$5t = 3(t + 2)$$
$$5t = 3t + 6$$
$$2t = 6$$

(a) $\begin{cases} t = 3 \text{ hr to hike from lake} \\ t + 2 = 5 \text{ hr to hike to lake} \end{cases}$

(b) Distance = (Rate)(Time) = $3 \cdot 5 = 15$ miles

Exercises 4–1 (page 113)

1. Prime **3.** Composite, because it has factors 2, 3, 4, and 6 other than 1 and 12. **5.** Composite

7. Composite **9.** Composite **11.** $2^2 \cdot 7$ **13.** 2^5 **15.** $2 \cdot 17$ **17.** $2^2 \cdot 3 \cdot 7$

19.
```
2 | 144
2 |  72
2 |  36
2 |  18
3 |   9
      3
```
$144 = 2^4 \cdot 3^2$

21.
```
2 | 156
2 |  78
3 |  39
     13
```
$156 = 2^2 \cdot 3 \cdot 13$

23. $3x(3x + 1)$ **25.** $5a^2(2a - 5)$ **27.** $2ab(a + 2b)$

29. $6c^2d^2(2c - 3d)$

31. GCF = $4x$
$$4x^3 - 12x - 24x^2$$
$$= 4x(\ \ - \ \ - \ \ )$$
$$= 4x(x^2 - 3 - 6x)$$

33. Cannot be factored

35. First combine the common terms
$6my + 15mz - 5n - 4n$
$= 6my + 15mz - 9n$
$= 3(2my + 5mz - 3n)$

37. GCF = Either $-2 \cdot 7xy^3$ or $+2 \cdot 7xy^3$

Then $\begin{Bmatrix} -14x^8y^9 + 42x^5y^4 - 28xy^3 \\ = -14xy^3(x^7y^6 - 3x^4y + 2) \end{Bmatrix}$ or $\begin{Bmatrix} -14x^8y^9 + 42x^5y^4 - 28xy^3 \\ = 14xy^3(-x^7y^6 + 3x^4y - 2) \end{Bmatrix}$

Both of these answers are correct; however, we will use whichever answer is most suitable for the purpose at hand. Usually, the first answer is more convenient.

39. $11a^{10}b^4(-4a^4b^3 - 3b + 2a)$ **41.** $2(9u^{10}v^5 + 12 - 7u^{10}v^6)$ **43.** $6x^2(3xy^4 - 2z^3 - 8x^2y^3)$

45. $-5r^7s^5(7t^4 + 11rs^4u^4 - 8p^8r^2s^3)$

47. GCF $= -12x^4y^3$

$$-24x^8y^3 - 12x^7y^4 + 48x^5y^5 + 60x^4y^6$$
$$= -12x^4y^3(\quad + \quad - \quad - \quad)$$
$$= -12x^4y^3(2x^4 + x^3y - 4xy^2 - 5y^3)$$

49. GCF $= 5(x - y)$

$$5(x - y) - 15(x - y)^2$$
$$= 5(x - y)[\quad - \quad]$$
$$= 5(x - y)[1 - 3(x - y)]$$
$$= 5(x - y)(1 - 3x + 3y)$$

51. $3(x - y)^2(1 + 3x - 3y)$ **53.** $3(a + b)(x - y)(1 + 2x - 2y)$

Exercises 4–2 (page 116)

1. $x^2 - 9$ **3.** $4u^2 - 25v^2$ **5.** $4x^4 - 81$ **7.** $x^{10} - y^{12}$ **9.** $49m^2n^2 - 4r^2s^2$ **11.** $144x^8y^6 - u^{14}v^2$

13. $[(a + b) + 2][(a + b) - 2] = (a + b)^2 - (2)^2$ **15.** $(2c + 1)(2c - 1)$ **17.** $2(x + 2y)(x - 2y)$

19. $(7u^2 + 6v^2)(7u^2 - 6v^2)$ **21.** $(ab + cd)(ab - cd)$

23. $x^4 - y^4 = (\sqrt{x^4} + \sqrt{y^4})(\sqrt{x^4} - \sqrt{y^4})$
$= (x^2 + y^2)(x^2 - y^2)$
$= (x^2 + y^2)(x + y)(x - y)$

25. $(2h^2k^2 + 1)(2h^2k^2 - 1)$ **27.** $(5a^2b + cd^2)(5a^2b - cd^2)$

29. $(x + y)^2 - (2)^2 = [(x + y) + 2][(x + y) - 2]$
$= (x + y + 2)(x + y - 2)$

31. $(a + b)x^2 - (a + b)y^2 = (a + b)(x^2 - y^2)$
$= (a + b)(x + y)(x - y)$

33. 6,293,441

Exercises 4–3A (page 119)

1. $a^2 + 7a + 10$ **3.** $y^2 - y - 72$ **5.** $x^2 + 6x + 9$ **7.** $b^2 - 8b + 16$ **9.** $6x^2 - 7x - 20$

11. $56z^2 + 5z - 6$ **13.** $2a^2 + 7ab + 5b^2$

15. $(4x - y)\quad(2x + 7y)$
$-2xy$
$+28xy$
$8x^2 + 26xy - 7y^2$

17. $9x^2 + 24x + 16$

19. $(7x - 10y)^2 = (7x)^2 - 2(7x)(10y) + (10y)^2$
$= 49x^2 - 140xy + 100y^2$

21. $4a^2 - 12ab + 9b^2$ **23.** $12x^4 + x^2y - 6y^2$

Exercises 4–3B (page 120)

1. $(k + 1)(k + 6)$ **3.** $(b - 2)(b - 7)$ **5.** Cannot be factored **7.** $(z + 2)(z - 3)$

9. $(m + 1)(m + 12)$ **11.** $(h - 2)(h + 4)$ **13.** $(t - 3)(t + 10)$

15. $u^2 + 12u - 64 = (u - 4)(u + 16)$
$+12 = -4 + +16$

$64 = 1 \cdot 64$
$2 \cdot 32$
$4 \cdot 16$
$8 \cdot 8$

17. First rearrange the terms
$x^2 + 3x + 2 = (x + 2)(x + 1)$
$3 = 2 + 1$

19. $(x + 2y)(x - y)$ **21.** $(z - 2a)(z - 10a)$ **23.** $(f + 2g)(f - 9g)$

25. $(a + b)^2 + 6(a + b) + 8 = [(a + b) + 2][(a + b) + 4]$
$+6 = +2 + +4$

$8 = 1 \cdot 8$
$= 2 \cdot 4$

$(a + b)^2 + 6(a + b) + 8 = (a + b + 2)(a + b + 4)$

27. $(x + y + 2)(x + y - 15)$

29. $(x + y + 3)(x + y - 5)$

Exercises 4–3C (page 123)

1. $(x + 2)(3x + 1)$ **3.** $(x + 1)(4x + 3)$ **5.** $(1 + 5x)(4 + x)$ **7.** $(a - 3)(5a - 1)$

9. $(7 - b)(1 - 3b)$ **11.** $(v - 3)(5v - 2)$ **13.** First rearrange the terms
$4 + 7h^2 - 11h = 7h^2 - 11h + 4$
$= (h - 1)(7h - 4)$
15. $(t + 6)(3t - 1)$

17. $8 + 12z - 8z^2 = 4(2 + 3z - 2z^2)$
$2 = 1 \cdot 2$ $4(1 + 2z)$ $(2 - z)$ $2 = 1 \cdot 2$
$+4z$
$-1z$
$+3z$
Therefore, $8 + 12z - 8z^2 = 4(1 + 2z)(2 - z)$

19. $2(2x - y)(3x + 4y)$ **21.** Cannot be factored

23. $2(v + 4w)(2v - w)$ **25.** $(2x + 3y)(4x - 5y)$ **27.** $(3a - 3b + 1)(a - b + 2)$

29. Let $(x + y) = a$
Then $5(x + y)^2 + 21(x + y) + 4 = 5a^2 + 21a + 4 = (\ a + \ )\ (a + \ )$
$5 = 1 \cdot 5$ $(5a + 1)(a + 4)$ $4 = 1 \cdot 4$
$2 \cdot 2$
$1a$
$20a$
$21a$
$21a$

Therefore, $5a^2 + 21a + 4 = (5a + 1)(a + 4)$
But, since $a = (x + y)$,
$5(x + y)^2 + 21(x + y) + 4 = [5(x + y) + 1][(x + y) + 4]$
$= (5x + 5y + 1)(x + y + 4)$

Exercises 4–3D (page 125)

1. $(2a + 1)^2$ **3.** $(6x + y)^2$ **5.** Cannot be factored **7.** $(3 - 2m)^2$ **9.** $(a^2 + b^2)^2$

11. $100m^2 + 180mn^2 + 81n^4$
$(10m)^2 + 2(10m)(9n^2) + (9n^2)^2$
$(10m + 9n^2)^2$

13. $2x^2y + 8xy^2 + 8y^3$
$2y(x^2 + 4xy + 4y^2)$
$2y[(x)^2 + 2(x)(2y) + (2y)^2]$
$2y(x + 2y)^2$

15. $3(mn - 1)^2$ **17.** $x^2(x - 3y)^2$

Exercises 4–4 (page 127)

1. $(x - 2)(x^2 + 2x + 4)$ **3.** $(4 + a)(16 - 4a + a^2)$ **5.** $(a - 1)(a^2 + a + 1)$

7. $1 - 27a^3b^3 = (1)^3 - (3ab)^3$
$= (1 - 3ab)[(1)^2 + (1)(3ab) + (3ab)^2]$
$= (1 - 3ab)(1 + 3ab + 9a^2b^2)$

9. $(2xy^2 + 3)(4x^2y^4 - 6xy^2 + 9)$ **11.** Cannot be factored

13. $a^4 + ab^3 = a[a^3 + b^3]$
$= a[(a)^3 + (b)^3]$
$= a(a + b)[(a)^2 - (a)(b) + (b)^2]$
$= a(a + b)(a^2 - ab + b^2)$

15. $2(2 - m)(4 + 2m + m^2)$ **17.** $(x + 2)(x^2 + x + 1)$

19. $a(a^2 - 6a + 12)$ **21.** $(x + 1)^3 - (x - 1)^3$
$= [(x + 1) - (x - 1)][(x + 1)^2 + (x + 1)(x - 1) + (x - 1)^2]$
$= (x + 1 - x + 1)(x^2 + 2x + 1 + x^2 - 1 + x^2 - 2x + 1)$
$= 2(3x^2 + 1)$

23. $2b(3a^2 + b^2)$

Exercises 4–5 (page 129)

1. $(a + b)(m + n)$ **3.** $(y + 1)(x - 1)$ **5.** $(h + k)(h - k + 2)$ **7.** Cannot be factored

9. $a(a + 6)(a^2 + 6)$ **11.** $(a + b)(a - b + a^2 - ab + b^2)$ **13.** $(x + y)(x^2 - xy + y^2)(x^3 - y^3 + 1)$

15. $a^2 + b^2 + 2ab + 2b + 2a$
$= a^2 + 2ab + b^2 + 2a + 2b$
$= (a + b)^2 + 2(a + b)$
$= (a + b)[(a + b) + 2]$
$= (a + b)(a + b + 2)$

17. $(3x + 2y)(x - y + 1)$

19. Rearrange terms
$x^3 + x^2 - 2xy + y^2 - y^3$
$= x^3 - y^3 + x^2 - 2xy + y^2$
$= (x - y)(x^2 + xy + y^2) + (x - y)^2$
$= (x - y)(x^2 + xy + y^2 + x - y)$

Exercises 4–6 (page 131)

1. $(x^2 + x + 2)(x^2 - x + 2)$ **3.** $(u^2 + 2u + 4)(u^2 - 2u + 4)$

5.
$$\begin{array}{l} 4m^4 + 5m^2 + 1 \\ \underline{\quad\;\; + \; m^2 \qquad - m^2} \\ 4m^4 + 4m^2 + 1 - m^2 \end{array}$$
$(2m^2 + 1)^2 - (m)^2$
$[(2m^2 + 1) + (m)][(2m^2 + 1) - (m)]$
$(2m^2 + m + 1)(2m^2 - m + 1)$

7.
$$\begin{array}{l} 64a^4 \qquad\qquad + b^4 \\ \underline{\qquad + 16a^2b^2 \qquad - 16a^2b^2} \\ 64a^4 + 16a^2b^2 + b^4 - 16a^2b^2 \end{array}$$
$(8a^2 + b^2)^2 - (4ab)^2$
$[(8a^2 + b^2) + (4ab)][(8a^2 + b^2) - (4ab)]$
$(8a^2 + 4ab + b^2)(8a^2 - 4ab + b^2)$
$(a + 4b)(a + b)(a - 4b)(a - b)$

9. $(2a^2 + 2a + 1)(2a^2 - 2a + 1)$ **11.** $(x^2 + 3x + 3)(x^2 - 3x + 3)$ **13.** $(a^2 + 5ab + 4b^2)(a^2 - 5ab + 4b^2)$

15. Cannot be factored **17.** $(3x^2 + 4xy - 2y^2)(3x^2 - 4xy - 2y^2)$ **19.** $(3x^2 + 2xy - y^2)(3x^2 - 2xy - y^2)$

21. $2(5x^2 + 4xy + y^2)(5x^2 - 4xy + y^2)$ **23.** $8m^4n + 2n^5 = 2n(4m^4 + n^4)$

$$\begin{array}{l} 4m^4 \qquad\qquad + n^4 \\ \underline{\quad + 4m^2n^2 \qquad - 4m^2n^2} \\ 4m^4 + 4m^2n^2 + n^4 - 4m^2n^2 \end{array}$$
$(2m^2 + n^2)^2 - (2mn)^2$
$(2m^2 + n^2 + 2mn)(2m^2 + n^2 - 2mn)$

Therefore, $8m^4n + 2n^5 = 2n(2m^2 + 2mn + n^2)(2m^2 - 2mn + n^2)$

Exercises 4–7A (page 134)

1. $x + 5 \quad$ R -3 **3.** $x^2 - x - 1 \quad$ R 10 **5.** $x^3 + 0x^2 + 0x - 1 \quad$ R 2 **7.** $x^3 + 2x^2 + 4x + 8$

9.
$$\begin{array}{r|rrrrrrr} & 1 & -3 & 0 & 0 & -2 & 3 & 5 \\ & & 3 & 0 & 0 & 0 & -6 & -9 \\ \hline 3 & 1 & 0 & 0 & 0 & -2 & -3 & -4 \end{array}$$
$x^5 + 0x^4 + 0x^3 + 0x^2 - 2x - 3 \quad$ R -4

11.
$$\begin{array}{r|rrrrr} & 3 & -1 & 9 & 0 & -1 \\ & & 1 & 0 & 3 & 1 \\ \hline \frac{1}{3} & 3 & 0 & 9 & 3 & 0 \end{array}$$
$3x^3 + 0x^2 + 9x + 3$

13.
$$\begin{array}{r|rrrr} & 2.6 & 0 & 1.8 & -6.4 \\ & & 3.9 & 5.85 & 11.475 \\ \hline 1.5 & 2.6 & 3.9 & 7.65 & 5.075 \end{array}$$
$2.6x^2 + 3.9x + 7.65 \quad$ R 5.075

Exercises 4–7B (page 136)

1. $(x - 1)(x^2 + 2x + 3)$ **3.** $(x - 2)(x - 3)(x + 2)$

5. $6x^3 - 13x^2 + 4$
Factors of the constant term are $\pm 1, \pm 2, \pm 4$

$$\begin{array}{r|rrrr} & 6 & -13 & 0 & 4 \\ & & 6 & -7 & -7 \\ \hline \text{Divide by } 1 & 6 & -7 & -7 & -3 \end{array}$$
Remainder is not zero; therefore $(x - 1)$ is not a factor

$$\begin{array}{r|rrrr} & 6 & -13 & 0 & 4 \\ & & 12 & -2 & -4 \\ \hline \text{Divide by } 2 & 6 & -1 & -2 & 0 \end{array}$$
Remainder is zero; therefore $(x - 2)$ is a factor

$6x^2 - x - 2$ — Quotient is another factor
$(2x + 1)(3x - 2)$ — Factors of quotient

Therefore, $6x^3 - 13x^2 + 4 = (x - 2)(2x + 1)(3x - 2)$

7. $(x+1)(x-2)(x-3)$ **9.** $(x+2)(x^3-2x^2+x+2)$ **11.** $(x-1)(x+2)(x-3)$

13. $(x-2)(2x^2-x-4)$ **15.** $(x+2)(x-2)(x+3)(x-3)$ **17.** $(x-y)(x^2+xy+y^2+1)$

19. $2x^4+3x^3+2x^2+x+1$
Factors of the constant term are ± 1

$$\begin{array}{rr|rrrrr} & & 2 & 3 & 2 & 1 & 1 \\ & & & -2 & -1 & -1 & 0 \\ \hline \text{Divide by} & -1 & 2 & 1 & 1 & 0 & 1 \end{array}$$
Remainder is not zero; therefore $(x+1)$ is not a factor

$$\begin{array}{rr|rrrrr} & & 2 & 3 & 2 & 1 & 1 \\ & & & 2 & 5 & 7 & 8 \\ \hline \text{Divide by} & 1 & 2 & 5 & 7 & 8 & 9 \end{array}$$
Remainder is not zero; therefore $(x-1)$ is not a factor

Since none of the possible factors work, the expression cannot be factored

21. $x^4-4x^3-7x^2+34x-24$
Factors of the constant term are $\pm 1, \pm 2, \pm 3, \pm 4, \pm 6, \pm 8, \pm 12, \pm 24$

$$\begin{array}{rr|rrrrr} & & 1 & -4 & -7 & 34 & -24 \\ & & & 1 & -3 & -10 & 24 \\ \hline \text{Divide by} & 1 & 1 & -3 & -10 & 24 & 0 \end{array}$$
Remainder is zero; therefore $(x-1)$ is a factor

$$\begin{array}{rr|rrrrr} & & 1 & -4 & -7 & 34 & -24 \\ & & & 2 & -4 & -22 & 24 \\ \hline \text{Divide by} & 2 & 1 & -2 & -11 & 12 & 0 \end{array}$$
Remainder is zero; therefore $(x-2)$ is a factor

$$\begin{array}{rr|rrrrr} & & 1 & -4 & -7 & 34 & -24 \\ & & & 4 & 0 & -28 & 24 \\ \hline \text{Divide by} & 4 & 1 & 0 & -7 & 6 & 0 \end{array}$$
Remainder is zero; therefore $(x-4)$ is a factor

$$\begin{array}{rr|rrrrr} & & 1 & -4 & -7 & 34 & -24 \\ & & & -3 & 21 & -42 & 24 \\ \hline \text{Divide by} & -3 & 1 & -7 & 14 & -8 & 0 \end{array}$$
Remainder is zero; therefore $(x+3)$ is a factor

Therefore, $x^4-4x^3-7x^2+34x-24=(x-1)(x-2)(x-4)(x+3)$
There can be no more linear factors, because if five or more linear factors are multiplied, the highest-degree term of their product is greater than fourth degree.

Exercises 4–8 (page 138)

1. $(4e-5)(3e+7)$ **3.** $4yz(2y+z)(2y-z)$ **5.** Cannot be factored **7.** $3(2my-3nz+5mz)$

9.
$6ac-6bd+6bc-6ad$ Common factor
$=6(ac-bd+bc-ad)$ Grouping
$=6[a(c-d)+b(c-d)]$ Common factor
$=6(a+b)(c-d)$

11. $2xy(y+3)(y-5)$ **13.** $3(x+2h)(x^2-2xh+4h^2)$

15. $2(x-1)(3x+5)$ **17.** $6a(b+c)$

19. $3x^3+x^2+3x+5$ Synthetic division

$$\begin{array}{r|rrrr} & 3 & 1 & 3 & 5 \\ & & -3 & 2 & -5 \\ \hline -1 & 3 & -2 & 5 & 0 \end{array}$$

$(x+1)(3x^2-2x+5)$

21. $(3e-5f)^2$

23.
$$\begin{array}{l} 4 \qquad\quad + k^4 \\ \underline{+\,4k^2 \qquad - 4k^2} \\ 4+4k^2+k^4-4k^2 \end{array}$$
Completing the square

$(2+k^2)^2-(2k)^2$ Difference of 2 squares
$[(2+k^2)+2k][(2+k^2)-2k]$
$(2+2k+k^2)(2-2k+k^2)$

25. $4(x+2y)(x^2-2xy+4y^2)$ **27.** $(x-1)(x+2)(x-3)$

Exercises 4–9 (page 141)

1.
$$\begin{aligned} 3x(x-4) &= 0 \\ 3x = 0 \quad &\mid \quad x-4=0 \\ x = 0 \quad &\mid \quad x = 4 \end{aligned}$$

3.
$$\begin{aligned} x(x-4) &= 12 \\ x^2-4x &= 12 \\ x^2-4x-12 &= 0 \\ (x-6)(x+2) &= 0 \\ x-6=0 \quad \mid \quad x+2 &= 0 \\ x=6 \quad \mid \quad x &= -2 \end{aligned}$$

5. $\{0, 2\}$

7.
$$\begin{aligned} x^2-x-12 &= 0 \\ (x-4)(x+3) &= 0 \\ x-4=0 \quad \mid \quad x+3 &= 0 \\ x=4 \quad \mid \quad x &= -3 \end{aligned}$$

9.
$$\begin{aligned} x^2-18 &= 9x \\ x^2-9x-18 &= 0 \end{aligned}$$
Cannot be solved by factoring

11. $\{-4, 3\}$ **13.** $\left\{-\frac{2}{3}, \frac{5}{2}\right\}$ **15.** $\left\{5, \frac{1}{3}\right\}$

17.
$$\begin{aligned} 5a^2 &= 16a-3 \\ 5a^2-16a+3 &= 0 \\ (1a-3)(5a-1) &= 0 \\ a-3=0 \quad \mid \quad 5a-1 &= 0 \\ a=3 \quad \mid \quad 5a &= 1 \\ a &= \frac{1}{5} \end{aligned}$$

19.
$$\begin{aligned} 4x^2+9 &= 12x \\ 4x^2-12x+9 &= 0 \\ (2x-3)^2 &= 0 \\ 2x-3 &= 0 \\ 2x &= 3 \\ x &= \frac{3}{2} \end{aligned}$$

21.
$$\begin{aligned} 4x(2x-1)(3x+7) &= 0 \\ 4x=0 \quad \mid \quad 2x-1=0 \quad \mid \quad 3x+7 &= 0 \\ x=0 \quad \mid \quad 2x=1 \quad \mid \quad 3x &= -7 \\ x=\frac{1}{2} \quad \mid \quad x &= -\frac{7}{3} \end{aligned}$$

23.
$$\begin{aligned} 2x^3+x^2 &= 3x \\ 2x^3+x^2-3x &= 0 \\ x(2x^2+x-3) &= 0 \\ x(1x-1)(2x+3) &= 0 \\ x=0 \quad \mid \quad x-1=0 \quad \mid \quad 2x+3 &= 0 \\ x=1 \quad \mid \quad 2x &= -3 \\ x &= -\frac{3}{2} \end{aligned}$$

25.
$$\begin{aligned} 21x^2+60x &= 18x^3 \\ 18x^3-21x^2-60x &= 0 \\ 3x(6x^2-7x-20) &= 0 \\ 3x(2x-5)(3x+4) &= 0 \\ 3x=0 \quad \mid \quad 2x-5=0 \quad \mid \quad 3x+4 &= 0 \\ x=0 \quad \mid \quad 2x=5 \quad \mid \quad 3x &= -4 \\ x=\frac{5}{2} \quad \mid \quad x &= -\frac{4}{3} \end{aligned}$$

27. $\{2, 1, -1\}$ **29.** $\{-3, -1, \frac{1}{2}\}$

Exercises 4–10 (page 144)

1. $\frac{-9}{3} = \frac{6}{-2}$ and $\frac{2}{3} = \frac{6}{9}$

3. There are two answers: $\{2, 7\}$ and $\{-7, -2\}$

5.
x = First integer
$x + 1$ = Second integer
$x + 2$ = Third integer

The product of the first two	is	10 less than	the product of the last two

$$\begin{aligned} x(x+1) &= (x+1)(x+2) - 10 \\ x^2+x &= x^2+3x+2-10 \\ 8 &= 2x \\ 4 &= x \\ 5 &= x+1 \\ 6 &= x+2 \end{aligned}$$

7. Length = 9 yd, Width = 5 yd

9. Side of larger square = 6 cm Side of smaller square = 3 cm

11. Length = 10 yd Width = 4 yd

13. (Length)(Width)(Depth) = Volume
$$\begin{aligned} (x+3)(x)(2) &= 80 \\ (x+3)(x) &= 40 \\ x^2+3x-40 &= 0 \\ (x+8)(x-5) &= 0 \\ x+8=0 \quad \mid \quad x-5 &= 0 \\ x=-8 \quad \mid \quad x &= 5 \text{ Width} \\ x+3 &= 8 \text{ Length} \end{aligned}$$
-8 has no meaning in this problem
(a) The dimensions of the metal sheet are 9 by 12 in.
(b) The dimensions of the box are:
Depth = 2 in., Width = 5 in., Length = 8 in.

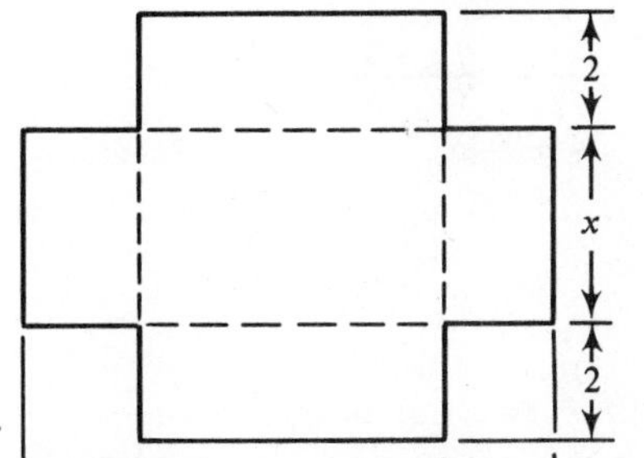

15. The number of children, n, must divide \$15 exactly.
If $n = 3$, each child would receive \$5. The products of the amounts they receive would be $5 \cdot 5 \cdot 5 = 125$, which is *not* divisible by 9 (the number of square feet in a square yard).
If $n = 5$, each child would receive \$3. $3 \cdot 3 \cdot 3 \cdot 3 \cdot 3 = 243$, which *is* exactly divisible by 9.
Solution: 5 children; each receives \$3.

Chapter Four Diagnostic Test or Review (page 145)

Following each problem number is the textbook section number (in parentheses) where that kind of problem is discussed.

1. (4–1)
$$\begin{array}{r|l} 2 & 126 \\ 3 & 63 \\ 3 & 21 \\ & 7 \end{array}$$
$126 = 2 \cdot 3 \cdot 3 \cdot 7$
$= 2 \cdot 3^2 \cdot 7$

2. (4–1) $2xy^3(4x^2y - 3xy - 5)$
$= (2xy^3)(4x^2y) + (2xy^3)(-3xy) + (2xy^3)(-5)$
$= 8x^3y^4 - 6x^2y^4 - 10xy^3$

3. (4–2) $(2x^4 + 3)(2x^4 - 3)$
$= (2x^4)^2 - (3)^2$
$= 4x^8 - 9$

4. (4–3A) $(5m - 2)\quad(3m + 4)$
Inner: $-6m$
Outer: $20m$
$15m^2 + 14m - 8$

5. (4–3A) $(3R^2 - 5)^2$
$= (3R^2)^2 + 2(3R^2)(-5) + (-5)^2$
$= 9R^4 - 30R^2 + 25$

6. (4–1, 4–2) $4x - 16x^3$
$= 4x(1 - 4x^2)$
$= 4x(1 + 2x)(1 - 2x)$

7. (4–4) $y^3 - 1 = (y)^3 - (1)^3$
$= (y - 1)(y^2 + y + 1)$

8. (4–3B) $x^2 - 2x - 15 = (x - 5)(x + 3)$
$-2 = -5 + +3$
$15 = 1 \cdot 15$
$3 \cdot 5$

9. (4–3C) $6x^2 - 5x - 6$
$6 = 1 \cdot 6$
$2 \cdot 3$
$(3x + 2)\quad(2x - 3)$
Inner: $+4x$
Outer: $-9x$
$-5x$
$6 = 1 \cdot 6$
$2 \cdot 3$

10. (4–5) $3ac + 6bc - 5ad - 10bd$
$= 3c(a + 2b) - 5d(a + 2b)$
$= (a + 2b)(3c - 5d)$

11. (4–6) $x^4 + 5x^2 + 9$ Factor by completing the square
$+ x^2 \quad - x^2$ Add and subtract x^2
$x^4 + 6x^2 + 9 - x^2$
$(x^2 + 3)^2 \quad - (x)^2$ Difference of two squares
$[(x^2 + 3) + x][(x^2 + 3) - x]$ or $[x^2 + x + 3][x^2 - x + 3]$

12. (4–7B) $x^3 - 8x + 3$ Use synthetic division
$1x^3 + 0x^2 - 8x + 3$
$$\begin{array}{r|rrrr} & 1 & 0 & -8 & 3 \\ & & -3 & 9 & -3 \\ \hline -3 & 1 & -3 & 1 & 0 \end{array}$$
$x^2 - 3x + 1$

Remainder is zero; therefore $(x + 3)$ is a factor.
Quotient is another factor.

Therefore, $x^3 - 8x + 3 = (x + 3)(x^2 - 3x + 1)$

13. (4–9)
$2x^2 + x = 15$
$2x^2 + x - 15 = 0$
$(x + 3)(2x - 5) = 0$
$x + 3 = 0 \quad | \quad 2x - 5 = 0$
$x = -3 \quad | \quad x = \frac{5}{2}$

14. (4–9)
$8y^2 = 4y$
$8y^2 - 4y = 0$
$4y(2y - 1) = 0$
$4y = 0 \quad | \quad 2y - 1 = 0$
$y = 0 \quad | \quad y = \frac{1}{2}$

15. (4–10) Let

$$
\begin{aligned}
x &= \text{Width}\\
x+11 &= \text{Length}\\
\text{Area} &= 60\\
(x+11)x &= 60\\
x^2+11x-60 &= 0\\
(x+15)(x-4) &= 0
\end{aligned}
$$

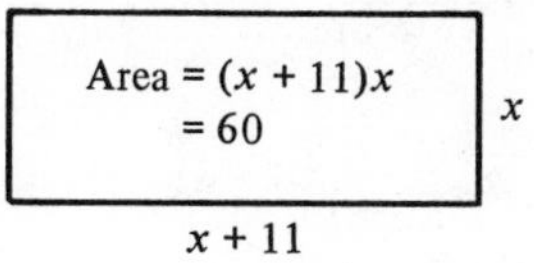

$$
\begin{array}{r|l}
x+15 = 0 & x-4 = 0 \\
x = -15 & x = 4 \quad \text{Width} \\
\text{Not possible} & x+11 = 15 \quad \text{Length}
\end{array}
$$

16. (4–10) Let x = Second term
$x + 7$ = Third term

Then $\dfrac{5}{x} = \dfrac{x+7}{6}$

$$
\begin{aligned}
x(x+7) &= 5(6)\\
x^2+7x &= 30\\
x^2+7x-30 &= 0\\
(x-3)(x+10) &= 0
\end{aligned}
$$

$$
\begin{array}{r|l}
x-3=0 & x+10 = 0\\
x=3 & x = -10
\end{array}
$$

If $x = 3$, then $x + 7 = 10$ and the proportion is $\dfrac{5}{3} = \dfrac{10}{6}$

If $x = -10$, then $x + 7 = -3$ and the proportion is $\dfrac{5}{-10} = \dfrac{-3}{6}$

Exercises 5–1 (page 150)

1. -4 **3.** None **5.** $5, -5$

7.

$$
\begin{aligned}
c^4-13c^2+36 &= 0\\
(c^2-4)(c^2-9) &= 0\\
(c+2)\ (c-2)\ (c+3)\ (c-3) &= 0\\
c=-2 \mid c=2 \mid c=-3 \mid c &= 3
\end{aligned}
$$

$-2, 2, -3$, and 3 must be excluded because they make the denominator $(c^4 - 13c^2 + 36)$ zero

9. -8

11. $\dfrac{8-y}{4y-7} = \dfrac{-(8-y)}{-(4y-7)} = \dfrac{y-8}{7-4y}$

13. $\dfrac{a-b}{(3a+2b)(a-5b)} = \dfrac{-(a-b)}{-(3a+2b)(a-5b)} = \dfrac{b-a}{(3a+2b)(5b-a)}$

15. $3m^2$ **17.** $\dfrac{15a^4b^3c^2}{-35ab^5c} = -\dfrac{3a^3c}{7b^2}$ **19.** $\dfrac{8(5-x)}{5(x+2)}$ **21.** Cannot be reduced because 4 is not a factor of the numerator.

23. $\dfrac{4x^2}{3w}$ **25.** $\dfrac{x+4}{x-5}$ **27.** $\dfrac{2x+y}{3x-y}$ **29.** $\dfrac{2x+3}{x-4}$ **31.** $-\dfrac{2x+y}{x+y}$

33. $\dfrac{a^3-1}{1-a^2} = \dfrac{(a-1)(a^2+a+1)}{(1-a)(1+a)} = \dfrac{(a-1)(a^2+a+1)}{(-1)(a-1)(a+1)} = -\dfrac{a^2+a+1}{a+1}$

Factored by completing the square

35. $\dfrac{4x^4+y^4}{6x^2+6xy+3y^2} = \dfrac{(2x^2+2xy+y^2)(2x^2-2xy+y^2)}{3(2x^2+2xy+y^2)} = \dfrac{2x^2-2xy+y^2}{3} = \dfrac{1}{3}(2x^2-2xy+y^2)$

37. $\dfrac{x^3+8}{x^3-3x^2+6x-4} = \dfrac{(x+2)(x^2-2x+4)}{(x-1)(x^2-2x+4)} = \dfrac{x+2}{x-1}$

Denominator factored by synthetic division

Exercises 5–2 (page 153)

1. $\dfrac{15x}{2y}$ **3.** $\dfrac{4n^2}{15}$ **5.** $\dfrac{9u^2}{14}$ **7.** $\dfrac{-3c^2}{8}$

9. $\dfrac{d^2e^2-d^3e}{12e^2d} \div \dfrac{d^2e^2-de^3}{3e^2d+3e^3} = \dfrac{d^2e(e-d)}{12e^2d} \cdot \dfrac{3e^2(d+e)}{de^2(d-e)} = -\dfrac{d+e}{4e}$

11. $\dfrac{5w^2}{6(w-5)}$

13. $\dfrac{4a^2+8ab+4b^2}{a^2-b^2} \div \dfrac{6ab+6b^2}{b-a} = \dfrac{\overset{2}{\cancel{4}}(\cancel{a+b})(\cancel{a+b})}{(\cancel{a+b})\underset{1}{(\cancel{a-b})}} \cdot \dfrac{\overset{-1}{(\cancel{b-a})}}{\underset{3}{\cancel{6}}b(\cancel{a+b})} = -\dfrac{2}{3b}$ **15.** $-\dfrac{a+1}{2(a^2+2a+4)}$

17. $\dfrac{(x+y)^2}{2}$ **19.** $\dfrac{m^2+2m+2}{m-2}$ **21.** $\dfrac{-h^2}{3k}$ **23.** $\dfrac{-5}{6z}$ **25.** $\dfrac{u+v-1}{v}$

27. $\dfrac{e^2+10ef+25f^2}{e^2-25f^2} \cdot \dfrac{3e-3f}{f-e} \div \dfrac{e+5f}{5f-e} = \dfrac{(\cancel{e+5f})(\cancel{e+5f})}{(\cancel{e+5f})\underset{-1}{(\cancel{e-5f})}} \cdot \dfrac{3\overset{-1}{(\cancel{e-f})}}{\underset{1}{(\cancel{f-e})}} \cdot \dfrac{\overset{1}{(\cancel{5f-e})}}{(\cancel{e+5f})} = 3$

29. $\dfrac{(x+y)^2+x+y}{(x-y)^2-x+y} \cdot \dfrac{x^2-2xy+y^2}{x^2+2xy+y^2} \cdot \dfrac{x+y}{x-y} = \dfrac{(\cancel{x+y})(x+y+1)}{(\cancel{x-y})(x-y-1)} \cdot \dfrac{(\cancel{x-y})(\cancel{x-y})}{(\cancel{x+y})(\cancel{x+y})} \cdot \dfrac{\cancel{x+y}}{\cancel{x-y}} = \dfrac{x+y+1}{x-y-1}$

Exercises 5–3 (page 158)

1. (1) $5^2 \cdot a^3;\ 3 \cdot 5 \cdot a$ Denominators in factored form
(2) $3, 5, a$ All the different factors
(3) $3^1, 5^2, a^3$ Highest power of each factor
(4) $\text{LCD} = 3^1 \cdot 5^2 \cdot a^3 = 75a^3$

3. (1) $2(w-5);\ 2^2w$
(2) $2,\ w,\ (w-5)$
(3) $2^2,\ w^1,\ (w-5)^1$
(4) $\text{LCD} = 2^2 \cdot w^1 \cdot (w-5)^1 = 4w(w-5)$

5. $(3b+c)(3b-c)^2$ **7.** $2(x-4)^2(x+5)$

9. (1) $2^2 \cdot 3 \cdot x^2 \cdot (x+2);\ (x-2)^2;\ (x+2)(x-2)$
(2) $2, 3, x, (x+2), (x-2)$
(3) $2^2, 3^1, x^2, (x+2)^1, (x-2)^2$
(4) $\text{LCD} = 2^2 \cdot 3 \cdot x^2 \cdot (x+2) \cdot (x-2)^2 = 12x^2(x+2)(x-2)^2$

11. 4 **13.** 3

15. $\dfrac{7z}{8z-4} + \dfrac{6-5z}{4-8z} = \dfrac{7z}{8z-4} + \dfrac{5z-6}{8z-4} = \dfrac{7z+5z-6}{8z-4} = \dfrac{12z-6}{8z-4} = \dfrac{\overset{3}{\cancel{6}}(\cancel{2z-1})}{\underset{2}{\cancel{4}}(\cancel{2z-1})} = \dfrac{3}{2}$ **17.** $\dfrac{27+35a^2}{75a^3}$

19. $\dfrac{147k^2-142h}{180h^2k^4}$ **21.** $\dfrac{x^2+6x-10}{2x^2}$ **23.** $\dfrac{3x^2-3}{x}$ **25.** $\dfrac{a^2+b^2}{a+b}$

27. $\dfrac{(x-3)^2}{x+3}$ **29.** $\dfrac{2t^2+5t-20}{t(t-4)}$ **31.** $\dfrac{9k^2-28k+14}{12k(2k-1)}$

33. $\text{LCD} = x(x-2)$

$$\frac{x}{1}+\frac{2}{x}+\frac{-3}{x-2} = \frac{x}{1}\cdot\frac{x(x-2)}{x(x-2)} + \frac{2}{x}\cdot\frac{(x-2)}{(x-2)} + \frac{-3}{x-2}\cdot\frac{x}{x} = \frac{x^2(x-2)}{x(x-2)} + \frac{2(x-2)}{x(x-2)} + \frac{-3x}{x(x-2)}$$
$$= \frac{x^3-2x^2+2x-4-3x}{x(x-2)} = \frac{x^3-2x^2-x-4}{x(x-2)}$$

35. $\dfrac{a^2}{b(a-b)}$

37. $\text{LCD} = (a+3)(a-1)$

$$\frac{2}{a+3}\cdot\frac{a-1}{a-1} + \frac{-4}{a-1}\cdot\frac{a+3}{a+3} = \frac{2a-2-4a-12}{(a+3)(a-1)} = \frac{-2a-14}{(a+3)(a-1)} \text{ or } \frac{14+2a}{(3+a)(1-a)}$$

39. $\dfrac{5}{(x-3)(x-2)}$ **41.** $-\dfrac{x+y}{xy}$ **43.** $\dfrac{x+4}{x^2-1}$ **45.** $\dfrac{12}{x-3}$ **47.** $\dfrac{2-7x}{(x+2)^2(x-2)}$

49. $\dfrac{4}{x^2+2x+4} + \dfrac{x-2}{x+2}$ $\text{LCD} = (x^2+2x+4)(x+2)$

$$\frac{4}{x^2+2x+4}\cdot\frac{x+2}{x+2} + \frac{x-2}{x+2}\cdot\frac{x^2+2x+4}{x^2+2x+4} = \frac{4x+8}{(x^2+2x+4)(x+2)} + \frac{x^3-8}{(x^2+2x+4)(x+2)}$$
$$= \frac{4x+x^3}{(x^2+2x+4)(x+2)}$$

51. $\dfrac{5}{2g^3}$ **53.** $\dfrac{10x^2-13x-81}{2(x-4)^2(x+5)}$

55. $\text{LCD} = 12e^2(e+3)(e-3)$

$$\frac{35}{3e^2}\cdot\frac{4(e+3)(e-3)}{4(e+3)(e-3)} - \frac{2e}{(e+3)(e-3)}\cdot\frac{12e^2}{12e^2} - \frac{3}{4(e-3)}\cdot\frac{3e^2(e+3)}{3e^2(e+3)} = \frac{140(e^2-9)-2e(12e^2)-9e^2(e+3)}{12e^2(e+3)(e-3)}$$
$$= \frac{140e^2-1260-24e^3-9e^3-27e^2}{12e^2(e+3)(e-3)} = \frac{-33e^3+113e^2-1260}{12e^2(e+3)(e-3)}$$

Exercises 5–4 (page 162)

1. $\frac{3m}{5}$ **3.** $\dfrac{\frac{15h-6}{18h}}{\frac{30h^2-12h}{8h}} = \frac{15h-6}{18h} \div \frac{30h^2-12h}{8h} = \frac{\cancel{3}(\cancel{5h-2})}{\underset{3}{\cancel{18}}h} \cdot \frac{\overset{2}{\cancel{8}}\cancel{h}}{\underset{3}{\cancel{6}}\cancel{h}(5h-2)} = \frac{2}{9h}$ **5.** $\frac{d}{c-2d}$ **7.** 1

9. $\frac{x^2}{y^2}$ **11.** $-\frac{x+2}{x}$ **13.** $\dfrac{\frac{x+4}{x} - \frac{3}{x-1}}{x+1+\frac{2x+1}{x-1}} = \dfrac{\frac{x(x-1)}{1}\cdot\frac{(x+4)}{x} + \frac{x(x-1)}{1}\cdot\left(-\frac{3}{x-1}\right)}{\frac{x(x-1)}{1}\cdot\frac{(x+1)}{1} + \frac{x(x-1)}{1}\cdot\left(\frac{2x+1}{x-1}\right)} = \frac{x^2+3x-4-3x}{x^3-x+2x^2+x}$

$$= \frac{x^2-4}{x^3+2x^2} = \frac{(\cancel{x+2})(x-2)}{x^2(\cancel{x+2})} = \frac{x-2}{x^2}$$

15. $\dfrac{4x^{-2}-y^{-2}}{2x^{-1}+y^{-1}} = \dfrac{\frac{4}{x^2}-\frac{1}{y^2}}{\frac{2}{x}+\frac{1}{y}} = \dfrac{\frac{x^2y^2}{1}\cdot\frac{4}{x^2} + \frac{x^2y^2}{1}\left(-\frac{1}{y^2}\right)}{\frac{x^2y^2}{1}\cdot\frac{2}{x} + \frac{x^2y^2}{1}\cdot\frac{1}{y}} = \frac{4y^2-x^2}{2xy^2+x^2y} = \frac{(\cancel{2y+x})(2y-x)}{xy(\cancel{2y+x})} = \frac{2y-x}{xy}$ **17.** $\frac{cd}{5d-2c}$

19. $\dfrac{\frac{x-2}{x+2}-\frac{x+2}{x-2}}{\frac{x-2}{x+2}+\frac{x+2}{x-2}} = \dfrac{\frac{(x+2)(x-2)}{1}\left(\frac{x-2}{x+2}\right) - \frac{(x+2)(x-2)}{1}\left(\frac{x+2}{x-2}\right)}{\frac{(x+2)(x-2)}{1}\left(\frac{x-2}{x+2}\right) + \frac{(x+2)(x-2)}{1}\left(\frac{x+2}{x-2}\right)} = \frac{(x-2)(x-2)-(x+2)(x+2)}{(x-2)(x-2)+(x+2)(x+2)}$

$$= \frac{x^2-4x+4-(x^2+4x+4)}{x^2-4x+4+x^2+4x+4} = \frac{-8x}{2x^2+8} = -\frac{4x}{x^2+4}$$

21. $\frac{-y}{x}$ **23.** $\frac{2x^2+1}{2x^3+3x}$ We leave the solution of this problem as a challenge to the student.

Exercises 5–5 (page 168)

1. $\left\{\frac{24}{13}\right\}$ **3.** $\left\{-\frac{1}{2}\right\}$ **5.** $\left\{\frac{25}{14}\right\}$ **7.** $\left\{\frac{23}{3}\right\}$ **9.** $\{19\}$ **11.** No solution; solution set = { }

13. $\frac{\cancel{2m-3}}{1}\cdot\frac{12m}{\cancel{2m-3}} = \frac{2m-3}{1}\cdot\frac{6}{1} + \frac{\cancel{2m-3}}{1}\cdot\frac{18}{\cancel{2m-3}}$

$$12m = (2m-3)6 + 18$$
$$12m = 12m - 18 + 18$$
$$12m = 12m \quad \text{Identity}$$

Solution set = R

15. Identity; solution set = R **17.** $\{-18\}$

19. $\left\{-\frac{14}{39}\right\}$ **21.** $\left\{-\frac{6}{5}\right\}$ **23.** $\left\{-\frac{1}{2}, \frac{5}{3}\right\}$ **25.**

$$\frac{3e-5}{4e} = \frac{e}{2e+3} \quad \text{This is a proportion}$$
$$(3e-5)(2e+3) = 4e(e) \quad \text{Product of means}$$
$$6e^2-e-15 = 4e^2 \quad = \text{Product of extremes}$$
$$2e^2-e-15 = 0$$
$$(e-3)(2e+5) = 0$$
$$e-3=0 \quad\Big|\quad 2e+5=0$$
$$e=3 \quad\Big|\quad e=-\frac{5}{2}$$

27. $\left\{2, -\frac{3}{4}\right\}$ **29.** $\{2\}$ **31.** $\left\{-\frac{1}{5}, 2\right\}$ **33.** $\{5, 9\}$

35. LCD $= x(x+3)(x+4)$

$$\frac{x(x+3)\cancel{(x+4)}}{1}\cdot\frac{6}{\cancel{x+4}} = \frac{x\cancel{(x+3)}(x+4)}{1}\cdot\frac{5}{\cancel{x+3}} + \frac{\cancel{x}(x+3)(x+4)}{1}\cdot\frac{4}{\cancel{x}}$$

$$\begin{aligned} x(x+3)6 &= x(x+4)5 + (x+3)(x+4)4 \\ 6x^2+18x &= 5x^2+20x+4x^2+28x+48 \\ 0 &= 3x^2+30x+48 \\ 0 &= 3(x^2+10x+16) \\ 0 &= 3(x+2)(x+8) \end{aligned}$$

$$x+2=0 \quad\Big|\quad x+8=0$$
$$x=-2 \quad\Big|\quad x=-8$$

37. $\left\{-\frac{7}{4}, 1\right\}$

39. $\left\{-3, \frac{6}{19}\right\}$

41. LCD $= 5(x+3)(x-3)$

$$\frac{5\cancel{(x+3)(x-3)}}{1}\cdot\frac{6}{\cancel{x^2-9}} + \frac{\cancel{5}(x+3)(x-3)}{1}\cdot\frac{1}{\cancel{5}} = \frac{\cancel{5}(x+3)\cancel{(x-3)}}{1}\cdot\frac{1}{\cancel{x-3}}$$

$$\begin{aligned} 30+x^2-9 &= 5x+15 \\ x^2-5x+6 &= 0 \\ (x-2)(x-3) &= 0 \end{aligned}$$

$$x-2=0 \quad\Big|\quad x-3=0$$
$$x=2 \quad\Big|\quad x=3$$

$x = 3$: Not a root (it makes a denominator zero and is therefore an excluded value)

Check
$$\frac{6}{x^2-9}+\frac{1}{5}=\frac{1}{x-3}$$
$$\frac{6}{2^2-9}+\frac{1}{5}\stackrel{?}{=}\frac{1}{2-3}$$
$$-\frac{6}{5}+\frac{1}{5}\stackrel{?}{=}-1$$
$$-1=-1$$

Exercises 5–6 (page 171)

1.
$$\begin{aligned} 2(3x-y) &= xy-12 \\ 6x-2y &= xy-12 \\ 6x+12 &= xy+2y \\ 6(x+2) &= (x+2)y \\ \frac{6\cancel{(x+2)}}{\cancel{(x+2)}} &= y \\ 6 &= y \end{aligned}$$

3. $x=\frac{5}{2}$

5. $m = x - zs$

7.
$$\begin{aligned} \frac{C}{1} &= \frac{5(F-32)}{9} \\ 9C &= 5(F-32) \\ 9C &= 5F-160 \\ 9C+160 &= 5F \\ \frac{9C+160}{5} &= F \end{aligned}$$

9. $s - a = c$

11. $r=\frac{A-P}{Pt}$

13. $b=\frac{32a}{43}$

15. $a=\frac{r}{2-rv^2}$

17. $r_1=\frac{Rr_2}{r_2-R}$

19. $A=\frac{Ca}{\pi(a-C)}$

21.
$$C=\frac{a}{1+\frac{a}{\pi A}}$$
$$C=\frac{\pi A\cdot a}{\pi A\cdot 1+\frac{\cancel{\pi A}}{1}\cdot\frac{a}{\cancel{\pi A}}}$$
$$\frac{C}{1}=\frac{\pi Aa}{\pi A+a} \quad \text{Proportion}$$
$$\begin{aligned} C(\pi A+a) &= \pi Aa \\ C\pi A+Ca &= \pi Aa \\ C\pi A &= \pi Aa - Ca \\ C\pi A &= a(\pi A - C) \\ \frac{C\pi A}{\pi A - C} &= a \end{aligned}$$

Exercises 5–7 (page 173)

1. $2\frac{2}{9}$ days **3.** $1\frac{5}{7}$ hr **5.** $4\frac{4}{5}$ hr

7. Machine A rate $= \dfrac{1 \text{ job}}{36 \text{ hr}} = \dfrac{1}{36}$ job per hr

Machine B rate $= \dfrac{1 \text{ job}}{24 \text{ hr}} = \dfrac{1}{24}$ job per hr

In 12 hr, machine A will do $12 \cdot \dfrac{1}{36} = \dfrac{1}{3}$ of job; therefore, $\dfrac{2}{3}$ of job is left to be done

Let $x =$ Hours needed for both machines to finish job

Then [Amount of job done by machine A] + [Amount of job done by machine B] = [$\frac{2}{3}$ of job]

$$\frac{1}{36} \cdot x + \frac{1}{24} \cdot x = \frac{2}{3}$$

LCD = 72

$$\frac{72}{1} \cdot \frac{x}{36} + \frac{72}{1} \cdot \frac{x}{24} = \frac{72}{1} \cdot \frac{2}{3}$$

$$2x + 3x = 48$$

$$5x = 48$$

$$x = \frac{48}{5} = 9\tfrac{3}{5} \text{ hr}$$

9. Let $x =$ Smaller number

$x + 8 =$ Larger number

[One-fourth the larger number] [is] [one more than] [one-third the smaller number]

$$\frac{1}{4}(x + 8) = 1 + \frac{1}{3}(x)$$

LCD = 12

$$\frac{12}{1} \cdot \frac{x + 8}{4} = \frac{12}{1} \cdot 1 + \frac{12}{1} \cdot \frac{x}{3}$$

$$3x + 24 = 12 + 4x$$

$$12 = x$$

$x = 12$ Smaller number

$x + 8 = 12 + 8 = 20$ Larger number

11.

	d =	r ·	t
Slow plane	d	400	$\frac{d}{400}$
Fast plane	d	500	$\frac{d}{500}$

[Slow plane's time] − [Fast plane's time] = [$\frac{1}{2}$ hr]

$$\frac{d}{400} - \frac{d}{500} = \frac{1}{2}$$

LCD = 2000

$$\frac{2000}{1} \cdot \frac{d}{400} - \frac{2000}{1} \cdot \frac{d}{500} = \frac{2000}{1} \cdot \frac{1}{2}$$

$$5d - 4d = 1000$$

$$d = 1000 \text{ miles}$$

13.

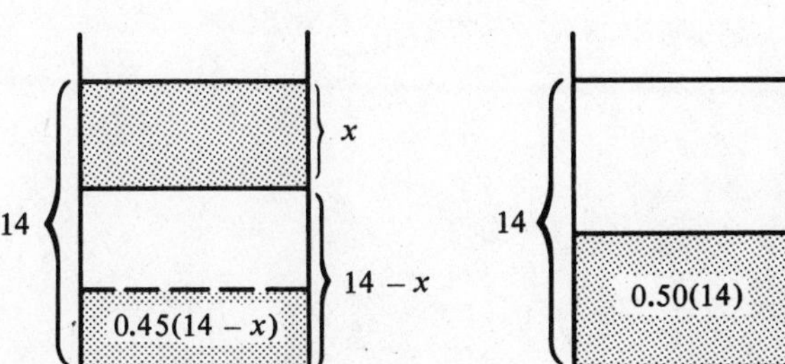

$$0.45(14 - x) + x = 0.50(14)$$

$$45(14 - x) + 100x = 50(14)$$

$$630 - 45x + 100x = 700$$

$$55x = 70$$

$$x = \frac{70}{55} = \frac{14}{11} = 1\frac{3}{11} \text{ qt}$$

15. 85%

17. Let u = Units digit
$u + 1$ = Tens digit

Product of digits	divided by	sum of digits	is	$\frac{6}{5}$
$(u+1)u$	$\div$	$(u+1+u)$	$=$	$\frac{6}{5}$

$$\frac{(u+1)u}{u+1+u} = \frac{6}{5}$$
$$\frac{u^2+u}{2u+1} = \frac{6}{5}$$
$$5(u^2+u) = 6(2u+1)$$
$$5u^2+5u = 12u+6$$
$$5u^2-7u-6 = 0$$
$$(u-2)(5u+3) = 0$$
$$u-2=0 \quad \Big| \quad 5u+3 = 0$$
$$u=2 \quad \Big| \quad u = -\frac{3}{5} \text{ Extraneous}$$
$$u+1=3$$

The number is 32

19. Let x = Hours for pipe 1 to fill tank
$x + 1$ = Hours for pipe 2 to fill tank

Pipe 1 rate $= \dfrac{1 \text{ tank}}{x \text{ hr}} = \dfrac{1}{x}$ tank per hr

Pipe 2 rate $= \dfrac{1 \text{ tank}}{(x+1)\text{hr}} = \dfrac{1}{x+1}$ tank per hr

Pipe 3 rate $= \dfrac{1 \text{ tank}}{2 \text{ hr}} = \dfrac{1}{2}$ tank per hr

Amount 1 does in 3 hr	+	Amount 2 does in 3 hr	−	Amount 3 does in 3 hr	=	1 full tank
$\frac{1}{x}(3)$	+	$\frac{1}{x+1}(3)$	−	$\frac{1}{2}(3)$	=	1

$$\frac{2x(x+1)}{1}\cdot\frac{3}{x} + \frac{2x(x+1)}{1}\cdot\frac{3}{x+1} - \frac{2x(x+1)}{1}\cdot\frac{3}{2} = \frac{2x(x+1)}{1}\cdot\frac{1}{1}$$
$$6(x+1)+6x-3x(x+1) = 2x(x+1)$$
$$6x+6+6x-3x^2-3x = 2x^2+2x$$
$$5x^2-7x-6 = 0$$
$$(x-2)(5x+3) = 0$$
$$x = 2\text{hr} \mid \text{Extraneous}$$

21. Let f = Unknown fraction

$$\left(3f+\frac{1}{3}f\right)f = f$$
$$\left(\frac{10}{3}f\right)f = f$$
$$\frac{10}{3}f^2 - f = 0$$
$$f\left(\frac{10}{3}f-1\right) = 0$$
$$f = \frac{0}{1} \quad \Big| \quad \frac{10}{3}f - 1 = 0$$
$$f = \frac{3}{10}$$

The fraction is $\frac{3}{1}$ or $\frac{0}{1}$

Chapter Five Diagnostic Test or Review (page 176)

Following each problem number is the textbook section number (in parentheses) where that kind of problem is discussed.

1. (5–1) (a) $\dfrac{2x+3}{x^2-4x}$

The values that make the denominator zero must be excluded:

$$x^2-4x=0$$
$$x(x-4)=0$$
$$x=0 \quad \Big| \quad x-4=0$$
$$x=4$$

Exclude: 0 and 4

(b)
$$3y^2-y-10=0$$
$$(3y+5)(y-2)=0$$
$$3y+5=0 \quad \Big| \quad y-2=0$$
$$y=-\frac{5}{3} \qquad y=2$$

Exclude: $-\dfrac{5}{3}$ and 2

2. (5–1) (a) Starting with $\dfrac{4}{-h}$, change two of its signs:

$$\frac{4}{-h}=\frac{-4}{-(-h)}=\frac{-4}{h}$$

h is the missing term

(b) Starting with $\dfrac{-3}{k-2}$, change two of its signs:

$$\frac{-3}{k-2}=\frac{-(-3)}{-(k-2)}=\frac{3}{2-k}$$

3 is the missing term

3. (5–1) (a) $\dfrac{4a^4b}{10a^2b^3}=\dfrac{2a^2}{5b^2}$

(b) $\dfrac{f^2+5f+6}{f^2-9}=\dfrac{(f+2)(f+3)}{(f-3)(f+3)}=\dfrac{f+2}{f-3}$

(c) $\dfrac{x^4-2x^3+5x^2-10x}{x^3-8}$ — Grouping (numerator); Difference of 2 cubes (denominator)

$$=\frac{x^3(x-2)+5x(x-2)}{(x-2)(x^2+2x+4)}$$
$$=\frac{\cancel{(x-2)}(x^3+5x)}{\cancel{(x-2)}(x^2+2x+4)}=\frac{x(x^2+5)}{x^2+2x+4}$$

4. (5–2) $\dfrac{z}{\cancel{(z-2)}}\cdot\dfrac{3\cancel{(z-2)}}{\underset{2}{6}z^2}=\dfrac{z}{2z^2}=\dfrac{1}{2z}$

5. (5–2) $\dfrac{3x^2}{\cancel{(x+2)}}\cdot\dfrac{\cancel{(x+2)}(x-2)}{\underset{2}{6x}}=\dfrac{x(x-2)}{2}$

6. (5–2) $\dfrac{3\cancel{(m+n)}}{(m-n)\cancel{(m^2+mn+n^2)}}\cdot\dfrac{\cancel{(m^2+mn+n^2)}}{\cancel{(m+n)}(m-n)}=\dfrac{3}{(m-n)^2}$

7. (5–3) $\dfrac{20a+27b}{12a-20b}+\dfrac{44a-13b}{20b-12a}=\dfrac{20a+27b}{12a-20b}+\dfrac{13b-44a}{12a-20b}$

$$=\frac{20a+27b+13b-44a}{12a-20b}=\frac{-24a+40b}{12a-20b}=\frac{-\overset{2}{\cancel{8}}\cancel{(3a-5b)}}{\cancel{4}\cancel{(3a-5b)}}=-2$$

8. (5–3) $\dfrac{y-2}{y}-\dfrac{y}{y+2}=\dfrac{y-2}{y}\cdot\dfrac{y+2}{y+2}-\dfrac{y}{y}\cdot\dfrac{y}{y+2}$

$$=\frac{(y-2)(y+2)}{y(y+2)}-\frac{y^2}{y(y+2)}=\frac{y^2-4-y^2}{y(y+2)}=\frac{-4}{y(y+2)}$$

9. (5–3) $\dfrac{x}{x+1}-\dfrac{x-1}{2(x-1)}=\dfrac{x}{x+1}\cdot\dfrac{2(x-1)}{2(x-1)}-\dfrac{x-1}{2(x-1)}\cdot\dfrac{x+1}{x+1}=\dfrac{2x^2-2x-x^2+1}{2(x+1)(x-1)}=\dfrac{x^2-2x+1}{2(x+1)(x-1)}$

$$=\frac{(x-1)\cancel{(x-1)}}{2(x+1)\cancel{(x-1)}}=\frac{x-1}{2(x+1)}$$

10. (5–4) $\dfrac{\dfrac{8h^4}{5k}}{\dfrac{4h^2}{15k^3}}=\dfrac{8h^4}{5k}\div\dfrac{4h^2}{15k^3}=\dfrac{\overset{2}{\cancel{8}}h^4}{\cancel{5}k}\cdot\dfrac{\overset{3}{\cancel{15}}k^3}{\cancel{4}h^2}=\dfrac{6h^2k^2}{1}=6h^2k^2$

11. (5–4) $\dfrac{6-\dfrac{4}{w}}{\dfrac{3w}{w-2}+\dfrac{1}{w}}\cdot\dfrac{w(w-2)}{w(w-2)}=\dfrac{\dfrac{6}{1}\cdot\dfrac{w(w-2)}{1}-\dfrac{4}{w}\cdot\dfrac{w(w-2)}{1}}{\dfrac{3w}{w-2}\cdot\dfrac{w(w-2)}{1}+\dfrac{1}{w}\cdot\dfrac{w(w-2)}{1}}=\dfrac{6w(w-2)-4(w-2)}{3w^2+w-2}$

$$=\frac{2(w-2)\cancel{(3w-2)}}{(w+1)\cancel{(3w-2)}}=\frac{2(w-2)}{w+1}$$

12. (5–5) $\frac{6}{2c-7} = \frac{9}{5-3c}$ Proportion

$$\begin{aligned} 6(5-3c) &= (2c-7)9 \\ 30-18c &= 18c-63 \\ 93 &= 36c \\ \frac{93}{36} &= c \\ \frac{31}{12} &= c \end{aligned}$$

13. (5–5) $\frac{\overset{2}{\cancel{6}}}{1}\cdot\frac{y+6}{\cancel{3}} = \frac{\overset{3}{\cancel{6}}}{1}\cdot\frac{y+4}{\cancel{2}} + \frac{\overset{2}{\cancel{6}}}{1}\cdot\frac{1}{\cancel{3}}$

$$\begin{aligned} 2(y+6) &= 3(y+4)+2 \\ 2y+12 &= 3y+12+2 \\ -2 &= y \end{aligned}$$

14. (5–5) $\frac{\cancel{(3a+5)}(a-2)}{1}\cdot\frac{2}{\cancel{3a+5}} - \frac{(3a+5)\cancel{(a-2)}}{1}\cdot\frac{6}{\cancel{a-2}} = \frac{(3a+5)(a-2)}{1}\cdot\frac{3}{1}$

$$\begin{aligned} (a-2)2-(3a+5)6 &= (3a^2-a-10)3 \\ 2a-4-18a-30 &= 9a^2-3a-30 \\ 0 &= 9a^2+13a+4 \\ 0 &= (9a+4)\ (a+1) \end{aligned}$$

$9a+4=0$ | $a+1=0$

$a=-\frac{4}{9}$ | $a=-1$

15. (5–5) $\frac{\overset{3}{\cancel{6}}z^2}{1}\cdot\frac{3}{\cancel{2}\cancel{z}} + \frac{6\cancel{z^2}}{1}\cdot\frac{3}{\cancel{z^2}} = \frac{\cancel{6}z^2}{1}\left(-\frac{1}{\cancel{6}}\right)$

$$\begin{aligned} 9z+18 &= -z^2 \\ z^2+9z+18 &= 0 \\ (z+3)(z+6) &= 0 \end{aligned}$$

$z+3=0$ | $z+6=0$

$z=-3$ | $z=-6$

16. (5–6) $\frac{I}{1} = \frac{E}{R+r}$ Proportion

$$\begin{aligned} I(R+r) &= E \\ IR+Ir &= E \\ Ir &= E-IR \\ r &= \frac{E-IR}{I} \end{aligned}$$

17. (5–7) Let x = Number of hours Ruben takes to make 8 bushings

Albert's rate $= \frac{24 \text{ bushings}}{8 \text{ hr}} = \frac{3}{1}$ bushings per hr

Ruben's rate $= \frac{8 \text{ bushings}}{x \text{ hr}} = \frac{8}{x}$ bushings per hr

$$\frac{3}{1}(4)+\frac{8}{x}(4) = 14$$

$$\frac{x}{1}\cdot\frac{12}{1}+\frac{x}{1}\cdot\frac{32}{x} = \frac{x}{1}\cdot\frac{14}{1}$$

$$\begin{aligned} 12x+32 &= 14x \\ 32 &= 2x \\ 16 &= x \end{aligned}$$

It takes Ruben 16 hr to make 8 bushings

Exercises 6–1 (page 180)

1. x^2 **3.** a **5.** $a^{1/4}$ **7.** $z^{1/6}$ **9.** $\frac{1}{w^{1/6}}$ **11.** $H^{3/2}$ **13.** $\frac{1}{x^{1/4}}$ **15.** $e^{1/5}$

17. $a^{1/4}$ **19.** $\frac{11}{P^{11/10}}$ **21.** $\frac{1}{z^{3/4}}$ **23.** $h^{1/6}$ **25.** $x^{7/6}$ **27.** $\frac{1}{z^{1/2}}$ **29.** $\frac{1}{x^{1/6}}$ **31.** $\frac{4}{u^{2/3}}$

33. $(16x^{-2/5}y^{4/9})^{3/2} = (2^4)^{3/2}(x^{-2/5})^{3/2}(y^{4/9})^{3/2}$
$= 2^6x^{-3/5}y^{2/3} = \frac{64y^{2/3}}{x^{3/5}}$
35. $\frac{R^{2/3}}{6S^{1/5}}$ **37.** $u^{3/4}v^{1/3}$ **39.** $\frac{3b^{1/6}}{2a^{1/6}}$ **41.** $\frac{R^{1/3}S^{1/3}}{4}$

43. $\dfrac{x^{3/5}}{y^{2/5}z^3}$ **45.** $\left(\dfrac{9x^{-2/3}y^{2/9}}{x^{-2}}\right)^{-3/2} = (3^2x^{-2/3-(-2)}y^{2/9})^{-3/2}$

$= 3^{2(-3/2)}x^{4/3(-3/2)}y^{2/9(-3/2)}$

$= 3^{-3}x^{-2}y^{-1/3} = \dfrac{1}{27x^2y^{1/3}}$

47. $\dfrac{1}{4}$ **49.** $\dfrac{2}{27}$

51. $(100)^{-1/2}(-8)^{1/3} = (10^2)^{-1/2}(-2^3)^{1/3}$

$= 10^{-1}(-2) = -\dfrac{2}{10} = -\dfrac{1}{5}$

53. 0.005 **55.** 1

Exercises 6–2 (page 183)

1. 4 **3.** 3 **5.** -4 **7.** $\sqrt{-25}$ Even index, negative radicand; roots are not real numbers. **9.** 3 **11.** 2

13. Not a real number **15.** 2 **17.** -10 **19.** $\sqrt[5]{a^3}$ **21.** $\sqrt{7}$ **23.** $\dfrac{1}{\sqrt{x}}$ **25.** $\sqrt[n]{7^m}$

27. $5^{1/2}$ **29.** $z^{1/3}$ **31.** $x^{3/4}$ **33.** $x^{2/3}$ **35.** $x^{4/3}$ **37.** $\sqrt[n]{x^{2n}} = x^{2n/n} = x^2$

39. 2 **41.** $(-27)^{2/3} = [(-3)^3]^{2/3}$

$= (-3)^{3(2/3)} = (-3)^2 = 9$

43. $4^{3/2} = (2^2)^{3/2} = 2^3 = 8$

45. $(-16)^{3/4} = (\sqrt[4]{-16})^3$ Even index, negative radicand; roots are not real numbers. **47.** 16 **49.** -1

Exercises 6–3A (page 187)

1. $2x$ **3.** $5x$ **5.** $2x$ **7.** $2xy^2$ **9.** $(2x-3)^2$ **11.** $\sqrt{(-2)^2} = |-2| = 2$ **13.** $4\sqrt{2}$

15. $-3\sqrt[3]{9}$ **17.** $8\sqrt{2}$ **19.** $m^2\sqrt{m}$ **21.** $-x\sqrt[5]{x^2}$ **23.** $2\sqrt[4]{2}$ **25.** $2\sqrt[4]{3}$ **27.** $2a^2b\sqrt{2}$

29. $3mn^2\sqrt{2mn}$ **31.** $-10a\sqrt[3]{3a^2b^2}$ **33.** $2m^2p^3\sqrt[5]{2mu}$ **35.** $2y^4z^2\sqrt{3xz}$

37. $\dfrac{3}{2abc}\sqrt[4]{2^5a^8b^9c^{10}} = \dfrac{3}{2abc}\sqrt[4]{2^4\cdot 2a^8b^8bc^8c^2}$

$= \dfrac{3\cdot 2a^2b^2c^2}{2abc}\sqrt[4]{2bc^2} = 3abc\sqrt[4]{2bc^2}$

39. $xy^2\sqrt[4]{2xy}$ **41.** $\sqrt[3]{8(a+b)^3} = \sqrt[3]{2^3(a+b)^3}$

$= 2(a+b)$

43. $\sqrt{4x^2+4x+1} = \sqrt{(2x+1)^2} = 2x+1$ **45.** $|2-y|$

Exercises 6–3B (page 191)

1. $\dfrac{4}{5}$ **3.** $-\dfrac{3}{4}$ **5.** $\dfrac{2}{3}$ **7.** $\dfrac{ab^2}{2}$ **9.** $\dfrac{2x}{y}$ **11.** $-\dfrac{x^3y^2}{3}$ **13.** $2\sqrt{5}$

15. $\dfrac{9}{\sqrt[3]{3}}\cdot\dfrac{\sqrt[3]{3^2}}{\sqrt[3]{3^2}} = \dfrac{9\sqrt[3]{3^2}}{\sqrt[3]{3^3}} = \dfrac{9\sqrt[3]{9}}{3} = 3\sqrt[3]{9}$ **17.** $-3c^2\sqrt[5]{d^4}$ **19.** $-\dfrac{m\sqrt[3]{9m^2}}{3}$ **21.** 1 **23.** $\dfrac{2\sqrt{3}}{b}$

25. $\dfrac{m\sqrt[4]{12p^2}}{2p}$ **27.** $\sqrt[5]{\dfrac{15x^4y^7}{24x^6y^2}} = \sqrt[5]{\dfrac{5y^5}{2^3x^2}\cdot\dfrac{2^2x^3}{2^2x^3}} = \dfrac{\sqrt[5]{20x^3y^5}}{\sqrt[5]{2^5x^5}} = \dfrac{y\sqrt[5]{20x^3}}{2x}$ **29.** $3xy\sqrt[3]{3x^2}$

31. $\frac{4x^2}{5y^3}\sqrt{\frac{3x^2y^3}{8x^5y}} = \frac{4x^2}{5y^3}\sqrt{\frac{3y^2}{8x^3}} = \frac{\overset{2}{\cancel{4}}x^2}{5y^3}\cdot\frac{y}{\cancel{2}x}\sqrt{\frac{3}{2x}\cdot\frac{2x}{2x}} = \frac{\cancel{2}x}{5y^2}\cdot\frac{1}{\cancel{2}x}\sqrt{6x} = \frac{1}{5y^2}\sqrt{6x}$

33. $\sqrt{\frac{a^2+2a-3}{a^2+4a+3}} = \sqrt{\frac{\cancel{(a+3)}(a-1)}{\cancel{(a+3)}(a+1)}} = \sqrt{\frac{a-1}{a+1}\cdot\frac{a+1}{a+1}} = \frac{\sqrt{a^2-1}}{\sqrt{(a+1)^2}} = \frac{\sqrt{a^2-1}}{a+1}$

35. $\frac{\sqrt{a^2-b^2}}{a-b}$ **37.** $\frac{2}{|x-2|}\sqrt{x^2-4}$

Exercises 6–4 (page 193)

1. 3 **3.** 8 **5.** 3 **7.** 3 **9.** $3x$ **11.** $10ab\sqrt{b}$ **13.** $2a\sqrt[5]{b^2}$ **15.** $2xy$ **17.** 175

19. $(2\sqrt[3]{4x^2})^2 = 2\cdot 2\cdot\sqrt[3]{4x^2}\cdot\sqrt[3]{4x^2}$
$= 4\sqrt[3]{4x^2\cdot 4x^2} = 4\sqrt[3]{2^4x^4}$
$= 4\sqrt[3]{2^3 2x^3x} = 4\cdot 2x\sqrt[3]{2x}$
$= 8x\sqrt[3]{2x}$

21. $2\sqrt{7x^3y^3}\cdot 5\sqrt{3xy}\cdot 2\sqrt{7x^3y}$
$= 2\cdot 5\cdot 2\sqrt{3\cdot 7^2x^6xy^4y}$
$= 20\cdot 7x^3y^2\sqrt{3xy}$
$= 140x^3y^2\sqrt{3xy}$

23. $120x^2\sqrt{x}$ **25.** $32x$

27. $(3\sqrt{2x}-5)^2 = 3^2(\sqrt{2x-5})^2$
$= 9(2x-5) = 18x-45$

29. $75x^3 - 100x^2$ **31.** 4 **33.** $\sqrt[3]{\frac{10}{2^3}} = \frac{\sqrt[3]{10}}{2}$ $\left(\text{or } \frac{1}{2}\sqrt[3]{10}\right)$

35. 6 **37.** $\frac{x^2\sqrt{5}}{5}$ **39.** $3\sqrt[4]{3}$

41. $\frac{\sqrt[5]{128z^7}}{\sqrt[5]{2z}} = \sqrt[5]{\frac{128z^7}{2z}} = \sqrt[5]{64z^6}$
$= \sqrt[5]{2^6z^6} = 2z\sqrt[5]{2z}$

43. $6x$

45. $\frac{6\sqrt[4]{2^5m^2}}{2\sqrt[4]{2m^3}} = 3\sqrt[4]{\frac{2^5m^2}{2m^3}} = 3\sqrt[4]{\frac{2^4}{m}\cdot\frac{m^3}{m^3}}$
$= 3\sqrt[4]{\frac{2^4m^3}{m^4}} = \frac{6\sqrt[4]{m^3}}{m}$

47. $\frac{\sqrt[3]{4B}}{3B}$ **49.** $\sqrt{x}$ **51.** $\sqrt[4]{a^3}$ **53.** $\sqrt[3]{x^2}$

55. $\sqrt[6]{27b^3} = (3^3b^3)^{1/6} = 3^{3/6}b^{3/6}$
$= 3^{1/2}b^{1/2} = (3b)^{1/2} = \sqrt{3b}$

57. m **59.** m **61.** $\sqrt[4]{a^3}$ **63.** $4\sqrt[6]{32}$ **65.** 2

67. $x\sqrt[12]{x^{11}}$

69. $\sqrt[3]{-8z^2}\,\sqrt[3]{-z}\,\sqrt[4]{16z^3} = (-2^3z^2)^{1/3}(-z)^{1/3}(2^4z^3)^{1/4}$
$= 2^1z^{2/3}z^{1/3}2^1z^{3/4} = 2^2z^{1+3/4} = 4z\sqrt[4]{z^3}$

71. x^2 **73.** $\sqrt[12]{G}$

75. $\frac{\sqrt[3]{-x^2}}{\sqrt[6]{x^5}} = \frac{-x^{2/3}}{x^{5/6}} = -x^{2/3-5/6}$
$= -x^{4/6-5/6} = -x^{-1/6} = -\frac{1}{x^{1/6}} = -\frac{1}{x^{1/6}}\cdot\frac{x^{5/6}}{x^{5/6}} = -\frac{x^{5/6}}{x^{6/6}} = -\frac{\sqrt[6]{x^5}}{x}$

77. $\frac{\sqrt[6]{x}}{x}$ **79.** $|k-4|$ or $|4-k|$

Exercises 6–5 (page 199)

1. $2\sqrt{5}$ **3.** $7\sqrt[4]{xy}$ **5.** $\frac{3\sqrt[3]{2}}{2}$ **7.** $6\sqrt{2}$ **9.** $10\sqrt{2x}$ **11.** $4\sqrt{5M}$ **13.** $4\sqrt[3]{x}$ **15.** $5a\sqrt[3]{a}$

17. $(y+x)\sqrt[5]{x^2y}$

19. $b\sqrt[4]{4a^4b} + ab\sqrt[4]{64b} - \sqrt[4]{4a^4b^5} = ab\sqrt[4]{4b} + ab\sqrt[4]{2^4\cdot 2^2b} - \sqrt[4]{4a^4b^4b}$
$= ab\sqrt[4]{4b} + 2ab\sqrt[4]{4b} - ab\sqrt[4]{4b}$
$= (ab+2ab-ab)\sqrt[4]{4b} = 2ab\sqrt[4]{4b}$

21. $-2\sqrt{6} + 2\sqrt{3}$

23. $-\dfrac{5\sqrt{2}}{12}$ **25.** $5a\sqrt[3]{2}$ **27.** $\dfrac{22}{5}\sqrt{5b}$ **29.** $4x\sqrt[3]{9x^2}$

31.
$$2k\sqrt[4]{\frac{3}{8k}} - \frac{1}{k}\sqrt[4]{\frac{2k^3}{27}} + 5k^2\sqrt[4]{\frac{6}{k^2}} = \frac{2k}{1}\sqrt[4]{\frac{3}{2^3k}\cdot\frac{2k^3}{2k^3}} - \frac{1}{k}\sqrt[4]{\frac{2k^3}{3^3}\cdot\frac{3}{3}} + \frac{5k^2}{1}\sqrt[4]{\frac{6}{k^2}\cdot\frac{k^2}{k^2}}$$
$$= \frac{2k}{1}\cdot\frac{1}{2k}\sqrt[4]{6k^3} - \frac{1}{k}\cdot\frac{1}{3}\sqrt[4]{6k^3} + \frac{5k^2}{1}\cdot\frac{1}{k}\sqrt[4]{6k^2}$$
$$= \sqrt[4]{6k^3} - \frac{1}{3k}\sqrt[4]{6k^3} + 5k\sqrt[4]{6k^2} = \left(1 - \frac{1}{3k}\right)\sqrt[4]{6k^3} + 5k\sqrt[4]{6k^2}$$
$$= \left(\frac{3k}{3k} - \frac{1}{3k}\right)\sqrt[4]{6k^3} + 5k\sqrt[4]{6k^2} = \frac{3k-1}{3k}\sqrt[4]{6k^3} + 5k\sqrt[4]{6k^2}$$

33. $4x$

35. $\dfrac{1-\sqrt{3}}{3}$ **37.** $2+\sqrt{2}$ **39.** $6+\sqrt{3}$ **41.** $x - 3\sqrt{x}$ **43.** $18x$

45.
$$\sqrt[3]{x}(2\sqrt[3]{x} - 3) = \sqrt[3]{x}\cdot 2\sqrt[3]{x} - \sqrt[3]{x}\cdot 3 = 2\sqrt[3]{x^2} - 3\sqrt[3]{x}$$

47. $13 + 5\sqrt{7}$ **49.** $7x - 16$ **51.** $12 - 20x\sqrt{3} + 25x^2$

53.
$$(\sqrt{ab} + 2\sqrt{a})^2 = (\sqrt{ab})^2 + 2(\sqrt{ab})(2\sqrt{a}) + (2\sqrt{a})^2$$
$$= ab + 4\sqrt{a^2b} + 4a = ab + 4a\sqrt{b} + 4a$$

55. $2 + 5\sqrt{2}$

57.
$$\frac{4\sqrt[3]{8x} + 6\sqrt[3]{32x^4}}{2\sqrt[3]{4x}} = \frac{4\sqrt[3]{8x}}{2\sqrt[3]{4x}} + \frac{6\sqrt[3]{32x^4}}{2\sqrt[3]{4x}}$$
$$= 2\sqrt[3]{\frac{8x}{4x}} + 3\sqrt[3]{\frac{32x^4}{4x}}$$
$$= 2\sqrt[3]{2} + 3\sqrt[3]{8x^3} = 2\sqrt[3]{2} + 3\cdot 2x$$
$$= 2\sqrt[3]{2} + 6x$$

59. $20a - 2\sqrt[4]{12}$ **61.** $2\sqrt{7}$ **63.** $\dfrac{a\sqrt[3]{a^2}}{2}$

65. $3(\sqrt{2}+1)$ **67.** $2(\sqrt{5}-\sqrt{2})$

69.
$$\frac{4\sqrt{3}-\sqrt{2}}{4\sqrt{3}+\sqrt{2}}\cdot\frac{4\sqrt{3}-\sqrt{2}}{4\sqrt{3}-\sqrt{2}} = \frac{(4\sqrt{3}-\sqrt{2})^2}{(4\sqrt{3})^2-(\sqrt{2})^2} = \frac{48 - 8\sqrt{6} + 2}{48-2}$$
$$= \frac{50 - 8\sqrt{6}}{46} = \frac{2(25 - 4\sqrt{6})}{46} = \frac{25 - 4\sqrt{6}}{23}$$

71. $2x + 1 - 2\sqrt{x^2 + x}$ **73.** 0

75. $\sqrt{2} + 2^{2/4} = \sqrt{2} + 2^{1/2} = \sqrt{2} + \sqrt{2} = 2\sqrt{2}$

Exercises 6–6 (page 204)

1. $\{8\}$ **3.** $\{2, 1\}$ **5.** $\{8\}$ **7.** $\{3\}$ **9.** $\left\{\dfrac{1}{2}\right\}$ **11.** $\{7\}$

13.
$$\sqrt{x+7} = 2x - 1$$
$$(\sqrt{x+7})^2 = (2x-1)^2$$
$$x + 7 = 4x^2 - 4x + 1$$
$$0 = 4x^2 - 5x - 6$$
$$0 = (x-2)(4x+3)$$
$$x - 2 = 0 \quad\Big|\quad 4x + 3 = 0$$
$$x = 2 \quad\Big|\quad x = -\frac{3}{4}$$

Check for $x = 2$
$$\sqrt{x+7} = 2x - 1$$
$$\sqrt{2+7} \stackrel{?}{=} 4 - 1$$
$$3 = 3$$

Check for $x = -\dfrac{3}{4}$
$$\sqrt{x+7} = 2x - 1$$
$$\sqrt{-\frac{3}{4} + 7} \stackrel{?}{=} 2\left(-\frac{3}{4}\right) - 1$$
$$\sqrt{\frac{25}{4}} \stackrel{?}{=} -\frac{3}{2} - 1$$
$$\frac{5}{2} \neq -\frac{5}{2}$$

15. No solution **17.** $\{4, 20\}$

19.
$$\begin{aligned} \sqrt[3]{2x+3} - 2 &= 0 \\ \left(\sqrt[3]{2x+3}\right)^3 &= (2)^3 \\ 2x + 3 &= 8 \\ 2x &= 5 \\ x &= \frac{5}{2} \end{aligned}$$

Check
$$\begin{aligned} \sqrt[3]{2\left(\frac{5}{2}\right) + 3} - 2 &\stackrel{?}{=} 0 \\ \sqrt[3]{5+3} - 2 &\stackrel{?}{=} 0 \\ \sqrt[3]{8} - 2 &\stackrel{?}{=} 0 \\ 2 - 2 &= 0 \end{aligned}$$

21. $\{4\}$ **23.** $\{25\}$ **25.** $\{32\}$ **27.** $\left\{\frac{1}{64}\right\}$ **29.** $\{9\}$

31.
$$\begin{aligned} \sqrt{4u+1} - \sqrt{u-2} &= \sqrt{u+3} \\ \left(\sqrt{4u+1} - \sqrt{u-2}\right)^2 &= \left(\sqrt{u+3}\right)^2 \\ 4u + 1 - 2\sqrt{4u+1}\sqrt{u-2} + u - 2 &= u + 3 \\ 4u - 4 &= 2\sqrt{(4u+1)(u-2)} \\ (2u-2)^2 &= \left(\sqrt{(4u+1)(u-2)}\right)^2 \\ 4u^2 - 8u + 4 &= 4u^2 - 7u - 2 \\ -u &= -6 \\ u &= 6 \end{aligned}$$

Check
$$\begin{aligned} \sqrt{4(6)+1} - \sqrt{(6)-2} &\stackrel{?}{=} \sqrt{(6)+3} \\ \sqrt{25} - \sqrt{4} &\stackrel{?}{=} \sqrt{9} \\ 5 - 2 &= 3 \end{aligned}$$

33. $\{14\}$

Exercises 6–7 (page 206)

1. 3 **3.** $3\sqrt{2}$ **5.** $2\sqrt{34}$

7.
$$\begin{aligned} (x+1)^2 + \left(\sqrt{20}\right)^2 &= (x+3)^2 \\ x^2 + 2x + 1 + 20 &= x^2 + 6x + 9 \\ 12 &= 4x \\ x &= 3 \end{aligned}$$

9. $4\sqrt{2}$ **11.** 7

13. Let $x =$ One leg
then $2x - 4 =$ Other leg

$$\begin{aligned} (10)^2 &= (2x-4)^2 + x^2 \\ 100 &= 4x^2 - 16x + 16 + x^2 \\ 0 &= 5x^2 - 16x - 84 \\ 0 &= (5x+14)(x-6) \end{aligned}$$

$5x + 14 = 0$, $x = -\frac{14}{5}$ Not a solution

$x - 6 = 0$, $x = 6$ One leg

$2x - 4 = 12 - 4 = 8$ Other leg

15. Sides 5 and 12. Hypotenuse 13.

17. (a) $AC = OB = 10$ (radius of circle)
(b) $AB = OC$

Let $x = OC = AB$
then
$$\begin{aligned} x^2 + 6^2 &= 10^2 \\ x^2 + 36 &= 100 \\ x^2 &= 64 \\ x &= 8 \end{aligned}$$
Therefore, $AB = 8$

Exercises 6–8A (page 210)

1. $3 + \sqrt{-16} = 3 + \sqrt{16}\sqrt{-1} = 3 + 4i$

3. $5 + 7i$

5. $5 + \sqrt{-32} = 5 + \sqrt{16}\sqrt{2}\sqrt{-1} = 5 + 4\sqrt{2}i$

7. $\sqrt{-36} + \sqrt{4} = \sqrt{36}\sqrt{-1} + 2 = 6i + 2 = 2 + 6i$

9. $3 + 2\sqrt{2}i$

11. $\sqrt{-64} = 0 + \sqrt{64}\sqrt{-1} = 0 + 8i$

13. $2i - \sqrt{9} = 2i - 3 = -3 + 2i$ **15.** $14 = 14 + 0i$

Exercises 6–8B (page 211)

1. $x = 3, y = -2$
Real parts equal and imaginary parts equal

3.
$$(4 + 3i) - (5 - i) = (3x + 2yi) + (2x - 3yi)$$
$$4 + 3i - 5 + i = 3x + 2yi + 2x - 3yi$$
$$-1 + 4i = 5x - yi$$
Therefore, $5x = -1$, $x = -\frac{1}{5}$ | $-y = 4$, $y = -4$

Real parts equal and imaginary parts equal

5. $x = \frac{7}{3}$, $y = -6$

7. $9 + 2i$ **9.** $2 - 6i$ **11.** $-13 + 0i$ **13.** $-15 - 48i$

15. $(2 + i) + (3i) - (2 - 4i)$
$= 2 + i + 3i - 2 + 4i$
$= 8i = 0 + 8i$

17. $-2 - 6i$

19. $(2 + 3i) - (x + yi) = 2 + 3i - x - yi$
$= (2 - x) + (3 - y)i$

21. $-\frac{2}{3}\sqrt{3} + 6i$

23. $(9 + \sqrt{-16}) + (2 + \sqrt{-25}) + (6 - \sqrt{-64})$
$= 9 + \sqrt{16}\sqrt{-1} + 2 + \sqrt{25}\sqrt{-1} + 6 - \sqrt{64}\sqrt{-1}$
$= 9 + 4i + 2 + 5i + 6 - 8i = (9 + 2 + 6) + (4 + 5 - 8)i = 17 + i$

25. $-12 + (3 - 2\sqrt{3})i$

Exercises 6–8C (page 213)

1. $2 + 0i$ **3.** $44 - 17i$ **5.** $14 + 5i$ **7.** $9 + 0i$ **9.** $-1 + 3i$ **11.** $-5 - 10i$

13. $-11 + 0i$ **15.** $-21 + 20i$ **17.** $7 - 24i$ **19.** $75 - 32i$ **21.** $(3i)^3 = 3^3 i^2 i = 27(-1)i = -27i$

23. 16

25. $(2 - \sqrt{-1})^2 = (2 - i)^2$
$= 4 - 4i + (i)^2$
$= 4 - 4i - 1 = 3 - 4i$

27. $15 - 8i$

29. $[3 + (-i)^6]^2 = [3 - 1]^2$
$= 2^2 = 4$

Exercises 6–8D (page 214)

1. $3 + 2i$ **3.** $-5 + 3i$ **5.** $-5i$ **7.** 10 **9.** $1 - 3i$

11. $\frac{1 + i}{1 - i} \cdot \frac{1 + i}{1 + i} = \frac{1 + 2i + i^2}{1 - i^2} = \frac{1 + 2i - 1}{1 - (-1)} = \frac{2i}{2}$
$= i = 0 + i$

13. $\frac{8 + i}{i} \cdot \frac{-i}{-i} = \frac{-8i - i^2}{-i^2} = \frac{-8i - (-1)}{-(-1)}$
$= 1 - 8i$

15. $0 - \frac{3}{2}i$

17. $\frac{4}{5} + \frac{3}{5}i$

19. $\frac{15i}{1 - 2i} \cdot \frac{1 + 2i}{1 + 2i} = \frac{15i(1 + 2i)}{1 - 4i^2} = \frac{\overset{3}{\cancel{15}}i(1 + 2i)}{\cancel{5}}$
$= 3i(1 + 2i) = 3i + 6i^2 = 3i + 6(-1)$
$= -6 + 3i$

21. $1 + 3i$ **23.** $\frac{19}{13} - \frac{4}{13}i$

25. $\frac{i(2 - \sqrt{-50})}{\sqrt{2}(2\sqrt{-1} + 5\sqrt{2})} = \frac{i(2 - 5\sqrt{2}i)}{\sqrt{2}(2i + 5\sqrt{2})} = \frac{\cancel{2i + 5\sqrt{2}}}{\sqrt{2}(\cancel{2i + 5\sqrt{2}})} = \frac{1}{\sqrt{2}} \cdot \frac{\sqrt{2}}{\sqrt{2}} = \frac{\sqrt{2}}{2} = \frac{\sqrt{2}}{2} + 0i$

Chapter Six Diagnostic Test or Review (page 217)

Following each problem number is the textbook section number (in parentheses) where that kind of problem is discussed.

1. (6–3A) $\sqrt[3]{54x^6y^7} = \sqrt[3]{2(27)x^6y^6y}$
$= 3x^2y^2\sqrt[3]{2y}$

2. (6–3B) $\frac{4xy}{\sqrt{2x}} = \frac{4xy}{\sqrt{2x}} \cdot \frac{\sqrt{2x}}{\sqrt{2x}} = \frac{4xy\sqrt{2x}}{2x} = 2y\sqrt{2x}$

3. (6–4) $\sqrt[6]{a^3} = a^{3/6} = a^{1/2} = \sqrt{a}$

4. (6–4) $(\sqrt[4]{z^2})^2 = (z^{2/4})^2 = z^{(2/4)2} = z$

5. (6–3B) $\dfrac{y}{2x}\sqrt[4]{\dfrac{16x^8y}{y^5}} = \dfrac{y}{2x}\sqrt[4]{\dfrac{16x^8}{y^4}} = \dfrac{y}{2x}\cdot\dfrac{2x^2}{y} = x$

6. (6–4) $\sqrt{x}\sqrt[3]{x} = x^{1/2}x^{1/3} = x^{1/2+1/3} = x^{5/6} = \sqrt[6]{x^5}$

7. (6–1) $16^{3/2} = (2^4)^{3/2} = 2^{4(3/2)} = 2^6 = 64$

8. (6–1) $(-27)^{2/3} = [(-3)^3]^{2/3} = (-3)^{3(2/3)} = (-3)^2 = 9$

9. (6–1) $x^{1/2}x^{-1/4} = x^{1/2-1/4} = x^{1/4}$

10. (6–1) $(R^{-4/3})^3 = R^{(-4/3)(3)} = R^{-4} = \dfrac{1}{R^4}$

11. (6–1) $\dfrac{a^{5/6}}{a^{1/3}} = a^{5/6-1/3} = a^{3/6} = a^{1/2}$

12. (6–1)
$$\left(\frac{x^{-2/3}y^{3/5}}{x^{1/3}y}\right)^{-5/2} = (x^{-2/3-1/3}y^{3/5-1})^{-5/2}$$
$$= (x^{-1}y^{-2/5})^{-5/2} = x^{(-1)(-5/2)}y^{(-2/5)(-5/2)}$$
$$= x^{5/2}y$$

13. (6–5)
$$4\sqrt{8y} + 3\sqrt{32y} = 4\sqrt{4\cdot 2y} + 3\sqrt{16\cdot 2y}$$
$$= 4(2)\sqrt{2y} + 3(4)\sqrt{2y}$$
$$= 20\sqrt{2y}$$

14. (6–5)
$$3\sqrt{\frac{5x^2}{2}} - 5\sqrt{\frac{x^2}{10}} = 3\sqrt{\frac{5x^2}{2}\cdot\frac{2}{2}} - 5\sqrt{\frac{x^2}{10}\cdot\frac{10}{10}}$$
$$= 3\sqrt{\frac{10x^2}{4}} - 5\sqrt{\frac{10x^2}{100}} = \frac{3x\sqrt{10}}{2} - \frac{5x\sqrt{10}}{10}$$
$$= \frac{3}{2}x\sqrt{10} - \frac{1}{2}x\sqrt{10} = x\sqrt{10}$$

15. (6–4) $\sqrt{2x^4}\sqrt{8x^3} = \sqrt{16x^6x} = 4x^3\sqrt{x}$

16. (6–5)
$$\sqrt{2x}(\sqrt{8x} - 5\sqrt{2}) = \sqrt{2x}\sqrt{8x} + \sqrt{2x}(-5\sqrt{2})$$
$$= \sqrt{16x^2} - 5\sqrt{4x} = 4x - 10\sqrt{x}$$

17. (6–5) $\dfrac{\sqrt{10x} + \sqrt{5x}}{\sqrt{5x}} = \dfrac{\sqrt{10x}}{\sqrt{5x}} + \dfrac{\sqrt{5x}}{\sqrt{5x}} = \sqrt{\dfrac{10x}{5x}} + 1 = \sqrt{2} + 1$

18. (6–5) $\dfrac{5}{\sqrt{7}+\sqrt{2}} = \dfrac{5}{\sqrt{7}+\sqrt{2}}\cdot\dfrac{\sqrt{7}-\sqrt{2}}{\sqrt{7}-\sqrt{2}} = \dfrac{5(\sqrt{7}-\sqrt{2})}{7-2} = \sqrt{7}-\sqrt{2}$

19. (6–8B)
$$(5 - \sqrt{-8}) - (3 - \sqrt{-18}) = 5 - \sqrt{4\cdot 2(-1)} - 3 + \sqrt{9\cdot 2(-1)}$$
$$= 5 - 2\sqrt{2}\,i - 3 + 3\sqrt{2}\,i = 2 + \sqrt{2}\,i$$

20. (6–8C)
$$(3 + i)(2 - 5i) = 6 - 15i + 2i - 5i^2$$
$$= 6 - 13i - 5(-1)$$
$$= 6 - 13i + 5 = 11 - 13i$$

21. (6–8D) $\dfrac{10}{1-3i} = \dfrac{10}{1-3i}\cdot\dfrac{1+3i}{1+3i} = \dfrac{10(1+3i)}{1-9i^2} = \dfrac{10(1+3i)}{10} = 1 + 3i$

22. (6–6)
$$x^{3/2} = 8$$
$$(x^{3/2})^{2/3} = 8^{2/3} = (2^3)^{2/3}$$
$$x = 2^2$$
$$x = 4$$

Check
$$x^{3/2} = 8$$
$$(4)^{3/2} \stackrel{?}{=} 2^3$$
$$(2^2)^{3/2} \stackrel{?}{=} 2^3$$
$$2^3 = 2^3$$

23. (6–6)

$$\sqrt{x-3}+5=x$$
$$(\sqrt{x-3})^2=(x-5)^2$$
$$x-3=x^2-10x+25$$
$$0=x^2-11x+28$$
$$0=(x-4)(x-7)$$
$$x-4=0 \quad | \quad x-7=0$$
$$x=4 \quad | \quad x=7$$

Check for $x = 4$

$$\sqrt{x-3}+5=x$$
$$\sqrt{4-3}+5\stackrel{?}{=}4$$
$$\sqrt{1}+5\stackrel{?}{=}4$$
$$1+5\neq 4$$

Therefore, 4 is not a solution

Check for $x = 7$

$$\sqrt{x-3}+5=x$$
$$\sqrt{7-3}+5\stackrel{?}{=}7$$
$$\sqrt{4}+5\stackrel{?}{=}7$$
$$2+5=7$$

Therefore, 7 is a solution

24. (6–8B)

$$(2+5i)-(x+i) = (4+2yi)-(3+yi)$$
$$(2-x)+(5-1)i = (4-3)+(2y-yi)$$
$$(2-x)+(4)i = (1)+(y)i$$
$$2-x = 1 \quad \text{Real parts equal}$$
$$x = 1$$
$$4 = y \quad \text{Imaginary parts equal}$$

25. (6–7)

$$(x+3)^2=(x+1)^2+(\sqrt{12})^2$$
$$x^2+6x+9=x^2+2x+1+12$$
$$4x=4$$
$$x=1$$

Exercises 7–1 (page 222)

1. (a) Let $P_1 = (-2,-2)$ and $P_2 = (2,1)$

Then
$$d=\sqrt{(x_2-x_1)^2+(y_2-y_1)^2}$$
$$=\sqrt{[2-(-2)]^2+[1-(-2)]^2}$$
$$=\sqrt{4^2+3^2}=\sqrt{16+9}=\sqrt{25}=5$$

(b) Let $P_1 = (-3,3)$ and $P_2 = (3,-1)$

Then
$$d=\sqrt{[3-(-3)]^2+[-1-3]^2}$$
$$=\sqrt{36+16}=\sqrt{52}=\sqrt{4\cdot 13}=2\sqrt{13}$$

(c) The distance formula will apply. However, notice that the ordinate (3) is the same for both points. A simple graph shows the distance by inspection. Distance = 7.

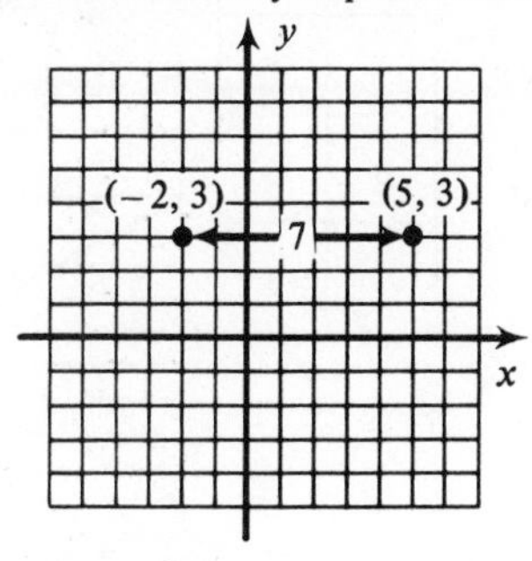

(d) The distance formula will apply. However, both points have the same x-coordinate. Therefore, a graph of the points shows the distance. Distance = 3.

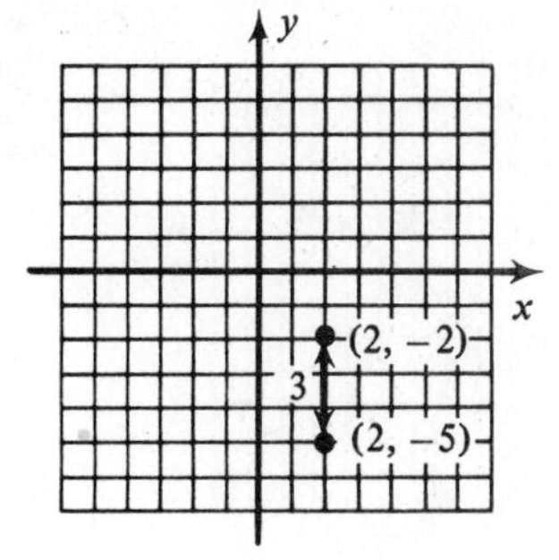

(e) $d=\sqrt{(4-0)^2+(6-0)^2}=\sqrt{16+36}=\sqrt{52}=\sqrt{4\cdot 13}=2\sqrt{13}$

3.
$$AB=\sqrt{[4-(-2)]^2+(2-2)^2}=\sqrt{6^2}=6$$
$$BC=\sqrt{(6-4)^2+(8-2)^2}=\sqrt{4+36}=\sqrt{40}$$
$$=\sqrt{4\cdot 10}=2\sqrt{10}$$
$$AC=\sqrt{[6-(-2)]^2+(8-2)^2}=\sqrt{64+36}$$
$$=\sqrt{100}=10$$

Perimeter $= 6+2\sqrt{10}+10=16+2\sqrt{10}$

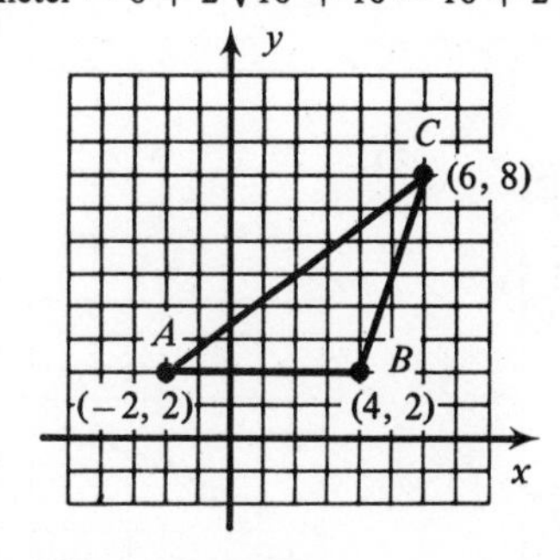

5.

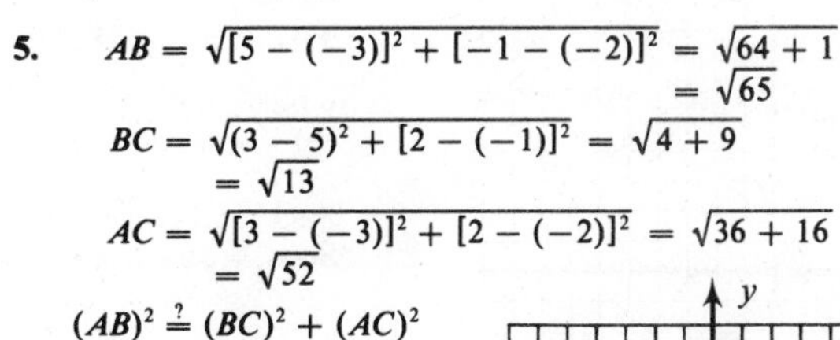

$$AB=\sqrt{[5-(-3)]^2+[-1-(-2)]^2}=\sqrt{64+1}$$
$$=\sqrt{65}$$
$$BC=\sqrt{(3-5)^2+[2-(-1)]^2}=\sqrt{4+9}$$
$$=\sqrt{13}$$
$$AC=\sqrt{[3-(-3)]^2+[2-(-2)]^2}=\sqrt{36+16}$$
$$=\sqrt{52}$$

$$(AB)^2\stackrel{?}{=}(BC)^2+(AC)^2$$
$$(\sqrt{65})^2\stackrel{?}{=}(\sqrt{13})^2+(\sqrt{52})^2$$
$$65=13+52$$

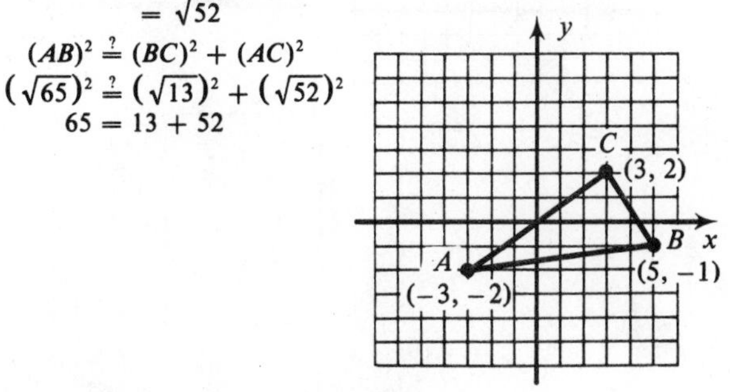

Therefore, the triangle is a right triangle

7. (a) $2\sqrt{10}$ (b) $2\sqrt{5}$

9. Yes, because $PQ + QR = PR$

Exercises 7–2 (page 225)

1.

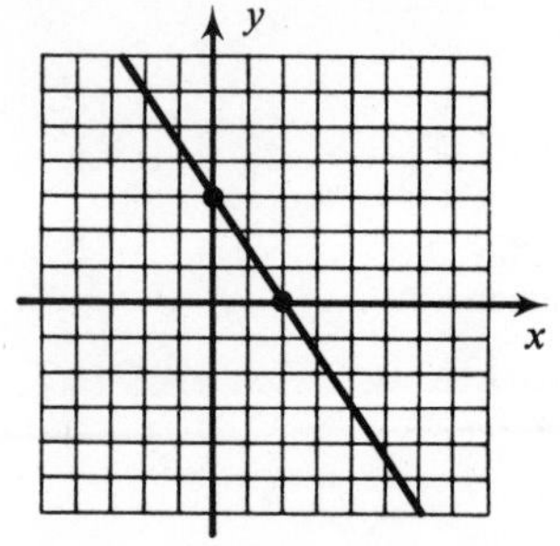

3.

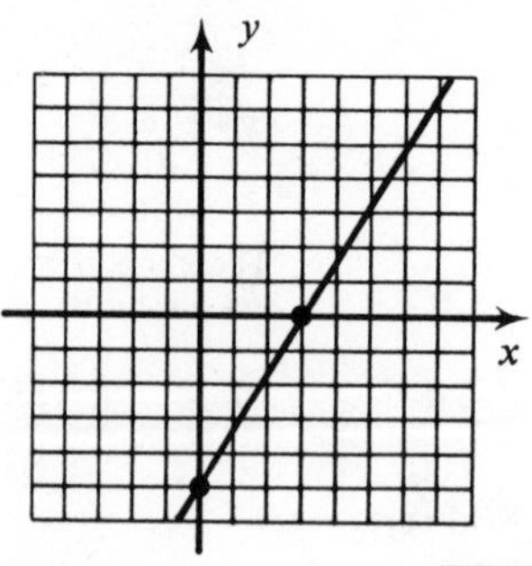

5.

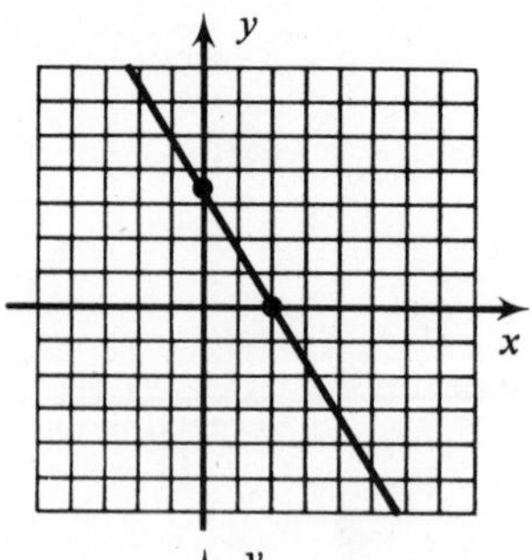

7.

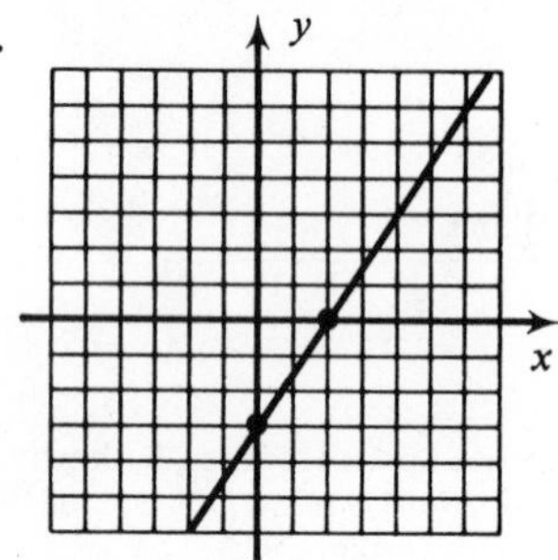

9. $9y = 25 - 7x$

x-intercept:

Set $y = 0$: $9(0) = 25 - 7x$

$x = 3\frac{4}{7}$

y-intercept:

Set $x = 0$: $9y = 25 - 7(0)$

$y = 2\frac{7}{9}$

x	y
$3\frac{4}{7}$	0
0	$2\frac{7}{9}$

11.

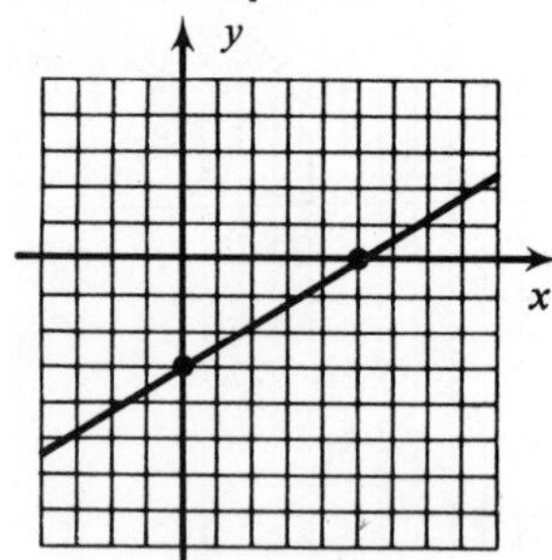

13. $6x + 11y = 0$

Both intercepts are (0,0)

Find another point:

Set $y = 3$: $6x + 11(3) = 0$

$6x + 33 = 0$

$x = -5\frac{1}{2}$

x	y
0	0
$-5\frac{1}{2}$	3

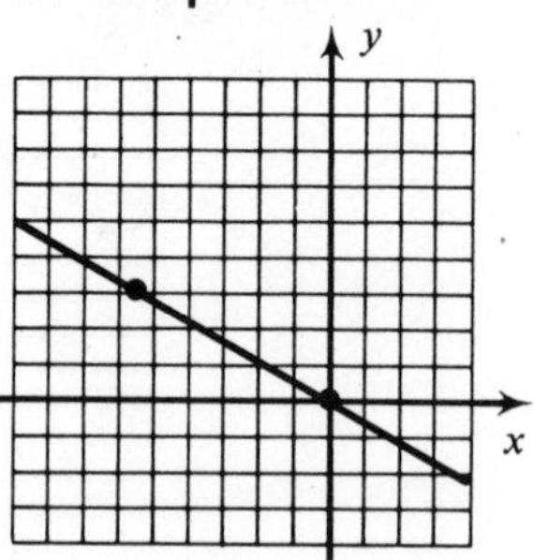

15.

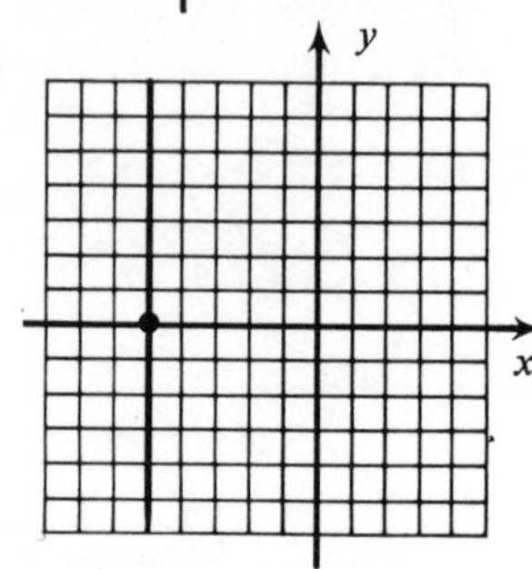

17. $7x + 5y = 2$

x-intercept:

Set $y = 0$: $7x + 5(0) = 2$

$x = \frac{2}{7}$

y-intercept:

Set $x = 0$: $7(0) + 5y = 2$

$y = \frac{2}{5}$

Find another point:

Set $x = 3$: $7(3) + 5y = 2$

$21 + 5y = 2$

$y = -3\frac{4}{5}$

x	y
$\frac{2}{7}$	0
0	$\frac{2}{5}$
3	$-3\frac{4}{5}$

19. $y = \frac{1}{2}x - 1$

$2y = x - 2$

x-intercept:

Set $y = 0$: $2(0) = x - 2$

$2 = x$

y-intercept:

Set $x = 0$: $2y = (0) - 2$

$y = -1$

x	y
2	0
0	-1

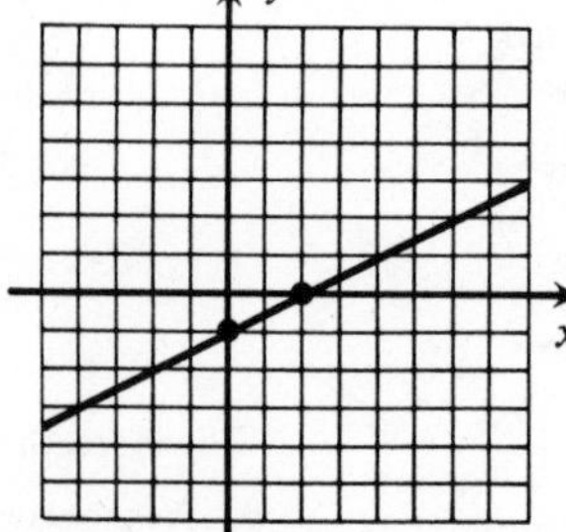

21.

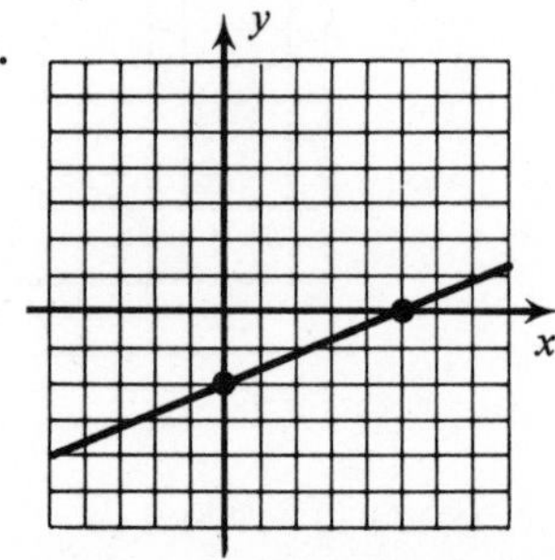

23.

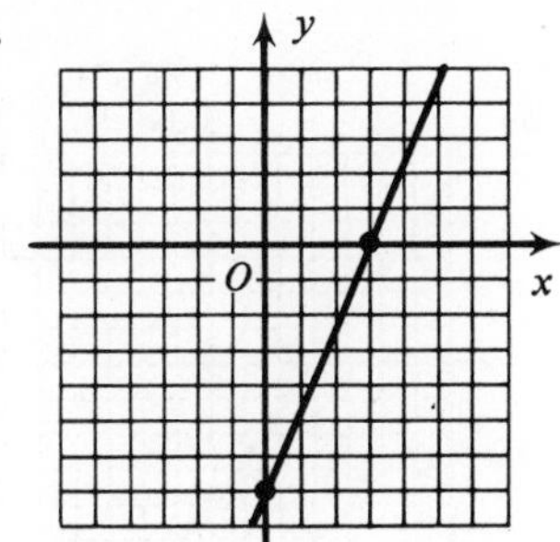

25.

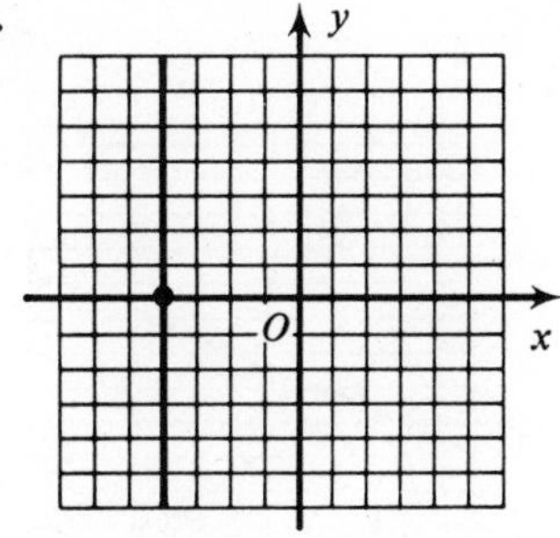

27.

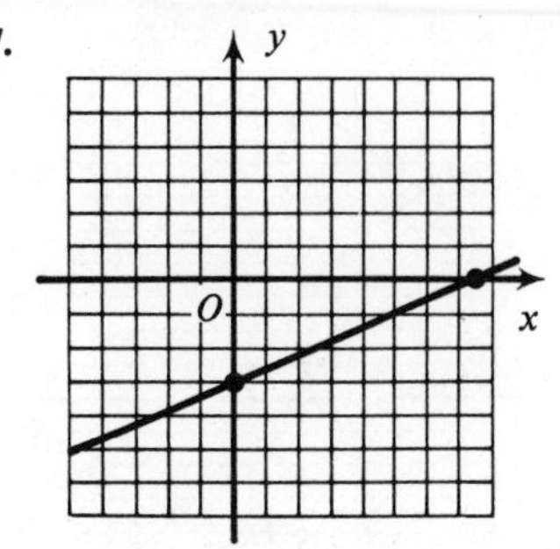

29. (a) $5x - 7y = 18$

x	y
$3\frac{3}{5}$	0
0	$-2\frac{4}{7}$

$2x + 3y = -16$

x	y
-8	0
0	$-5\frac{1}{3}$

(b) $(-2,-4)$

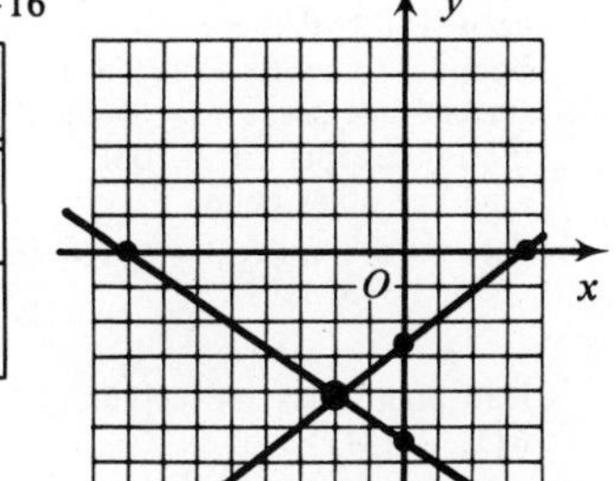

Exercises 7–3 (page 232)

1. $\frac{2}{9}$ **3.** $-\frac{1}{3}$ **5.** $\frac{0}{9} = 0$ **7.** $m = \frac{y_2 - y_1}{x_2 - x_1} = \frac{(-2) - (3)}{(-4) - (-4)} = \frac{-5}{0}$
Not a real number; m does not exist

9. $5x - 3y + 7 = 0$

11. $5x - 2y - 10 = 0$

13. $\frac{3}{1}\cdot\frac{y}{1} = \frac{\cancel{3}}{1}\left(-\frac{5}{\cancel{3}}x\right) + \frac{3}{1}\cdot\frac{4}{1}$
$3y = -5x + 12$
$5x + 3y - 12 = 0$

15. $6x + 2y - 11 = 0$ **17.** $y - 3 = 0$

19. $x - 7 = 0$ **21.** $5x - 7y - 21 = 0$ **23.** $8x + 6y - 3 = 0$

25. $y = mx + b$
$y = 0x + 5$
$y = 5$ or $y - 5 = 0$

27. $x - 5y - 19 = 0$ **29.** $x - 4y + 26 = 0$

31. (a) $y = \frac{4}{5}x + 4$
(b) $\frac{4}{5}$
(c) 4

33. (a) $y = -\frac{2}{9}x - \frac{5}{3}$
(b) $-\frac{2}{9}$
(c) $-\frac{5}{3}$

35. $5x + 2y - 38 = 0$ **37.** $4x + 3y - 40 = 0$ **39.** $2x + 3y + 9 = 0$

41. It must have the same slope as $3x - 5y = 6$
$-5y = -3x + 6$
$y = \frac{3}{5}x - \frac{6}{5}$
$m = \frac{3}{5}$

$y - y_1 = m(x - x_1)$
$y - 7 = \frac{3}{5}[x - (-4)]$
$5y - 35 = 3x + 12$
$0 = 3x - 5y + 47$

43. $3x + 5y - 12 = 0$

45. x-intercept $= (-6,0)$
y-intercept $= (0,4)$
$m = \frac{(4) - (0)}{(0) - (-6)} = \frac{4}{6} = \frac{2}{3}$
$y - y_1 = m(x - x_1)$
$y - 4 = \frac{2}{3}(x - 0)$
$3y - 12 = 2x$
$0 = 2x - 3y + 12$

Exercises 7–4 (page 239)

1.

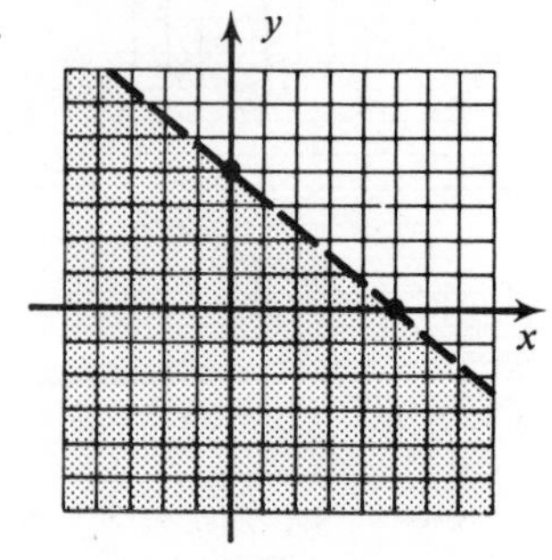

3.

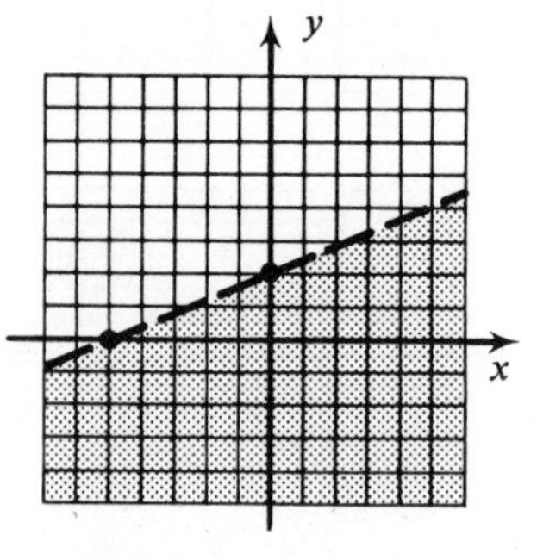

5.

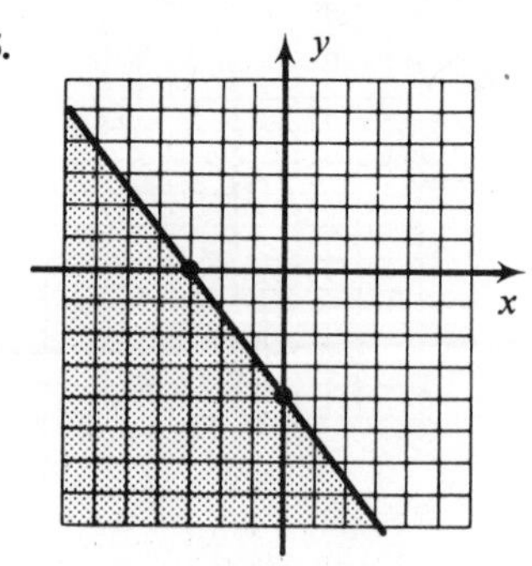

7.

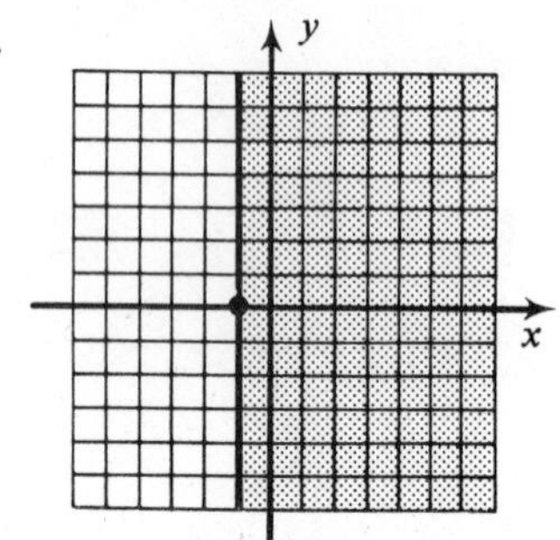

9. $6x - 13y > 0$

Boundary line: $6x - 13y = 0$

Dashed line because $=$ is not included in $>$

Both intercepts are at $(0,0)$

Find one other point:

Set $x = 6$: $\quad 6(6) - 13y = 0$

$$y = 2\frac{10}{13}$$

x	y
0	0
6	$2\frac{10}{13}$

Half-plane does not include $(1,1)$ because:

$6(1) - 13(1) > 0$

$-7 > 0$ False

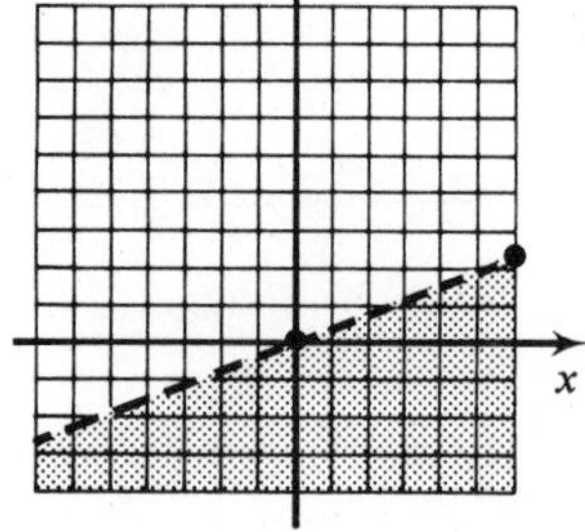

11.

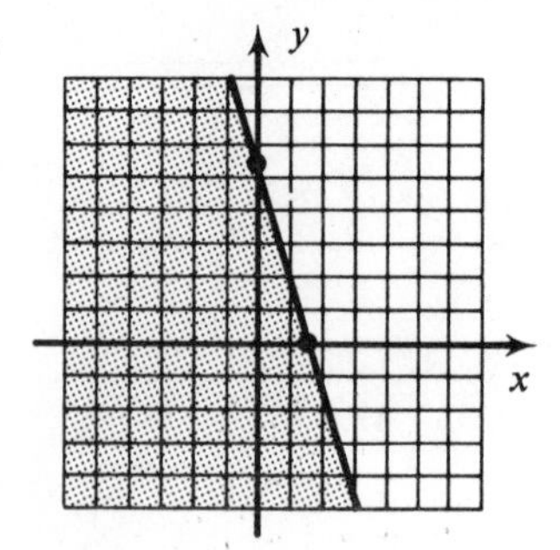

13.

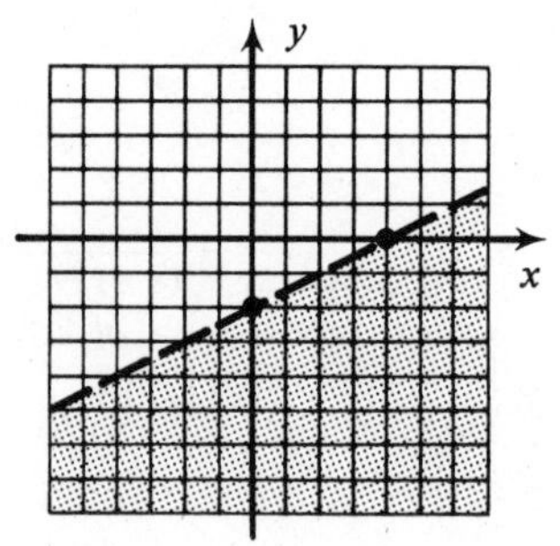

15. $4(x + 2) + 7 \leq 3(5 - 2x)$

$$4x + 8 + 7 \leq 15 - 6x$$

$$10x \leq 0$$

$$x \leq 0$$

Boundary line: $x = 0$

Solid line because $=$ is included in $\leq$

Half-plane does not include $(1,0)$ because:

$1 \leq 0$ False

17. $\dfrac{2x + y}{3} - \dfrac{x - y}{2} \geq \dfrac{5}{6}$

LCD $= 6$

$$\overset{2}{\cancel{6}}\left(\frac{2x + y}{\cancel{3}}\right) + \overset{3}{\cancel{6}}\left(\frac{-(x - y)}{\cancel{2}}\right) \geq \cancel{6}\left(\frac{5}{\cancel{6}}\right)$$

$$4x + 2y - 3x + 3y \geq 5$$

$$x + 5y \geq 5$$

x	y
0	1
5	0

Boundary line: $x + 5y = 5$

Solid line because $=$ is included in $\geq$

Half-plane does not include $(0,0)$ because:

$(0) + 5(0) \geq 5$

$0 \geq 5$ False

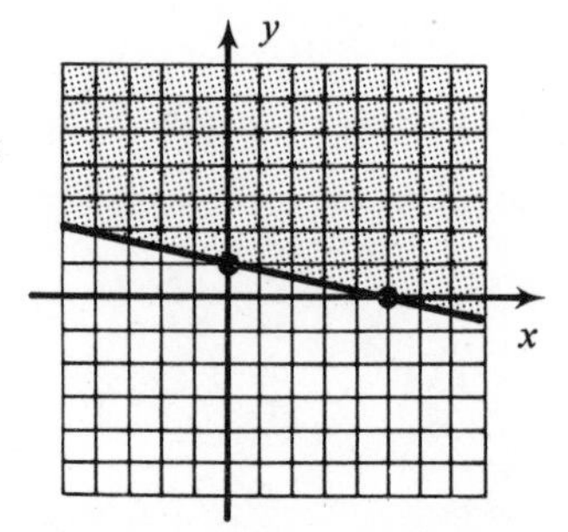

Exercises 7–5 (page 245)

1. Domain $= \{2, 3, 0, -3\}$
Range $= \{-1, 4, 2, -2\}$

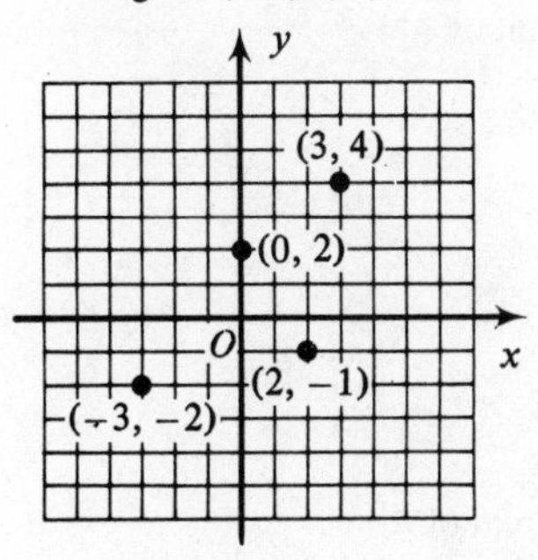

3. Range $= \left\{4, \frac{12}{5}, 2, \frac{6}{5}\right\}$

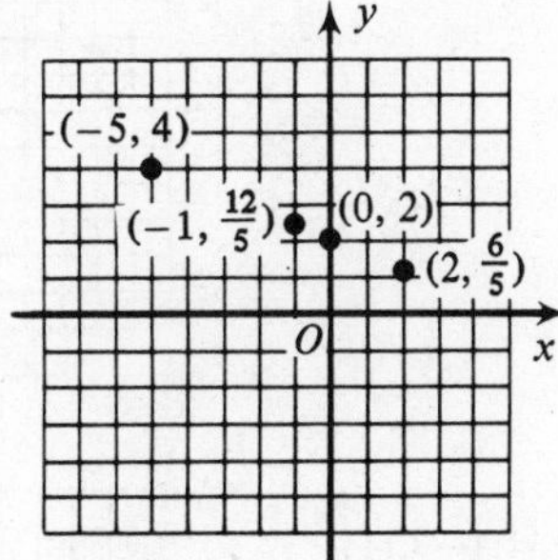

5. $y = 2x - 3$

x	y
0	−3
3	3

7. (b) and (c), because any vertical line meets the graph at no more than one point

9. (a) 5
(b) −1
(c) $3a - 7$
(d) $3x + 5$

11. 7

13.
$$f(x) = 3(x)^2 - 2(x) + 4$$
$$f(3) = 3(3)^2 - 2(3) + 4 = 27 - 6 + 4 = 25$$
$$f(1) = 3(1)^2 - 2(1) + 4 = 3 - 2 + 4 = 5$$
$$f(0) = 3(0)^2 - 2(0) + 4 = 4$$
Therefore, $2f(3) + 4f(1) - 3f(0) = 2(25) + 4(5) - 3(4)$
$= 50 + 20 - 12 = 58$

15.
$$f(x) = (x)^3$$
$$f(-3) = (-3)^3 = -27$$
$$6g(x) = 6\left(\frac{1}{x}\right)$$
$$6g(2) = 6\left(\frac{1}{2}\right) = 3$$
Therefore, $f(-3) - 6g(2)$
$= -27 - 3 = -30$

17. 42

19. $D_f = \{x | x \geq 4\}$ because $\sqrt{x - 4}$ will not have a real value when $x - 4 < 0$ or $x < 4$
$R_f = \{y | y \geq 0\}$ because $y = \sqrt{x - 4}$, which is the principal square root and therefore cannot be negative

21.
$$f(x) = (x)^2 - (x)$$
$$f(x + h) = (x + h)^2 - (x + h)$$
$$= x^2 + 2xh + h^2 - x - h$$
$$\frac{f(x + h) - f(x)}{h} = \frac{x^2 + 2xh + h^2 - x - h - x^2 + x}{h}$$
$$= \frac{2xh + h^2 - h}{h} = \frac{2xh}{h} + \frac{h^2}{h} + \frac{-h}{h} = 2x + h - 1$$

23. $3r - 4$

25.
$$z = g(x,y) = 5x^2 - 2y^2 + 7x - 4y$$
$$g(3,-4) = 5(3)^2 - 2(-4)^2 + 7(3) - 4(-4)$$
$$= 45 - 32 + 21 + 16 = 50$$

27.
$$A = f(P,i,n) = P(1 + i)^n$$
$$f(100, 0.08, 12) = 100(1 + 0.08)^{12} = 100(1.08)^{12} \doteq 251.82$$
Found by calculator *or* by using logarithms

Exercises 7–6 (page 250)

1. $\mathcal{R}^{-1} = \{(7,-10), (-8,3), (-4,-5), (9,3)\}$
$D_{\mathcal{R}^{-1}} = \{7, -8, -4, 9\} = R_{\mathcal{R}}$
$R_{\mathcal{R}^{-1}} = \{-10, 3, -5, 3\} = D_{\mathcal{R}}$
$\mathcal{R}$ is *not* a function
$\mathcal{R}^{-1}$ *is* a function

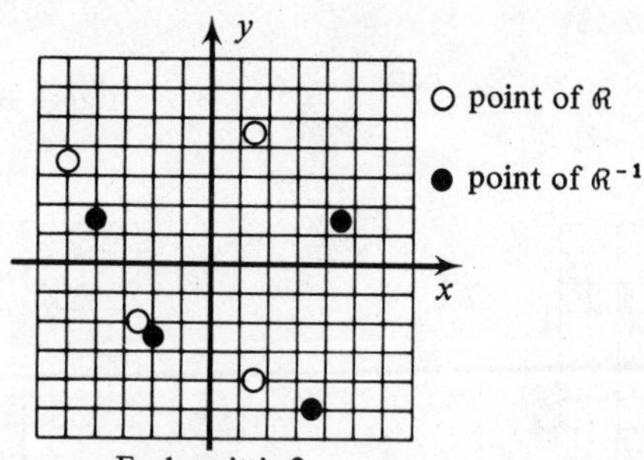

Each unit is 2

3. $y = f(x) = 2x^2 - 7$
$f(-3) = 2(-3)^2 - 7 = 11; (-3,11)$
$f(-2) = 2(-2)^2 - 7 = 1; (-2,1)$
$f(0) = 2(0)^2 - 7 = -7; (0,-7)$
$f(1) = 2(1)^2 - 7 = -5; (1,-5)$
$\mathcal{R}^{-1} = \{(11,-3), (1,-2), (-7,0), (-5,1)\}$
$D_{\mathcal{R}^{-1}} = \{11, 1, -7, -5\} = R_{\mathcal{R}}$
$R_{\mathcal{R}^{-1}} = \{-3, -2, 0, 1\} = D_{\mathcal{R}}$
Both $\mathcal{R}$ and $\mathcal{R}^{-1}$ are functions.

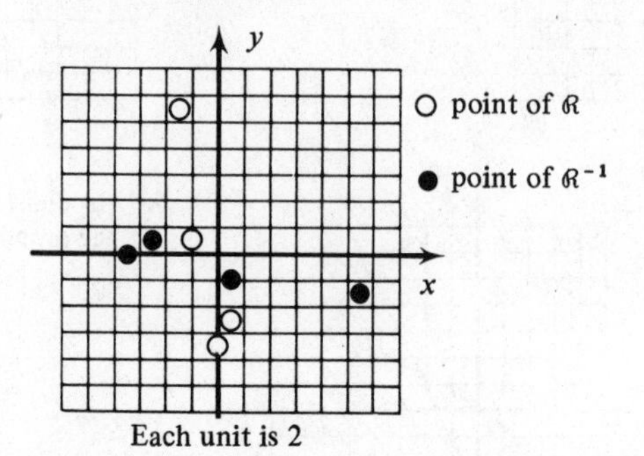

Each unit is 2

5. $\mathcal{R}^{-1} = \{(-6,-4), (1,-3), (10,0), (6,2)\}$
$D_{\mathcal{R}^{-1}} = \{-6, 1, 10, 6\} = R_{\mathcal{R}}$
$R_{\mathcal{R}^{-1}} = \{-4, -3, 0, 2\} = D_{\mathcal{R}}$
Both $\mathcal{R}$ and $\mathcal{R}^{-1}$ are functions.

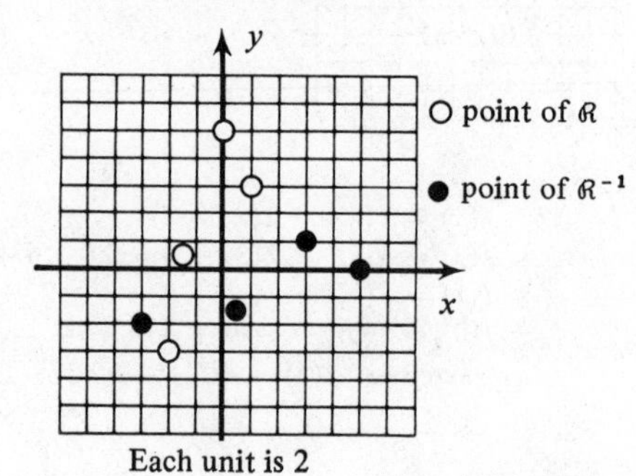

Each unit is 2

7. $y = f(x) = 5 - 2x$
$x = 5 - 2y$
$y = \dfrac{5-x}{2} = f^{-1}(x)$

$y = 5 - 2x$

x	y
$2\frac{1}{2}$	0
0	5

$y = \dfrac{5-x}{2}$
$2y = 5 - x$

x	y
5	0
0	$2\frac{1}{2}$

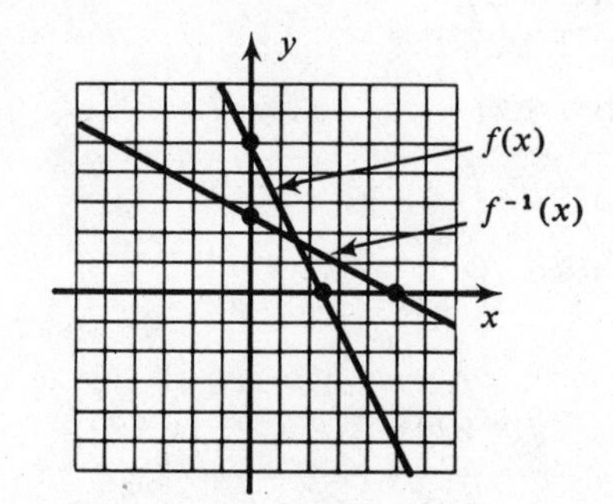

9. $y = f(x) = \dfrac{4x-3}{5}$
$x = \dfrac{4y-3}{5}$
$5x = 4y - 3$
$y = \dfrac{5x+3}{4} = f^{-1}(x)$

11. $y = \dfrac{4x+8}{3} = f^{-1}(x)$

13. $y = f(x) = \dfrac{5}{x+2}$
$\dfrac{x}{1} = \dfrac{5}{y+2}$
$x(y+2) = 5$
$xy + 2x = 5$
$xy = 5 - 2x$
$y = \dfrac{5-2x}{x} = f^{-1}(x)$

Exercises 7–7 (page 254)

1.

x	y
-2	4
-1	1
0	0
1	1
2	4

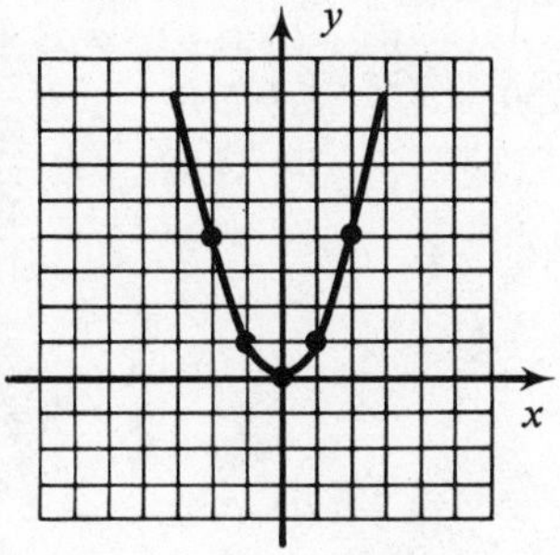

3.

x	y
-2	8
-1	3
0	0
1	-1
2	0
3	3
4	8

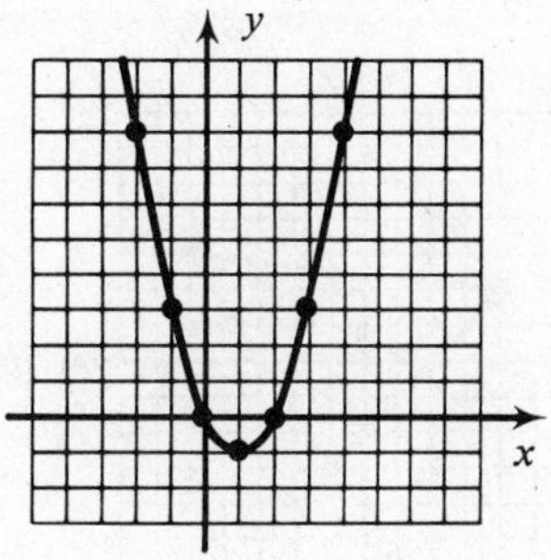

5.

x	y
-2	-8
-1	-3
0	0
1	1
2	0
3	-3
4	-8

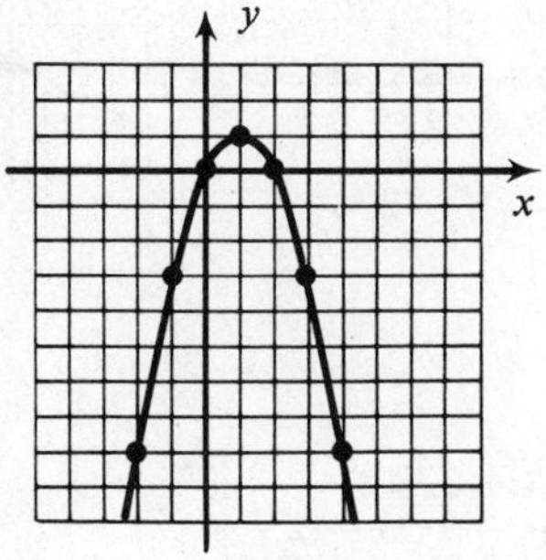

7. $y = x^3$

$y = (-2)^3 = -8$

$y = (-1)^3 = -1$

$y = 0^3 = 0$

$y = 1^3 = 1$

$y = 2^3 = 8$

x	y
-2	-8
-1	-1
0	0
1	1
2	8

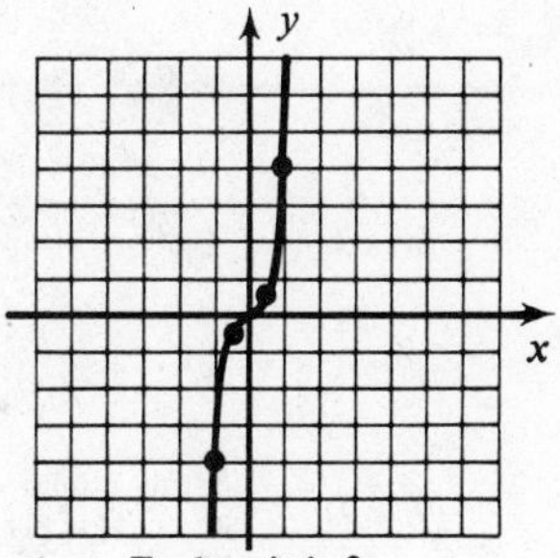

Each unit is 2

9. $f(x) = x^3 - 2x^2 - 13x + 20$

1 − 2 − 13 20

x				$f(x)$
-4	1	-6	11	-24
-3	1	-5	2	14
-2	1	-4	-5	30
0	1	-2	-13	20
2	1	0	-13	-6
3	1	1	-10	-10
4	1	2	-5	0
5	1	3	2	30

Range = $\{-24, 14, 30, 20, -6, -10, 0, 30\}$

11. $R = P(1) = (1)^9 - 15 = -14$

Chapter Seven Diagnostic Test or Review (page 257)

Following each problem number is the textbook section number (in parentheses) where that kind of problem is discussed.

1. (a) (7–1)

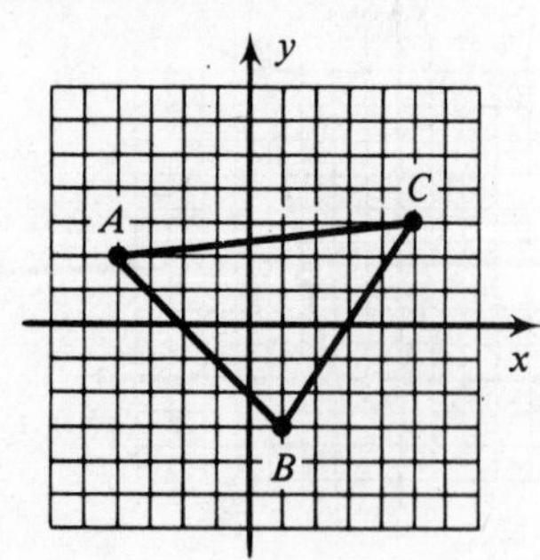

(b) (7–1) $A(-4,2)$, $B(1,-3)$

$$(AB)^2 = [1-(-4)]^2 + (-3-2)^2 = 25 + 25 = 50$$
$$AB = \sqrt{50} = \sqrt{25 \cdot 2} = 5\sqrt{2}$$

(c) (7–3) $B(1,-3)$, $C(5,3)$

$$m = \frac{3-(-3)}{5-1} = \frac{6}{4} = \frac{3}{2}$$

(d) (7–3) $A(-4,2)$, $B(1,-3)$

$$m = \frac{-3-2}{1-(-4)} = \frac{-5}{5} = -1$$
$$y - y_1 = m(x - x_1)$$
$$y - (-3) = -1(x-1)$$
$$y + 3 = -x + 1$$
$$x + y + 2 = 0$$

(e) (7–2)
If $y = 0$: $x + y + 2 = 0$
$x + 0 + 2 = 0$
$x = -2$

(f) (7–2)
If $x = 0$: $x + y + 2 = 0$
$0 + y + 2 = 0$
$y = -2$

2. (7–3) $y = mx + b$

$$y = \frac{6}{5}x - 4$$
$$5y = 6x - 20$$
$$0 = 6x - 5y - 20$$

3. (7–2) $x - 2y = 6$

If $x = 0$: $0 - 2y = 6$
$y = -3$

If $y = 0$: $x - 2(0) = 6$
$x = 6$

x	y
0	−3
6	0

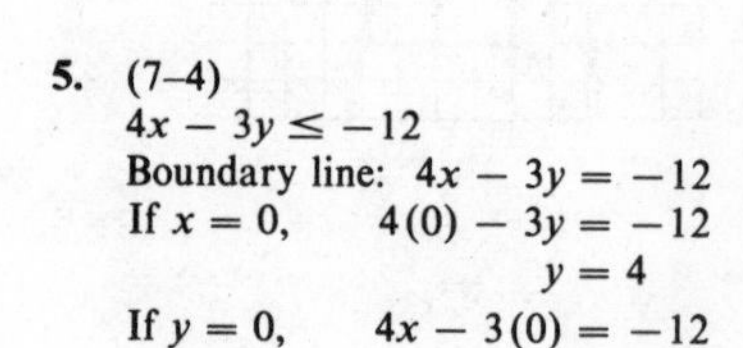

4. (7–7)
$y = 1 + x - x^2$

	x	y
$y = 1 + (-2) - (-2)^2 = -5$	−2	−5
$y = 1 + (-1) - (-1)^2 = -1$	−1	−1
$y = 1 + (0) - (0)^2 = 1$	0	1
$y = 1 + (1) - (1)^2 = 1$	1	1
$y = 1 + (2) - (2)^2 = -1$	2	−1
$y = 1 + (3) - (3)^2 = -5$	3	−5

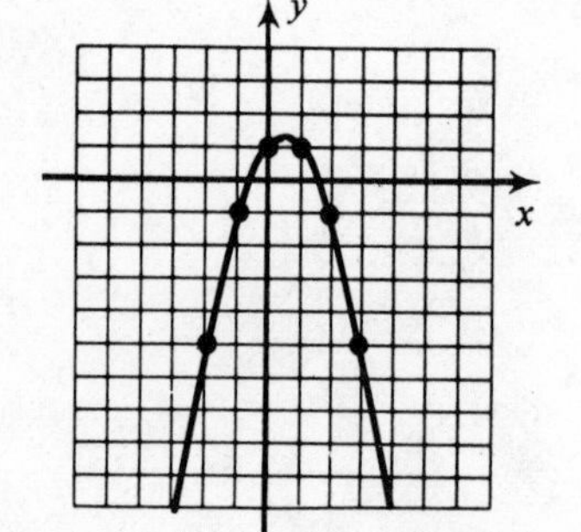

5. (7–4)
$4x - 3y \leq -12$
Boundary line: $4x - 3y = -12$
If $x = 0$, $4(0) - 3y = -12$
$y = 4$
If $y = 0$, $4x - 3(0) = -12$
$x = -3$

x	y
0	4
−3	0

Boundary line is solid because equality is included
Test point: (0,0) $4x - 3y \leq -12$
$4(0) - 3(0) \leq -12$
$0 \leq -12$ False

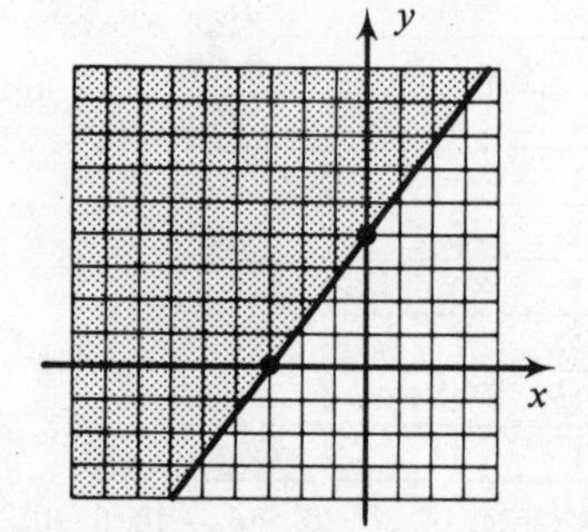

6. (7–5) (a) $\{(-4, -5), (2, 4), (4, -2), (-2, 3), (2, -1)\}$
Domain $= \{-4, 2, 4, -2\}$
(b) $\{(-4, -5), (2, 4), (4, -2), (-2, 3), (2, -1)\}$
Range $= \{-5, 4, -2, 3, -1\}$
(c)

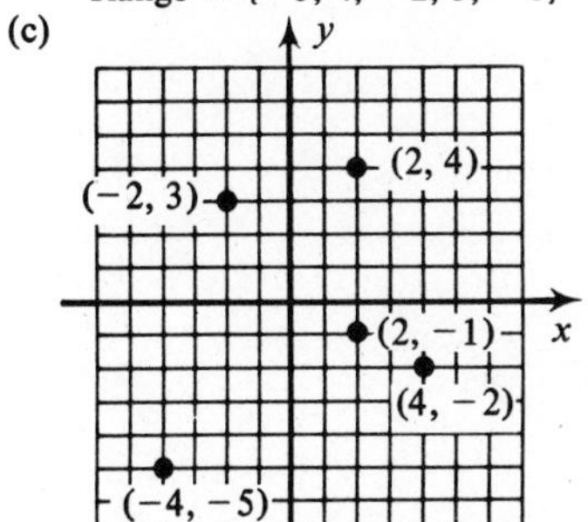

(d) The relation *is not* a function because (2, 4) and (2, −1) have the same first coordinate and different second coordinates

7. (7–5) $f(x) = 3(x)^2 - 5$

(a) $f(-2) = 3(-2)^2 - 5 = 12 - 5 = 7$

(b) $f(4) = 3(4)^2 - 5 = 48 - 5 = 43$

(c) $\dfrac{f(4) - f(-2)}{6} = \dfrac{43 - 7}{6} = \dfrac{36}{6} = 6$

8. (7–6) $\mathcal{R} = \{(1,4), (-5,-3), (4,-2), (-5,2)\}$
$\mathcal{R}^{-1} = \{(4,1), (-3,-5), (-2,4), (2,-5)\}$
$D_{\mathcal{R}^{-1}} = \{4, -3, -2, 2\} = R_{\mathcal{R}}$
$R_{\mathcal{R}^{-1}} = \{1, -5, 4\} = D_{\mathcal{R}}$
$\mathcal{R}$ *is not* a function
$\mathcal{R}^{-1}$ *is* a function

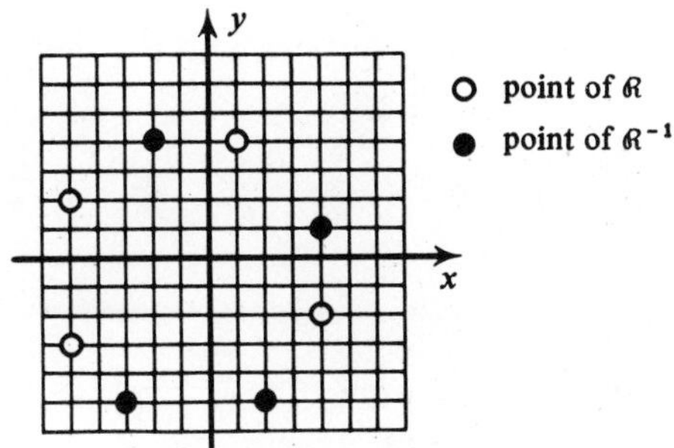

9. (7–6) $y = f(x) = -\dfrac{3}{2}x + 1$

$x = -\dfrac{3}{2}y + 1$

$2x = -3y + 2$
$3y = 2 - 2x$
$y = \dfrac{2 - 2x}{3} = f^{-1}(x)$

$y = -\dfrac{3}{2}x + 1$

$2y = -3x + 2$

x	y
$\frac{2}{3}$	0
0	1
4	−5

$y = \dfrac{2 - 2x}{3}$

$3y = 2 - 2x$

x	y
1	0
0	$\frac{2}{3}$
−5	4

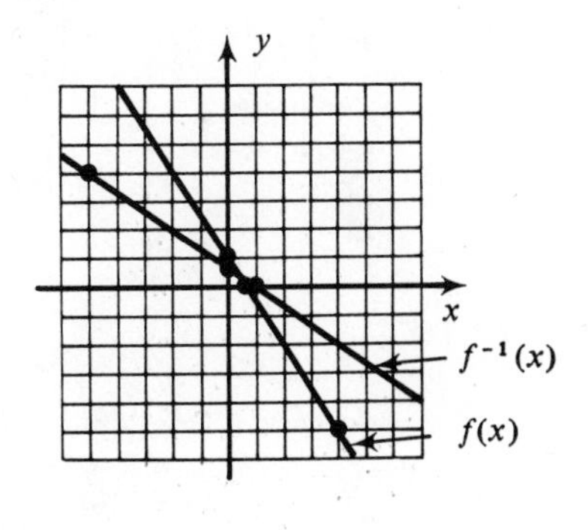

10. (7–6) $y = f(x) = \dfrac{15}{4(2 - 5x)}$

$x = \dfrac{15}{4(2 - 5y)}$

$4x(2 - 5y) = 15$
$8x - 20xy = 15$
$y = \dfrac{8x - 15}{20x} = f^{-1}(x)$

Exercises 8–1 (page 263)

1. $a = 3$, $b = 5$, $c = -2$

3. $3x^2 + 0x - 4 = 0$ $\begin{cases} a = 3 \\ b = 0 \\ c = -4 \end{cases}$

5. $3x^2 - 4x + 12 = 0$ $\begin{cases} a = 3 \\ b = -4 \\ c = 12 \end{cases}$

7. $3x(x - 2) = (x + 1)(x - 5)$
$3x^2 - 6x = x^2 - 4x - 5$
$2x^2 - 2x + 5 = 0$ $\begin{cases} a = 2 \\ b = -2 \\ c = 5 \end{cases}$

9. $\{1, 4\}$

11.
$$\frac{x-1}{4} + \frac{6}{x+1} = 2$$

LCD $= 4(x+1)$

$$\frac{4(x+1)}{1} \cdot \frac{x-1}{4} + \frac{4(x+1)}{1} \cdot \frac{6}{(x+1)} = 4(x+1) \cdot 2$$
$$(x+1)(x-1) + 24 = 8(x+1)$$
$$x^2 - 1 + 24 = 8x + 8$$
$$x^2 - 8x + 15 = 0$$
$$(x-3)(x-5) = 0$$
$$x - 3 = 0 \quad | \quad x - 5 = 0$$
$$x = 3 \quad | \quad x = 5$$

13. $\left\{\frac{1}{2}, -\frac{2}{3}\right\}$

15.
$$\frac{3x-2}{6} = \frac{2}{x+1}$$
$$(3x-2)(x+1) = 6(2) \quad \text{Product of extremes}$$
$$3x^2 + x - 2 = 12 \quad = \text{Product of means}$$
$$3x^2 + x - 14 = 0$$
$$(x-2)(3x+7) = 0$$
$$x - 2 = 0 \quad | \quad 3x + 7 = 0$$
$$x = 2 \quad | \quad x = -\frac{7}{3}$$

17.
$$z^{-4} - 4z^{-2} = 0$$

Let $a = z^{-2}$
$a^2 = z^{-4}$
$$a^2 - 4a = 0$$
$$a(a-4) = 0$$
$$a = 0 \quad | \quad a - 4 = 0$$
$$\quad | \quad a = 4$$
$$z^{-2} = 0 \quad | \quad z^{-2} = 4$$
$$(z^{-2})^{-1/2} = 0^{-1/2} \quad | \quad (z^{-2})^{-1/2} = 4^{-1/2}$$
$$z = 0^{-1/2} \quad | \quad z = \pm\frac{1}{\sqrt{4}} = \pm\frac{1}{2}$$

Does not exist

19. Let $a = K^{-1/3}$
$a^2 = K^{-2/3}$
$$K^{-2/3} + 2K^{-1/3} + 1 = 0$$
$$a^2 + 2a + 1 = 0$$
$$(a+1)(a+1) = 0$$

Same answer for both factors
$$a + 1 = 0$$
$$a = -1$$
$$K^{-1/3} = -1$$
$$(K^{-1/3})^{-3} = (-1)^{-3} = \frac{1}{(-1)^3} = \frac{1}{-1} = -1$$
$$K = -1$$

21. Let $a = (x^2 - 4x)$
$a^2 = (x^2 - 4x)^2$
$$(x^2 - 4x)^2 - (x^2 - 4x) - 20 = 0$$
$$a^2 - a - 20 = 0$$
$$(a+4)(a-5) = 0$$
$$a + 4 = 0 \quad | \quad a - 5 = 0$$
$$x^2 - 4x + 4 = 0 \quad | \quad x^2 - 4x - 5 = 0$$
$$(x-2)(x-2) = 0 \quad | \quad (x-5)(x+1) = 0$$
$$x - 2 = 0 \quad | \quad x - 5 = 0 \quad | \quad x + 1 = 0$$
$$x = 2 \quad | \quad x = 5 \quad | \quad x = -1$$

Solution set: $\{-1, 2, 5\}$ (all check)

23. The area plus the perimeter is 80
$$2W^2 + 6W = 80$$

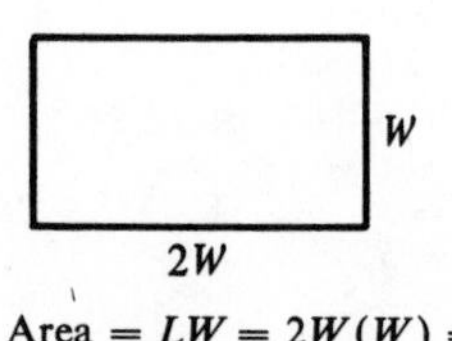

$$2W^2 + 6W - 80 = 0$$
$$2(W^2 + 3W - 40) = 0$$
$$2(W+8)(W-5) = 0$$
$$W + 8 = 0 \quad | \quad W - 5 = 0$$
$$W = -8 \quad | \quad W = 5$$
$$\text{No meaning} \quad | \quad 2W = 2(5)$$
$$2W = 10$$

Width = 5; Length = 10

Area $= LW = 2W(W) = 2W^2$
$p = 2(L) + 2(W)$
$p = 2(2W) + 2(W)$
$p = 6W$

25.

	d	r	t
Going	120	r	$\frac{120}{r}$
Returning	120	$r+20$	$\frac{120}{r+20}$

$$\text{Time to go} + \text{Time to return} = 5 \text{ hr}$$
$$\frac{120}{r} + \frac{120}{r+20} = 5$$
$$\frac{24}{r} + \frac{24}{r+20} = 1$$

LCD $= r(r+20)$

$$\frac{\cancel{r}(r+20)}{1}\cdot\frac{24}{\cancel{r}} + \frac{r\cancel{(r+20)}}{1}\cdot\frac{24}{\cancel{r+20}} = \frac{r(r+20)}{1}\cdot 1$$
$$24r + 480 + 24r = r^2 + 20r$$
$$r^2 - 28r - 480 = 0$$
$$(r+12)(r-40) = 0$$

$r + 12 = 0$ | $r - 40 = 0$
$r = -12$ | $r = 40$ mph
No meaning | Rate going

27. 8 and 10, or -10 and -8 **29.** Width = 5, length = 12

Exercises 8–2 (page 267)

1. $\{4, -4\}$ **3.** $\left\{\pm\frac{2\sqrt{5}}{5}i\right\}$ **5.** $\{0, 6\}$ **7.** $\left\{0, \frac{1}{6}\right\}$

9.
$$\frac{x+2}{3x} = \frac{x+1}{x}$$
$$3x(x+1) = x(x+2) \quad \text{Product of means} = \text{Product of extremes}$$
$$3x^2 + 3x = x^2 + 2x$$
$$2x^2 + x = 0$$
$$x(2x+1) = 0$$

$x = 0$ | $2x + 1 = 0$
Not possible | $x = -\frac{1}{2}$

11. $\{0, 5\}$

13.
$$12x = 8x^3$$
$$0 = 8x^3 - 12x$$
$$0 = 4x(2x^2 - 3)$$

$4x = 0$ | $2x^2 - 3 = 0$
$x = 0$ | $2x^2 = 3$
| $x^2 = \frac{3}{2}$
| $x = \pm\sqrt{\frac{3}{2}}$
| $x = \pm\sqrt{\frac{3}{2}\cdot\frac{2}{2}} = \pm\frac{\sqrt{6}}{2}$

15. $\{0, \sqrt{7}, -\sqrt{7}\}$

17.

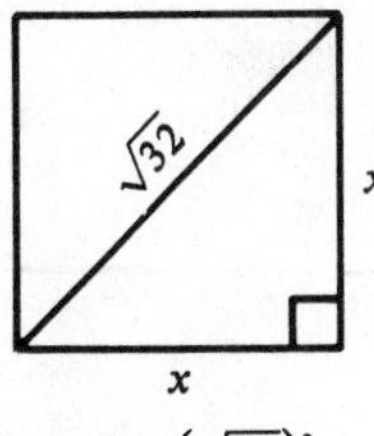

$$x^2 + x^2 = (\sqrt{32})^2$$
$$2x^2 = 32$$
$$x^2 = 16$$
$$x = 4$$
(-4 has no meaning in this problem)

19. $\doteq$ 14.42 cm

21. Let x = Hours for Merwin Merwin does $\frac{1}{x}$ of the work each hour

$x + 3$ = Hours for Mina Mina does $\frac{1}{x+3}$ of the work each hour

Mina works a total of 8 hr and Merwin works 3 hr to do the complete job

$$\text{Work done by Mina} + \text{Work done by Merwin} = \text{Total work done (the complete job)}$$
$$8\left(\frac{1}{x+3}\right) + 3\left(\frac{1}{x}\right) = 1$$

(Problem 21 continued on next page)

$$\text{LCD} = x(x + 3)$$

$$\frac{x\cancel{(x+3)}}{1} \cdot \frac{8}{\cancel{x+3}} + \frac{\cancel{x}(x+3)}{1} \cdot \frac{3}{\cancel{x}} = \frac{x(x+3)}{1} \cdot 1$$

$$8x + 3x + 9 = x^2 + 3x$$
$$x^2 - 8x - 9 = 0$$
$$(x - 9)(x + 1) = 0$$

$x + 1 = 0$	$x - 9 = 0$
$x = -1$	$x = 9$ Hours for Merwin to do the job
No meaning	$x + 3 = 12$ Hours for Mina to do the job

Exercises 8–3 (page 271)

1.
$$x^2 = 6x + 11$$
$$x^2 - 6x = 11$$
$$x^2 - 6x + 9 = 9 + 11$$
$$(x - 3)^2 = 20$$
$$\sqrt{(x-3)^2} = \pm\sqrt{20}$$
$$x - 3 = \pm\sqrt{4 \cdot 5}$$
$$x = 3 \pm 2\sqrt{5}$$

3. $\{2 \pm \sqrt{17}\}$ **5.** $\{1 \pm \sqrt{3}\}$ **7.** $\{3 \pm \sqrt{14}\}$ **9.** $\left\{1, -\frac{2}{3}\right\}$

11. $\{2 \pm \sqrt{3}\}$ **13.** $\{2 \pm \sqrt{2}\}$

15. $3x^2 + 2x + 1 = 0 \quad \begin{cases} a = 3 \\ b = 2 \\ c = 1 \end{cases}$

$$x = \frac{-(2) \pm \sqrt{(2)^2 - 4(3)(1)}}{2(3)}$$
$$= \frac{-2 \pm \sqrt{4 - 12}}{6} = \frac{-2 \pm \sqrt{-8}}{6}$$
$$= \frac{-2 \pm 2\sqrt{2}\,i}{6} = \frac{-1 \pm \sqrt{2}\,i}{3}$$

17. $\left\{\frac{4 \pm \sqrt{2}\,i}{2}\right\}$ **19.** $\left\{\frac{1}{2}, -3\right\}$

21. $\left\{\frac{5 \pm \sqrt{23}\,i}{2}\right\}$

23. $x^2 - 4x + 5 = 0 \quad \begin{cases} a = 1 \\ b = -4 \\ c = 5 \end{cases}$

$$x = \frac{-(-4) \pm \sqrt{(-4)^2 - 4(1)(5)}}{2(1)}$$
$$= \frac{4 \pm \sqrt{16 - 20}}{2} = \frac{4 \pm \sqrt{-4}}{2}$$
$$= \frac{4 \pm 2i}{2} = 2 \pm i$$

Check for $x = 2 - i$:

$$x^2 - 4x + 5 = 0$$
$$(2 - i)^2 - 4(2 - i) + 5 \stackrel{?}{=} 0$$
$$4 - 4i + i^2 - 8 + 4i + 5 \stackrel{?}{=} 0$$
$$4 - 4i - 1 - 8 + 4i + 5 \stackrel{?}{=} 0$$
$$9 - 9 + 4i - 4i \stackrel{?}{=} 0$$
$$0 = 0$$

25. $\left\{\frac{3 \pm \sqrt{2}}{2}\right\}$

27.

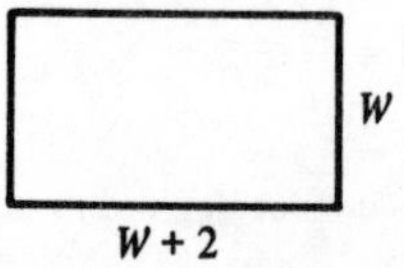

$$\text{Area} = LW = (W + 2)W = 2$$
$$W^2 + 2W = 2$$
$$W^2 + 2W - 2 = 0 \quad \begin{cases} a = 1 \\ b = 2 \\ c = -2 \end{cases}$$

$$W = \frac{-(2) \pm \sqrt{(2)^2 - 4(1)(-2)}}{2(1)} = \frac{-2 \pm \sqrt{4 + 8}}{2}$$
$$= \frac{-2 \pm \sqrt{12}}{2} = \frac{-2 \pm 2\sqrt{3}}{2}$$
$$= -1 \pm \sqrt{3} = \begin{cases} -1 + \sqrt{3} \doteq -1 + 1.732 \\ -1 - \sqrt{3} \doteq -1 - 1.732 \end{cases}$$
$$\doteq \begin{cases} 0.732 \doteq 0.73 & \text{Width; 2.73 Length} \\ -2.732 \doteq -2.73 & \text{Not possible} \end{cases}$$

29. 2

31. $x^3 - 8x^2 + 16x - 8$

	1	−8	16	−8
1	1	−7	9	1
−1	1	−9	25	−33
2	1	−6	4	0

$(x-2)(x^2-6x+4)=0$

$x = 2$

$$x = \frac{-(-6) \pm \sqrt{(-6)^2 - 4(1)(4)}}{2(1)} = \frac{6 \pm \sqrt{36-16}}{2} = \frac{6 \pm \sqrt{20}}{2} = \frac{6 \pm 2\sqrt{5}}{2} = 3 \pm \sqrt{5} \doteq 3 \pm 2.236 = \begin{cases} 3 + 2.236 = 5.236 \doteq 5.24 \\ 3 - 2.236 = 0.764 \doteq 0.76 \end{cases}$$

$\{2, 5.24, 0.76\}$

Exercises 8–4 (page 275)

1. Real, rational, and unequal **3.** Real, irrational, and unequal **5.** Real, rational, and unequal **7.** Roots are complex conjugates **9.** One real, rational root **11.** $x^2 - 2x - 8 = 0$ **13.** $x^2 - 5x = 0$ **15.** $6x^2 - 7x + 2 = 0$ **17.** $x^2 - 4x + 1 = 0$ **19.** $x^2 - 18 = 0$

21.
$$x = \frac{1+\sqrt{3}\,i}{2} \qquad\qquad x = \frac{1-\sqrt{3}\,i}{2}$$
$$2x = 1 + \sqrt{3}\,i \qquad\qquad 2x = 1 - \sqrt{3}\,i$$
$$2x - 1 - \sqrt{3}\,i = 0 \qquad\qquad 2x - 1 + \sqrt{3}\,i = 0$$
$$[(2x-1) - \sqrt{3}\,i][(2x-1) + \sqrt{3}\,i] = 0$$
$$(2x-1)^2 - (\sqrt{3}\,i)^2 = 0$$
$$4x^2 - 4x + 1 + 3 = 0$$
$$4x^2 - 4x + 4 = 0$$
$$x^2 - x + 1 = 0$$

23. $x^3 - 8x^2 + 19x - 12 = 0$

25. $15x^3 - 26x^2 + 13x - 2 = 0$

27.
$$(x-3)(x+2i)(x-2i) = 0$$
$$(x-3)(x^2+4) = 0$$
$$x^3 - 3x^2 + 4x - 12 = 0$$

Exercises 8–5 (page 282)

1. $-1, 3$ **3.** $-2, 1$

5. $y = f(x) = x^2 - 2x - 13$
Set $y = 0$ Then $x^2 - 2x - 13 = 0$
Use formula:
$$x = \frac{-(-2) \pm \sqrt{(-2)^2 - 4(1)(-13)}}{2(1)} = \frac{2 \pm \sqrt{4+52}}{2} = \frac{2 \pm \sqrt{56}}{2} = \frac{2 \pm \sqrt{4 \cdot 14}}{2} = \frac{2 \pm 2\sqrt{14}}{2} = 1 \pm \sqrt{14}$$
Therefore, the x-intercepts are $1 + \sqrt{14}$ and $1 - \sqrt{14}$

7. $y = f(x) = x^2 - 6x + 10$
Set $y = 0$ Then $x^2 - 6x + 10 = 0$
Use formula:
$$x = \frac{-(-6) \pm \sqrt{(-6)^2 - 4(1)(10)}}{2(1)} = \frac{6 \pm \sqrt{36-40}}{2} = \frac{6 \pm \sqrt{-4}}{2} = \frac{6 \pm 2i}{2} = 3 \pm i$$
This means that the curve does not cross the x-axis

9. $x = 2$ **11.** $x = -\frac{3}{2}$ **13.** $x = 0$

15. Minimum $= -2$
Vertex: $(3, -2)$

17. $f(x) = 8x - 5x^2 - 3$
$= -5x^2 + 8x - 3$

$a < 0$

Therefore, $f(x)$ has a maximum

$$f\left(-\frac{b}{2a}\right) = f\left(-\frac{8}{2(-5)}\right) = f\left(\frac{4}{5}\right)$$

$$f\left(\frac{4}{5}\right) = -5\left(\frac{4}{5}\right)^2 + 8\left(\frac{4}{5}\right) - 3$$

$$= -\frac{16}{5} + \frac{32}{5} - \frac{15}{5}$$

$$= \frac{1}{5} \quad \text{maximum}$$

Vertex: $\left(\frac{4}{5}, \frac{1}{5}\right)$

19. Maximum $= 2$ Vertex: $(1,2)$

21.

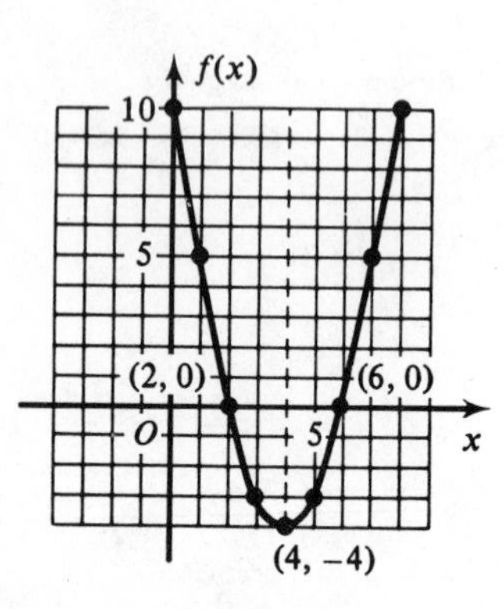

23.

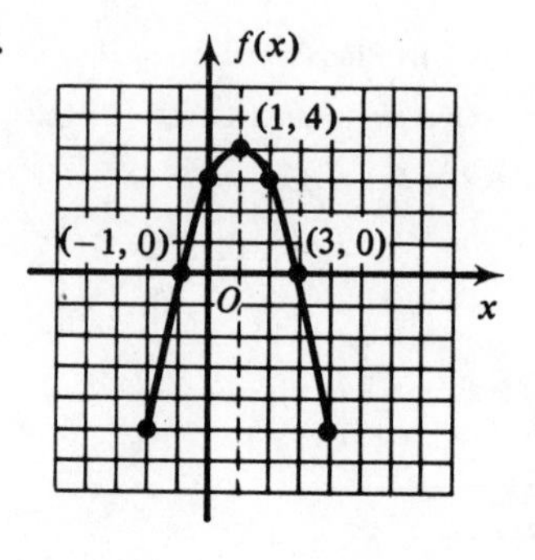

25. $f(x) = 6 + x^2 - 4x$
$= x^2 - 4x + 6$

1. Axis of symmetry: $x = -\frac{-4}{2} = 2$
2. Minimum $(a > 0)$: $f(2) = 4 - 8 + 6 = 2$
3. Vertex: $(2,2)$
4. Find points:
   $f(x) = x^2 - 4x + 6$
   $f(1) = 1 - 4 + 6 = 3$
   $f(0) = 0 - 0 + 6 = 6$
   No x-intercepts

x	$f(x)$
1	3
0	6

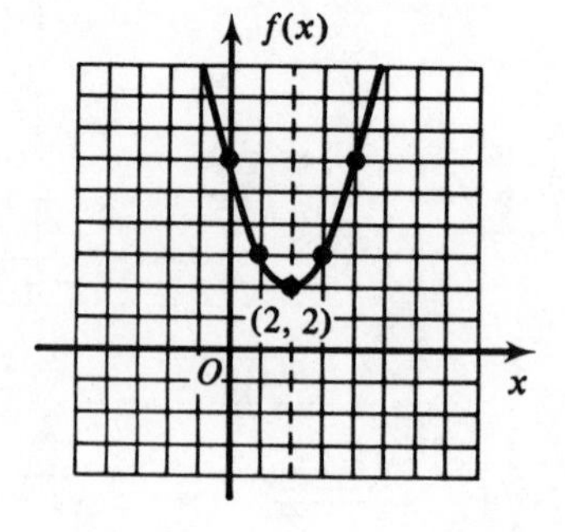

5. Plot symmetrical points and draw graph

27.

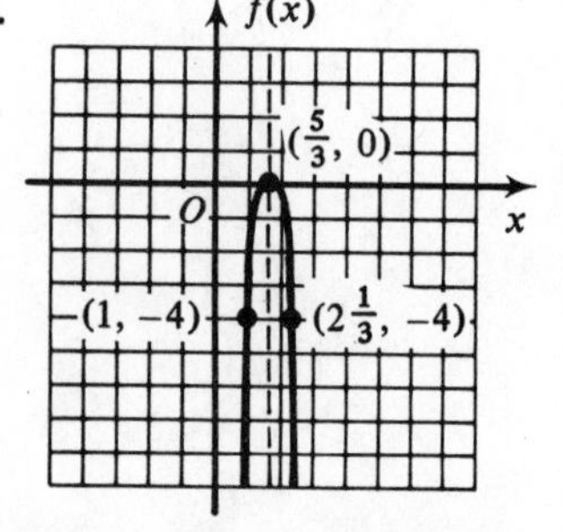

29. (a) $x = f(0) = (0)^2 + 0 - 2 = -2$

(b) $0 = y^2 + y - 2$
$= (y + 2)(y - 1)$
$y = -2 \mid y = 1$

(c) $y = \dfrac{-b}{2a} = \dfrac{-1}{2} = -\dfrac{1}{2}$

(d) $y = -\dfrac{1}{2}$

$x = f\left(-\frac{1}{2}\right) = \left(-\frac{1}{2}\right)^2 + \left(-\frac{1}{2}\right) - 2 = \frac{1}{4} - \frac{1}{2} - 2 = -\frac{9}{4}$

Vertex: $\left(-\frac{9}{4}, -\frac{1}{2}\right)$

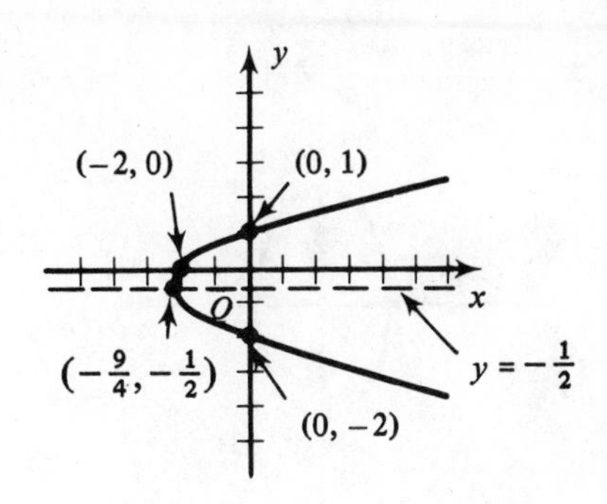

31.

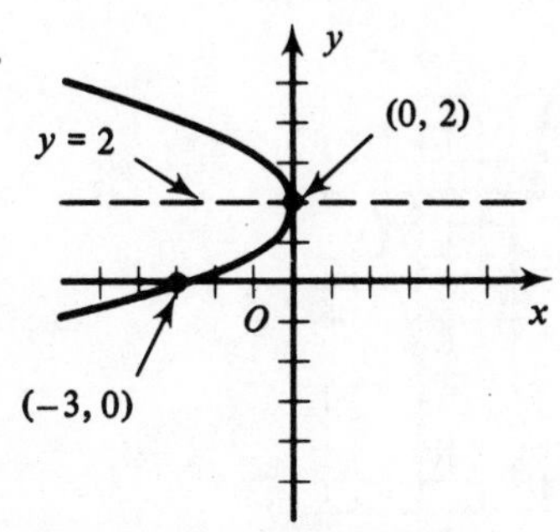

Exercises 8–6 (page 288)

1.

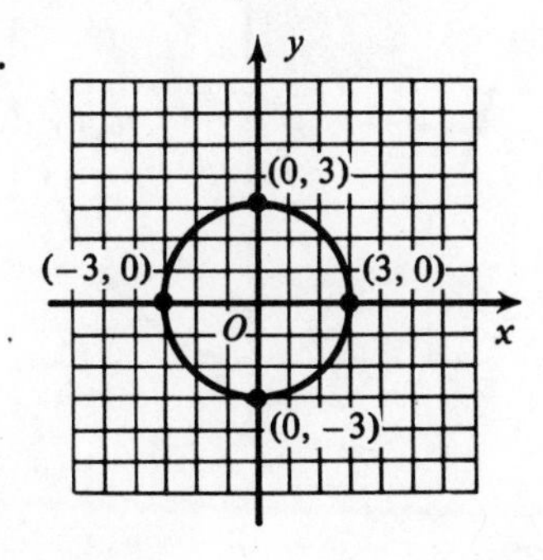

3.

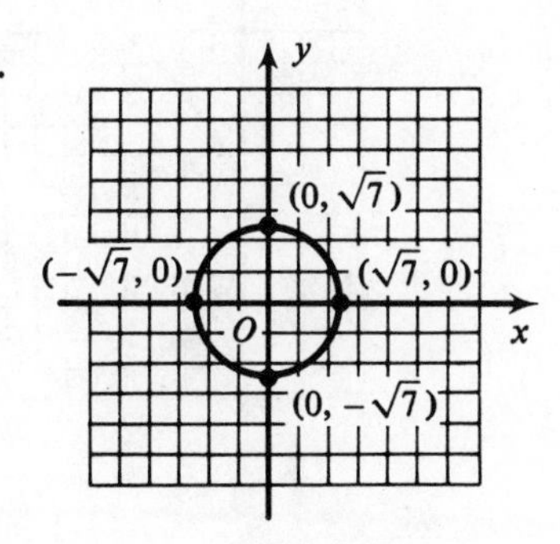

5. $4x^2 + 9y^2 = 36$

Ellipse

x-intercepts: Set $y = 0$ in $4x^2 + 9y^2 = 36$

$$4x^2 + 9(0)^2 = 36$$
$$4x^2 = 36$$
$$x^2 = 9$$
$$x = \pm 3$$

y-intercepts: Set $x = 0$ in $4x^2 + 9y^2 = 36$

$$4(0)^2 + 9y^2 = 36$$
$$9y^2 = 36$$
$$y^2 = 4$$
$$y = \pm 2$$

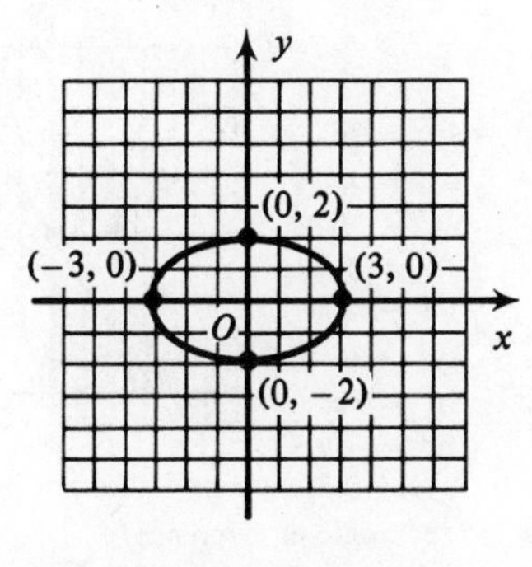

7. Ellipse

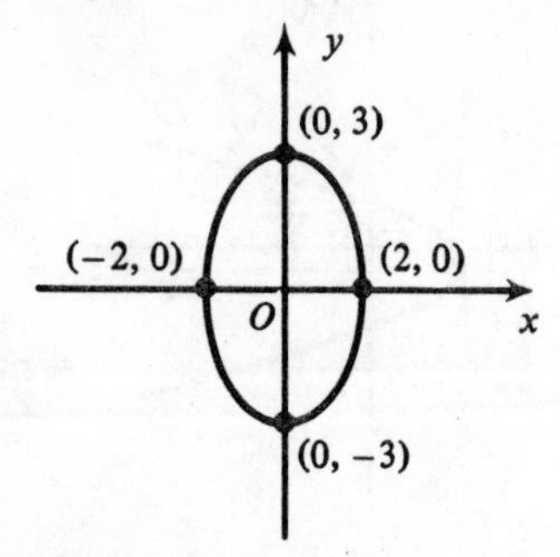

9. $y = x^2 + 2x - 3$

Parabola

x-intercepts: Set $y = 0$ in $y = x^2 + 2x - 3$

$$0 = x^2 + 2x - 3$$
$$0 = (x + 3)(x - 1)$$
$$x = -3 \mid x = 1$$

y-intercept: Set $x = 0$ in $y = x^2 + 2x - 3$

$$= 0 + 0 - 3$$
$$= -3$$

Axis of symmetry: $x = -\dfrac{b}{2a} = -\dfrac{2}{2} = -1$

Minimum point: When $x = -1$. $y = x^2 + 2x - 3$

$$= (-1)^2 + 2(-1) - 3$$
$$= -4$$

Therefore, *vertex* is at $(-1, -4)$

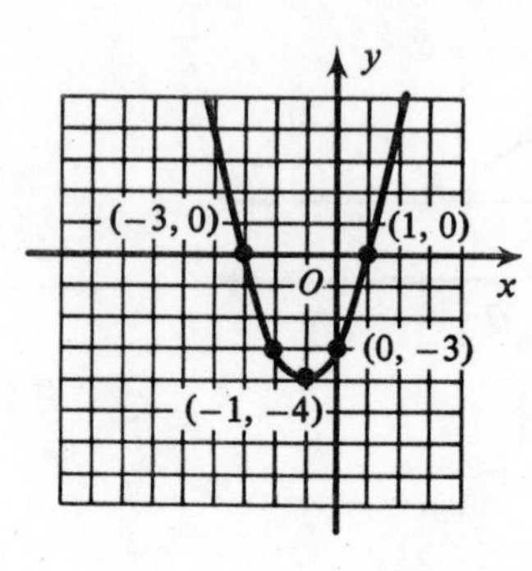

11. $y^2 = 8x$

Parabola

$x = \frac{1}{8}y^2 + 0y + 0$

x-intercept: $x = \frac{1}{8}(0)^2 + 0(0) + 0 = 0$

y-intercept: $0 = \frac{1}{8}y^2 \Rightarrow y = 0$

Axis of symmetry: $y = \dfrac{-b}{2a} = \dfrac{-0}{2\left(\frac{1}{8}\right)} = 0$

Vertex: (0,0)

Other points: When $x = 2$. $y^2 = 16$

$$y = \pm 4$$

(2,4), (2,−4)

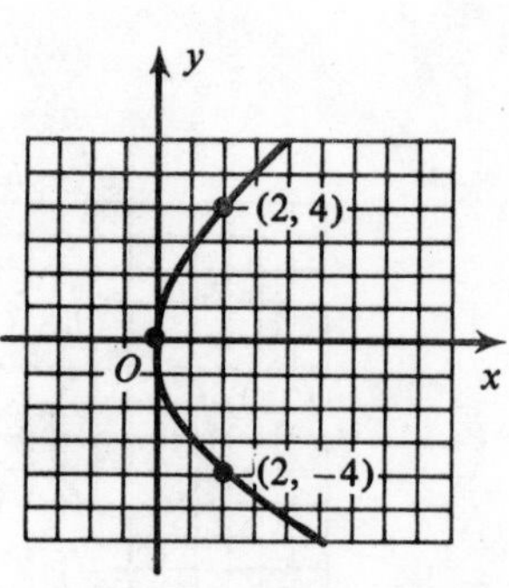

13. Parabola

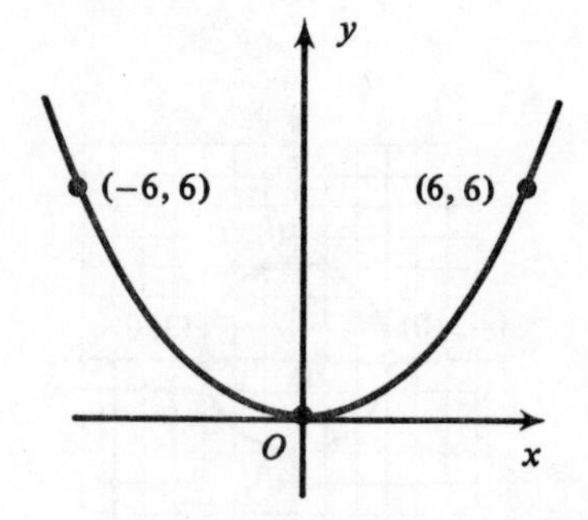

15. $9x^2 - 4y^2 = 36$

Hyperbola

x-intercepts: Set $y = 0$ in $9x^2 - 4y^2 = 36$

$$9x^2 - 4(0)^2 = 36$$

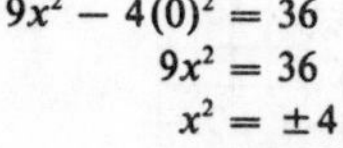

$$9x^2 = 36$$
$$x^2 = \pm 4$$
$$x = \pm 2$$

y-intercepts: Set $x = 0$ in $9x^2 - 4y^2 = 36$

$$9(0)^2 - 4y^2 = 36$$
$$y^2 = -9$$

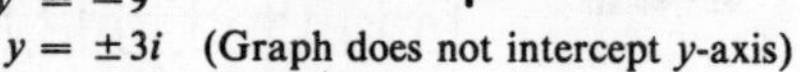

$y = \pm 3i$ (Graph does not intercept y-axis)

Lines $y = \pm 3$ and $x = \pm 2$ form the rectangle of reference

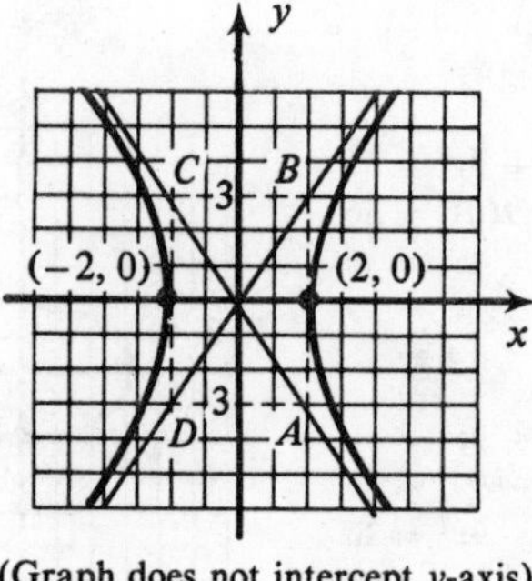

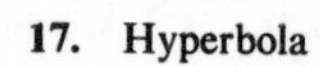

17. Hyperbola

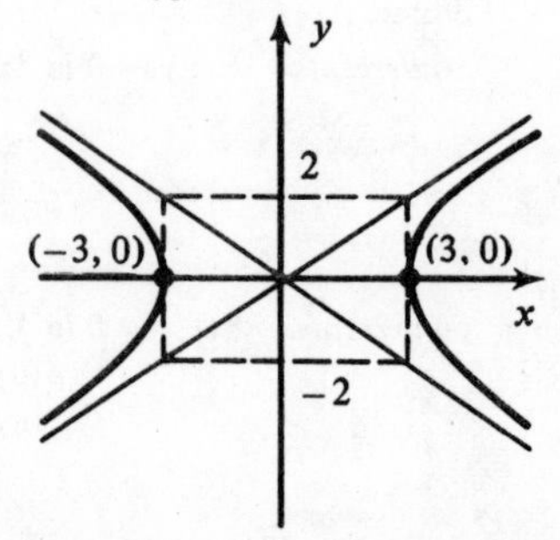

Exercises 8–7 (page 292)

1. $\{x \mid -1 < x < 2\}$

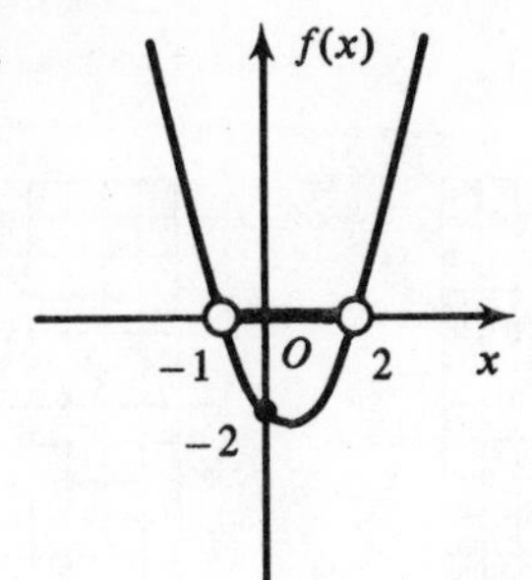

3. $\{x \mid x < -5\} \cup \{x \mid x > 1\}$

5.
$$x^2 + 7 > 6x$$
$$x^2 - 6x + 7 > 0$$
Let $f(x) = x^2 - 6x + 7 = 0$
$$x = \frac{-(-6) \pm \sqrt{36 - 28}}{2} = \frac{6 \pm \sqrt{8}}{2}$$
$$= \frac{6 \pm 2\sqrt{2}}{2} = 3 \pm \sqrt{2}$$
x-intercepts: $3 - \sqrt{2}$ and $3 + \sqrt{2} \doteq 1.6$ and 4.4
$f(0) = 0 + 7 = 7$
Therefore, the graph goes through (0,7)
Solution set: $\{x|x < 3 - \sqrt{2}\} \cup \{x|x > 3 + \sqrt{2}\}$

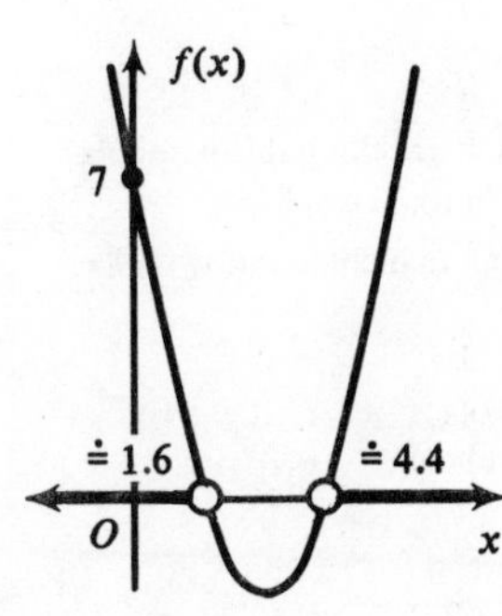

7. $\{x|0 \leq x \leq 5\}$

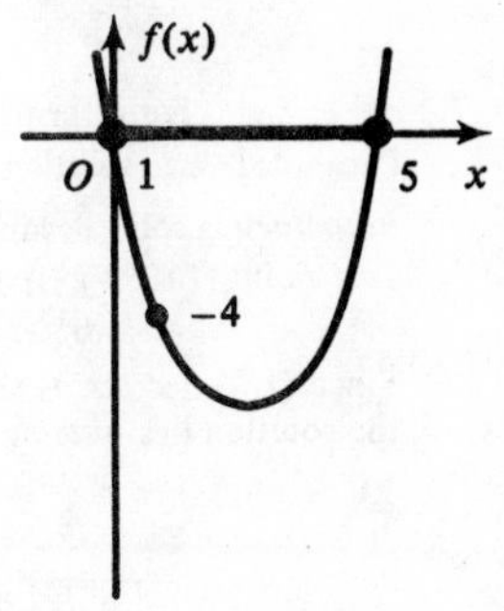

9. $\left\{x \mid -4 \leq x \leq \frac{3}{2}\right\}$ **11.** Solution set is empty: { }

13. Multiply by $(x + 1)^2$, which is positive. This clears fractions and does not change the sense of the inequality
$$\frac{(x+1)^2}{1} \cdot \frac{x}{1} < \frac{(x+1)^2}{1} \cdot \frac{2}{x+1}$$
$$(x + 1)^2 x < 2(x + 1)$$
$$(x + 1)^2 x - 2(x + 1) < 0$$
$$(x + 1)[(x + 1)x - 2] < 0$$
$$(x + 1)(x^2 + x - 2) < 0$$
$$(x + 1)(x + 2)(x - 1) < 0$$
Let $f(x) = (x + 1)(x + 2)(x - 1) = 0$
x-intercepts: $-2, -1$, and 1
$f(0) = (1)(2)(-1) = -2$
Therefore, the curve goes through $(0, -2)$
Solution set: $\{x|x < -2\} \cup \{x|-1 < x < 1\}$

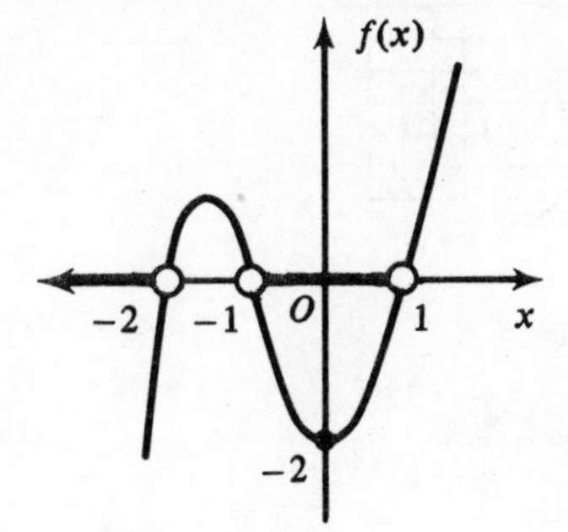

15. $(x + 3)(x + 2)(x - 1)(x + 1)(x - 3) > 0$
Let $f(x) = (x + 3)(x + 2)(x - 1)(x + 1)(x - 3)$
From this function we see that the x-intercepts are $-3, -2, -1, 1$, and 3
$f(0) = (3)(2)(-1)(1)(-3) = 18$
Knowing that $f(0)$ is positive tells us that the curve is above the x-axis when $x = 0$. Then, starting at a point above the $x =$ axis at $x = 0$, we draw a smooth curve through the x-intercepts. From the drawing we observe the intervals where the function is positive.

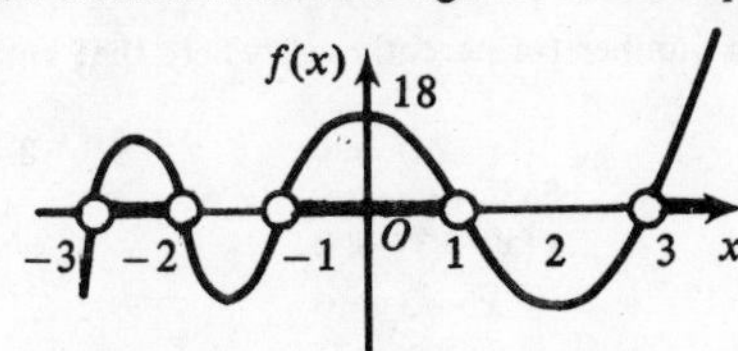

Solution set: $\{x|-3 < x < -2\} \cup \{x|-1 < x < 1\} \cup \{x|x > 3\}$

Exercises 8–8 (page 294)

1. Circle

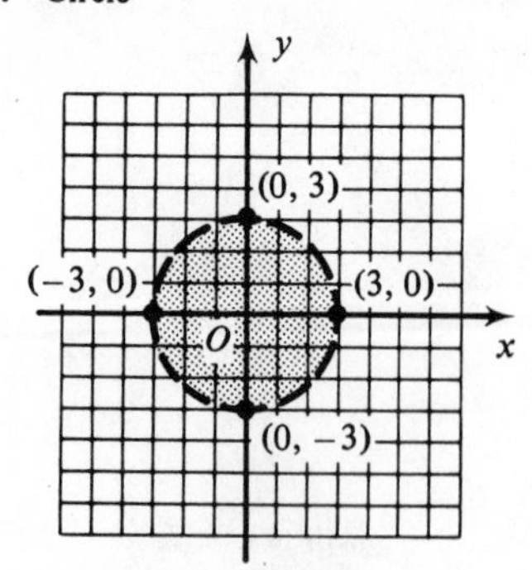

3. Ellipse

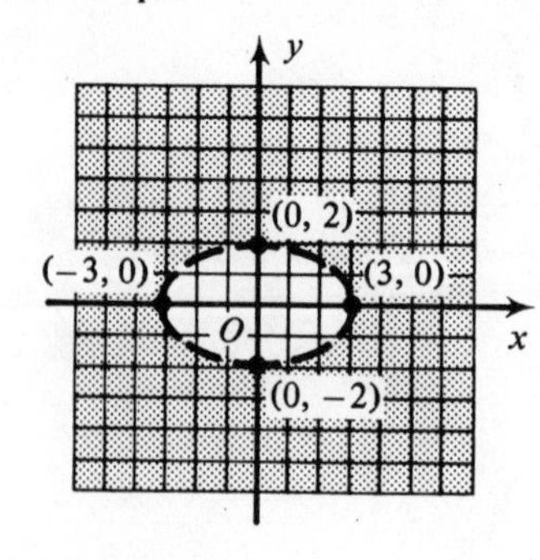

5. Parabola

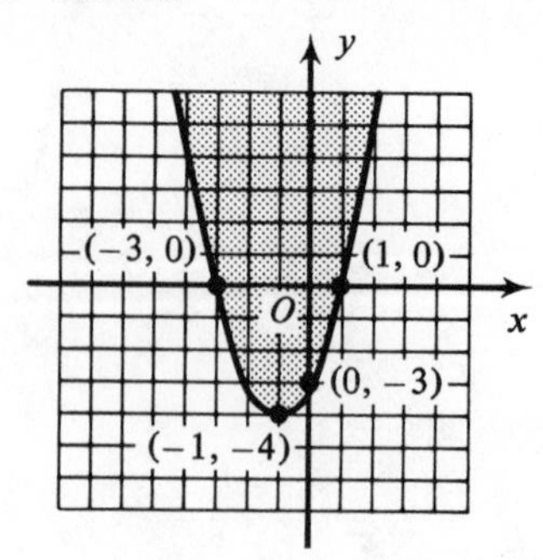

7. $y^2 \geq 8x$ Parabola } For information about graphing this parabola, see solution of #11 in Exercises 8–6.

Boundary is solid because = sign *is* included in $y^2 \geq 8x$.

Test Point (1,0): $y^2 \geq 8x$

$0^2 \geq 8(1)$ False

Since (1,0) lies *inside* the parabola, the solution lies *outside* the parabola.

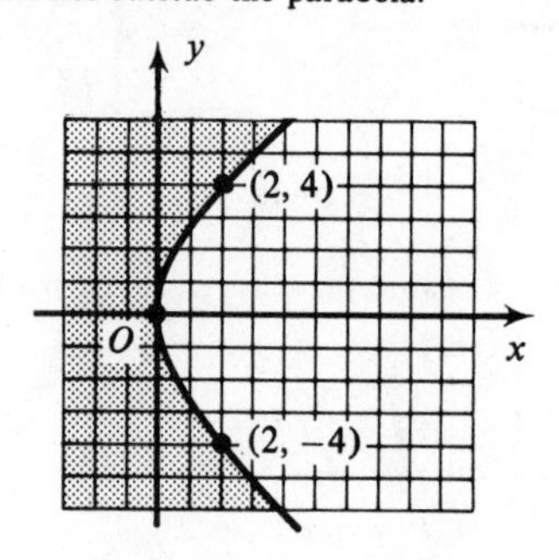

9. $9x^2 - 4y^2 = 36$
Hyperbola

x-intercepts: Set $y = 0$ in $9x^2 - 4y^2 = 36$

$$9x^2 - 4(0)^2 = 36$$
$$9x^2 = 36$$
$$x^2 = \pm 4$$
$$x = \pm 2$$

y-intercepts: Set $x = 0$ in $9x^2 - 4y^2 = 36$

$$9(0)^2 - 4y^2 = 36$$
$$y^2 = -9$$
$$y = \pm 3i$$

(Graph does not intercept y-axis)

Lines $y = \pm 3$ and $x = \pm 2$ form the rectangle of reference

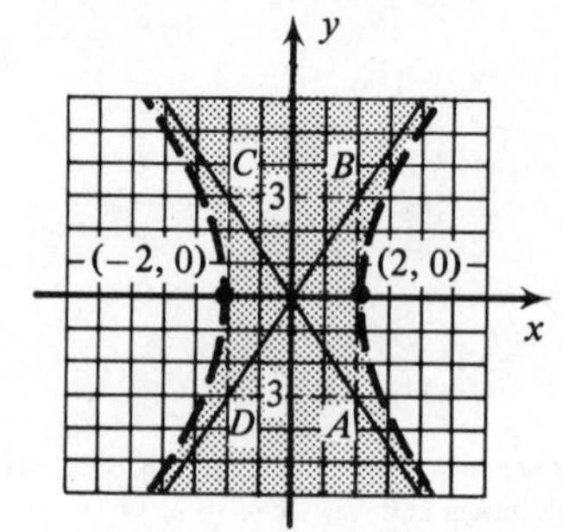

Chapter Eight Diagnostic Test or Review (page 295)

Following each problem number is the textbook section number (in parentheses) where that kind of problem is discussed.

1. (8–1)

$$4x^2 = 3 + 4x$$
$$4x^2 - 4x - 3 = 0$$
$$(2x + 1)(2x - 3) = 0$$

$2x + 1 = 0$, $x = -\frac{1}{2}$ | $2x - 3 = 0$, $x = \frac{3}{2}$

2. (8–2)

$$2x^2 = 6x$$
$$2x^2 - 6x = 0$$
$$2x(x - 3) = 0$$

$2x = 0$, $x = 0$ | $x - 3 = 0$, $x = 3$

3. (8–1, 8–2)

$$2x^2 = 18$$
$$2x^2 - 18 = 0$$
$$2(x^2 - 9) = 0$$
$$2(x + 3)(x - 3) = 0$$

$x + 3 = 0$, $x = -3$ | $x - 3 = 0$, $x = 3$

4. (8–1) $\frac{x+1}{2} + \frac{4}{x+1} = 2 \qquad \text{LCD} = 2(x+1)$

$$\frac{2(x+1)}{1} \cdot \frac{x-1}{2} + \frac{2(x+1)}{1} \cdot \frac{4}{x+1} = \frac{2(x+1)}{1} \cdot \frac{2}{1}$$

$$(x+1)(x-1) + 2 \cdot 4 = 4(x+1)$$
$$x^2 - 1 + 8 = 4x + 4$$
$$x^2 - 4x + 3 = 0$$
$$(x-1)(x-3) = 0$$
$$x - 1 = 0 \quad | \quad x - 3 = 0$$
$$x = 1 \quad | \quad x = 3$$

Check for x = 1	*Check for x = 3*
$\frac{x-1}{2} + \frac{4}{x+1} = 2$	$\frac{x-1}{2} + \frac{4}{x+1} = 2$
$\frac{(1)-1}{2} + \frac{4}{(1)+1} \stackrel{?}{=} 2$	$\frac{(3)-1}{2} + \frac{4}{(3)+1} \stackrel{?}{=} 2$
$0 + 2 = 2$	$1 + 1 = 2$

Therefore, both 1 and 2 are solutions

5. (8–3) $x^2 = 6x - 7$

$$x^2 - 6x + 7 = 0$$
$$x = \frac{-(-6) \pm \sqrt{(-6)^2 - 4(7)}}{2(1)}$$
$$= \frac{6 \pm \sqrt{8}}{2} = \frac{6 \pm 2\sqrt{2}}{2} = 3 \pm \sqrt{2}$$

6. (8–1) (1) $2x^{2/3} + 3x^{1/3} = 2$

Let $z = x^{1/3}$; then $z^2 = x^{2/3}$

Equation (1) becomes $2z^2 + 3z - 2 = 0$

$$(z+2)(2z-1) = 0$$

$z + 2 = 0$	$2z - 1 = 0$
$z = -2$	$z = \frac{1}{2}$
$x^{1/3} = -2$	$x^{1/3} = \frac{1}{2}$
$(x^{1/3})^3 = (-2)^3$	$(x^{1/3})^3 = \left(\frac{1}{2}\right)^3$
$x = -8$	$x = \frac{1}{8}$

Check for x = −8	*Check for x = 1/8*
$2(-8)^{2/3} + 3(-8)^{1/3} \stackrel{?}{=} 2$	$2\left(\frac{1}{8}\right)^{2/3} + 3\left(\frac{1}{8}\right)^{1/3} \stackrel{?}{=} 2$
$2(4) + 3(-2) \stackrel{?}{=} 2$	$2\left(\frac{1}{4}\right) + 3\left(\frac{1}{2}\right) \stackrel{?}{=} 2$
$8 - 6 = 2$ *True*	$\frac{1}{2} + \frac{3}{2} = 2$ *True*

Therefore, both -8 and $\frac{1}{8}$ are solutions

7. (8–4) $25x^2 - 20x + 7 = 0 \quad \begin{cases} a = 25 \\ b = -20 \\ c = 7 \end{cases}$

$$b^2 - 4ac = (-20)^2 - 4(25)7$$
$$= 400 - 700 = -300$$

Therefore, roots are complex conjugates

8. (8–4)

$x = 2 + \sqrt{3}\,i$	$x = 2 - \sqrt{3}\,i$
$x - (2 + \sqrt{3}\,i) = 0$	$x - (2 - \sqrt{3}\,i) = 0$
$(x - 2 - \sqrt{3}\,i) = 0$	$(x - 2 + \sqrt{3}\,i) = 0$

$$(x - 2 - \sqrt{3}\,i)(x - 2 + \sqrt{3}\,i) = 0$$
$$[(x-2) - \sqrt{3}\,i][(x-2) + \sqrt{3}\,i] = 0$$
$$(x-2)^2 - (\sqrt{3}\,i)^2 = 0$$
$$x^2 - 4x + 4 - 3i^2 = 0$$
$$x^2 - 4x + 4 - 3(-1) = 0$$
$$x^2 - 4x + 4 + 3 = 0$$
$$x^2 - 4x + 7 = 0$$

9. (8–8) $25x^2 + 16y^2 > 400$

Boundary is ellipse $25x^2 + 16y^2 = 400$

x-intercepts: Set $y = 0$ in $25x^2 + 16y^2 = 400$

$$25x^2 + 16(0)^2 = 400$$
$$25x^2 = 400$$
$$x^2 = 16$$
$$x = \pm 4$$

y-intercepts: Set $x = 0$ in $25x^2 + 16y^2 = 400$

$$25(0)^2 + 16y^2 = 400$$
$$16y^2 = 400$$
$$y^2 = 25$$
$$y = \pm 5$$

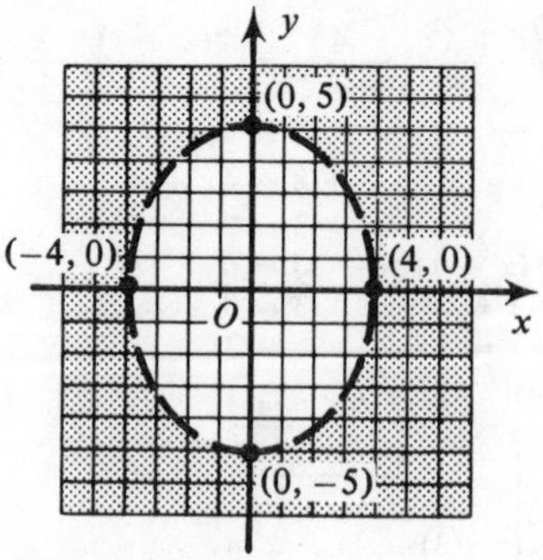

Boundary is dashed because $=$ sign is not included in $25x^2 + 16y^2 > 400$.

Test point (0,0) in $25x^2 + 16y^2 > 400$

$25(0)^2 + 16(0)^2 > 400$ False

Therefore, the solution lies outside the ellipse, because (0,0) lies inside the ellipse.

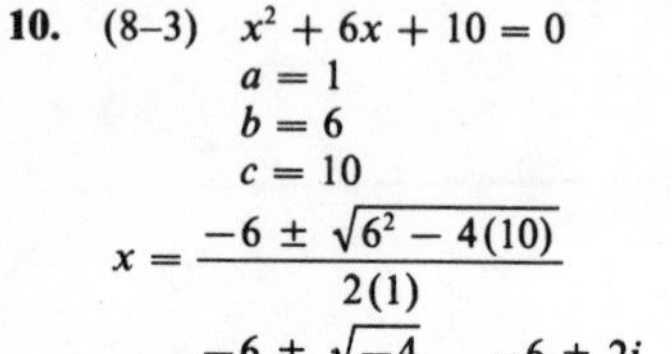

10. (8–3) $x^2 + 6x + 10 = 0$

$a = 1$

$b = 6$

$c = 10$

$$x = \frac{-6 \pm \sqrt{6^2 - 4(10)}}{2(1)}$$
$$= \frac{-6 \pm \sqrt{-4}}{2} = \frac{-6 \pm 2i}{2}$$
$$= -3 \pm i$$

11. (8–7) $x^2 + 5 < 6x$

$x^2 - 6x + 5 < 0$

$(x - 1)(x - 5) < 0$

Let $f(x) = (x - 1)(x - 5) = 0$

Then $x - 1 = 0 \mid x - 5 = 0$

$x = 1 \mid x = 5$

Therefore, the x-intercepts are 1 and 5

$f(x) = (x - 1)(x - 5)$

$f(0) = (-1)(-5) = 5$

Therefore, the curve goes through (0,5)

Solution set: $\{x \mid 1 < x < 5\}$

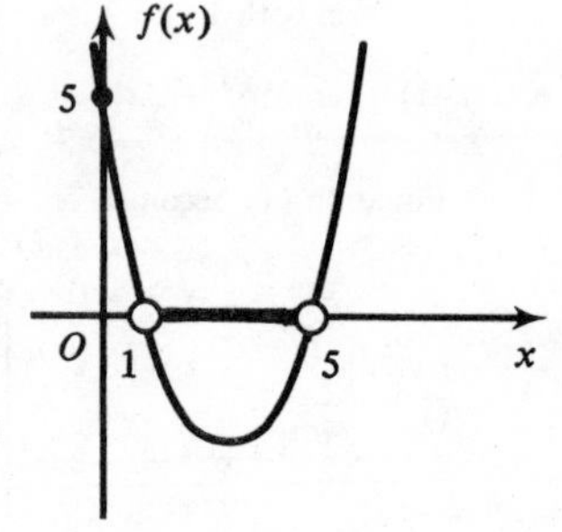

12. (8–7)

$x^2 + 2x \geq 8$

$x^2 + 2x - 8 \geq 0$

$(x + 4)(x - 2) \geq 0$

Let $f(x) = (x + 4)(x - 2) = 0$

Then $x + 4 = 0 \mid x - 2 = 0$

$x = -4 \mid x = 2$

Therefore, the x-intercepts are -4 and 2.

$f(x) = (x + 4)(x - 2)$

$f(0) = (4)(-2) = -8$

Therefore, the curve goes through $(0, -8)$

Solution set: $\{x \mid x \leq -4\} \cup \{x \mid x \geq 2\}$

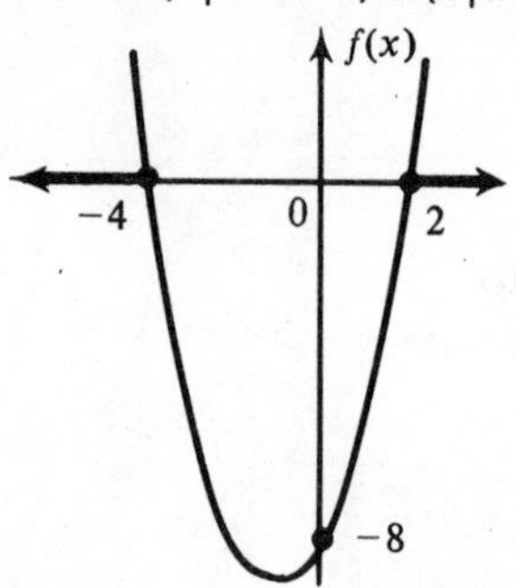

13. (8–5) $f(x) = x^2 - 4x$

(a) Equation of axis of symmetry: $x = -\dfrac{b}{2a} = -\dfrac{-4}{2(1)} = 2$

(b) Vertex: $f(2) = (2)^2 - 4(2)$
$= 4 - 8 = -4$

Vertex: $(2, -4)$

(c) x-intercepts: $f(x) = x(x - 4) = 0$

$x = 0 \mid x - 4 = 0$

$x = 4$

x-intercepts: 0 and 4

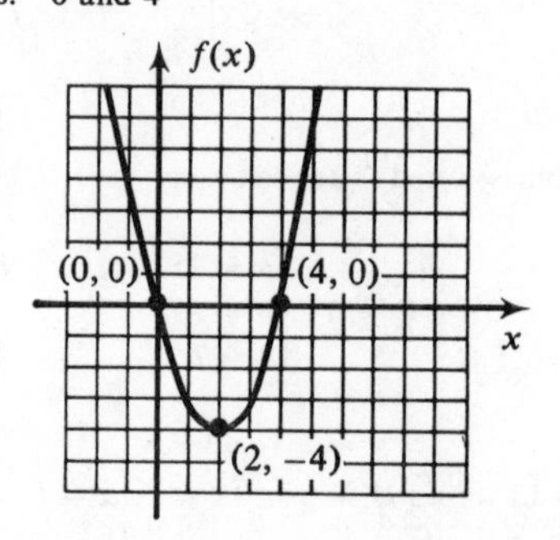

14. (8–1) Let x = Number of hours for Oscar to do the job

Oscar does $\dfrac{1}{x}$ of the work each hour

$x + 2$ = Number of hours for Jan to do the job

Jan does $\dfrac{1}{x+2}$ of the work each hour

Jan works a total of 4 hr and Oscar works 3 hr to do the complete job

(Problem 14 continued on next page)

Work done by Jan	+	Work done by Oscar	=	Total work done (the complete job)
$4\left(\dfrac{1}{x+2}\right)$	+	$3\left(\dfrac{1}{x}\right)$	=	1

$$\frac{4}{x+2} + \frac{3}{x} = 1 \qquad \text{LCD} = x(x+2)$$

$$\frac{x(\cancel{x+2})}{1} \cdot \frac{4}{\cancel{x+2}} + \frac{\cancel{x}(x+2)}{1} \cdot \frac{3}{\cancel{x}} = \frac{x(x+2)}{1} \cdot \frac{1}{1}$$

$$4x + (x+2)3 = x(x+2)$$
$$4x + 3x + 6 = x^2 + 2x$$
$$0 = x^2 - 5x - 6$$
$$0 = (x-6)(x+1)$$
$$x - 6 = 0 \quad \Big| \quad x + 1 = 0$$
$$x = 6 \quad \Big| \quad x = -1 \quad \text{No meaning}$$

Therefore, $x = 6$ hours for Oscar to do the job
$x + 2 = 8$ hours for Jan to do the job

Exercises 9–1 (page 302)

1.

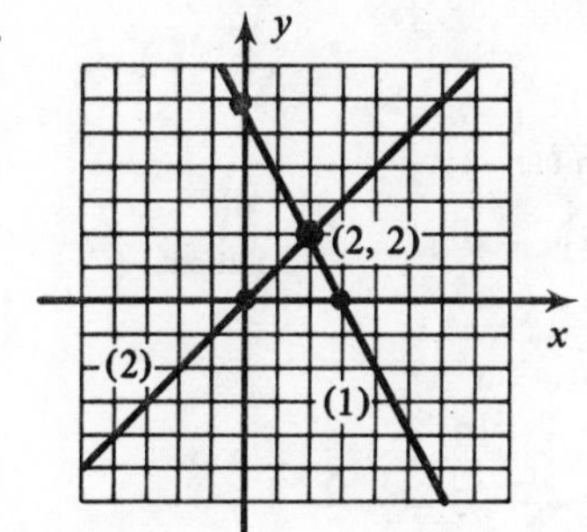

3.

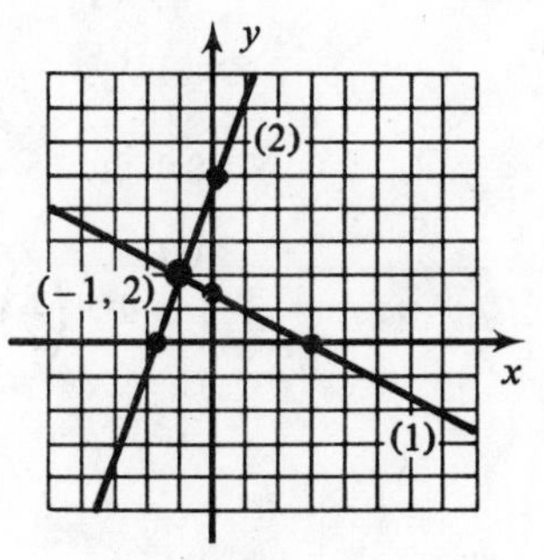

5.

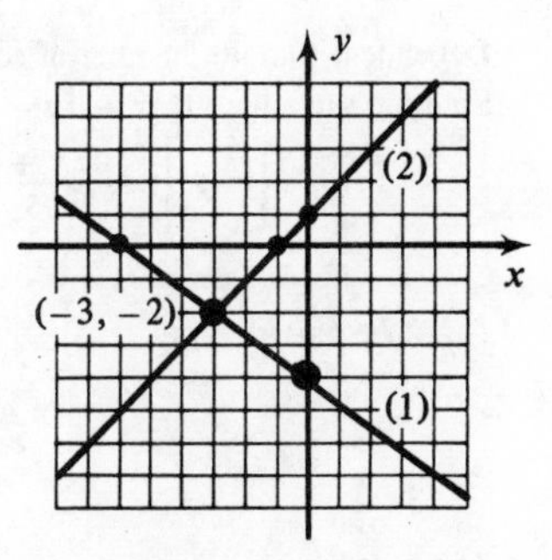

7.

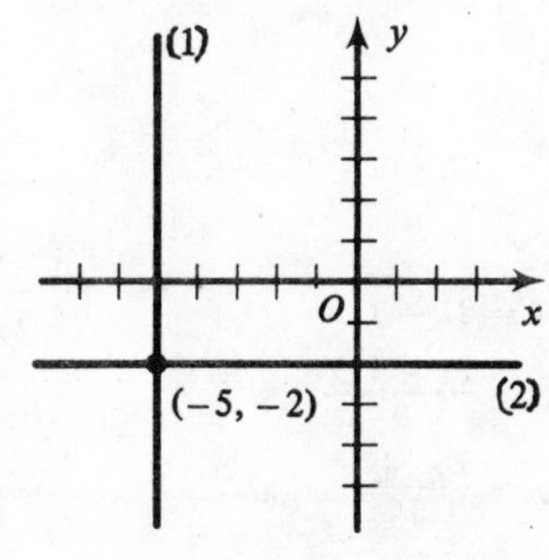

9.

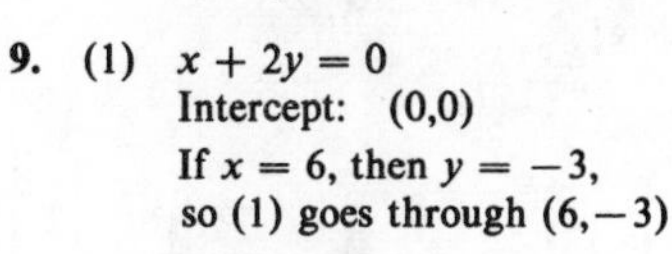

(1) $x + 2y = 0$
Intercept: (0,0)
If $x = 6$, then $y = -3$,
so (1) goes through $(6, -3)$
(2) $2x - y = 0$
Intercept: (0,0)
If $x = 3$, then $y = 6$,
so (2) goes through $(3,6)$

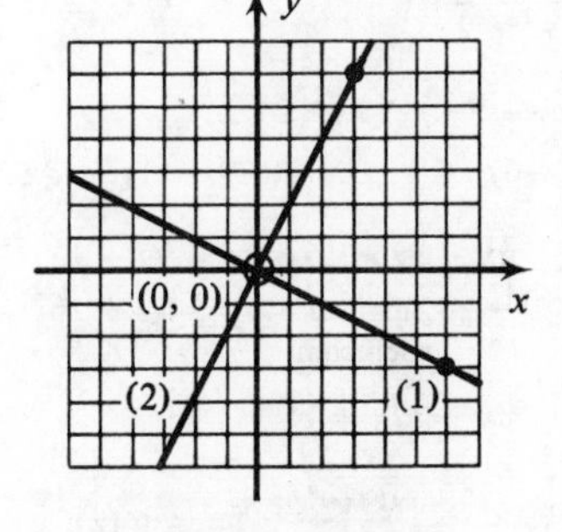

11.

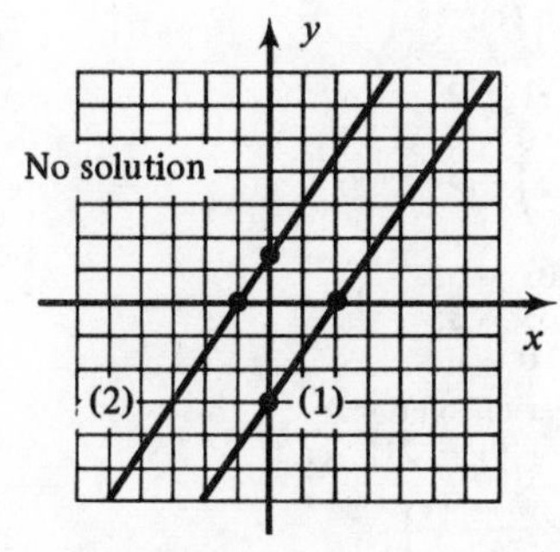

13. Since both lines have the same intercepts, they are the same line

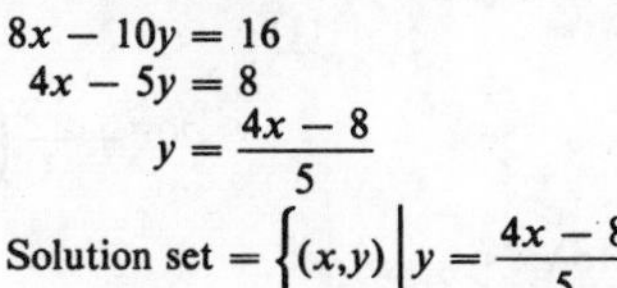

$$8x - 10y = 16$$
$$4x - 5y = 8$$
$$y = \frac{4x - 8}{5}$$

Solution set $= \left\{(x,y) \,\middle|\, y = \dfrac{4x - 8}{5}\right\}$

Infinite number of solutions

(1)

(2)

15.

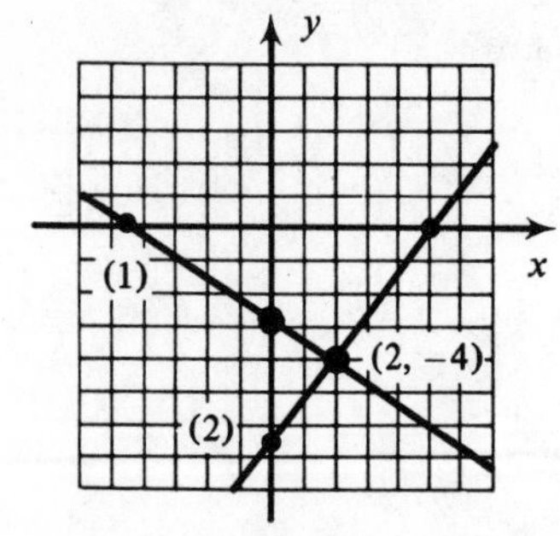

Exercises 9–2 (page 306)

1. (2,−5) 3. (7,−3) 5. (−3,−8) 7. (5,−6)

9.
$$\begin{array}{ll} 9]\ 3] & 6x - 10y = 6 \Rightarrow 18x - 30y = 18 \\ 6]\ 2] & 9x - 15y = -4 \Rightarrow \underline{18x - 30y = -8} \end{array}$$
Inconsistent (no solution) $\qquad 0 \neq 26$

11.
$$\begin{array}{ll} 6]\ 2] & 3x - 5y = -2 \Rightarrow 6x - 10y = -4 \\ 3]\ 1] & -6x + 10y = 4 \Rightarrow \underline{-6x + 10y = 4} \\ & 0 = 0 \end{array}$$
Dependent (infinite number of solutions)
Solution set: $\{(x,y)|3x - 5y = -2\}$ or
$$\left\{(x,y)\,\middle|\, y = \frac{3x+2}{5}\right\}$$

13.
$$\begin{array}{ll} 6]\ 3] & 9x + 4y = -4 \Rightarrow 27x + 12y = -12 \\ 4]\ 2] & 15x - 6y = 25 \Rightarrow \underline{30x - 12y = 50} \\ & 57x = 38 \\ & x = \frac{38}{57} = \frac{2}{3} \end{array}$$
Substitute $x = \frac{2}{3}$ in (1):
$$\begin{array}{ll} (1) & 9x + 4y = -4 \\ & \overset{3}{\cancel{9}}\left(\frac{2}{\cancel{3}}\right) + 4y = -4 \\ & 6 + 4y = -4 \\ & 4y = -10 \\ & y = \frac{-10}{4} = -\frac{5}{2} \end{array}$$
Solution: $\left(\frac{2}{3}, -\frac{5}{2}\right)$

15. $\left(\frac{14}{3}, -\frac{9}{2}\right)$ 17. $\doteq (-2.35, -5.81)$

Exercises 9–3 (page 310)

1. (4,−6) 3. (−9,−11) 5. (8,12)

7. (1) $6x - 9y = 16 \Rightarrow 6x = 9y + 16$
Smallest coefficient $\qquad x = \frac{9y+16}{6}$

(2) $8x - 12y = 16$

Substitute $\frac{9y+16}{6}$ for x in (2):
$$\begin{array}{ll} (2) & 8x - 12y = 16 \\ & \frac{\overset{4}{\cancel{8}}}{1}\left(\frac{9y+16}{\underset{3}{\cancel{6}}}\right) - 12y = 16 \end{array}$$
LCD = 3
$$\frac{\cancel{3}}{1}\cdot\frac{4}{1}\left(\frac{9y+16}{\cancel{3}}\right) - \frac{3}{1}\left(\frac{12y}{1}\right) = \frac{3}{1}\left(\frac{16}{1}\right)$$
$$36y + 64 - 36y = 48$$
$$64 \neq 48$$
Inconsistent (no solution)

9. (1) $20x + 15y = 35$
(2) $12x + 9y = 21 \Rightarrow 9y = 21 - 12x$
Smallest coefficient
$$y = \frac{21 - 12x}{9}$$
$$y = \frac{7 - 4x}{3}$$
Substitute $\frac{7-4x}{3}$ for y in (1):
$$\begin{array}{ll} (1) & 20x + 15y = 35 \\ & 20x + \frac{\overset{5}{\cancel{15}}}{1}\left(\frac{7-4x}{\cancel{3}}\right) = 35 \\ & 20x + 35 - 20x = 35 \\ & 35 = 35 \\ & \textit{or}\ \ 0 = 0 \end{array}$$
Dependent (infinite number of solutions)
Solution set: $\{(x,y)|12x + 9y = 21\}$ or
$$\left\{(x,y)\,\middle|\, y = \frac{7-4x}{3}\right\}$$

11. $\left(\frac{1}{2}, \frac{3}{4}\right)$

13. (1) $4x + 4y = 3 \Rightarrow 4x = 3 - 4y$

(2) $6x + 12y = -6 \qquad x = \dfrac{3 - 4y}{4}$

Substitute $\dfrac{3 - 4y}{4}$ for x in (2):

$$
\begin{aligned}
(2) \qquad 6x + 12y &= -6 \\
\overset{3}{\not 6}\left(\frac{3 - 4y}{\underset{2}{\not 4}}\right) + 12y &= -6
\end{aligned}
$$

LCD = 2

$$
\begin{aligned}
\frac{\not 2}{1} \cdot 3\left(\frac{3 - 4y}{\not 2}\right) + 2(12y) &= 2(-6) \\
9 - 12y + 24y &= -12 \\
12y &= -21 \\
y = -\frac{21}{12} = -\frac{7}{4} &= -1\tfrac{3}{4}
\end{aligned}
$$

Substitute $y = -\dfrac{7}{4}$ in $x = \dfrac{3 - 4y}{4}$

$$x = \frac{3 - 4\left(-\frac{7}{4}\right)}{4} = \frac{3 + 7}{4} = \frac{10}{4} = 2\tfrac{1}{2}$$

Solution: $(2\tfrac{1}{2}, -1\tfrac{3}{4})$

15. $\left(\dfrac{8}{3}, -\dfrac{9}{2}\right)$

Exercises 9–4 (page 313)

1.

$$
\begin{array}{lrcl}
(1) & 2x + y + z &=& 4 \\
(2) & x - y + 3z &=& -2 \\
(3) & x + y + 2z &=& 1 \\
\hline
(1) + (2): & 3x \quad + 4z &=& 2 \\
(2) + (3): & 2x \quad + 5z && -1
\end{array}
$$

Next, eliminate x:

$$
\begin{aligned}
2]\ 3x + 4z &= 2 \Rightarrow 6x + 8z = 4 \\
3]\ 2x + 5z &= -1 \Rightarrow \underline{6x + 15z = -3} \\
&\qquad\qquad -7z = 7 \\
&\qquad\qquad\ \ z = -1
\end{aligned}
$$

Substitute $z = -1$ in $3x + 4z = 2$

$$
\begin{aligned}
3x - 4 &= 2 \\
3x &= 6 \\
x &= 2
\end{aligned}
$$

Substitute $x = 2$ and $z = -1$ in (3):

$$
\begin{aligned}
(3) \qquad x + y + 2z &= 1 \\
2 + y - 2 &= 1 \\
y &= 1
\end{aligned}
$$

Therefore, the solution is $(2, 1, -1)$

3. $(2, 3, -4)$ **5.** $(-1, 3, 2)$

7.

$$
\begin{array}{lrcll}
(1) & x \quad + 2z &=& 7 & \\
(2) & 2x - y \qquad &=& 5 & \\
(3) & 2y + z &=& 4 & \\
\hline
2(2) + (3): & 4x \quad + z &=& 14 & (4) \\
(1): & x \quad + 2z &=& 7 &
\end{array}
$$

$$
\begin{aligned}
2]\ 4x + 1z = 14 &\Rightarrow 8x + 2z = 28 \\
(1) &\quad\ \ \underline{x + 2z = 7} \\
&\quad 7x \qquad = 21 \\
&\qquad\ \ x = 3
\end{aligned}
$$

Substitute $x = 3$ in (4):

$$
\begin{aligned}
(4) \qquad 4x + z &= 14 \\
4(3) + z &= 14 \\
z &= 2
\end{aligned}
$$

Substitute $z = 2$ in (3):

$$
\begin{aligned}
(3) \qquad 2y + z &= 4 \\
2y + 2 &= 4 \\
2y &= 2 \\
y &= 1
\end{aligned}
$$

Therefore, the solution is $(3, 1, 2)$

9. $\left(\dfrac{1}{2}, \dfrac{2}{3}, 4\right)$

11.

$$
\begin{array}{lrcll}
(1) & x + y + z + w &=& 5 & \\
(2) & 2x - y + 2z - w &=& -2 & \\
(3) & x + 2y - z - 2w &=& -1 & \\
(4) & -x + 3y + 3z + w &=& 1 & \\
\hline
(1) + (2): & 3x \quad + 3z &=& 3 & (5) \\
2(1) + (3): & 3x + 4y + z &=& 9 & (6) \\
(2) + (4): & x + 2y + 5z &=& -1 & (7) \\
\hline
(6): & 3x + 4y + z &=& 9 & \\
2(7): & 2x + 4y + 10z &=& -2 & \\
\hline
 & x \quad - 9z &=& 11 & \\
\frac{1}{3}(5): & x \quad + z &=& 1 & (8) \\
\hline
 & -10z &=& 10 & \\
 & z &=& -1 &
\end{array}
$$

Substitute $z = -1$ in (8):

$$
\begin{aligned}
(8) \qquad x + z &= 1 \\
x + (-1) &= 1 \\
x &= 2
\end{aligned}
$$

Substitute $x = 2$ and $z = -1$ in (7):

$$
\begin{aligned}
(7) \qquad x + 2y + 5z &= -1 \\
2 + 2y - 5 &= -1 \\
2y &= 2 \\
y &= 1
\end{aligned}
$$

Substitute $x = 2$, $y = 1$, and $z = -1$ in (1):

$$
\begin{aligned}
(1) \qquad x + y + z + w &= 5 \\
2 + 1 - 1 + w &= 5 \\
w &= 3
\end{aligned}
$$

Therefore, the solution is $(2, 1, -1, 3)$

13. $(-2,3,-5,-1)$ **15.** $\left(\frac{1}{2}, -\frac{3}{4}, \frac{2}{3}, \frac{3}{5}\right)$

Exercises 9–5 (page 318)

1. 7 **3.** $\begin{vmatrix} 2 & -4 \\ 5 & -3 \end{vmatrix} = (2)(-3) - (5)(-4) = -6 - (-20) = -6 + 20 = 14$ **5.** -41 **7.** $(2,3)$ **9.** $(1,-1)$ **11.** $(-1,1)$

13. $3x - 4y = 5$
$9x + 8y = 0$

$$x = \frac{\begin{vmatrix} 5 & -4 \\ 0 & 8 \end{vmatrix}}{\begin{vmatrix} 3 & -4 \\ 9 & 8 \end{vmatrix}} = \frac{(5)(8) - (0)(-4)}{(3)(8) - (9)(-4)} = \frac{40}{60} = \frac{2}{3}$$

$$y = \frac{\begin{vmatrix} 3 & 5 \\ 9 & 0 \end{vmatrix}}{\begin{vmatrix} 3 & -4 \\ 9 & 8 \end{vmatrix}} = \frac{(3)(0) - (9)(5)}{(3)(8) - (9)(-4)} = \frac{-45}{60} = -\frac{3}{4}$$

Therefore, the solution is $\left(\frac{2}{3}, -\frac{3}{4}\right)$

15. $(0,0)$ **17.** $x + 3y = 1$
$2x + 6y = 3$

$$x = \frac{\begin{vmatrix} 1 & 3 \\ 3 & 6 \end{vmatrix}}{\begin{vmatrix} 1 & 3 \\ 2 & 6 \end{vmatrix}} = \frac{6 - 9}{6 - 6} = \frac{-3}{0} \quad \text{Not possible}$$

$$y = \frac{\begin{vmatrix} 1 & 1 \\ 2 & 3 \end{vmatrix}}{\begin{vmatrix} 1 & 3 \\ 2 & 6 \end{vmatrix}} = \frac{3 - 2}{6 - 6} = \frac{1}{0} \quad \text{Not possible}$$

Inconsistent (no solution)

19. $2x - 4y = 6$
$3x - 6y = 9$

$$x = \frac{\begin{vmatrix} 6 & -4 \\ 9 & -6 \end{vmatrix}}{\begin{vmatrix} 2 & -4 \\ 3 & -6 \end{vmatrix}} = \frac{(6)(-6) - (9)(-4)}{(2)(-6) - (3)(-4)} = \frac{-36 + 36}{-12 + 12} = \frac{0}{0} \quad \text{Cannot be determined}$$

$$y = \frac{\begin{vmatrix} 2 & 6 \\ 3 & 9 \end{vmatrix}}{\begin{vmatrix} 2 & -4 \\ 3 & -6 \end{vmatrix}} = \frac{(2)(9) - (3)(6)}{(2)(-6) - (3)(-4)} = \frac{18 - 18}{-12 + 12} = \frac{0}{0} \quad \text{Cannot be determined}$$

Dependent (an infinite number of solutions)

Solution set $= \{(x,y) \mid 2x - 4y = 6\}$

or $\left\{(x,y) \mid y = \frac{x - 3}{2}\right\}$

21. $\begin{vmatrix} 2 & -4 \\ 3 & x \end{vmatrix} = 20$

$2x - (3)(-4) = 20$
$2x + 12 = 20$
$2x = 8$
$x = 4$

Exercises 9–6 (page 323)

1. $\begin{vmatrix} 4 & -1 \\ -3 & 0 \end{vmatrix}$ **3.** $\begin{vmatrix} 1 & 2 & 1 \\ 3 & 1 & 2 \\ 4 & 2 & 0 \end{vmatrix} = (1)\begin{vmatrix} 3 & 1 \\ 4 & 2 \end{vmatrix} - (2)\begin{vmatrix} 1 & 2 \\ 4 & 2 \end{vmatrix} + (0)\begin{vmatrix} 1 & 2 \\ 3 & 1 \end{vmatrix}$

$= 1(6 - 4) - 2(2 - 8) + 0 = 2 + 12 = 14$

5. 25 **7.** -88

9. -168 **11.** -17

13. $2x + y + z = 4$
$x - y + 3z = -2$
$x + y + 2z = 1$

$$D = \begin{vmatrix} 2 & 1 & 1 \\ 1 & -1 & 3 \\ 1 & 1 & 2 \end{vmatrix} = (2)\begin{vmatrix} -1 & 3 \\ 1 & 2 \end{vmatrix} - (1)\begin{vmatrix} 1 & 3 \\ 1 & 2 \end{vmatrix} + (1)\begin{vmatrix} 1 & -1 \\ 1 & 1 \end{vmatrix}$$

$= 2(-5) - 1(2 - 3) + 1(1 + 1)$
$= -10 + 1 + 2 = -7$

$$D_x = \begin{vmatrix} 4 & 1 & 1 \\ -2 & -1 & 3 \\ 1 & 1 & 2 \end{vmatrix} = (4)\begin{vmatrix} -1 & 3 \\ 1 & 2 \end{vmatrix} - (-2)\begin{vmatrix} 1 & 1 \\ 1 & 2 \end{vmatrix} + (1)\begin{vmatrix} 1 & 1 \\ -1 & 3 \end{vmatrix}$$

$$= 4(-5) + 2(1) + 1(4)$$
$$= -20 + 2 + 4 = -14$$

$$D_y = \begin{vmatrix} 2 & 4 & 1 \\ 1 & -2 & 3 \\ 1 & 1 & 2 \end{vmatrix} = +(2)\begin{vmatrix} -2 & 3 \\ 1 & 2 \end{vmatrix} - (4)\begin{vmatrix} 1 & 3 \\ 1 & 2 \end{vmatrix} + (1)\begin{vmatrix} 1 & -2 \\ 1 & 1 \end{vmatrix}$$

$$= 2(-7) - 4(-1) + 1(3)$$
$$= -14 + 4 + 3 = -7$$

$$D_z = \begin{vmatrix} 2 & 1 & 4 \\ 1 & -1 & -2 \\ 1 & 1 & 1 \end{vmatrix} = (2)\begin{vmatrix} -1 & -2 \\ 1 & 1 \end{vmatrix} - (1)\begin{vmatrix} 1 & -2 \\ 1 & 1 \end{vmatrix} + (4)\begin{vmatrix} 1 & -1 \\ 1 & 1 \end{vmatrix}$$

$$= 2(1) - 1(3) + 4(2)$$
$$= 2 - 3 + 8 = 7$$

$$x = \frac{D_x}{D} = \frac{-14}{-7} = 2, \quad y = \frac{D_y}{D} = \frac{-7}{-7} = 1, \quad z = \frac{D_z}{D} = \frac{7}{-7} = -1$$

Therefore, the solution is $(2,1,-1)$

15. $\left(\frac{1}{2}, \frac{2}{3}, 4\right)$ **17.** $(3,1,2)$ **19.**

$$\begin{vmatrix} x & 0 & 1 \\ 0 & 2 & 3 \\ 4 & -1 & -2 \end{vmatrix} = 6$$

$$x\begin{vmatrix} 2 & 3 \\ -1 & -2 \end{vmatrix} + 1\begin{vmatrix} 0 & 2 \\ 4 & -1 \end{vmatrix} = 6$$

$$x(-1) + 1(-8) = 6$$
$$-x - 8 = 6$$
$$x = -14$$

Exercises 9–7 (page 328)

1. $(4,8)$ and $(-2,2)$ **3.** (1) $x^2 = 4y$
(2) $x - y = 1$

Solve (2) for y:

$$x - y = 1$$
$$y = x - 1$$

Substitute $x - 1$ for y in (1):

(1) $x^2 = 4y$

$$x^2 = 4(x - 1)$$
$$x^2 = 4x - 4$$
$$x^2 - 4x + 4 = 0$$
$$(x - 2)(x - 2) = 0$$

Therefore, $x = 2$

Substitute $x = 2$ in (2):

(2) $x - y = 1$

$$2 - y = 1 \Rightarrow y = 1$$

Therefore, there is only one solution, $(2,1)$

5. $(-5,0)$ and $(4,3)$ **7.** $(-2,-2)$ and $(4,1)$

9. $(6,5)$, $(6,-5)$, $(-6,5)$ and $(-6,-5)$

11. (1) $2x^2 + 3y^2 = 21$
(2) $x^2 + 2y^2 = 12$

1] $2x^2 + 3y^2 = 21 \Rightarrow 2x^2 + 3y^2 = 21$
2] $1x^2 + 2y^2 = 12 \Rightarrow 2x^2 + 4y^2 = 24$

$$y^2 = 3$$
$$y = \pm\sqrt{3}$$

Substitute $y = \pm\sqrt{3}$ in (2):

(2) $x^2 + 2y^2 = 12$

$$x^2 + 2(\pm\sqrt{3})^2 = 12$$
$$x^2 + 6 = 12$$
$$x^2 = 6$$
$$x = \pm\sqrt{6}$$

Therefore, the four solutions are $(\sqrt{6},\sqrt{3}), (\sqrt{6},-\sqrt{3}), (-\sqrt{6},\sqrt{3}), (-\sqrt{6},-\sqrt{3})$

13. (3,0), (−3,0), (0,−2)

15. (1,3), (−1,−3), (3,1), (−3,−1)

17. (1) $xy = -4$
(2) $y = 6x - 9 - x^2$

Solve (1) for y:

(1) $xy = -4$

$$y = \frac{-4}{x}$$

Substitute $\frac{-4}{x}$ for y in (2):

(2)
$$y = 6x - 9 - x^2$$
$$\frac{-4}{x} = 6x - 9 - x^2$$
$$-4 = 6x^2 - 9x - x^3$$
$$x^3 - 6x^2 + 9x - 4 = 0$$

	1	−6	9	−4	Factor using synthetic division
		1	−5	4	
1	1	−5	4	0	

$$(x-1)(x^2 - 5x + 4) = 0$$

$x - 1 = 0$, $x = 1$ | $x^2 - 5x + 4 = 0$, $(x-1)(x-4) = 0$

$x = 1$ | $x - 4 = 0$, $x = 4$

When $x = 1$ in (1):
(1) $xy = -4$
$(1)y = -4$
$y = -4$

When $x = 4$ in (1):
(1) $xy = -4$
$(4)y = -4$
$y = -1$

Therefore, there are two solutions, (1,−4) and (4,−1)

Exercises 9–8 (page 330)

1. 9 and 21

3. 5 quarters, 10 nickels

5. x = Smaller number
y = Larger number

	Twice the smaller	plus	three times the larger	is	34
(1)	$2x$	+	$3y$	=	34
	Five times the smaller	minus	twice the larger	is	9
(2)	$5x$	−	$2y$	=	9

Use addition-subtraction:

(1) 2] $2x + 3y = 34 \Rightarrow 4x + 6y = 68$
(2) 3] $5x - 2y = 9 \Rightarrow 15x - 6y = 27$

$$19x = 95$$
$$x = 5 \quad \text{Smaller number}$$

Substitute $x = 5$ in (1):

(1)
$$2x + 3y = 34$$
$$2(5) + 3y = 34$$
$$10 + 3y = 34$$
$$3y = 24$$
$$y = 8 \quad \text{Larger number}$$

7. $x =$ Number of 15¢ stamps
$y =$ Number of 20¢ stamps

(1) $x + y = 22 \Rightarrow x = 22 - y$
(2) $15x + 20y = 370$

Substitute $22 - y$ for x in (2):

$$\begin{aligned} (2)\quad 15x + 20y &= 370 \\ 15(22 - y) + 20y &= 370 \\ 330 - 15y + 20y &= 370 \\ 5y &= 40 \\ y &= 8 \quad \text{20¢ stamps} \end{aligned}$$

Substitute $y = 8$ in $x = 22 - y$:
$x = 22 - 8 = 14$ 15¢ stamps

11. Let $u =$ Units digit
$t =$ Tens digit
$h =$ Hundreds digit

	The sum of digits	is	20
(1)	$u + t + h$	$=$	20
	Tens digit	is	3 more than units digit
(2)	t	$=$	$3 + u$
	Sum of hundreds digit and tens digit	is	15
(3)	$h + t$	$=$	15

$$\begin{aligned} (1)\quad h + t + u &= 20 \\ (2)\quad t - u &= 3 \\ (3)\quad h + t &= 15 \\ \hline (1) + (2)\quad h + 2t &= 23 \quad (4) \\ (3)\quad h + t &= 15 \\ \hline (4) - (3)\quad t &= 8 \end{aligned}$$

Substitute $t = 8$ in (3) $\Rightarrow h = 7$
Substitute $t = 8$ in (2) $\Rightarrow u = 5$
Therefore, the number is 785

13. 540 mph speed of plane   60 mph speed of wind

15. Let $a =$ Hours for A to do job
$b =$ Hours for B to do job
$c =$ Hours for C to do job

Then in 1 hr A does $\frac{1}{a}$ of the job, B does $\frac{1}{b}$ of the job, and C does $\frac{1}{c}$ of the job

$$\begin{aligned} (1)\quad \frac{2}{a} + \frac{2}{b} + \frac{2}{c} &= 1 \\ (2)\quad \frac{3}{b} + \frac{3}{c} &= 1 \\ (3)\quad \frac{4}{a} + \frac{4}{b} &= 1 \end{aligned}$$

$$\begin{aligned} (1)\quad \frac{2}{a} + \frac{2}{b} + \frac{2}{c} &= 1 \\ \tfrac{1}{2}(3)\quad \frac{2}{a} + \frac{2}{b} &= \frac{1}{2} \\ \hline (1) - \tfrac{1}{2}(3)\quad \frac{2}{c} &= \frac{1}{2} \Rightarrow c = 4 \text{ hr} \end{aligned}$$

Substitute $c = 4$ in (2) $\Rightarrow b = 12$ hr
Substitute $b = 12$ in (3) $\Rightarrow a = 6$ hr

9. $x =$ Numerator
$y =$ Denominator

	A fraction	has a value of	$\frac{2}{3}$
(1)	$\frac{x}{y}$	$=$	$\frac{2}{3}$
	10 is added to its numerator;	its value becomes	1
(2)	$\frac{x + 10}{y - 5}$	$=$	1
	5 is subtracted from its denominator		

(1) $\frac{x}{y} = \frac{2}{3} \Rightarrow 3x = 2y$

(2) $\frac{x + 10}{y - 5} = 1 \Rightarrow x + 10 = y - 5 \Rightarrow x = y - 15$

Substitute $y - 15$ for x in

$$\begin{aligned} 3x &= 2y \\ 3(y - 15) &= 2y \\ 3y - 45 &= 2y \\ y &= 45 \end{aligned}$$

Substitute $y = 45$ in $x = y - 15$
$x = 45 - 15 = 30$

Original fraction was $\frac{30}{45}$

17.

	10 yrs ago	3 yrs ago	Now	15 yrs in future
Al:	$a - 10$	$a - 3$	a	$a + 15$
Chet:	$c - 10$	$c - 3$	c	$c + 15$
Muriel:	$m - 10$	$m - 3$	m	$m + 15$

$$\begin{array}{lll} (1) & a - 10 = 2(c - 10) & \text{Ten years ago Al was twice as old as Chet.} \\ (2) & m - 3 = \frac{3}{4}(c - 3) & \text{Three years ago Muriel was } \tfrac{3}{4} \text{ Chet's age.} \\ (3) & a + 15 = m + 15 + 8 & \text{In 15 years Al will be 8 yrs older than Muriel.} \\ (4) & a - 2c \qquad = -10 & \text{by simplifying (1)} \\ (5) & \quad -3c + 4m = 3 & \text{by simplifying (2)} \\ (6) & a \qquad - m = 8 & \text{by simplifying (3)} \end{array}$$

$$\begin{array}{ll} (4)\text{–}(6): & 4] \; -2c + m = -18 \Rightarrow -8c + 4m = -72 \\ (5): & \quad -3c + 4m = 3 \Rightarrow -3c + 4m = 3 \\ & \qquad 5c = 75 \\ & \qquad c = 15 \end{array}$$

Substituting $c = 15$ *in (4):*

$$\begin{aligned} a - 2(15) &= -10 \\ a &= 20 \end{aligned}$$

Substituting $c = 15$ *in (5):*

$$\begin{aligned} -3(15) + 4m &= 3 \\ 4m &= 48 \\ m &= 12 \end{aligned}$$

Therefore,

$$\begin{aligned} \text{Al} &= 20 \text{ yr} \\ \text{Chet} &= 15 \text{ yr} \\ \text{Muriel} &= 12 \text{ yr} \end{aligned}$$

19. $-3, 7$

21.

$$\begin{array}{lrl} (1) & xy = 12 & A = LW \\ (2) & (x + 2)^2 = x^2 + y^2 & \text{Pythagorean Theorem} \\ & x^2 + 4x + 4 = x^2 + y^2 & \\ (3) & 4x + 4 = y^2 & \end{array}$$

Solving (1) for x: $x = \frac{12}{y}$

Substituting $x = \frac{12}{y}$ in (3):

$$4\left(\frac{12}{y}\right) + 4 = y^2$$

$$\begin{aligned} 48 + 4y &= y^3 \\ 0 &= y^3 - 4y - 48 \end{aligned}$$

$$\begin{array}{r|rrrr} & 1 & 0 & -4 & -48 \\ & & 4 & 16 & 48 \\ \hline 4 & 1 & 4 & 12 & 0 \end{array}$$

$y = 4$ | $y^2 + 4y + 12 = 0$

$x = \frac{12}{y} = \frac{12}{4} = 3$ | $y = \frac{-4 \pm \sqrt{4^2 - 4(1)(12)}}{2(1)} = \frac{-4 \pm \sqrt{16 - 48}}{2}$ Complex roots

Therefore, length = 4
width = 3

23. Tie costs \$1.05; pin costs \$0.05.

25.

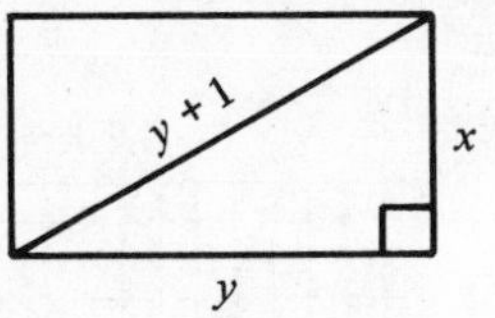

$$
\begin{array}{lrll}
(1) & xy = 60 & & A = LW \\
 & (y+1)^2 = x^2 + y^2 & & \text{Pythagorean Theorem} \\
 & y^2 + 2y + 1 = x^2 + y^2 & & \\
(2) & 2y + 1 = x^2 & &
\end{array}
$$

Solving (1) for y, we have: $y = \dfrac{60}{x}$ Substituting $y = \dfrac{60}{x}$ in (2), we have:

$$2\left(\frac{60}{x}\right) + 1 = x^2$$

$$120 + x = x^3 \quad \text{Multiplying each term by LCD } x$$

$$0 = x^3 - x - 120$$

$$
\begin{array}{r|rrrr}
 & 1 & 0 & -1 & -120 \\
 & & 5 & 25 & 120 \\
\hline
5 & 1 & 5 & 24 & 0
\end{array}
$$

$$
\begin{array}{r|l}
x = 5 & x^2 + 5x + 24 = 0 \\
y = \dfrac{60}{x} = \dfrac{60}{5} = 12 & x = \dfrac{-5 \pm \sqrt{5^2 - 4(1)(24)}}{2} \quad \text{Quadratic formula} \\
\begin{cases} \text{Length} = 12 \\ \text{Width} = 5 \end{cases} & \;\;= \dfrac{-5 \pm \sqrt{25 - 96}}{2} \quad \text{Complex roots}
\end{array}
$$

27. Let D = Original amount of Dan's money
F = Original amount of Fran's money

	D	F
First step:	$D + 2D = 3D$	$F - 2D$
Second step:	$3D - \frac{1}{4}(F - 2D)$	$F - 2D + \frac{1}{4}(F - 2D) = \frac{5}{4}(F - 2D)$
Third step:	$3D - \frac{1}{4}(F - 2D) + 3D - \frac{1}{4}(F - 2D) = 100$	$\frac{5}{4}(F - 2D) - \left[3D - \frac{1}{4}(F - 2D)\right] = 75$
	$6D - \frac{1}{2}(F - 2D) = 100$	$\frac{5}{4}(F - 2D) - 3D + \frac{1}{4}(F - 2D) = 75$
	$12D - (F - 2D) = 200$	$\frac{3}{2}(F - 2D) - 3D = 75$
	$12D - F + 2D = 200$	$3(F - 2D) - 6D = 150$
		$3F - 6D - 6D = 150$
	(1) $14D - 200 = F$	(2) $F = 4D + 50$

Equating the values of F, we have:

$14D - 200 = 4D + 50$

$10D = 250$

$D = \$25$ Amount of Dan's money

$F = 4D + 50 = 4(25) + 50 = \$150$ Amount of Fran's money

28. Let B = Length of body
T = Length of tail

Then $B = 4 + T$ (1)

and $T = 4 + \frac{1}{2}B$ (2)

Substitute $B = 4 + T$ in (2):

$T = 4 + \frac{1}{2}(4 + T)$

$T = 4 + 2 + \frac{1}{2}T$

$T = 12$ in.

$B = 4 + 7 = 4 + 12 = 16$ in.

Therefore, the length of the fish $= T + B + 4 = 12 + 16 + 4 = 32$ in.

Exercises 9–9 (page 335)

1.

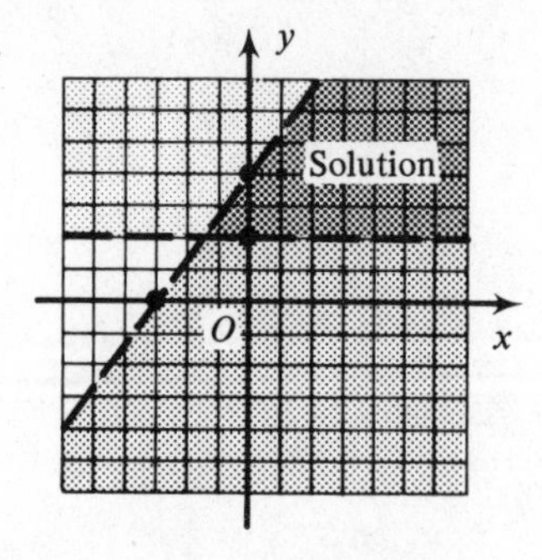

Both lines are dashed lines
because equality is not included

3.

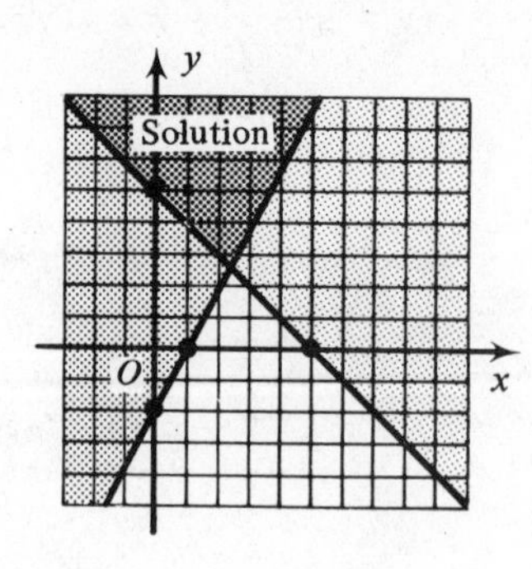

5.

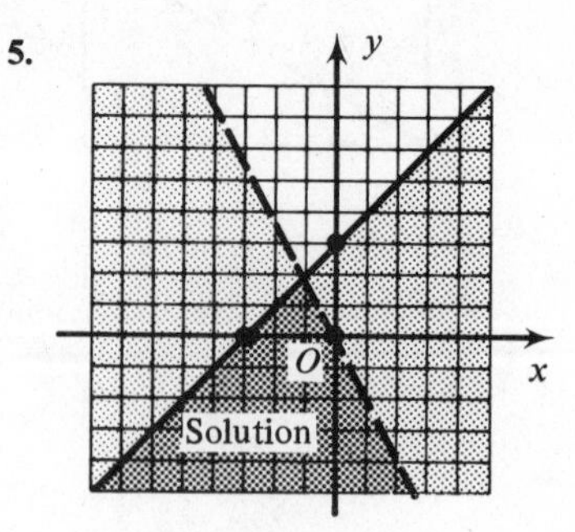

7.

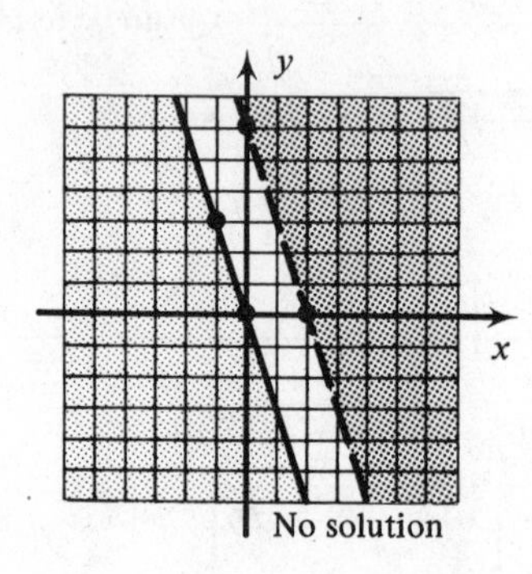

9.

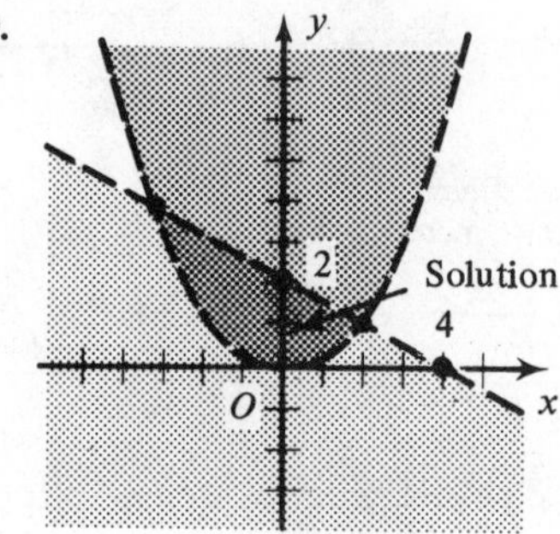

11. 1. *Graph inequality (1):* $x^2 + 4y^2 < 4$

Boundary: $x^2 + 4y^2 = 4$

x-intercepts: $y = 0$: $x^2 + 4(0)^2 = 4$

$x = \pm 2$

y-intercepts: $x = 0$: $0^2 + 4y^2 = 4$

$y = \pm 1$

Therefore, the boundary is an ellipse.
The ellipse is dashed because $=$ is *not* included.

Test point (0,0) in $x^2 + 4y^2 < 4$

$0^2 + 4(0)^2 < 4$ True

The solutions of (1) lie inside the ellipse along with (0,0).

2. *Graph inequality (2):* $x^2 - y^2 \geq 1$

Boundary: $x^2 - y^2 = 1$

x-intercepts: $y = 0$: $x^2 - (0)^2 = 1$

$x = \pm 1$

y-intercepts: $x = 0$: $0^2 - y^2 = 1$

$y = \pm\sqrt{-1} = \pm i$

Therefore, the graph does not cross the *y*-axis.
The boundary is an hyperbola.

Test point (0,0) in $x^2 - y^2 \geq 1$

$0^2 - (0)^2 \geq 1$ False

The solutions of (2) lie on the opposite side of the hyperbola from (0,0).
The solution of the *system* is shown in the figure.

y
1
−2 −1 1 2 x
−1
Solution

13. (1) $3x - 2y < 6$
(2) $x + 2y \leq 4$
(3) $6x + y > -6$

Boundary line for (1): $3x - 2y = 6$

Intercepts:

x	y
2	0
0	−3

The half-plane includes (0, 0) because $3(0) - 2(0) < 6$

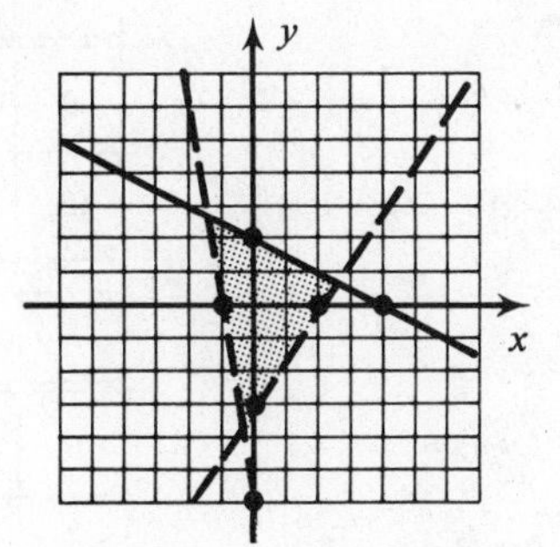

Shaded area is solution. The boundary lines for (1) and (3) are dashed because equality is not included; the boundary line for (2) is solid because equality is included.

Boundary line for (2): $x + 2y = 4$

Intercepts:

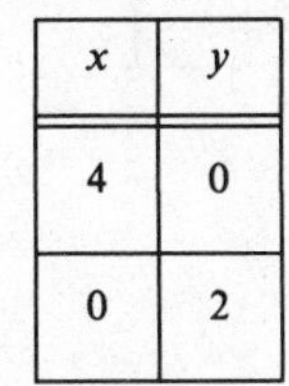

x	y
4	0
0	2

The half-plane includes (0, 0) because $(0) + 2(0) \leq 4$

Boundary line for (3): $6x + y = -6$

Intercepts:

x	y
−1	0
0	−6

The half-plane includes (0, 0) because $6(0) + (0) > -6$

15. 1. *Graph inequality (1):* $v > x^2 - 2x$

Boundary: Parabola $y = x^2 - 2x$

x-intercepts: $y = 0$: $0 = x^2 - 2x$
$0 = x(x - 2)$
$x = 0 \mid x = 2$

y-intercept: $x = 0$: $y = 0^2 - 2(0) = 0$

Axis of symmetry: $y = \dfrac{x_1 + x_2}{2} = \dfrac{0 + 2}{2} = 1$

Parabola is dashed because = is *not* included

Test point (1, 0) in $y > x^2 - 2x$
$0 > 1^2 - 2(1)$
$0 > 1 - 2$ True

Therefore, solution of (1) lies inside parabola, along with (1, 0).

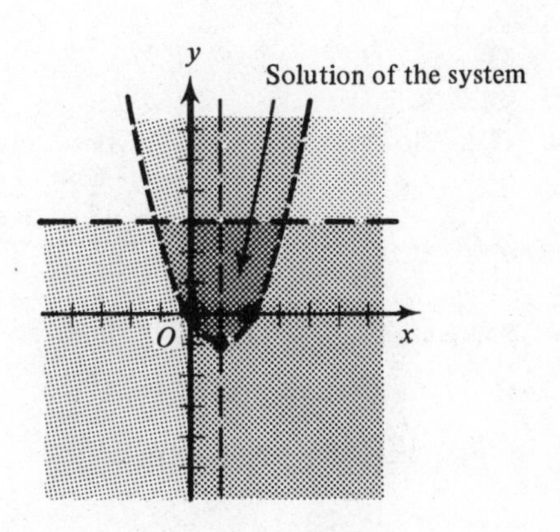

2. *Graph inequality (2):* $x \geq 0$

Solution of (2) includes the y-axis and all points to the right of it.

3. *Graph inequality (3):* $y < 3$

Solution of (3) is all points *below* the line $y = 3$.
The solution of the *system* is shown in the figure.

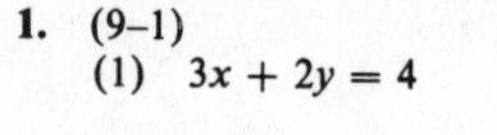

Chapter Nine Diagnostic Test or Review (page 337)

Following each problem number is the textbook section number (in parentheses) where that kind of problem is discussed.

1. (9–1)

(1) $3x + 2y = 4$

If $x = 0$, $3(0) + 2y = 4 \Rightarrow y = 2$

If $y = 0$, $3x + 2(0) = 4 \Rightarrow x = \frac{4}{3}$

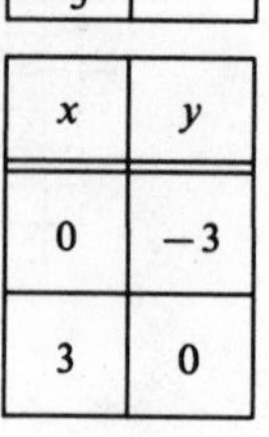

x	y
0	2
$1\frac{1}{3}$	0

(2) $x - y = 3$

If $x = 0$, $0 - y = 3 \Rightarrow y = -3$

If $y = 0$, $x - 0 = 3 \Rightarrow x = 3$

x	y
0	−3
3	0

(1) (2) (2, −1) x y

Solution $(2, -1)$

2. (9–2)

$$\begin{aligned} 2]\ 4x - 3y &= 13 \Rightarrow & 8x - 6y &= 26 \\ 3]\ 5x - 2y &= 4 \Rightarrow & 15x - 6y &= 12 \\ & & 7x &= -14 \\ & & x &= -2 \end{aligned}$$

Substitute $x = -2$ into (2):

$$\begin{aligned} (2) \quad 5x - 2y &= 4 \\ 5(-2) - 2y &= 4 \\ -10 - 2y &= 4 \\ -2y &= 14 \\ y &= -7 \end{aligned}$$

Solution $(-2, -7)$

3. (9–3) (1) $5x + 4y = 23$

(2) $3x + 2y = 9 \Rightarrow 2y = 9 - 3x$

$$y = \frac{9 - 3x}{2}$$

Substitute $\frac{9 - 3x}{2}$ in place of y in (1):

$$\begin{aligned} (1) \quad 5x + 4y &= 23 \\ 5x + \frac{\overset{2}{\cancel{4}}}{1}\left(\frac{9 - 3x}{\cancel{2}}\right) &= 23 \\ 5x + 2(9 - 3x) &= 23 \\ 5x + (18 - 6x) &= 23 \\ -x &= 5 \\ x &= -5 \end{aligned}$$

Substitute $x = -5$ in $y = \frac{9 - 3x}{2} = \frac{9 - 3(-5)}{2}$

$$= \frac{9 + 15}{2} = \frac{24}{2} = 12$$

Solution: $(-5, 12)$

4. (9–2)

$$\begin{aligned} 16]\ 2]\ 15x + 8y &= -18 \Rightarrow & 30x + 16y &= -36 \\ 8]\ 1]\ 9x + 16y &= -8 \Rightarrow & 9x + 16y &= -8 \\ & & 21x &= -28 \\ & & x &= \frac{-28}{21} = \frac{-4}{3} \end{aligned}$$

Substitute $x = \frac{-4}{3}$ in (1):

$$\begin{aligned} (1) \quad 15x + 8y &= -18 \\ \frac{\overset{5}{\cancel{15}}}{1}\left(\frac{-4}{\cancel{3}}\right) + 8y &= -18 \\ -20 + 8y &= -18 \\ 8y &= 2 \\ y &= \frac{2}{8} = \frac{1}{4} \end{aligned}$$

Solution: $\left(\frac{-4}{3}, \frac{1}{4}\right)$

5. (9–2) 4] 2] 10] 5]

$$\begin{array}{rcrcrcl} -10x + 35y &=& -18 &\Rightarrow& -20x + 70y &=& -36 \\ 4x - 14y &=& 8 &\Rightarrow& 20x - 70y &=& 40 \\ \hline & & & & 0 &\neq& 4 \end{array}$$

No solution

6. (9–8) Let x = Larger number
y = Smaller number

The sum of two numbers is 12

(1) $x + y = 12$

Their difference is 34

(2) $x - y = 34$

$$\begin{array}{lrcl} (1) & x + y &=& 12 \\ (2) & x - y &=& 34 \\ \hline & 2x &=& 46 \\ & x &=& 23 \end{array}$$

Substitute $x = 23$ in (1):

$$\begin{array}{lrcl} (1) & x + y &=& 12 \\ & 23 + y &=& 12 \\ & y &=& -11 \end{array}$$

The numbers are 23 and -11.

7. (9–4)

$$\begin{array}{lrcl} (1) & x + y + z &=& 0 \\ (2) & 2x \quad\quad - 3z &=& 5 \\ (3) & 3y + 4z &=& 3 \\ \hline 2(1) - (2) & 2y + 5z &=& -5 \quad (4) \\ (3) & 3y + 4z &=& 3 \\ \hline 3(4) - 2(3) & 7z &=& -21 \\ & z &=& -3 \end{array}$$

Substitute $z = -3$ in (2):
$2x - 3(-3) = 5 \Rightarrow x = -2$

Substitute $z = -3$ in (3):
$3y + 4(-3) = 3 \Rightarrow y = 5$

Solution: $(-2, 5, -3)$

8. (9–5, 9–6) (a) $\begin{vmatrix} 8 & -9 \\ 5 & -3 \end{vmatrix} = (8)(-3) - (5)(-9) = -24 + 45 = 21$

(b) $\begin{vmatrix} 2 & 0 & -1 \\ -3 & 1 & 4 \\ -1 & 5 & 6 \end{vmatrix} = +(2)\begin{vmatrix} 1 & 4 \\ 5 & 6 \end{vmatrix} - (0)\begin{vmatrix} -3 & 4 \\ -1 & 6 \end{vmatrix} + (-1)\begin{vmatrix} -3 & 1 \\ -1 & 5 \end{vmatrix}$

$= 2(6 - 20) - 0 - 1(-15 + 1) = -14$

9. (9–6)

$$\begin{aligned} 1x - 1y - 1z &= 0 \\ 1x + 3y + 1z &= 4 \\ 7x - 2y - 5z &= 2 \end{aligned}$$

$$D = \begin{vmatrix} 1 & -1 & 1 \\ 1 & 3 & 1 \\ 7 & -2 & -5 \end{vmatrix}$$

$$y = \frac{\begin{vmatrix} 1 & 0 & -1 \\ 1 & 4 & 1 \\ 7 & 2 & -5 \end{vmatrix}}{\begin{vmatrix} 1 & -1 & -1 \\ 1 & 3 & 1 \\ 7 & -2 & -5 \end{vmatrix}} = \frac{D_y}{D}$$

10. (9–7) (1) $y^2 = 8x$
(2) $3x + y = 2 \Rightarrow y = 2 - 3x$

Substitute $2 - 3x$ for y in (1):

$$\begin{array}{lrcl} (1) & y^2 &=& 8x \\ & (2 - 3x)^2 &=& 8x \\ & 4 - 12x + 9x^2 &=& 8x \\ & 9x^2 - 20x + 4 &=& 0 \\ & (x - 2)(9x - 2) &=& 0 \end{array}$$

$x - 2 = 0$ | $9x - 2 = 0$

$x = 2$ | $x = \frac{2}{9}$

If $x = 2$, $y = 2 - 3x = 2 - 3(2) = -4$
Therefore, one solution is $(2, -4)$.

If $x = \frac{2}{9}$, $y = 2 - 3x = 2 - 3\left(\frac{2}{9}\right) = \frac{4}{3}$.

Therefore, a second solution is $\left(\frac{2}{9}, \frac{4}{3}\right)$.

11. (9–9)
(1) $2x + 3y \leq 6$
Boundary line $2x + 3y = 6$ is solid because equality is included

If $y = 0$, $2x + 3(0) = 6 \Rightarrow x = 3$

If $x = 0$, $2(0) + 3y = 6 \Rightarrow y = 2$

x	y
3	0
0	2

(Problem 11 is continued on next page)

Substitute (0,0) in (1):
$2(0) + 3(0) \le 6$
$0 \le 6$ *True*
Therefore, the half-plane containing (0,0) is the solution of (1)

(2) $y - 2x < 2$
Boundary line $y - 2x = 2$ is dashed because equality is *not* included
If $y = 0,\ 0 - 2x = 2 \Rightarrow x = -1$

If $x = 0,\ y - 2(0) = 2 \Rightarrow y = 2$

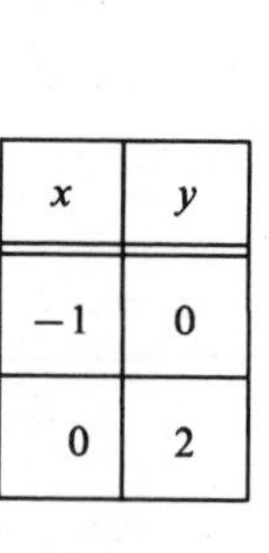

x	y
-1	0
0	2

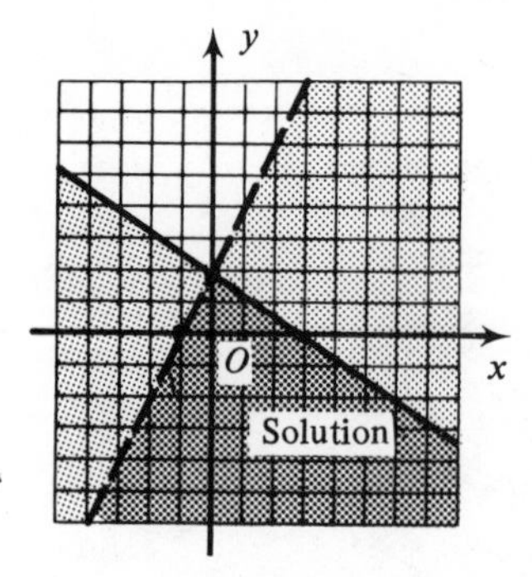

Substitute (0,0) in (2):
$0 - 2(0) < 2$
$0 < 2$ *True*
Therefore, the half-plane containing (0,0) is a solution of (2)

12. (9–9)
Graph (1) $\qquad 4x^2 + y^2 < 16$
Boundary: $\qquad 4x^2 + y^2 = 16$
x-intercepts: $y = 0$: $\qquad 4x^2 + (0)^2 = 16$
$x = \pm 2$
y-intercepts: $x = 0$: $\qquad 4(0)^2 + y^2 = 16$
$y = \pm 4$
Therefore, the boundary is an ellipse.
The ellipse is dashed because = is *not* included.
Test (0, 0) in $\qquad 4x^2 + y^2 < 16$
$4(0)^2 + (0)^2 < 16$ *True*
The solutions of (1) lie inside the ellipse along with (0, 0).

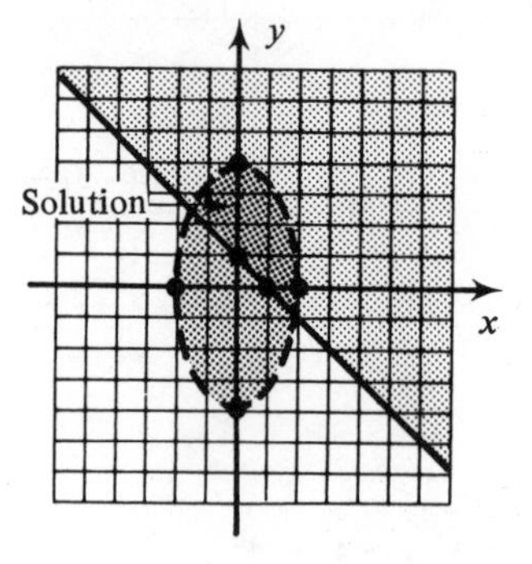

Graph (2) $\qquad x + y \ge 1$
Boundary: $\qquad x + y = 1$
x-intercept: $y = 0$ $\qquad x + (0) = 1 \Rightarrow x = 1$

y-intercept: $x = 0$ $\qquad (0) + y = 1 \Rightarrow y = 1$

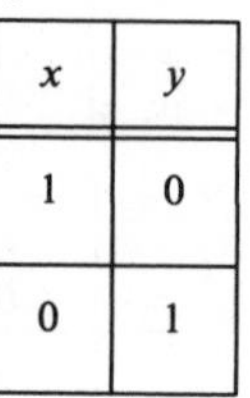

x	y
1	0
0	1

Boundary line is solid because = is included.
Test (0, 0) in $\qquad x + y \ge 1$
$(0) + (0) \ge 1$ *False*
Therefore, the solution of (2) is the half-plane *not* containing (0, 0).

Exercises 10–1 (page 342)

1. $\log_3 9 = 2$ **3.** $\log_{10} 1000 = 3$ **5.** $\log_2 16 = 4$ **7.** $\log_3\left(\frac{1}{9}\right) = -2$ **9.** $12^0 = 1 \Longleftrightarrow \log_{12} 1 = 0$

11. $16^{1/2} = 4 \Longleftrightarrow \log_{16} 4 = \frac{1}{2}$ **13.** $\log_2 8 = 3$ **15.** $\log_4 \frac{1}{16} = -2$ **17.** $8^2 = 64$ **19.** $7^2 = 49$

21. $5^0 = 1$ **23.** $\log_{10} 100 = 2 \Longleftrightarrow 10^2 = 100$ **25.** $10^{-3} = 0.001$ **27.** $16^{1/2} = 4$ **29.** $10^5 = 100{,}000$

31. *Logarithmic form* $x = \log_{10}10{,}000$ $\iff$ *Exponential form* $10^x = 10{,}000 = 10^4$, $x = 4$

Therefore, $\log_{10}10{,}000 = 4$

33. *Logarithmic form* $x = \log_4 8$ $\iff$ *Exponential form*

$$4^x = 8$$
$$(2^2)^x = 2^3$$
$$2^{2x} = 2^3$$
$$2x = 3$$
$$x = \frac{3}{2}$$

Therefore, $\log_4 8 = \frac{3}{2}$

35. 4 **37.** 1 **39.** 0 **41.** $\frac{1}{2}$ **43.** 25 **45.** 2 **47.** 3

49. *Logarithmic form* $\log_9\left(\frac{1}{3}\right) = x$ $\iff$ *Exponential form*

$$9^x = \frac{1}{3}$$
$$(3^2)^x = 3^{-1}$$
$$3^{2x} = 3^{-1}$$
$$2x = -1$$
$$x = -\frac{1}{2}$$

51. *Logarithmic form* $\log_{10}10^{-4} = x$ $\iff$ *Exponential form* $10^x = 10^{-4}$, $x = -4$

53. *Logarithmic form* $\log_{3/2}N = 2$ $\iff$ *Exponential form* $N = \left(\frac{3}{2}\right)^2 = \frac{9}{4}$

55. *Logarithmic form* $\log_5 125 = x$ $\iff$ *Exponential form* $5^x = 125 = 5^3$, $x = 3$

57. *Logarithmic form* $\log_b 8 = 1.5$ $\iff$ *Exponential form*

$$b^{1.5} = 8 = 2^3$$
$$b^{3/2} = 2^3$$
$$(b^{3/2})^{2/3} = (2^3)^{2/3}$$
$$b = 2^2 = 4$$

59. *Logarithmic form* $\log_2 N = -2$ $\iff$ *Exponential form*

$$N = 2^{-2} = \frac{1}{2^2}$$
$$N = \frac{1}{4}$$

61. 125 **63.** $\frac{5}{4}$ **65.** $\frac{3}{2}$ **67.** 8

Exercises 10–2 (page 346)

1.

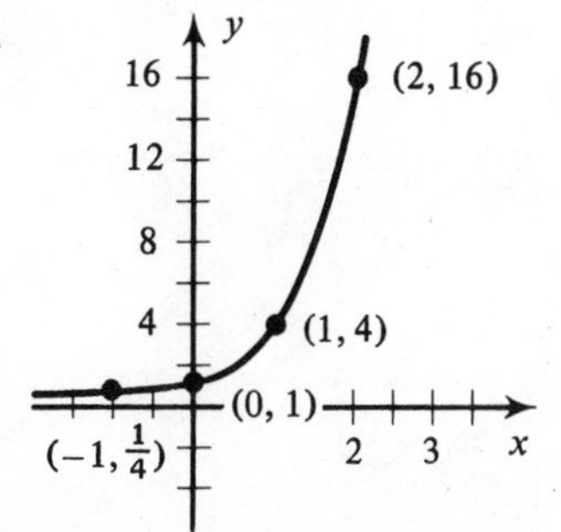

3.

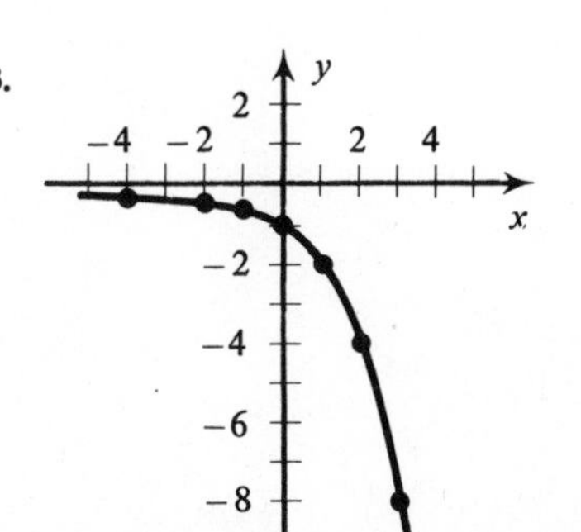

5.

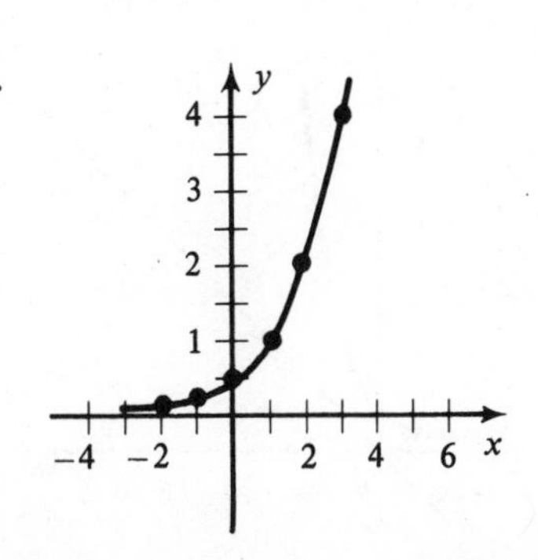

7. $y = \left(\frac{3}{2}\right)^x$

Give x-values and solve for y:

$$y = \left(\frac{3}{2}\right)^{-3} = \left(\frac{2}{3}\right)^3 = \frac{8}{27}$$
$$y = \left(\frac{3}{2}\right)^{-2} = \left(\frac{2}{3}\right)^2 = \frac{4}{9}$$
$$y = \left(\frac{3}{2}\right)^{-1} = \frac{2}{3}$$
$$y = \left(\frac{3}{2}\right)^0 = 1$$
$$y = \left(\frac{3}{2}\right)^1 = \frac{3}{2}$$
$$y = \left(\frac{3}{2}\right)^2 = \frac{9}{4}$$
$$y = \left(\frac{3}{2}\right)^3 = \frac{27}{8}$$
$$y = \left(\frac{3}{2}\right)^4 = \frac{81}{16} \doteq 5$$
$$y = \left(\frac{3}{2}\right)^5 = \frac{243}{32} \doteq 7.6$$

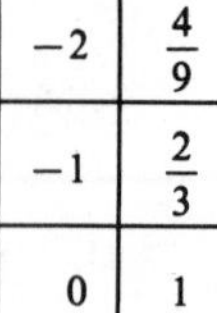

x	y
-3	$\frac{8}{27}$
-2	$\frac{4}{9}$
-1	$\frac{2}{3}$
0	1
1	$\frac{3}{2}$
2	$\frac{9}{4}$
3	$\frac{27}{8}$
4	$\frac{81}{16}$
5	$\frac{243}{32}$

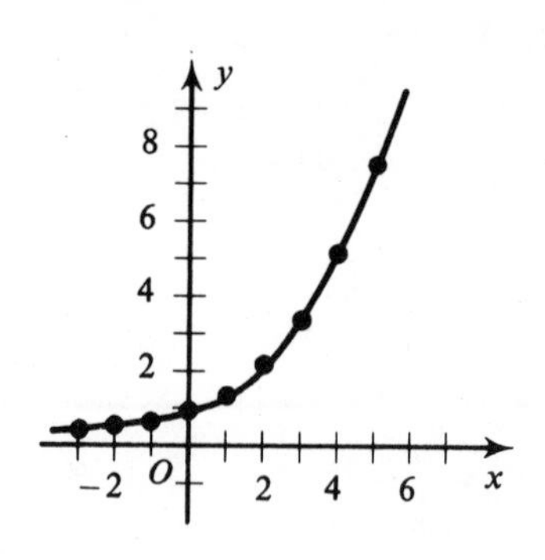

9. The exponential form for $y = \log_2 x$ is $x = 2^y$
The exponential form for $y = \log_{10} x$ is $x = 10^y$

$x = 2^y$

x	y
$\frac{1}{4}$	-2
$\frac{1}{2}$	-1
1	0
2	1
4	2
8	3

$x = 10^y$

x	y
0.01	-2
0.1	-1
1	0
10	1

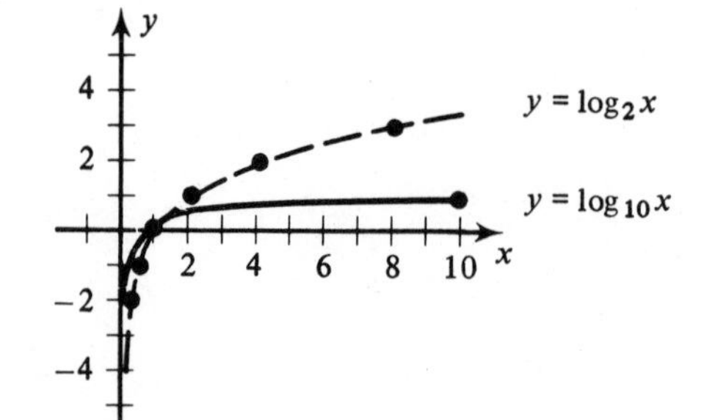

11.

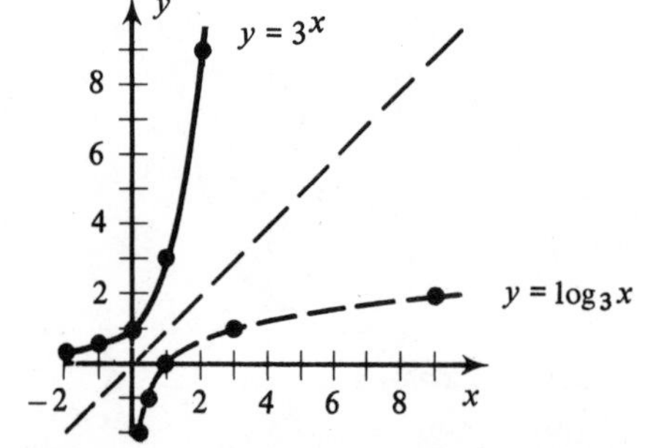

Exercises 10–3 (page 348)

1. $\log_{10}31 + \log_{10}7$ **3.** $\log_{10}41 - \log_{10}13$ **5.** $3\log_{10}19$ **7.** $\log_{10}17 + \log_{10}31 - \log_{10}29$

9. $\log_{10}\sqrt[5]{75} = \log_{10}75^{1/5} = \frac{1}{5}\log_{10}75$ **11.** $\log_{10}53 - \log_{10}11 - 2\log_{10}19$

13. $\log_{10}\left[\frac{35\sqrt{73}}{(1.06)^8}\right] = \log_{10}35\sqrt{73} - \log_{10}(1.06)^8 = \log_{10}35 + \frac{1}{2}\log_{10}73 - 8\log_{10}1.06$ **15.** 1.146 **17.** 0.109

19.
$$\begin{aligned}\log_{10}\sqrt{27} &= \log_{10}(27)^{1/2} = \log_{10}(3^3)^{1/2}\\ &= \log_{10}3^{3/2} = \frac{3}{2}\log_{10}3\\ &= (1.5)(0.477) = 0.716\end{aligned}$$

21. 1.954 **23.** 3.112 **25.** 0.119

27.
$$\begin{aligned}\log_{10}6000 &= \log_{10}(2)(3)(10^3)\\ &= \log_{10}2 + \log_{10}3 + 3\log_{10}10\\ &= 0.301 + 0.477 + 3 = 3.778\end{aligned}$$

29. $\log_b xy$ **31.** $\log_b\left(\frac{x^2}{y^3}\right)$ **33.** $\log_b\left(\frac{x^3}{y^6}\right)$ **35.** $\log_e v^4$

37.
$$\begin{aligned}\log_a(x^2 - y^2) - 3\log_e(x+y) &= \log_a(x^2 - y^2) - \log_a(x+y)^3\\ &= \log_a\left[\frac{x^2 - y^2}{(x+y)^3}\right] = \log_a\left[\frac{\cancel{(x+y)}(x-y)}{(x+y)^{\cancel{3}2}}\right] = \log_a\left[\frac{(x-y)}{(x+y)^2}\right]\end{aligned}$$

39.
$$\begin{aligned}2\log_b 2xy - \log_b 3xy^2 + \log_b 3x &= \log_b(2xy)^2 - \log_b 3xy^2 + \log_b 3x = \log_b 4x^2y^2 - \log_b 3xy^2 + \log_b 3x\\ &= \log_b\left(\frac{4x^2y^2 3x}{3xy^2}\right) = \log_b 4x^2 = \log_b(2x)^2 = 2\log_b 2x\end{aligned}$$

Exercises 10–4A (page 352)

1. 2.856×10^1 **3.** $0.06184 = 0.06\,184 = 6.184 \times 10^{-2}$ **5.** $3\,700.5 = 3.7005 \times 10^3$ **7.** 2 **9.** 0

11. -1, written as $9 - 10$ **13.** 9 3000000 Characteristic is 7 **15.** 0.00008 06 Characteristic is -5, written as $5 - 10$

17. 4 **19.** 5 (same as exponent of 10) **21.** -3 (same as exponent of 10), written as $7 - 10$

Exercises 10–4B (page 353)

1. 2.8774 **3.** 1.2304 **5.** 3.5250 **7.** 3.8451 **9.** $8.7810 - 10$ **11.** $9.9566 - 10$

13. 1.7701 **15.** $6.7612 - 10$ **17.** 3.7513

Exercises 10–4C (page 355)

1. 1.3683 **3.** 0.4860 **5.** $8.8224 - 10$ **7.** 2.1781 **9.** 5.2702 **11.** $9.9034 - 10$

13. $7.6024 - 10$ **15.** 6.9230 **17.** 2.5111

Exercises 10–5 (page 358)

1. 3530 **3.** 9.13 **5.** 0.179 **7.** 3200 **9.** 0.00409 **11.** 11,450

13. 0.0009773 **15.** 14.74 **17.** 0.04907

Exercises 10–6 (page 360)

1. 45.92 **3.** 26.26 **5.** 1.538 **7.** 0.7630 **9.** 0.3259 **11.** $2.863 + \log 38.46 = 2.863 + 1.5850 = 4.4480$

13. $N = \dfrac{\log 7.86}{\log 38.4}$

$\log N = \log(\log 7.86) - \log(\log 38.4)$
$= \log(0.8954) - \log(1.5843)$
$= (9.9520 - 10) - (0.1999)$
$\log N = 9.7521 - 10$
$N = .5\,651 = 0.5651$ (decimal point moved -1)

15. 16,890

17. Let $N = \sqrt[5]{\dfrac{(5.86)(17.4)}{\sqrt{450}}}$

$\log 5.86 = 0.7679$
$\log 17.4 = 1.2405$ (+)
2.0084
$\log 450 = 2.6532$
$\frac{1}{2}\log 450 = 1.3266$ (−)

$\log N = \frac{1}{5}(0.6818)$
$= 0.1364$
$N = 1.369$

19. $335.68

Exercises 10–7 (page 364)

1. $-\frac{2}{3}$ **3.** $-\frac{3}{2}$ **5.** 1

7.
$25^{2x+3} = 5^{x-1}$
$(5^2)^{2x+3} = 5^{x-1}$
$5^{4x+6} = 5^{x-1}$
$4x + 6 = x - 1$
$3x = -7$
$x = -\frac{7}{3}$

9.
$2^x = 3$
$\log 2^x = \log 3$
$x \log 2 = \log 3$
$x = \dfrac{\log 3}{\log 2} = \dfrac{0.4771}{0.3010}$
$\doteq 1.585$

11. $\doteq 0.125$

13. $\doteq 2.018$

15.
$(8.71)^{2x+1} = 8.57$
$\log(8.71)^{2x+1} = \log 8.57$
$(2x+1)\log 8.71 = \log 8.57$
$2x + 1 = \dfrac{\log 8.57}{\log 8.71} = \dfrac{0.9330}{0.9400}$
$2x + 1 \doteq 0.9926$
$x \doteq -0.0037$

17.
$\log(3x-1) + \log 4 = \log(9x+2)$
$\log(3x-1)4 = \log(9x+2)$
$(3x-1)4 = 9x + 2$
$12x - 4 = 9x + 2$
$3x = 6$
$x = 2$

Check for x = 2
$\log[3(2) - 1] + \log 4 \overset{?}{=} \log[9(2) + 2]$
$\log 5 + \log 4 \overset{?}{=} \log 20$
$\log(5)(4) \overset{?}{=} \log 20$
$\log 20 = \log 20$

Therefore, 2 is a solution

19. 5 **21.** 6 **23.** 2, 5

25.
$\log x + \log(x-3) = \log 4$
$\log x(x-3) = \log 4$
$x(x-3) = 4$
$x^2 - 3x - 4 = 0$
$(x-4)(x+1) = 0$
$x - 4 = 0 \quad | \quad x + 1 = 0$
$x = 4 \quad | \quad x = -1$

Check for x = 4
$\log 4 + \log(4-3) \overset{?}{=} \log 4$
$\log 4 + \log 1 \overset{?}{=} \log 4$
$\log 4 + 0 \overset{?}{=} \log 4$
$\log 4 = \log 4$

Therefore, 4 is a solution

Check for x = −1
$\log(-1) + \log(-1-3) \overset{?}{=} \log 4$
(Not real numbers)

Therefore, −1 *is not* a solution

27.
$\log(5x+2) - \log(x-1) = 0.7782$
$\log\left(\dfrac{5x+2}{x-1}\right) = 0.7782$

Let $N = \dfrac{5x+2}{x-1}$

Then $\log N = 0.7782$
$N = 6.000$

Therefore, $\dfrac{5x+2}{x-1} \doteq 6$
$5x + 2 \doteq 6x - 6$
$8 \doteq x$

Check for x = 8
$\log(5x+2) - \log(x-1) = 0.7782$
$\log[5(8) + 2] - \log(8-1) \overset{?}{=} 0.7782$
$\log 42 - \log 7 \overset{?}{=} 0.7782$
$\log\left(\dfrac{42}{7}\right) \overset{?}{=} 0.7782$
$\log 6 \overset{?}{=} 0.7782$
$0.7782 = 0.7782$

Therefore, 8 is a solution

29. 2802

Exercises 10–8 (page 366)

1. $\log_2 156 = \dfrac{\log_{10}156}{\log_{10}2} = \dfrac{2.1931}{0.3010} \doteq 7.286$ **3.** $\doteq 2.109$ **5.** $\log_e 3.04 = \dfrac{\log_{10}3.04}{\log_{10}2.718} = \dfrac{0.4829}{0.4343} \doteq 1.112$

7. $\doteq 3.983$ **9.** $\doteq 2.757$

Chapter Ten Diagnostic Test or Review (page 368)

Following each problem number is the textbook section number (in parentheses) where that kind of problem is discussed.

1. (10–1) *Logarithmic form* $\log_2 16 = 4 \iff$ *Exponential form* $2^4 = 16$

2. (10–1) *Logarithmic form* $\log_{2.5} 6.25 = 2 \iff$ *Exponential form* $(2.5)^2 = 6.25$

3. (10–1) $\log_4 N = 3 \iff 4^3 = N$
Therefore, $N = 64$

4. (10–1) $\log_{10} 10^{-2} = x \iff 10^x = 10^{-2}$
Therefore, $x = -2$

5. (10–1) $\log_b 6 = 1 \iff b^1 = 6$
Therefore, $b = 6$

6. (10–1) $\log_5 1 = x \iff 5^x = 1 = 5^0$
Therefore, $x = 0$

7. (10–1) $\log_{0.5} N = -2 \iff (0.5)^{-2} = N$

$$\left(\frac{1}{2}\right)^{-2} = N$$
$$\frac{1^{-2}}{2^{-2}} = N$$
$$2^2 = N$$
$$N = 4$$

8. (10–3) $\log x + \log y - \log z = \log\left(\dfrac{xy}{z}\right)$

9. (10–3) $\dfrac{1}{2}\log x^4 + 2\log x = \log(x^4)^{1/2} + 2\log x$
$= \log x^2 + 2\log x$
$= 2\log x + 2\log x = 4\log x$

10. (10–3) $\log(x^2 - 9) - \log(x - 3) = \log(x+3)(x-3) - \log(x-3)$
$= \log(x+3) + \log(x-3) - \log(x-3)$
$= \log(x+3)$

11. (10–7) $\log(3x+5) - \log 7 = \log(x-1)$

$$\log\left[\frac{(3x+5)}{7}\right] = \log(x-1)$$
$$\frac{3x+5}{7} = \frac{x-1}{1}$$

LCD = 7

$$\frac{\not{7}}{1}\cdot\frac{3x+5}{\not{7}} = \frac{7}{1}\cdot\frac{x-1}{1}$$
$$3x + 5 = 7x - 7$$
$$12 = 4x$$
$$x = 3$$

Check

$\log(3x+5) - \log 7 = \log(x-1)$
$\log[3(3)+5] - \log 7 \stackrel{?}{=} \log(3-1)$
$\log 14 - \log 7 \stackrel{?}{=} \log 2$
$\log(2\cdot 7) - \log 7 \stackrel{?}{=} \log 2$
$\log 2 + \log 7 - \log 7 \stackrel{?}{=} \log 2$
$\log 2 = \log 2$

12. (10–7)

$$x^{1.24} = 302$$
$$\log x^{1.24} = \log 302$$
$$1.24\log x = \log 302$$
$$\log x = \frac{\log 302}{1.24} = \frac{2.4800}{1.24}$$
$$\log x = 2.0000$$
$$x = 100$$

13. (10–6) Let $N = (1.71)^4$
$\log N = \log(1.71)^4$
$\log N = 4\log 1.71 = 4(0.2330)$
$\log N = 0.9320$
$N = 8.55$

14. (10–6) Let $N = \dfrac{\sqrt[3]{8.84}}{30.3}$

then $\log N = \log \sqrt[3]{8.84} - \log 30.3$

$\frac{1}{3}\log 8.84 = \frac{1}{3}(0.9465) = 10.3155 - 10$

$\log 30.3 = 1.4814$ (−)

$\log N = 8.8341 - 10$

$N = 0.06825$

15. (10–8) Find $\log_2 718$

$$\log_2 N = \frac{\log_{10} N}{\log_{10} 2}$$

$$\log_2 718 = \frac{\log 718}{\log 2} = \frac{2.8561}{0.3010} \doteq 9.489$$

16. (10–2) The inverse exponential function of $y = 7^x$ is $x = 7^y$
The inverse logarithmic function of $y = 7^x$ is $y = \log_7 x$

We find points on the curve $y = 7^x$ by giving x-values and solving for y:

$y = 7^x$

$y = 7^{-2} = \frac{1}{49}$

$y = 7^{-1} = \frac{1}{7}$

$y = 7^0 = 1$

$y = 7^1 = 7$

$y = 7^2 = 49$

x	y
-2	$\frac{1}{49}$
-1	$\frac{1}{7}$
0	1
1	7
2	49

We find points on the curve $y = \log_7 x$ by plotting points on the equivalent exponential curve $x = 7^y$
In this case, we give y values and solve for x:

$x = 7^y$

$x = 7^{-2} = \frac{1}{49}$

$x = 7^{-1} = \frac{1}{7}$

$x = 7^0 = 1$

$x = 7^1 = 7$

$x = 7^2 = 49$

x	y
$\frac{1}{49}$	-2
$\frac{1}{7}$	-1
1	0
7	1
49	2

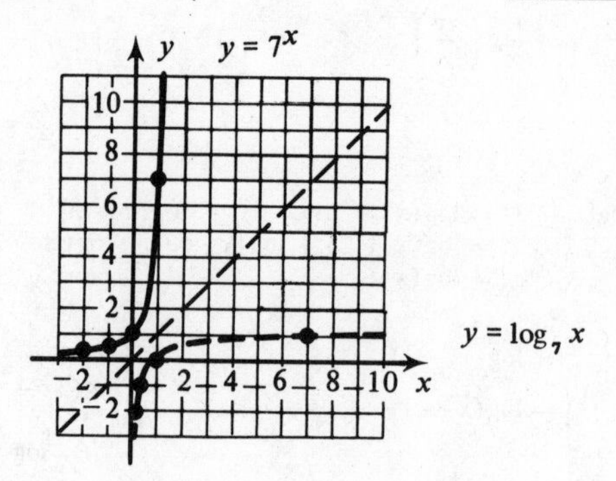

Exercises 11–1 (page 372)

1. 25, 30, 35 **3.** 9, 7, 5 **5.** 8, 4, 0 **7.** 5, 6, 7

9. $a_n = \dfrac{1-n}{n}$

$a_1 = \dfrac{1-(1)}{(1)} = 0$

$a_2 = \dfrac{1-(2)}{(2)} = -\dfrac{1}{2}$

$a_3 = \dfrac{1-(3)}{(3)} = -\dfrac{2}{3}$

$a_4 = \dfrac{1-(4)}{(4)} = -\dfrac{3}{4}$

The first four terms are $0, -\frac{1}{2}, -\frac{2}{3}, -\frac{3}{4}$

11. $-1, -\frac{2}{3}, -\frac{3}{5}$ **13.** 0, 3, 8 **15.** 8 **17.**
$$a_n = \frac{n-1}{n+1}$$
$$a_1 = \frac{1-1}{1+1} = 0$$
$$a_2 = \frac{2-1}{2+1} = \frac{1}{3}$$
$$a_3 = \frac{3-1}{3+1} = \frac{1}{2}$$
$$S_3 = 0 + \frac{1}{3} + \frac{1}{2} = \frac{5}{6}$$
19. 45

Exercises 11–2 (page 376)

1. AP, $d = 5$ **3.** 7, 4, 1, −2

$-2 - 1 = -3$
$1 - 4 = -3$
$4 - 7 = -3$

AP, because all the differences are −3

5. Not an AP **7.** Not an AP

9. AP, $d = -x + 1$ **11.** 5, −2, −9, −16 **13.** 19, −5, −29, −53 **15.** 7, 13, 19, 25, 31 **17.** 172

19. 86 **21.** $x, 2x + 1, 3x + 2, \ldots$
$$d = (2x + 1) - x = x + 1$$
$$a_{11} = a_1 + (11 - 1)d = x + 10(x + 1)$$
$$a_{11} = 11x + 10$$
23. 2550 **25.** 7,650 **27.** $d = 4$, $a_1 = -5$

29.
$$a_n = a_1 + (n - 1)d$$
$$16 = -5 + (n - 1)(3)$$
$$21 = 3n - 3 \Rightarrow 3n = 24 \Rightarrow n = 8$$
$$S_8 = \frac{8}{2}(-5 + 16) = 4(11) = 44$$
31. $n = 11$, $S_n = -473$ **33.** $n = 4$, $d = 4$

35.
$$a_n = a_1 + (n - 1)d$$
$$a_n = a_1 + 8\left(\frac{3}{2}\right) \Rightarrow a_n = a_1 + 12$$
$$S_n = \frac{n}{2}(a_1 + a_n)$$
$$-\frac{9}{4} = \frac{9}{2}(a_1 + a_n)$$
Substitute $a_1 + 12$ for a_n
$$-\frac{9}{4} = \frac{9}{2}(a_1 + a_1 + 12)$$
$$-\frac{9}{4} = \frac{9}{2}(2a_1 + 12)$$
$$-\frac{9}{4} = 9a_1 + 54$$
$$-9 = 36a_1 + 216 \Rightarrow a_1 = -\frac{25}{4}$$
Substitute $a_1 = -\frac{25}{4}$ in $a_n = a_1 + 12$
$$a_n = -\frac{25}{4} + \frac{48}{4} = \frac{23}{4}$$

37. $a_1 = -\frac{7}{3}$, $a_n = a_{15} = 7$

39.
$a_1 = 16$, $a_2 = 48$ } $d = 48 - 16 = 32$
$a_3 = 80$
.
.
.
$$a_{10} = a_1 + 9d = 16 + 9(32)$$
$$a_{10} = 16 + 288 = 304 \text{ ft}$$
$$a_{12} = 304 + 64 = 368 \text{ ft}$$
$$S_{12} = \frac{12}{2}(16 + 368) = 6(384) = 2304 \text{ ft}$$

Exercises 11–3 (page 380)

1. GP, $r = 3$

3. GP, $r = -3$

5. $2, \frac{1}{2}, \frac{1}{8}, \frac{1}{16}$

$\frac{1}{16} \div \frac{1}{8} = \frac{1}{16} \cdot \frac{8}{1} = \frac{1}{2}$

$\frac{1}{8} \div \frac{1}{2} = \frac{1}{8} \cdot \frac{2}{1} = \frac{1}{4}$

Not GP, because all ratios are not equal

7. Not a GP

9. GP, $r = 2y$

11. $12, 4, \frac{4}{3}, \frac{4}{9}, \frac{4}{27}$

13. $-9, 3, -1, \frac{1}{3}$

15.
$$a_5 = a_1 r^4$$
$$54 = \frac{2}{3} r^4$$
$$81 = r^4$$
$$r = \pm 3$$
Therefore, there are two answers

One answer: $r = 3$

$\frac{2}{3}, 2, 6, 18, 54$

Second answer: $r = -3$

$\frac{2}{3}, -2, 6, -18, 54$

17. $-\frac{64}{81}$

19. $-\frac{27}{1250}$

21. $16x, 8xy, 4xy^2, \ldots$

$$\frac{8xy}{16x} = \frac{y}{2} = r$$
$$a_8 = a_1 r^7 = 16x\left(\frac{y}{2}\right)^7 = \frac{xy^7}{8}$$

23.
$$a_5 = a_1 r^4$$
$$80 = a_1\left(\frac{2}{3}\right)^4$$
$$80 = \frac{16}{81}a_1$$
$$a_1 = 405$$
$$S_5 = \frac{a_1(1 - r^5)}{1 - r}$$
$$= \frac{405\left[1 - \left(\frac{2}{3}\right)^5\right]}{1 - \frac{2}{3}} = \frac{405\left[1 - \frac{32}{243}\right]}{\frac{1}{3}} = \frac{405}{1}\left(\frac{211}{243}\right)\frac{3}{1} = 1055$$

25. $a_1 = \frac{405}{2} \qquad S_5 = \frac{275}{2}$

27.
$$\left.\begin{array}{ll}(2) & a_5 = a_1 r^4 = \frac{112}{9} \\ (1) & a_3 = a_1 r^2 = 28\end{array}\right\} \quad r^2 = \frac{112}{9} \div 28 = \frac{112}{9} \cdot \frac{1}{28} = \frac{4}{9}$$
$$r = \pm\frac{2}{3}$$

$r_1 = \frac{2}{3}$:

(1) $a_1\left(\frac{2}{3}\right)^2 = 28$

$$\frac{4}{9}a_1 = 28$$
$$a_1 = 63$$
$$S_5 = \frac{63\left[1 - \left(\frac{2}{3}\right)^5\right]}{1 - \frac{2}{3}} = \frac{63 - \frac{224}{27}}{\frac{1}{3}}$$
$$= \frac{1477}{27} \cdot \frac{3}{1} = \frac{1477}{9}$$

$r_2 = -\frac{2}{3}$:

(1) $a_1\left(-\frac{2}{3}\right)^2 = 28$

$$a_1 = 63$$
$$S_5 = \frac{63\left[1 - \left(-\frac{2}{3}\right)^5\right]}{1 - \left(-\frac{2}{3}\right)} = \frac{63 - \left(-\frac{224}{27}\right)}{\frac{5}{3}}$$
$$= \frac{1925}{27} \cdot \frac{3}{5} = \frac{385}{9}$$

29.

$$a_n = a_1 r^{n-1}$$
$$3 = a_1\left(\frac{1}{2}\right)^{n-1}$$
$$(1) \quad \frac{3}{2} = a_1\left(\frac{1}{2}\right)^{n}$$

$$S_n = \frac{a_1(1-r^n)}{1-r}$$
$$189 = \frac{a_1\left[1-\left(\frac{1}{2}\right)^n\right]}{1-\frac{1}{2}} = \frac{a_1 - a_1\left(\frac{1}{2}\right)^n}{\frac{1}{2}}$$
$$189 = \frac{a_1 - \frac{3}{2}}{\frac{1}{2}} = 2a_1 - 3 \quad \leftarrow \text{From (1)}$$
$$192 = 2a_1$$
$$a_1 = 96$$

Substitute $a_1 = 96$ in (1):
$$\frac{3}{2} = 96\left(\frac{1}{2}\right)^n$$
$$\left(\frac{1}{2}\right)^6 = \left(\frac{1}{2}\right)^n$$
$$6 = n$$

31.
$$a_1 + a_1\left(\frac{6}{5}\right) + a_1\left(\frac{6}{5}\right)^2 = 22{,}750$$
$$a_1\left(1+\frac{6}{5}+\frac{36}{25}\right) = 22{,}750$$
$$\frac{91}{25}a_1 = 22{,}750$$
$$a_1 = \$6250$$

$$a_5 = a_1 r^4$$
$$= 6250\left(\frac{6}{5}\right)^4 = 6250\left(\frac{1296}{625}\right)$$
$$= \$12{,}960$$

33. $\doteq 12$ sec

35.
$$a_n = a_1 r^{n-1}$$
(a) $a_{10} = 1(2)^9 = 512¢ = \$5.12$

(b) $a_{31} = 1(2)^{30} \doteq 1{,}073{,}000{,}000¢ = \$10{,}730{,}000$

(c) $S_{31} = \frac{a_1(1-r^n)}{1-r} = \frac{1(1-2^{31})}{1-2} = 2^{31} - 1 \doteq 2{,}147{,}000{,}000¢ = \$21{,}470{,}000$

Exercises 11–4 (page 383)

1. $\frac{9}{2}$

3. $\frac{4}{3} - 1 + \frac{3}{4}$

$$\frac{\frac{3}{4}}{-1} = -\frac{3}{4} = r$$
$$S_\infty = \frac{a_1}{1-r} = \frac{\frac{4}{3}}{1-\left(-\frac{3}{4}\right)} = \frac{\frac{4}{3}}{\frac{7}{4}} = \frac{16}{21}$$

5. $\frac{1}{9}$

7. -18

9. $\frac{2}{9}$

11.
$$0.05454\ldots$$
$$= 0.054 + 0.00054 + \cdots$$
$$\frac{0.00054}{0.054} = 0.01 = r$$
$$S_\infty = \frac{a_1}{1-r} = \frac{0.054}{1-0.01} = \frac{0.054}{0.99} = \frac{3}{55}$$

13.
$$8.6444\ldots$$
$$= 8.6 + 0.04 + 0.004 + 0.0004 + \cdots$$
$$r = 0.1, \quad S_\infty = \frac{0.04}{1-0.1} = \frac{0.04}{0.9} = \frac{2}{45}$$
$$8.6444\ldots = \frac{8}{1} + \frac{6}{10} + \frac{2}{45}$$
$$= \frac{720}{90} + \frac{54}{90} + \frac{4}{90} = \frac{778}{90} = \frac{389}{45}$$

15. $\frac{1864}{495}$

17.

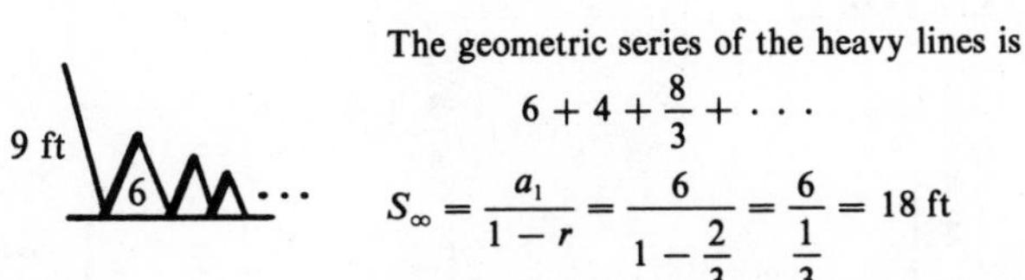

The geometric series of the heavy lines is
$$6 + 4 + \frac{8}{3} + \cdots$$
$$S_\infty = \frac{a_1}{1-r} = \frac{6}{1-\frac{2}{3}} = \frac{6}{\frac{1}{3}} = 18 \text{ ft}$$
This distance is doubled to include all but the first drop: $2(18) = 36$

The total distance traveled $= 9 + 36 = 45$ ft

19. From the sketch, you see that we have a situation that is different from the bouncing ball described in Exercise 17. This is a single geometric series.

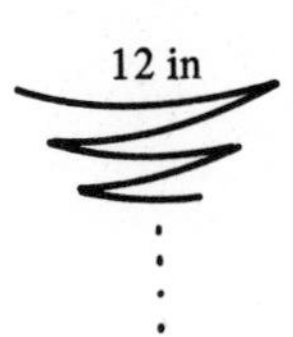

$$S_\infty = \frac{12}{1-\frac{9}{10}} = \frac{12}{\frac{1}{10}} = 120 \text{ in.}$$

21. 32 in.

23. One cut $\Rightarrow$ 0.002 in. high $= 10^{-3} \times 2^1$ in. high
Two cuts $= 10^{-3} \times 2^2$ in. high
Fifty cuts $= 10^{-3} \times 2^{50}$ in. high

$$H = \text{Height} = 10^{-3}(2^{50}) \text{ in.} = \frac{10^{-3}(2^{50})}{12(5280)} \text{ miles}$$

Use logs: $\log 2^{50} = 50 \log 2 = 50(0.3010) = 15.0500$
$\log 10^{-3} = -3.0000$, sum 12.0500

$\log 12 = 1.0792$
$\log 5280 = 3.7226$ (+)
4.8018

$12.0500 - 4.8018$ (−): $\log H = 7.2482$

$\log H = 7.2482 \Rightarrow H \doteq 17{,}710{,}000$ miles

Exercises 11–5 (page 388)

1. $x^4 + 8x^3 + 24x^2 + 32x + 16$

3. $729r^6 + 1458r^5s + 1215r^4s^2 + 540r^3s^3 + 135r^2s^4 + 18rs^5 + s^6$

5. $32x^5 - 40x^4 + 20x^3 - 5x^2 + \frac{5}{8}x - \frac{1}{32}$

7. $\frac{x^4}{81} + \frac{2x^3}{9} + \frac{3x^2}{2} + \frac{9x}{2} + \frac{81}{16}$

9. $(x + x^{-1})^4$
$= (x)^4 + 4(x)^3(x^{-1}) + \frac{4 \cdot 3}{1 \cdot 2}(x)^2(x^{-1})^2 + 4(x)(x^{-1})^3 + (x^{-1})^4$
$= x^4 + 4x^2 + 6 + 4x^{-2} + x^{-4}$

11. $x^2 + 4x^{3/2}y^{1/2} + 6xy + 4x^{1/2}y^{3/2} + y^2$

13. $(1)^3 + 3(1)^2(-\sqrt[3]{x}) + 3(1)(-\sqrt[3]{x})^2 + (-\sqrt[3]{x})^3$
$= 1 - 3\sqrt[3]{x} + 3\sqrt[3]{x^2} - x$

15. $x^{10} + 20x^9y^2 + 180x^8y^4 + 960x^7y^6 + \cdots$

17. $2048h^{11} - 11{,}264h^{10}k^2 + 28{,}160h^9k^4 - 42{,}240h^8k^6$

19. $(1 + 0.02)^3 = (1)^3 + 3(1)^2(0.02)^1 + 3(1)^1(0.02)^2 + (0.02)^3 = 1 + 0.06 + 0.0012 + 0.000008 = 1.061208$

Check $(1 + 0.02)^3 = (1.02)^3$

```
  1.02          1.0404
  1.02            1.02
  ----          ------
   204           20808
  102          10404
  ------       --------
  1.0404       1.061208
```

Do by calculator

Chapter Eleven Diagnostic Test or Review (page 390)

Following each problem number is the textbook section number (in parentheses) where that kind of problem is discussed.

1. (11–1) $a_n = \frac{2n-1}{n}$

$$S_4 = \frac{2(1)-1}{1} + \frac{2(2)-1}{2} + \frac{2(3)-1}{3} + \frac{2(4)-1}{4} = 1 + \frac{3}{2} + \frac{5}{3} + \frac{7}{4} = \frac{71}{12}$$

2. (11–2, 11–3) (a) $8, -20, 50, - + \cdots$

$\frac{50}{-20} = -\frac{5}{2}$, $\frac{-20}{8} = -\frac{5}{2}$ — GP, $r = -\frac{5}{2}$

(b) $\frac{1}{2}, \frac{3}{4}, 1, \frac{5}{4}, \frac{3}{2}, \ldots$

$1 - \frac{3}{4} = \frac{1}{4}$, $\frac{3}{4} - \frac{1}{2} = \frac{1}{4}$ — AP, $d = \frac{1}{4}$

(c) $2x - 1, 3x, 4x + 2, \ldots$

$4x + 2 - 3x = x + 2$
$3x - (2x - 1) = x + 1$ } Therefore, *not* an AP

$\frac{4x+2}{3x} \neq \frac{3x}{2x-1}$ } Therefore, *not* a GP

(d) $\frac{c^4}{16}, -\frac{c^3}{8}, \frac{c^2}{4}, - + \cdots$

$\frac{c^2}{4} \div \left(-\frac{c^3}{8}\right) = \frac{c^2}{4} \cdot \left(-\frac{8}{c^3}\right) = -\frac{2}{c}$
$-\frac{c^3}{8} \div \frac{c^4}{16} = -\frac{c^3}{8} \cdot \frac{16}{c^4} = -\frac{2}{c}$ } GP $r = -\frac{2}{c}$

3. (11–2) $x + 1, (x + 1) + (x - 1), (x + 1) + 2(x - 1), (x + 1) + 3(x - 1), (x + 1) + 4(x - 1)$
$x + 1, \quad 2x, \quad 3x - 1, \quad 4x - 2, \quad 5x - 3$

4. (11–2)
$$\begin{aligned} a_5 &= a_1 + 4d \\ -2 &= 2 + 4d \\ -4 &= 4d \\ -1 &= d \end{aligned}$$
$2, \; 2 + (-1), \; 2 + 2(-1), \; 2 + 3(-1), \; 2 + 4(-1)$
$2, \; 1, \; 0, \; -1, \; -2,$

5. (11–2) $1 - 6h, 2 - 4h, 3 - 2h, \ldots$
$(2 - 4h) - (1 - 6h) = 1 + 2h = d$
$a_{15} = a_1 + 14d = (1 - 6h) + 14(1 + 2h) = 15 + 22h$

6. (11–2)
$$\begin{aligned} a_7 = a_1 - 6d \Rightarrow 2 &= a_1 - 6d \quad (1) \\ a_3 = a_1 - 2d \Rightarrow 1 &= a_1 - 2d \quad (2) \\ 1 &= \quad - 4d \quad \text{Subtract (2) from (1)} \\ -\frac{1}{4} &= d \end{aligned}$$

Substitute $d = -\frac{1}{4}$ in (2):
$$\begin{aligned} 1 &= a_1 - 2\left(-\frac{1}{4}\right) \\ 1 &= a_1 + \frac{1}{2} \\ \frac{1}{2} &= a_1 \end{aligned}$$

7. (11–2)
$$\begin{aligned} S_n &= \frac{n}{2}(a_1 + a_n) \\ 7 &= \frac{n}{2}(10 - 8) \\ 7 &= n \end{aligned}$$
$$\begin{aligned} a_n &= a_1 + (n - 1)d \\ -8 &= 10 + 6d \\ -18 &= 6d \\ -3 &= d \end{aligned}$$

8. (11–3) $\frac{c^4}{16}, \; \frac{c^4}{16}\left(-\frac{2}{c}\right), \; \frac{c^4}{16}\left(-\frac{2}{c}\right)^2, \; \frac{c^4}{16}\left(-\frac{2}{c}\right)^3, \; \frac{c^4}{16}\left(-\frac{2}{c}\right)^4$
$\frac{c^4}{16}, \; -\frac{c^3}{8}, \; \frac{c^2}{4}, \; -\frac{c}{2}, \; 1$

9. (11–3) $8, -20, 50, - + \cdots$
$\frac{-20}{8} = -\frac{5}{2} = r$
$a_7 = a_1 r^6 = 8\left(-\frac{5}{2}\right)^6 = \frac{5^6}{2^3} = \frac{15{,}625}{8}$

10. (11–3)

$$a_4 = a_1r^3 \Rightarrow -8 = a_1r^3 \quad (1)$$
$$a_2 = a_1r \Rightarrow -18 = a_1r \quad (2)$$
$$\frac{8}{18} = r^2 \quad \text{Divide (1) by (2)}$$
$$\pm\frac{2}{3} = r$$

If $r = \frac{2}{3}$:

Substitute $r = \frac{2}{3}$ in (2):

$$-18 = a_1\left(\frac{2}{3}\right) \Rightarrow a_1 = -27$$

$$S_4 = \frac{a_1(1 - r^4)}{1 - r} = \frac{-27\left(1 - \frac{16}{81}\right)}{1 - \frac{2}{3}}$$

$$S_4 = \frac{-27\left(\frac{65}{81}\right)}{\frac{1}{3}} = -65$$

If $r = -\frac{2}{3}$:

Substitute $r = -\frac{2}{3}$ in (2):

$$-18 = a_1\left(-\frac{2}{3}\right) \Rightarrow a_1 = 27$$

$$S_4 = \frac{a_1(1 - r^4)}{1 - r} = \frac{27\left(1 - \frac{16}{81}\right)}{1 - \left(-\frac{2}{3}\right)}$$

$$S_4 = \frac{27\left(\frac{65}{81}\right)}{\frac{5}{3}} = \frac{65}{5} = 13$$

11. (11–3)

$$a_n = a_1r^{n-1}$$
$$a_nr = a_1r^n$$
$$-16\left(-\frac{2}{3}\right) = a_1\left(-\frac{2}{3}\right)^n$$
$$(1) \quad \frac{32}{3} = a_1\left(-\frac{2}{3}\right)^n$$
$$S_n = \frac{a_1(1 - r^n)}{1 - r}$$
$$26 = \frac{a_1\left[1 - \left(-\frac{2}{3}\right)^n\right]}{1 - \left(-\frac{2}{3}\right)}$$

From (1)

$$26 = \frac{a_1 - a_1\left(-\frac{2}{3}\right)^n}{\frac{5}{3}} = \frac{a_1 - \frac{32}{3}}{\frac{5}{3}}$$
$$26 = \frac{3a_1 - 32}{5}$$
$$130 = 3a_1 - 32$$
$$54 = a_1$$

Substitute $a_1 = 54$ in (1):

$$\frac{32}{3} = 54\left(-\frac{2}{3}\right)^n$$
$$\frac{16}{81} = \left(-\frac{2}{3}\right)^n \Rightarrow \left(\frac{2}{3}\right)^4 = \left(-\frac{2}{3}\right)^n$$
$$4 = n$$

12. (11–4) $8, -4, 2, -1, \ldots$

$$\frac{-1}{2} = -\frac{1}{2}$$
$$\frac{2}{-4} = -\frac{1}{2}$$
$$\frac{-4}{8} = -\frac{1}{2}$$

GP, $r = -\frac{1}{2}$

$$S_\infty = \frac{a_1}{1 - r} = \frac{8}{1 - \left(-\frac{1}{2}\right)} = \frac{8}{\frac{3}{2}} = \frac{16}{3}$$

13. (11–4) $3.0333\ldots = 3 + 0.0333\ldots$

$$0.0333\ldots = 0.03 + \underbrace{0.003 + 0.0003}_{} + \cdots \qquad \frac{0.0003}{0.003} = \frac{1}{10} = r$$

$$0.0333\ldots = \frac{0.03}{1 - \frac{1}{10}} = \frac{0.03}{0.9} = \frac{1}{30}$$

Therefore, $3.0333\ldots = 3 + \frac{1}{30} = \frac{91}{30}$

14. (11–4)

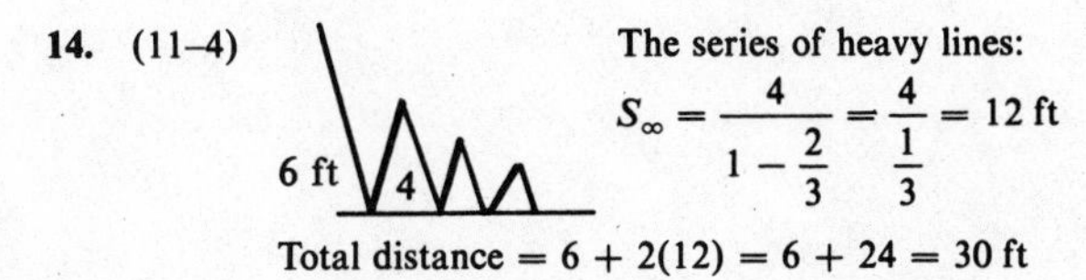

The series of heavy lines:

$$S_\infty = \frac{4}{1 - \frac{2}{3}} = \frac{4}{\frac{1}{3}} = 12 \text{ ft}$$

Total distance $= 6 + 2(12) = 6 + 24 = 30$ ft

15. (11–2) $9, 27, 45, \ldots$

$$\left.\begin{array}{l} 45 - 27 = 18 \\ 27 - 9 = 18 \end{array}\right\} \begin{array}{l} \text{AP} \\ d = 18 \end{array}$$

$$a_{10} = a_1 + 9d = 9 + 9(18) = 171 \text{ ft}$$

$$S_{12} = \frac{n}{2}(a_1 + a_{12}) = \frac{12}{2}[9 + 9 + 11(18)] = 1296 \text{ ft}$$

16. (11–5)

$$\left(\frac{1}{5} - 5y\right)^4 = \left(\frac{1}{5}\right)^4 + 4\left(\frac{1}{5}\right)^3(-5y)^1 + \frac{4 \cdot 3}{1 \cdot 2}\left(\frac{1}{5}\right)^2(-5y)^2 + 4\left(\frac{1}{5}\right)^1(-5y)^3 + (-5y)^4$$

$$= \frac{1}{625} - \frac{4}{25}y + 6y^2 - 100y^3 + 625y^4$$

17. (11–5)

$$\left(2x - \frac{1}{2}y\right)^8 = (2x)^8 + 8(2x)^7\left(-\frac{1}{2}y\right)^1 + \frac{8 \cdot 7}{1 \cdot 2}(2x)^6\left(-\frac{1}{2}y\right)^2 + \cdots$$

$$= 256x^8 - 512x^7y + 448x^6y^2 + \cdots$$

INDEX

STUDENT QUESTIONNAIRE

*Your chance to rate **Intermediate Algebra** (Willis/Johnston).*

In order to keep this text responsive to your needs, it would help us to know what you, the student, thought of this text. We would appreciate it if you would answer the following questions. Then cut out the page, fold, seal, and mail it; no postage is required. Thank you for your help.

Which chapters did you cover? (circle) 1 2 3 4 5 6 7 8 9 10 11 All

Does the book have enough worked-out examples? Yes ____ No ____

enough exercises? Yes ____ No ____

Which helped most?

Explanations ____ Examples ____ Exercises ____ All three ____ Other ________ (fill in)

Were the answers at the back of the book helpful? Yes ____ No ____

Did the answers have any typos or misprints? If so, where?

For you, was the course elective? ____ Required? ____

Do you plan to take more mathematics courses? Yes ____ No ____

If yes, which ones?

Finite mathematics	____	Precalculus	____
Statistics	____	Calculus (for engineering and physics)	____
Trigonometry	____	Calculus (for business and social science)	____
Analytic geometry	____	Other	________
College algebra	____		

How much algebra did you have before this course? Terms in high school (circle) 1 2 3 4

Courses in college 1 2 3

If you had algebra before, how long ago?

Last 2 years ____ 3–5 years ago ____ 5 years or longer ____

What is your major or your career goal? ____________ Your age? ____

What did you like the most about ***Intermediate Algebra?***

Can we quote you? Yes ______ No ______

What did you like least about the book?

College ________________________________ State ______________

--FOLD HERE--

First Class
PERMIT NO. 34
Belmont Ca.

BUSINESS REPLY MAIL
No postage necessary if mailed in United States

Postage will be paid by
WADSWORTH PUBLISHING COMPANY, INC.
10 Davis Drive
Belmont, California 94002
ATTN: Rich Jones, Mathematics editor

Table I Square Roots

N	$\sqrt{N}$	N	$\sqrt{N}$	N	$\sqrt{N}$	N	$\sqrt{N}$
1	1.000	51	7.141	101	10.050	151	12.288
2	1.414	52	7.211	102	10.100	152	12.329
3	1.732	53	7.280	103	10.149	153	12.369
4	2.000	54	7.348	104	10.198	154	12.410
5	2.236	55	7.416	105	10.247	155	12.450
6	2.449	56	7.483	106	10.296	156	12.490
7	2.646	57	7.550	107	10.344	157	12.530
8	2.828	58	7.616	108	10.392	158	12.570
9	3.000	59	7.681	109	10.440	159	12.610
10	3.162	60	7.746	110	10.488	160	12.649
11	3.317	61	7.810	111	10.536	161	12.689
12	3.464	62	7.874	112	10.583	162	12.728
13	3.606	63	7.937	113	10.630	163	12.767
14	3.742	64	8.000	114	10.677	164	12.806
15	3.873	65	8.062	115	10.724	165	12.845
16	4.000	66	8.124	116	10.770	166	12.884
17	4.123	67	8.185	117	10.817	167	12.923
18	4.243	68	8.246	118	10.863	168	12.961
19	4.359	69	8.307	119	10.909	169	13.000
20	4.472	70	8.367	120	10.954	170	13.038
21	4.583	71	8.426	121	11.000	171	13.077
22	4.690	72	8.485	122	11.045	172	13.115
23	4.796	73	8.544	123	11.091	173	13.153
24	4.899	74	8.602	124	11.136	174	13.191
25	5.000	75	8.660	125	11.180	175	13.229
26	5.099	76	8.718	126	11.225	176	13.266
27	5.196	77	8.775	127	11.269	177	13.304
28	5.292	78	8.832	128	11.314	178	13.342
29	5.385	79	8.888	129	11.358	179	13.379
30	5.477	80	8.944	130	11.402	180	13.416
31	5.568	81	9.000	131	11.446	181	13.454
32	5.657	82	9.055	132	11.489	182	13.491
33	5.745	83	9.110	133	11.533	183	13.528
34	5.831	84	9.165	134	11.576	184	13.565
35	5.916	85	9.220	135	11.619	185	13.601
36	6.000	86	9.274	136	11.662	186	13.638
37	6.083	87	9.327	137	11.705	187	13.675
38	6.164	88	9.381	138	11.747	188	13.711
39	6.245	89	9.434	139	11.790	189	13.748
40	6.325	90	9.487	140	11.832	190	13.784
41	6.403	91	9.539	141	11.874	191	13.820
42	6.481	92	9.592	142	11.916	192	13.856
43	6.557	93	9.644	143	11.958	193	13.892
44	6.633	94	9.695	144	12.000	194	13.928
45	6.708	95	9.747	145	12.042	195	13.964
46	6.782	96	9.798	146	12.083	196	14.000
47	6.856	97	9.849	147	12.124	197	14.036
48	6.928	98	9.899	148	12.166	198	14.071
49	7.000	99	9.950	149	12.207	199	14.107
50	7.071	100	10.000	150	12.247	200	14.142

Table II Common Logarithms

N	0	1	2	3	4	5	6	7	8	9
1‸0	.0000	.0043	.0086	.0128	.0170	.0212	.0253	.0294	.0334	.0374
1‸1	.0414	.0453	.0492	.0531	.0569	.0607	.0645	.0682	.0719	.0755
1‸2	.0792	.0828	.0864	.0899	.0934	.0969	.1004	.1038	.1072	.1106
1‸3	.1139	.1173	.1206	.1239	.1271	.1303	.1335	.1367	.1399	.1430
1‸4	.1461	.1492	.1523	.1553	.1584	.1614	.1644	.1673	.1703	.1732
1‸5	.1761	.1790	.1818	.1847	.1875	.1903	.1931	.1959	.1987	.2014
1‸6	.2041	.2068	.2095	.2122	.2148	.2175	.2201	.2227	.2253	.2279
1‸7	.2304	.2330	.2355	.2380	.2405	.2430	.2455	.2480	.2504	.2529
1‸8	.2553	.2577	.2601	.2625	.2648	.2672	.2695	.2718	.2742	.2765
1‸9	.2788	.2810	.2833	.2856	.2878	.2900	.2923	.2945	.2967	.2989
2‸0	.3010	.3032	.3054	.3075	.3096	.3118	.3139	.3160	.3181	.3201
2‸1	.3222	.3243	.3263	.3284	.3304	.3324	.3345	.3365	.3385	.3404
2‸2	.3424	.3444	.3464	.3483	.3502	.3522	.3541	.3560	.3579	.3598
2‸3	.3617	.3636	.3655	.3674	.3692	.3711	.3729	.3747	.3766	.3784
2‸4	.3802	.3820	.3838	.3856	.3874	.3892	.3909	.3927	.3945	.3962
2‸5	.3979	.3997	.4014	.4031	.4048	.4065	.4082	.4099	.4116	.4133
2‸6	.4150	.4166	.4183	.4200	.4216	.4232	.4249	.4265	.4281	.4298
2‸7	.4314	.4330	.4346	.4362	.4378	.4393	.4409	.4425	.4440	.4456
2‸8	.4472	.4487	.4502	.4518	.4533	.4548	.4564	.4579	.4594	.4609
2‸9	.4624	.4639	.4654	.4669	.4683	.4698	.4713	.4728	.4742	.4757
3‸0	.4771	.4786	.4800	.4814	.4829	.4843	.4857	.4871	.4886	.4900
3‸1	.4914	.4928	.4942	.4955	.4969	.4983	.4997	.5011	.5024	.5038
3‸2	.5051	.5065	.5079	.5092	.5105	.5119	.5132	.5145	.5159	.5172
3‸3	.5185	.5198	.5211	.5224	.5237	.5250	.5263	.5276	.5289	.5307
3‸4	.5315	.5328	.5340	.5353	.5366	.5378	.5391	.5403	.5416	.5428
3‸5	.5441	.5453	.5465	.5478	.5490	.5502	.5514	.5527	.5539	.5551
3‸6	.5563	.5575	.5587	.5599	.5611	.5623	.5635	.5647	.5658	.5670
3‸7	.5682	.5694	.5705	.5717	.5729	.5740	.5752	.5763	.5775	.5786
3‸8	.5798	.5809	.5821	.5832	.5843	.5855	.5866	.5877	.5888	.5899
3‸9	.5911	.5922	.5933	.5944	.5955	.5966	.5977	.5988	.5999	.6010
4‸0	.6021	.6031	.6042	.6053	.6064	.6075	.6085	.6096	.6107	.6117
4‸1	.6128	.6138	.6149	.6160	.6170	.6180	.6191	.6201	.6212	.6222
4‸2	.6232	.6243	.6253	.6263	.6274	.6284	.6294	.6304	.6314	.6325
4‸3	.6335	.6345	.6355	.6365	.6375	.6385	.6395	.6405	.6415	.6425
4‸4	.6435	.6444	.6454	.6464	.6474	.6484	.6493	.6503	.6513	.6522
4‸5	.6532	.6542	.6551	.6561	.6571	.6580	.6590	.6599	.6609	.6618
4‸6	.6628	.6637	.6646	.6656	.6665	.6675	.6684	.6693	.6702	.6712
4‸7	.6721	.6730	.6739	.6749	.6758	.6767	.6776	.6785	.6794	.6803
4‸8	.6812	.6821	.6830	.6839	.6848	.6857	.6866	.6875	.6884	.6893
4‸9	.6902	.6911	.6920	.6928	.6937	.6946	.6955	.6964	.6972	.6981
5‸0	.6990	.6998	.7007	.7016	.7024	.7033	.7042	.7050	.7059	.7067
5‸1	.7076	.7084	.7093	.7101	.7110	.7118	.7126	.7135	.7143	.7152
5‸2	.7160	.7168	.7177	.7185	.7193	.7202	.7210	.7218	.7226	.7235
5‸3	.7243	.7251	.7259	.7267	.7275	.7284	.7292	.7300	.7308	.7316
5‸4	.7324	.7332	.7340	.7348	.7356	.7364	.7372	.7380	.7388	.7396
N	0	1	2	3	4	5	6	7	8	9